CRC Handbook of Nuts

Author

James A. Duke
Agricultural Research Service
United States Department of Agriculture
Beltsville, Maryland

CRC Press
Taylor & Francis Group
Boca Raton London New York

CRC Press is an imprint of the
Taylor & Francis Group, an **informa** business

First published 1989 by CRC Press
Taylor & Francis Group
6000 Broken Sound Parkway NW, Suite 300
Boca Raton, FL 33487-2742

Reissued 2018 by CRC Press

Library of Congress Cataloging-in-Publication Data

Duke, James A., 1929-
CRC handbook of nuts / author, James A. Duke.
p. cm.
Bibliography: p.
Includes index.
ISBN 0-8493-3636-8
1. Nuts--Handbooks, manuals, etc. I. Title. II. Title: Handbook of nuts.
SB401.A4D84 1989
634'.5--dc19 88-14586

A Library of Congress record exists under LC control number: 88014586

Publisher's Note
The publisher has gone to great lengths to ensure the quality of this reprint but points out that some imperfections in the original copies may be apparent.

Disclaimer
The publisher has made every effort to trace copyright holders and welcomes correspondence from those they have been unable to contact.

ISBN 13: 978-1-315-89203-0 (hbk)
ISBN 13: 978-1-351-07113-0 (ebk)

Visit the Taylor & Francis Web site at http://www.taylorandfrancis.com and the
CRC Press Web site at http://www.crcpress.com

INTRODUCTION*

Appropriately, one might commence a Handbook of Nuts with a definition of nut. But, if you'll pardon the jargon, that's a "tough nut to crack." To drive home my definition, I'll here recount an anecdote. For several years I was peripherally embroiled in a controversy over that definition. Various people interested in the jojoba (*Simmondsia chinensis*) would call or write, hoping my definition would support their contention that the jojoba was or was not a nut.

Finally, lawyers from the Internal Revenue Service (IRS) called and asked me to send my definition, in writing! I quote my cautious letter to the IRS:

> I understand that the IRS has special treatment for certain farmers raising fruits and nuts.
> I quote definitions from my favorite glossary, B. D. Jackson's *A Glossary of Botanic Terms*, 4th ed., Hafner Publishing Company, New York, 1928, reprinted 1953:
>
> *Nut*: a hard indehiscent one-seeded fruit.
> *Fruit*: (1) Strictly, the pericarp and its seeds, the fertilized and developed ovary.
> I think the jojoba "beans" would qualify just as well as the pecan as both a nut and a fruit, botanically speaking. There are popular concepts of the fruit as fleshy and/or wet, the nut as nonfleshy and/or dry. Relatively speaking, the jojoba is as dry as a pecan and popularly considered a nut. But botanically, a nut is just one kind of fruit. Hence, I conclude the jojoba bean is both a nut and a fruit from a botanical point of view.

Ironically, IRS definitions may make or break a nut species. Vietmeyer[329] shows how an IRS ruling in 1969 withdrew a number of nut species, especially almond, from its list of speculative agricultural investments. "Pistachios, however, remained an allowable tax write-off. Suddenly alone it became a hot investment." Vietmeyer calls this the unexpected source for the real advance into commercialization; e.g., Getty Oil, Superior Oil, and Tenneco West then invested in pistachios. By 1984 we had some 20,000 hectares pistachios and pushed Turkey out of the number 2 production spot. In 1985, Vietmeyer went even farther with his NRC report on jojoba,[233] perhaps giving the jojoba more momentum. Tax advantages to the jojoba may hinge on whether or not it is defined as a nut. Who knows? Perhaps the future of the jojoba as a new crop may hinge on its definition.

In 1985, I received a hasty call from an official of the Jojoba Grower's Association, distressed by the IRS interpretation of my letter quoted previously. The official enticed me to agree that, in common, if not botanical parlance, the words fruit and nut implied edibility.

Here I quote the letter drafted (but never typed) to that official. Following conversations with other jojoba fans in the government, I feared the last half of the letter might jeopardize the future of the jojoba as a "new crop". So few "new crops" break through the economic catch 22 here in the U.S.! The farmer won't grow it until industry provides a guaranteed market, and industry won't generate a market until there's a guaranteed source (the farmer).

> Enclosed is a copy of my letter of July 5, 1983 to the IRS re jojoba. When asked by legal types how to define something, I like to quote published definitions, rather than inventing my own. Trained as a taxonomist, I resorted to Jackson's glossary.[147] Horticulturists might resort to other sources.
> It is true that the popular concepts of fruits and nuts with most people may imply edibility. Few, if any, modern Americans eat jojoba "nuts." I would have to agree with you, Dr. M. Faust, of USDA, and J. Janick, of Purdue University, that, if edibility is a prerequisite part of the definition of fruit or nut, jojoba is best not considered a nut.

* Expanded from talks presented at the Agricultural Marketing Workshop for the Caribbean Basin, Miami, Florida, September 24-27, 1984; and New Orleans, Louisiana, September 16-20, 1985.

I realize that paragraph two is what you wanted to hear. Hence, I separate it from the following paragraph which, being something you may not wish to hear, can be extricated from the rest of my letter.

Two books which I procured in preparing a draft *Handbook of Nuts* are Rosengarten's *The Book of Edible Nuts* (Walker and Company, New York, 1984),[283] and Menninger's *Edible Nuts of the World* (Horticultural Books, Inc., Stuart, Florida, 1977).[209] Menninger, who defines nut as "any hardshelled fruit or seed of which the kernel is eaten by mankind", treats the jojoba as an edible nut, noting the Indian consumption and the vulgar names "goat nut" and "deer nut." Rosengarten employs the word "nut" "in the broad and popular sense, covering a wide variety of fruits or seeds, some of which would not be classified as nuts according to strict botanical definition." He groups jojoba among "Thirty Other Edible Nuts", adding as common names "sheep nut" and "pignut". He says "Its fruits and leaves are devoured with avidity by goats, sheep, and deer. Indians of the desert Southwest gathered jojoba nuts and ate them, raw or roasted; their flavor is reminisent of the hazelnut, but more bitter . . . Today the use of jojoba nuts for human consumption is mainly of historic interest." This paragraph of my letter reinforces my reluctance to disqualify jojoba, even in the popular concept "nut." I have tasted them raw, and find them about as unappealing as most acorns I have tried.

Those seeking to exclude jojoba from the staid society of nuts might say that jojoba, an American species, is, with good reason, excluded from Woodruff's *Tree Nuts* (2 vols., AVI Publishing Company, Westport, Connecticut, 1967[341] and Jaynes *Handbook of North American Nut Trees* (NNGA, Knoxville, Tennessee, 1969).[148] My superficial examination of these revealed no definition of nut.

For the record, I did send the following letter and poem that encapsulated my seedy feelings.

Thanks for your letter of April 25, re the jojoba.

While not fully understanding the tax implication of the Jackson (botanical) interpretation of the word "nut" and "fruit", I surely agree with you that in common, rather than botanical parlance, the words fruit and nut imply edibility. Hence, the common parlance for an orchard of nuts would be a cultivated grove of trees or shrubs for their edible nuts. I don't frankly believe that jojoba falls into that common concept.

Hence, the botanical definition of nut is at variance with the popular definition of nut. I think jojoba is a fruit and/or nut according to Jackson's technical definition, but not according to common parlance.

Not a Nut?
(The incredible inedible nut!!!)

The Jojoba Growers' Association
Wishes, to my consternation,
That I'd retract a note,
That long ago I wrote
For IRS edification.

I sent Jackson's definition
To the IRS Commission
I resorted to quote, but,
Jojoba's both fruit and nut,
Which promotes the Growers' dissension.

I find it perfectly credible
To define a "nut" as an edible,
But even that caveat
Won't change the fact that
Its edible uses are negligible.

Poets sometimes get in a rut,
Nonpoetic lines, dry and cut.
No amount of stink
Will lead me to think
The jojoba nut's not a nut.

My interpretation of the facts is
Jojoba's not good for the gut,
And when you tally your taxes,
The jojoba nut's not a nut.

Anonpoet
April 29, 1985

I have included in this book many species which are not true nuts. Unlike a one-seeded peanut, a peanut with two or three seeds in the indehiscent pod is disqualified because it has more than one seed. But I excluded many nuts treated in my *Handbook of Legumes of World Economic Importance.*[83] There are many seeds in the Brazil nut pod, which rules them out (as one-seeded fruits). Similarly, there are many "nuts" in the colas, included in this book, and many "beans" in the cacao pod of the same family. Cacao is no further from the definition of nut than is cola. Cacao will be considered in the volume on Money Crops. As a matter of fact, nearly half the species in this book are *not* nuts in the narrowest sense: "one-seeded indehiscent fruits, the kernels of which are edible."

In 1984, I addressed the Agricultural Marketing Workshop (Miami) on subtropical and tropical nuts. The feedback I got from that first meeting suggested that I may have overdosed the audience with suggestions of nuts that might be grown in the tropics. There are hundreds of species that can be called nuts, by any of several possible definitions. And due to the overall higher species diversity in the tropics, there is a concomitant higher number of nut species available for consideration in the tropical environments with which we were concerned.

During that same year, CRC Press published Martin's *Handbook of Tropical Food Crops*[203] just before I attended the Miami conference that presaged the New Orleans conference of the Agricultural Marketing Workshop. Carl Campbell's[60] excellent chapter on Fruits and Nuts gave a good overview of the cultivation of fruits and nuts, and included short treatises on the cashew, pili nut, brazil nut, breadfruit, coconut, oil palm, and lychee (really a fruit). In a summary table, he listed a few others, the monkey pot (*Lecythis elliptica*), the paradise nut (*Lecythis zabucajo*), the jackfruit (*Artocarpus heterophylla*), the salak (*Salacca edulis*), the peach palm (*Bactris gasipaes*), the macadamia (*Macadamia integrifolia*), and the jujube (*Ziziphus mauritiana*). Certain virtues were suggested for nut trees:

- Dietary diversity
- High oil content
- Luxury long-distance commercial markets
- Important to subsistence farmers
- Everbearing
- Low maintenance
- Intercropping potential
- Wood as byproduct
- Land stabilization

Following my presentations, CRC advised me that they would publish this *Handbook of Nuts.* It was designed to contain information summaries on about 100 nut species, in the same format as my Plenum *Handbook of Legumes of World Economic Importance* (Plenum Press[83]) with succinct paragraphs on Uses, Folk Medicine, Chemistry, Description, Germplasm, Distribution, Ecology, Cultivation, Harvesting, Yields and Economics, Energy, Biotic Factors.

The following recommendations seem germaine to potential nut producers.

1. Understand the crop and its requirements — take the principles of production and do good, replicated, semi-commercial research to adapt the crop to your own situations.
2. Select growing areas where good production of a crop can be concentrated — secure large quantities of nuts to make an impact on the export market.
3. Develop or choose the best varieties and disease-free planting stocks.
4. Concentrate on producing high quality produce to ensure repeat sales.
5. Time production so that it will not overlap competitor production, if any.
6. Practice insect, disease, and pest control — consider quarantine and import regulations for the crop.
7. Develop attractive and protective packaging that is distinctive and makes your product recognizable.
8. Do not plant a tree until you've tentatively contracted a market. Many advanced technological studies concern temperate nuts and oil seeds.

Chemical Business (CB) ran an article on Oleochemicals (Research Sparks Oleochemical Hopes).[320] Oleochemicals are defined as the industrial products based on animal fats and vegetable oils, a $1.2 billion segment of the U.S. chemical industry (cf. nut imports worth ca. $300 million, 200 in brazil nuts, 50 in cashews).

Unlike nuts, oleochemicals find their way into:

1. The personal care product market (20%)
2. Industrial lubricants and related products (14%)
3. Coatings (10%)
4. Detergent intermediates (10%)
5. Plastics, alkyds, urethanes, cellophane, cleaners, detergents (18%)
6. Textiles, emulsion, polymerization, rubber, asphalt, mining, miscellaneous

In this handbook I treat both kinds of nuts, (1) the familiar nuts that we eat and (2) a few oleochemical or chemurgic nuts. Some of the chemurgic nuts of the tropics are tung and its relatives, purging nut, marking nut, jojoba, and some even more obscure species. I suspect more technological advances are emerging with oil palms than with edible nuts.

Lauric acid is now obtained mainly from coconut oil and secondarily from palm kernel oil. Finding an alternative source of lauric acid has sparked much industry interest. Henkel Corporation is betting on palm kernel oil in the short run, "in about 5 years, lauric acid from palm kernel oil will add about 75% to current supplies."[320] We use about 2 billion pounds of oleochemicals, which include fatty acids, surfactants, and other esters, amines, natural glycerins, natural alkanoamides, and primary amides and bisamides, at only $0.60 per pound = 1% of U.S. Chemical Revenues.

Exciting new technologies are being explored in the search for alternative sources of lauric acid. In the continental U.S., the technologies are directed more to temperate annuals than to tropical perennials, but potential is probably greater among tropical perennials which need not contend with winter. Some of the technologies *do* relate to tropical nuts. The kernel of the oil palm is a nut. Britain's Unilever, and others, are propagating high-yielding oil palms and these are showing up in palm plantations. Such palms can produce more than ten times as much oil as the temperate soybean. Elsewhere I have speculated that 2 billion ha oilpalm yielding 25 barrels oil/ha could, with transesterification, support the world's requirements for 50 billion barrels oil.[320]

Meanwhile, back in the temperate zone, Calgene[320] is looking at *Cuphea*, an oilseed with low yields and other agronomic problems, but a crop which produces lauric acid, a short-chain fatty acid. "Most oilseeds, including rapeseed, make long chain fatty acids (C-18 and up) . . . but . . . because the plants do not 'know how' to stop molecule chain growth, no midchain fatty acids, such as lauric acid, are produced by the plants . . . Some oilseed species such as cuphea do know how . . . Calgene scientists plan to isolate the gene or genes responsible and transfer them to rapeseed. Calgene has already overcome difficulties in introducing foreign genes into rapeseed and making the transformed rapeseed plants grow . . . Calgene researchers may be able to modify plants to produce whatever fatty acid is desired."[320] They "expect to have a series of genes cloned and to be able to 'mix-and-match' genes in a low-cost production plant to produce custom-designed plants that produce specialty oils." One potential of this research is the possibility of finding plant sources that can compete with petroleum feedstocks. Some *Cuphea* species synthesize the C-8 and C-10 fatty acids that could potentially replace petroleum based C-7 and C-9 fatty acids.

An Ohio subsidiary of Lubrizol has developed a high-oleic acid sunflower with 80% oleic acid, up from the traditional 40%. They put in a 20,000 hectare crop of high-oleic acid sunflowers. Perhaps those interested in tropical nuts should look more to the pataua, *Jessenia bataua*, a tropical perennial producing perhaps 3 to 6 MT of oil with 80% oleic acid according to some authors. This oil has been favorably compared with olive oil, at a much lower price.[32]

So much for the annual cupheas, brassicas, and sunflower, the latter treated as a nut by both Menninger[209] and Rosengarten.[283] None of the biotechnologies mentioned are unique to annuals; they can apply just as well to perennials. But it is easier to keep an annual proprietary. Perennials, once given to the world, can usually be clonally reproduced ad infinitum. Hence, I speculate that the world at large, especially the tropical world, would fare better if the new technologies were developed for perennial species, while the seed salesmen and gene-grabbers might fare better with annuals.

Whether annual or perennial, plant sources of oleochemicals, or proteins, or pesticides, or drugs, always seem to suffer from one valid criticism. As Tokay (1985) notes, " . . . the use of natural raw materials that are often inconsistent in composition from batch to batch causes processing headaches. In addition, most fractionation processes produce many by-products and co-products, which are often difficult to sell."[320] Contrastingly, we read in *Science*, September 13, 1985, "Whole plant utilization—extracting medicines, leaf proteins, vitamins, polyphenols, essential oils, and chemurgics, and using the residues for alcohol production for energy—could move us from the petrochemical to the phytochemical era, with the possible fringe benefits of slowing the 'greenhouse effect' and making us more self-sufficient."[88]

Balandrin and Klocke[28] indicate that much evidence shows that natural product research is still potentially less expensive and more fruitful (in terms of new prototype compounds discovered) than are large chemical synthesis programs.

New technologies for better extraction of main products and co-products and by-products are rapidly coming on line. Work goes on with the transesterification of palm oil, which could effectively fuel the diesel needs of the world. In "A Green World Instead of the Greenhouse",[87] one finds scores (yields in barrels oil equivalent per hectare) for several energy plants.

Babassu 5-60	Peachpalm 35—105
Cassava 6,11,15—45	Peanut 4,5,13
Castor 13	Purging nut 18
Coconut 11,25	Rape 4,5
Cottonseed 1	Sesame 8

Date palm 10—20	Soybean 2,2,6
Eucalypt 76	Sugarcane 13,15
Melaleuca 76	Sunflower 4,6
Nypa 30—90	Sweet potato 30—90
Oilpalm 24—58	Tung 17

Coconut is just one of the hundreds of palms which can be termed a nut. Oil-palms are also considered nuts, even by Menninger,[209] if their seeds are edible. Botanically, many familiar palms might better be classified as drupes, but their energetic potential is noteworthy. In his survey of "Amazonian Oil Palms of Promise", Balick[29] notes that most oil palms have a high yield and produce one or both of the basic types of oil (kernel and pulp). These two types usually differ chemically. More importantly, "Most of the palms would seem to be well adapted to underused agricultural lands in tropical areas, where climate or other factors preclude the cultivation of the more common oleaginous plants."[29]

The palms on people's minds today include, besides the conventional coconut and oil palm, both sources of lauric acid, the babassu and the pataua. And then there's another tree, the inche, not even a palm, attracting the attention of the oil-palm people.

The jojoba, with which I opened my talk in 1984, is not even an oilseed, but a "waxseed". Since it is so important to my introduction, I have left it in this CRC Handbook of Nuts.

Menninger, in his *Edible Nuts of the World*[209] after noting that "A thousand kinds of nuts in this world are hunted and eaten by hungry people" defines nut as "any hard shelled fruit or seed of which the kernel is eaten by man-kind." He purports to exclude those nuts that never see the interior of the human stomach in his chapter, "Not Nuts."

Rosengarten, in *The Book of Edible Nuts,*[283] is more cautious, like me, mostly quoting other definitions. Then he selects twelve important edible nuts and discusses their relation to the various definitions. That discussion bears repeating:

> Few botanical terms are used more loosely than the word 'nut'. Technically, according to Funk & Wagnalls Standard Encyclopedic College Dictionary (1968), a nut is '1. A dry fruit consisting of a kernel or seed enclosed in a woody shell; the kernel of such fruit, especially when edible, as of the peanut, walnut, or chestnut; *Bot.* A hard, indehiscent, one-seeded pericarp generally resulting from a compound ovary, as the chestnut or acorn.' (Indehiscent means that the seedcase does not split open spontaneously when ripe.) The nut has also been described as a one-celled, one-seeded, dry fruit with a hard pericarp (shell); and, more simply, as the type of fruit that consists of one edible, hard seed covered with a dry, woody shell that does not split open at maturity. Only a fraction of so-called nuts—for example, chestnuts, filberts, and acorns—answer this description. The peanut is not really a nut; it is a legume or pod, like the split pea, lentil, or bean—but an indehiscent one because the pod does not split open upon maturing. The shelled peanut is a seed or bean. The edible seeds of almonds, walnuts, pecans, pistachios, hickory nuts, and macadamia nuts are enclosed in the hard stones of a drupe—like the stones of peaches, cherries, or plums. A drupe is a soft, fleshy fruit with a spongy or fibrous husk, which may or may not split free from the inner hard-shelled stone containing the seed. In plums and peaches, we eat the fleshy parts and throw away the stones; but the fleshy part of the walnut, for example, is removed and discarded, while the kernel of the stone—the nut —is eaten. The shell of a drupe nut, like the walnut, corresponds to the hard, outer layer of the peach stone. The coconut is the seed of a fibrous drupe. The Brazil nut is a seed with a hard seed coat, as is the pinon nut. Another dry, indehiscent fruit type is the achene—a small, thin shell containing one seed, attached to the outer layer at one point only—as in the dandelion and buttercup. The sunflower seed is an achene. A true nut resembles an achene, but it develops from more than one carpel (female reproductive structure), is usually larger and has a harder, woody wall; e.g., the difference between the filbert nut and the sunflower achene.[283]

In 1983/84, the U.S. imported nuts worth \$305 million per year, with \$216 million in brazil nuts, and \$55 million in cashews, cf. \$233 with \$159 and \$46, respectively, in 1982/83[119] (Table 1).

Table 1
DATA ON AN IMPORTANT DOZEN NUTS

	U.S. production 1980 (1,000 tons)[a]	Import costs annual 1983—84 (million $)[b]	Per cap. consumption (shelled) 1960—1979 (lbs)[a]	Price per pound ($)[a]	Oil percentage (APB) (%)[a]
Almonds	260		0.45	1.75	54.2
Brazil nuts		216		1.65	66.9
Cashew		55		2.15	45.7
Chestnut				1.65	1.5
Coconut			0.50	.65	35.3
Filbert	15		0.08	1.40	62.4
Macadamia	15		0.033	5.50	71.6
Peanut			7.1	.65	47.5
Pecan	92		0.30	2.75	71.2
Pistachio	14			3.30	53.7
Sunflower				.55	47.3
Walnut (Persian)	197		0.50	2.00	64.0

[a] Rosengarten.[283]
[b] Gyawa.[119]

In 1980, the U.S. produced on an in-shell basis, ca.260,000 tons almonds, 197,000 tons walnuts, 92,000 tons pecans, 15,000 tons filberts, 15,000 tons macadamia, and 14 tons pistachios, for a total approximating 600,000 MT nuts production. Of these, it might be noted that only 92,000 (the pecans) were from a native American species.

Here we see a parallel with the other major groups of crops; North America has not contributed much to America's foodbasket. "Of all the horticultural products given by our continent to civilization, none are of more importance than the pecan, nor destined to play a more vital role in our pomological future." Moreover, a great slave, Antoine, of the Oak Valley Plantation, in Louisiana, is accredited with our most important contribution to the nut basket. "The slave Antoine had thus laid the foundation upon which was to be erected a great industry . . . "[341]

Mostly maturing in fall, the temperate zone nuts are extremely rich in calories. Rosengarten notes that one pound of nut kernels (assuming 3,000 calories of fuel value per pound) is equivalent in energy value to about 2.4 lbs breads, 3.2 lbs steak, 8 lbs potato, or 10.4 lbs apple.[283] Oils of the temperate zone are higher in unsaturated fatty acids in general, than oils from the tropics like the palm oils, brazil nut, cashew, etc. It is rather well known that the unsaturated fats are more healthy than the saturated. It is not so well known that you could clone a pecan, grow it in a cold and a hot locale, and have a higher unsaturated profile at the colder locale. In other words, the oils from the tropics will, in general, be less healthy than those from the temperate zone. Perhaps we should raise our edible oils in the temperate zone and our fuel oils in the tropics. But save the pilis, cashews, and brazil nuts for the palates they please so well.

Rosengarten adds that "Most nuts are an excellent source of calcium, phosphorus, iron, potassium, and the B vitamins."[283] This is true on an as-purchased basis, because nuts contain so little water. On a zero-moisture basis (Table 2), the nuts do not seem particularly outstanding with these nutrients. Some of the more familiar nuts are compared in Table 3.

Table 2
AVERAGES OF PROXIMATE ANALYSIS (OF CONVENTIONAL FOODS)

Per 100 g

FOOD		(No.)	H_2O (g)	Cal (g)	Prot (g)	Fat (g)	Total carb (g)	Fiber (g)	Ash (g)	Ca (mg)	P (mg)	Fe (mg)	Na (mg)	K (mg)	B-Car (μg)	Thia (mg)	Rib (mg)	Nia (mg)	Vit ((mg
Aerial	(APB)	(5)	92.4	25	1.4	.2	5.4	.9	.6	31	31	.6	4	224	271	.08	.09	.8	4
veg.	(ZMB)		0	331	18.5	2.6	71.4	11.9	7.9	410	410	7.9	53	2,963	3,585	1.06	1.19	10.6	55
Tame	(APB)	(5)	90.5	27	2.7	.4	5.0	1.0	1.5	106	46	2.4	89	440	3,568	.08	.16	.5	5
greens	(ZMB)			289	28.4	4.0	54.7	10.5	15.8	1,116	484	25.3	937	4,633	37,571	.84	1.68	4.7	60
Shoots	(APB)	(5)	89.9	31	3.5	.5	5.2	.6	.9	31	59	1.1	4	345	1,166	.15	.19	1.0	4
	(ZMB)		0	307	34.6	5.0	51.5	5.9	8.9	307	584	10.9	40	3,416	11,543	1.48	1.88	9.9	40
Wild	(APB)	(5)	88	35	2.8	.5	6.5	1.3	2.1	186	57	2.7	40	382	5,652	.11	.24	.9	6
greens	(ZMB)			292	23.7	4.5	54.0	11.2	17.5	1,553	475	22.2	337	3,183	47,433	.9	2.0	7.5	56
Fruits	(APB)	(10)	84.9	54	.8	.4	13.4	1.1	.5	19	17	.6	1	178	280	.04	.04	.4	2
	(ZMB)		0	361	5.2	2.6	88.7	7.3	3.3	128	117	4.0	8	1,181	1,854	.28	.28	2.7	15
Subter.	(APB)	(5)	84.1	55	2.3	.2	12.5	.9	1.0	28	68	.9	31	337	1,328	.08	.05	.4	1
veg.	(ZMB)		0	346	14.5	1.0	78.6	5.7	6.0	176	428	5.5	194	2,119	8,352	.50	.31	2.5	8
Roots	(APB)	(10)	79.7	74	1.8	.3	17.3	1.0	.9	32	54	1.0	14	399	568	.09	.05	.8	1
	(ZMB)		0	365	8.9	1.5	85.2	4.9	4.4	158	266	4.9	69	1,966	2,798	.44	.25	3.9	8
Cereals	(APB)	(10)	11.1	342	11.0	2.3	74.0	2.6	1.6	35	305	3.4	4	339	0	.45	.20	3.8	[illegible]
	(ZMB)		0	385	12.4	2.6	83.2	2.9	1.8	39	343	3.8	4	381	0	.51	.22	4.3	[illegible]
Pulses	(APB)	(10)	10.3	372	24.4	8.0	54.0	4.5	3.3	106	395	6.5	17	1,056	34	.66	.22	3.8	[illegible]
	(ZMB)		0	415	27.2	8.9	60.2	5.0	3.7	118	440	7.2	19	1,177	37	.73	.25	4.2	[illegible]
Nuts	(APB)	(10)	4.2	639	14.4	62.3	16.6	1.7	2.5	86	461	4.1	4	592	55	.65	.24	1.8	Trac
	(ZMB)		0	667	15.0	65.0	17.3	1.8	2.6	90	481	4.3	4	618	57	.68	.25	1.9	Trac
Overall	(APB)	(75)	63.5	165	6.5	7.5	21.0	1.6	1.5	66	149	2.3	21	429	1,292	.24	.15	1.4	2
average	(ZMB)		0	376	18.8	9.8	64.4	6.7	7.2	410	403	9.6	166	2,164	11,323	.74	.83	5.2	24

From Duke, J. A. and Atchley, A. A., *Proximate Analysis Tables of Higher Plants*, CRC Press, Boca Raton, 1986.

Table 3
PROXIMATE ANALYSES OF TEN IMPORTANT NUTS
(per 100g)

Nuts

Food item	H_2O (g)	Cal	Prot (g)	Fat (g)	Total carb (g)	Fiber (g)	Ash (g)	Ca (mg)	P (mg)	Fe (mg)	Na (mg)	K (mg)	β-Car (μg)	Thia (mg)	Rib (mg)	Nia (mg)	Vit C (mg)
mond	4.7	598	18.6	54.2	19.5	2.6	3.0	234	504	4.7	4	773	0	.24	.92	3.5	Tr
azil nut	4.6	654	14.3	66.9	10.9	3.1	3.3	186	693	3.4	1	715	Tr	.96	.12	1.6	—
ıshew	5.2	561	17.2	45.7	29.3	1.4	2.6	38	373	3.8	15	464	60	.43	.25	1.8	—
ckory	3.3	673	13.2	68.7	12.8	1.9	2.0	Tr	360	2.4	—	—	—	—	—	—	—
acadamia	3.0	691	7.8	71.6	15.9	2.5	1.7	48	161	2.0	—	264	0	.34	.11	1.3	0
can	3.4	687	9.2	71.2	14.6	2.3	1.6	73	289	2.4	Tr	603	78	.86	.13	.9	2
linut	6.3	669	11.4	71.1	8.4	2.7	2.8	140	554	3.4	3	489	24	.88	.09	.5	Tr
ıon	3.1	635	13.0	60.5	20.5	1.1	2.9	12	604	5.2	—	—	18	1.28	.23	4.5	Tr
ıtachio	5.3	594	19.3	53.7	19.0	1.9	2.7	131	500	7.3	—	972	138	.67	—	1.4	0
alnut	3.1	628	20.5	59.3	14.8	1.7	2.3	Tr	570	6.0	3	460	180	.22	.11	.7	—
ı. (APB)	4.2	639	14.4	62.3	16.6	1.7	2.5	86	461	4.1	4	592	55	.65	.24	1.8	Tr
ı. (ZMB)	0	667	15.0	65.0	17.3	1.8	2.6	90	481	4.3	4	618	57	.68	.25	1.9	Tr

ıte: APB, Average on an "as-purchased basis"; ZMB, Average on a zero-moisture basis; tr, Trace.

om Duke, J. A., *CRC Handbook of Medicinal Herbs*, CRC Press, Boca Raton, Fla., 1985.

THE EDITOR

Born in Birmingham, Alabama in 1929, James A. (''Jim'') Duke is a Phi Beta Kappa graduate of the University of North Carolina, where he took his AB (1952) and MS (1955) in Botany. Following a 2 1/2 year tour of military duty at Fort Dietrick, Maryland, Duke took his PhD in Botany in 1961 at the University of North Carolina. Then he undertook postdoctoral activities at Washington University and the Missouri Botanical Garden in St. Louis, Missouri, where he began studies of neotropical ethnobotany, his overriding interest to this day. From 1963 to 1965, Duke served with the USDA at Beltsville, Maryland, devoting much of his time to neotropical ecology, especially seedling ecology. In 1965, he joined Battelle Columbus Laboratories, for ecological and ethnological studies in Panama and Colombia to determine the ethnological feasibility of a Sea Level Canal. He spent long weeks in the forests with the various ethnic groups there, learning which plants they used as food, fiber, fuel, and medicine. Duke counts these years in Panama and Colombia as the most interesting in his career. He rejoined the USDA in 1971 for crop diversification and medicinal plant studies in developing countries. He has taped dozens of TV and radio shows on various aspects of economic botany. He is a popular lecturer on the subjects of ethnobotany, herbs, medicinal plants, and new crops, and their ecology. He grows dozens of interesting plants on his six acre farmette (Herbal Vineyard) with his wife and illustrator, Peggy. Fluent in Spanish, Duke has studied and/or lectured in Argentina, Belize, Bolivia, Brazil, Burma, Chile, China, Colombia, Costa Rica, Dominican Republic, Ecuador, Egypt, England, France, Guatemala, Honduras, India, Iran, Italy, Jamaica, Korea, Laos, Mexico, Panama, Peru, Puerto Rico, Singapore, Syria, Thailand, and Vietnam. Considered a key figure in the ''Herbal Renaissance'', Duke received the Cutty Sark Science Award in 1981. Currently, he is with the Germplasm Introduction and Evaluation Laboratory, USDA, Beltsville, Maryland, working on an encyclopedia of economic plants.

Dr. Duke is an active member in the American Herb Association (Life), American Society of Pharmacognosy, Association for Tropical Biology, Council for Agricultural Science and Technology (Cornerstone Life), Friends of the National Arboretum, Herb Research Foundation, International Association of Plant Taxonomists (Life), International Platform Association, International Society for Tropical Ecology, International Society for Tropical Root Crops (Life), International Weed Science Society (Life), Organization for Tropical Studies (Life), Oriental Healing Arts Society, Programa Interciencia de Recursos Biologicos (Advisory Council), Sigma Xi (Life), Society for Economic Botany (Life), Southern Appalachian Botanical Club (Life), Tri-State Bluegrass Association (Life), and Weed Science Society of America.

Having published several scientific papers and book chapters, Duke is sole or senior author of several books, including *Isthmian Ethnobotanical Dictionary*, Harrod and Company, Baltimore, 1972 (being reissued with illustrations by Scientific Publishers, Jodphur, India, 1986); *Handbook of Legumes of World Economic Importance*, Plenum Press, New York, 1981 (Japanese translation, 1986); *Medicinal Plants of the Bible*, Trado-Medic Books, Buffalo, New York, 1983; with E. S. Ayensu, *Medicinal Plants of China*, 2 volumes, Reference Publications, Algonac, Michigan, 1985; *Culinary Herbs: A Potpourri*, Trado-Medic Books, Buffalo, New York, 1985; *CRC Handbook of Medicinal Herbs*, CRC Press, Boca Raton, Florida, 1986; *Handbook of Northeastern Amerindian Medicinal Plants*, Quarterman Press, Lincoln, Massachusetts, 1986; with A. A. Atchley, *CRC Handbook of Proximate Analysis Tables of Higher Plants*, CRC Press, Boca Raton, Florida, 1986; and with A. A. Atchley, K. T. Ackerson, and P. K. Duke, *CRC Handbook of Agricultural Energy Potential of Developing Countries*, CRC Press, Boca Raton, Florida, 1987. Duke is a contributing editor to the following periodicals: *Bu$iness of Herbs, Coltsfoot, Herbalgram*, and the *International Permaculture Species Yearbook*; and he frequently writes for *American*

Health, *Economic Botany*, and *Organic Gardening*. In 1985, he published the *Herbalbum: an Anthology of Varicose Verse*, parts of which were recorded in 1986 as an LP album (bluegrass), the *Herbalbum*, by a studio band in Nashville, Tennessee. In 1987, Quarterman Press published *Living Liqueurs*.

ACKNOWLEDGMENTS (Conceptualization)

Herb Strum, Agricultural Marketing Specialist, USDA, triggered all this when he called and asked if I knew anyone who could address his Agricultural Marketing Workshops on tropical nuts. The next thing you know, I became the speaker without portfolio. Since these nuts are high-priced, light-weight, often labor-intensive crops, it was only natural that I should view the nuts as possible alternative crops for narcotics. For their support in my alternative crops program, I am indebted to the USDA's Dr. T. J. Army, Deputy Administrator, National Program Staff, Beltsville, Maryland; Dr. W. A. Gentner, Research Leader, Weed Science Laboratory, ARS, BARC, Beltsville, Maryland, and Quentin Jones, Assistant to Deputy Administrator for Germplasm (now retired), National Program Staff, Beltsville, Maryland.

In the preparation for these talks, I called on those more knowledgeable to help me decide what should be discussed in papers on Tropical Nuts. I sent these fine correspondents the crude check list, as follows:

TROPICAL NUTS

Anacardium occidentale	Cashew
Artocarpus altilis (A. communis)	Breadfruit
Bauhinia esculenta (Tylosema esculentum)	Marama nut or bean
Bertholettia excelsa	Brazil nut
Buchanania latifolia (lanza)	Cudapah almond or cuddapaha almond or Chironji nut
Canarium indicum	Java almond
Caryocar nuciferum	Suari nut
Caryodendron	Inchi nut
Cordeauxia edulis	Jeheb nut, ye-eb nut
Irvingia gabonensis	Dika nut
Lecythis ollaria	Sapucaja nut
Lecythis minor	—
Lecythis zabucajo	Paradise nut
Licania rigida	Oiticica
Macadamia spp.	Macadamia nut
Omphalea megacarpa	Hunter's nut
Ongokea klaineana	Isano nut
Palaquium burukii	Siak illipe nut
Pangium edule	Pangi nut
Poga oleosa	Oboga nut
Ricinodendron heudelotii	Essang nut
Sclerocarya caffra	Marula nut
Sterculia chicha	Maranhao nut
Terminalia catappa	Indian almond
Terminalia okara	Okari nut
Telfairia pedata	Oyster nut

Omit:

Jojoba
Coconut
Chestnut, water
Cola nut
Chufa or Tiger nut
Peanut
Groundnut
Litchi nut

Frank Martin added the jackfruit (*Artocarpus integer*) and the champedak (*Artocarpus heterophylla*), emphasizing that they were distinct species. He also added *Aleurites triloba* Forst, one of the many candle nuts, stating that it is edible when roasted. Further, he added the palmyra palm (*Borassus flabellifer* L.), *Gnetum gnemon* (adding that it is excellent), and *Telfairia occidentalis*, another oysternut. He challenged my exclusion of the coconut, and cautioned that *Sterculia chicha* contains a poisonous cyclopropenoid fatty acid.

Gerardo Budowski, of CATIE, added *Salacca edulis*, which is very important in Indonesia, often served at receptions. After he consulted Menninger, he queried how worthwhile are some of the nuts. If you listen to Menninger, all kinds of things are nuts, and may be delicious — to some, such as a large group of palms.

So I wrote to palm specialist Dennis Johnson, and sent him the list of the more than 50 genera of palms that Menninger had included in this books. Dennis seemed comfortable with leaving these in a talk on nuts and added *Areca*, which Menninger excluded because it was not ingested, and the Pacific ivory nut, *Coelococcus*.

Harold Winters, retired USDA author of Kennard and Winters, *Some Fruits and Nuts for the Tropics*,[160] also added several species to my list.

Bob Knight, of the USDA Station at Miami, reminded me of the double meaning of breadnut (1) as the nut of *Brosimum alicastrum*, also known as Maya Breadnut, and (2) as a seeded breadfruit. Bob also added *Aleurites moluccana* as a chemurgic nut, and *Castanopsis* as an edible. He reminded me of the unfortunate consequences of overeating seleniferous species of the Lecithidaceae.

Carl Campbell, also of Florida, added *Brosimum* too, with the Pili Nut (*Canarium ovatum*) and the Malabar chestnut (*Pachira macrocarpa*). He reminded me, as did Julia Morton and Bob Knight, that the pangi nut and the oiticica were "toxic(?)" and "hardly edible", respectively. They are right.

Ernie Imle, retired USDA cacao specialist, sent literature ranking the pili nut, *Canarium ovatum*, up with the macadam and cashew. He mentioned that several lines of pili were established at La Zamorana, Honduras.

Julia Morton added the jackfruit *Artocarpus heterophylla*, the breadnut, *Brosimum alicastrum*, the quandong, *Fusanus acuminatus*, the Tahiti chestnut, *Inocarpus edulis*, and the Saba nut, or *Pachira aquatica*, and included data on these and other nut species which I have incorporated in my write-ups on these species. She also added her acuminate capsular reviews of Menninger's and Rosengarten's books and equally acuminate warnings on other of our nut species.

I also acknowledge the help of Jayne Maclean, National Agricultural Library, who went through a list of tropical nuts to check how many citations there were in her computerized search. The tabulation which follows, with the number of "hits", might suggest the relative importance of the tropical nuts in the literature:

162 *Anacardium occidentale*	0 *Omphalea megacarpa*
22 *Artocarpus altilis* or *communis*	0 *Ongokea klaineana* (= *O. gore*)
2 *Bauhinia esculenta*	1 *Palaquium*
24 *Bertholettia excelsa*	0 *Pangium edule*
2 *Buchanania latifolia*	0 *Pogo oleosa*
0 *Canarium indicum*	2 *Licania rigida*
0 *Caryocar nuciferum*	0 *Ricinodendron* sp.
0 *Caryodendron* sp.	0 *Sclerocarya caffra*
1 *Irvingia gabonensis*	1 *Sterculia chicha*
3 *Cordeauxia edulis*	8 *Terminalia catappa*
0 *Lecithis ollaria*	0 *Telfairia pedata*
0 *Lecithis zabucajo*	

My wife, Peggy, has helped in gathering and touching up illustrations, some in the public domain, some being redrawn and reproduced here with the permission of the artist and/or publishers. She has gone to the libraries and herbaria around Washington to seek out illustrations, or specimens with which to improve on the quality of the illustrations herein. She is responsible for those drawings bearing her name. Last and most, my thanks go to my program assistant, Judy duCellier, who helped compile information into format from several disparate sources. Not only has she learned to read my handwritten annotations and seek out data from obscure sources, she has been good enough to type the manuscript as well. In the civil service system, the very fact that she types the data she helped gather may jeopardize her promotion potential. Take this as my letter of recommendation.

ACKNOWLEDGMENTS (Procedure and format)

For conventional nut species, I was immensely aided by a USDA contract with Dr. C. F. Reed,[278] who prepared rough drafts on description, uses, varieties, distribution, ecology, cultivation, economics, yields, and biotic factors of 1000 economic species. I was responsible for the drafts of the nonconventional species reviewed herein as nuts, and Judy duCellier and I edited, updated, and augmented the Reed drafts on the conventional species. Certain major sources constituted the major documentation for Dr. Reed's early drafts and my final drafts.

For the Use paragraph, the major references were Bailey,[25] Bogdan,[43] Brown,[51] Brown and Merrill,[52] Burkill,[55] C.S.I.R.,[70] Dalziel,[71] Duke,[81,86] *Hortus III,*[139] MacMillan,[196] Martin and Ruberte,[202] Uphof,[324] and many others. Often in this or other paragraphs I have internally cited the Chemical Marketing Reporter,[351] a weekly tabloid with much useful information.

For the Folk Medicine paragraphs, primary resources were Boulos,[45] C.S.I.R.,[70] Duke,[80] Duke and Ayensu,[90] Duke and Wain,[93] Hartwell,[126] Kirtikar and Basu,[165] List and Horhammer,[187] Morton,[224] Perry,[249] and Watt and Breyer-Brandwijk.[332]

For the Chemistry paragraph, the major references were C.S.I.R.,[70] Duke,[86] Duke and Atchley,[89] Gibbs,[109] Gohl,[110] Leung et al.,[181] List and Horhammer,[187] and Morton.[224]

For the Description paragraph, various floras were consulted in addition to the prime references, Kirtikar and Basu,[165] Little,[188] Ochse,[238] Radford, Ahles, and Bell,[276] and Reed.[278]

For the Germplasm paragraph, the major references were Duke,[82] Reed,[278] and Zeven and Zhukhovsky;[350] for the Distribution paragraph, various floras, Holm et al.,[134] Little,[188] and NAS;[231,231] for the Ecology paragraph, C.S.I.R.,[70] Duke,[82] Holm et al.,[135] Little,[188] and NAS.[231,232] (While ecological amplitudes were available for many of these nuts from Duke,[821] in other cases I amplified the Duke data from other sources. For yet other species with no hard data, I estimated ecological magnitudes.)

For the Cultivation and Harvesting paragraphs, C.S.I.R.,[70] NAS,[231,232] Purseglove,[272] Reed[70] were consulted; for the Yields and Economics paragraph, Bogdan,[43] Duke,[82] FAO,[98] and Reed;[70] for the Energy paragraph, Channel,[61] Duke,[87] NAS,[232] and Westlake;[334] for the Biotic Factors paragraph, Browne,[53] and Agriculture Handbook No. 165,[4] were the primary references. Dr. C. F. Reed went through some USDA mycology files[186] for those on which he cooperated. These names have not all been verified. In the Biotic Factor or Cultivation paragraph, there may be bibliographic mention of pesticides. In no way do I imply acceptance or rejection of a pesticide by inclusion or omission. I have merely recited items that may be of interest to those seeking information on pesticides.

I have omitted several "nuts" included in my *Handbook of Legumes of World Economic Importance.*[83] I have added other legumes, e.g., the groundnut, *Apios* (not really a nut, but a root), and the yeheb, the tallownut, which were not covered in the handbook. I rank *Apios* with the promising, yet still undeveloped, new crops of the New World.[40]

Warning — Although I have compiled from the literature folk medicinal applications for some of these nut species, neither I nor my publishers endorse or even suggest self diagnosis or herbal medication. The folk medicinal information was compiled from open literature, and I cannot vouch for its safety nor efficacy. As a matter of fact, I suspect some folk medicinal applications are both dangerous and inefficacious.

TABLE OF CONTENTS

ACROCOMIA SCLEROCARPA Mart. (ARCEACEAE) — Gru-Gru Nut, Coco de Catarro, Macauba, Mucaja
Syn.: *Acrocomia aculeata* (Jacq.) Lodd.

Uses — The slimy, soft external tissue (mesocarp) and the seed yield oil. The mesocarp oil can be used as cooking oil, without refining, if extracted from fresh or properly stored fruits. The mesocarp oil is also used for soaps. The kernel oil, with a sweet taste like coconut oil, is used as an edible oil, e.g., in the preparation of margarine.[29,152]

Folk medicine — Sometimes used as a purgative and vermifuge.

Chemistry — Seed contains 60% fat with 17% saturated fatty acids (74.6% oleic acid and 8% linoleic acid). Fruit contains 4.58 mg carotene per 100 g fresh weight. Flowers contain 2.1% gallic acid and tannin.[187] According to Balick,[29] air-dried kernels yield 53 to 65 (to 69.4%), pulp up to 63.7% fat. The yellow pulp oil is softer and has a higher iodine value than palm oil, but, unfortunately, hydrolyzes rapidly after harvested, especially if damaged, like the oil palm. Johnson[152] says that fresh fruits contain 35% moisture; dry fruit mesocarp yields 33% oil, the kernel 53.75%.

Description — Armed palm to 11 m tall. Leaves pinnate, armed, like the trunk. Inflorescence with very sharp fine spines. Fruit a reddish-yellow edible drupe surrounded by a tough woody kernel. Dry fruits weigh about 18 g, with 19.8% outer shell, 41.1% mesocarp pulp, 29.0% inner shell, and 10.1% kernel.

Germplasm — Reported from the South American Center of Diversity, gru-gru is reported to tolerate drought.

Distribution — Widely dispersed in Brazil, especially in Minas Gerais, where it grows in dense groves. Ranging into Paraguay.

Ecology — Estimated to range from Tropical Wet to Dry through Subtropical Wet to Dry Forest Life Zones, gru-gru nut is estimated to tolerate annual precipitation of 10 to 40 dm, annual temperature of 22 to 28°C, and pH of 6 to 8. Sometimes gregarious in dense groves. In Johnson,[152] Balick notes this palm occurs in drier regions than most palms, and therefore might be a useful economic plant in the dry areas.

Cultivation — Usually not cultivated.

Harvesting — Balick[29] notes the following, for oil palms in general, not necessarily for this species. "In commercial production, palm fruits first are harvested and removed from the panicles upon which they are formed. Sterilization is next, to inactivate the enzymes present in the mesocarp. These enzymes can cause deterioration of the oil through lipolysis, an increase of the free fatty acid content known commercially as rancidity. A so-called "hard oil", with up to 94.5% free fatty acids, is made by fermenting rather than sterilizing the ripe palm fruits. Sterilization also stops oxidation, which lowers the bleachability of the oil and makes it less valuable for commercial use. The fruits are then macerated to separate the oily pulp from the kernels. In small-scale, local production, natives may pound the fruits with a log or stone to release the pulp. On a large plantation, special machinery is used. To release the oil, this pulpy mass is pressed with a hand press, if primitively processed, or with heavy mechanical presses if on an industrial scale. Clarification follows: in a small operation, the oil is allowed to rise through a layer of boiling water and is then skimmed off. Large processing factories use a settling and centrifuge process. For commercial use, the oil is usually bleached, removing certain natural red or green pigments. These colors may lower the monetary value of the oil.

Kernels of some species of palms are often saved for their oil as a by-product of primitive fruit processing. These are then shipped to mills located in central areas, where heavier equipment is used for extraction. Natives in the past and today extract palm kernel oil by baking the kernels in an oven and pounding them in hollow logs. The resulting mash is boiled in pots with water, and the oil is collected as it rises to the top. Palm kernel cake, a product of the extraction, is a good protein source and may be used for either human consumption or as an animal feed.[29]

Yields and economics — Small local Brazilian establishments develop the oils, which are little known in the world market. Brazil produced small quantities of the oil before and during World War II. In 1980, Brazilian production was limited to three States: Maranhao, Ceara, and Minas Gerais, producing only 190 tons.[152]

Energy — The oil could be used like that of other oil palms for energy, the press-cake for alcohol production or animal feed. An 18 g fruit would yield ca. 2.4 g mesocarp oil and 1 g kernel oil.[152]

Biotic factors — No data available.

ACROCOMIA TOTAI Mart. (ARECACEAE) — Gru-Gru Nut, Paraguay Coco-Palm, Mbocaya

Uses — Since pre-Colombian times, this palm has, with *Copernicia australis* (most abundant palm in Paraguay), supplied food, shelter, and the raw material for fabrication of soaps, hats, ropes, baskets, bags, hammocks, and mats. In Argentina, it is regarded as an ornamental palm with edible nuts. Leaves are sometimes lopped for fodder in the dry season. The "cabbage" and base of the involucral leaves are eaten in salads. Ripe fruits are edible and tasty. Five industrially useful products are obtainable: pulp oil, kernel oil, kernel meal, kernel cake, and extracted pulp. The kernel oil is most valuable and abundant, usable for soap and food.[29]

Folk medicine — No data available.

Chemistry — Per 100 g, the mesocarp is reported to contain 4.3 g H_2O, 4.2 g protein, 27.9 g fat, 4.8 g total sugars, 8.8 g fiber, 10.32 g ash, 90 mg Ca, 120 mg P, and 2,180 mg K. Other data are tabulated in Markley.[199] (See Tables 1 and 2).

Description — Monoecious palm to 15 (to 20) m tall, the stipe provided with stout spines, some 7.5 to 12.5 (to 17) cm long. Leaves pinnate, 2 to 3 m long, individual leaflets 50 to 70 cm long; petiole with spines on the dorsal surface. Spadix interfoliar, 1 m long, like the inner spathe densely spinose. Fruits yellow, rounded, ca. 3 to 4 cm diam. with dark orange oily pulp, rich in carotene.

Germplasm — Reported from the South American (Paraguayan) Center of Diversity, mbocaya, or cvs thereof, is reported to tolerate savannas. Some trees are almost devoid of spines, except just below the crown.

Distribution — Higher altitude savannas in Argentina and Paraguay.

Ecology — Estimated to range from Tropical Very Dry to Wet through Subtropical Wet to Dry Forest Life Zones, mbocaya is estimated to tolerate annual precipitation of 8 to 35 dm, annual temperature of 22 to 28°C, and pH of 6 to 8.

Cultivation — Markley[199] calculates yield, at 10 × 4 m spacing (250 trees/ha) at 640 kg oil/ha, at 10 × 6 (166 trees) at 424 kg/ha, at 10 × 8 (125 trees) at 320 kg/ha, and at 10 × 10 (100 trees/ha) at 256 kg oil/ha. Markley's information suggests that the seeds might be as recalcitrant as those of oil palms.[199]

Harvesting — Humans usually eat only the pulp of freshly fallen fruits owing to the difficulty of extracting the kernels. Nature (decay and/or defecation, followed by rains) often leaves clean nuts lying on the ground, to be harvested by humans. Leaves are sometimes lopped to leave only two in the dry season.

Yields and Economics — The mbocaya palm is of greater economic importance to Paraguay than any other indigenous palm. Between 1940 and 1951, Paraguay produced 883 to 2,849 MT of kernel oil annually, exporting 13 to 2,588, and 170 to 1,125 MT pulp oil, exporting 109 to 2,074 MT. In 1971, Paraguay exported 7,400 MT, up from 2300 tons in 1964.[152] Commenting on comparative yields of oil per ha, Markley[199] shows only 96 to 640 kg/ha for this species, compared to 2,790 for oil palm, 818 for coconut, 420 for sesame, 392 for rapeseed, 308 for sunflower, 230 for peanuts, 193 for flaxseeds, and 190 for soybeans.

Energy — The oil could be used like that of other oil palms for energy, the press-cake for alcohol production or animal feed. Brazil is now studying this plant as a renewable source of fuel oil.[152]

Biotic factors — A highly destructive stem borer or snout beetle (*Rhyna barbirostris*) attacks the palm. Larvae may devour the whole interior, except for the long cellulose fiber. A fungus, probably *Phaecophora acrocomiac*, may cause yellow blotches with black centers on the leaves. Ruminants may eat the whole fruit, regurgitating or even defecating entire kernels ("nuts"). Seedlings may be devoured by insects, birds, or other animals, as well as attacked by microorganisms.

Table 1
COMPOSITION (%) OF COMMERCIAL SAMPLES OF *A. TOTAI* PRODUCTS[199]

Constituent	Outer hull (epicarp)	Pulp (mesocarp)	Pulp, expeller cake	Shell (endocarp)	Kernel	Kernel, expeller cake
Moisture (H_2O)	6.65	4.31	5.26	6.84	3.17	7.44
Lipides (oil)	3.88	27.94	6.26	2.46	66.75	7.22
Nitrogen	0.74	0.67	0.98	0.31	2.02	5.50
Protein ($N \times 6.25$)	4.62	4.18	6.12	1.94	12.62	34.38
Crude fiber	36.00	8.82	6.83	49.69	8.60	11.65
Sugars (total)	—	4.85	5.16	—	1.28	2.80
Ash	5.82	10.32	9.16	3.26	1.98	5.37
Potassium	2.18	2.18	2.75	1.02	1.36	1.55
Phosphorus	0.10	0.12	0.16	0.04	0.42	1.14
Calcium	0.07	0.09	0.10	0.04	0.08	0.27

Table 2
CHARACTERISTICS AND COMPOSITION OF THE PULP OILS OF *A. TOTAI* AND *E. GUINEENSIS*[199]

Characteristic	*A. totai*[14]	*A. totai*[13]	*E. guineensis*
Specific gravity (40°C)	—	0.9240	0.898—0.901
Refractive index (40°C)	1.4615	1.4582—1.4607	1.453—1.456
Titer value (°C)	—	26.1—33.2	40—47
Iodine value	68.4	54.5—66.7	44—58
Unsaponifiable matter (%)	0.81	0.27—0.55	<0.8
Saponification value	197.0	200—209	195—205
Free fatty acids (% palmitic)	41.2	1—?	
Total fatty acids			
Iodine value	69.7	—	—
Thiocyanogen value	66.6	—	—
Saturated (%)	20.0	—	39—50
Oleic (%)	80.0	—	38—52
Linoleic (%)	0.0	—	6—10

ADHATODA VASICA (L.) Nees (ACANTHACEAE) — Malabar Nut, Adotodai, Pavettia, Wanepala, Basak
Syn.: ***Justicia adhatoda*** L.

Uses — Plants grown for reclaiming waste lands. Because of its fetid scent, it is not eaten by cattle and goats. Leaves and twigs commonly used in Sri Lanka as green manure for field crops, and elsewhere in rice fields. Leaves, on boiling in water, give durable yellow dye used for coarse cloth and skins; in combination with indigo, cloth takes a greenish-blue to dark green color. Also used to impart black color to pottery. Stems and twigs used as supports for mud-walls. Wood makes good charcoal for gunpowder, and used as fuel for brick-making. Ashes used in place of crude carbonate of soda for washing clothes. In Bengal, statue heads are carved from the wood. Leaves also used in agriculture as a weedicide, insecticide, and fungicide, as they contain the alkaloid, vasicine. As a weedicide, it is used against aquatic weeds in rice-fields; as insecticide, used in same way tobacco leaves; as fungicide, they prevent growth of fungi on fruits which are covered with vasica leaves. Market gardeners place layers of leaves over fruit, like mangoes, plantains, and custard-apples, which have been picked in immature state to hasten ripening and to ensure development of natural color in these fruits without spoilage.[86]

Folk medicine — Plant has many medicinal uses. Whole plant used in Sri Lanka for treatment of excessive phlegm, and in menorrhagia. Leaves are source of an expectorant drug used to relieve coughs. Plants are used in folk remedies for glandular tumors in India. Leaf used for asthma, bronchitis, consumption, cough, fever, jaundice, tuberculosis; smoked for asthma; prescribed as a mucolytic, antitussive, antispasmodic, expectorant. Ayurvedics[165] use the root for hematuria, leucorrhea, parturition, and strangury, the plant for asthma, blood impurities, bronchitis, consumption, fever, heart disease, jaundice, leucoderma, loss of memory (amnesia), stomatosis, thirst, tumors, and vomiting. Yunani use the fruit for bronchitis, the flowers for jaundice, poor circulation, and strangury; the emmenagogue leaves in gonorrhea, and the diuretic root in asthma, bilious nausea, bronchitis, fever, gonorrhea, and sore eyes.[86]

Chemistry — Used in Indian medicine for more than 2000 years, adhatoda now has a whole book dedicated to only one of its active alkaloids.[22] In addition to antiseptic and insecticidal properties, vasicine produces a slight fall of blood pressure, followed by rise to the original level, and an increase in the amplitude of heart beats and a slowing of the rhythm. It has a slight but persistent bronchodilator effect. With a long history as an expectorant in India, vasicine has recently been modified to form the derivative bromhexine, a mucolytic inhalant agent, which increases respiratory fluid volume, diluting the mucus, and reduces its viscosity. Fluid extract of leaves liquifies sputum, relieving coughs and bronchial spasms. The plant also contains an unidentified principle agent active against the tubercular bacillus. Adhatodine, anisotinine, betaine, vasakin, vasicine, vasicinine, vasicinol, vasicinone, vasicoline, vasicolinone, are reported. Deoxyvasicine is a highly effective antifeedant followed by vasicinol and vasicine. These plant products as antifeedants could be safely used for controlling pests on vegetable crops.''[14,86] Atal[22] devoted a whole book to the chemistry and pharmacology of Vasicine-A. At the Regional Research Laboratory (RRL), in Jammu, vasicine showed a definite bronchodilatory effect, comparable to that of theophylline, as well as hypotensive, respiratory stimulant, and uterotonic activities.[13] The total alkaloid content is up to 0.4%, of which 85 to 90% is vasicine.

Toxicity — Vasicine is toxic to cold-blooded creatures (including fish) but not to mammals. Although it is not listed in many poisonous plant books, the fact that it is not grazed suggests that it could well be poisonous.[22] Vasicine and vasicinol exhibit potential to reduce fertility in insects. ''Vasicine is also likely to replace the abortifacient drugs in current use as its abortifacient activity is comparable to prostaglandins.''[22] In large doses the leaves cause diarrhea and nausea.[86]

Description — A gregarious, evergreen, densely branched shrub 1.5 to 3 (to 6) m tall; bark smooth, ash-colored; branches softly hairy, internodes short; leaves opposite, elliptic, ovate or elliptic-lanceolate, pointed at both ends, acuminate, entire, minutely pubescent, 12.5 to 20 cm. long, 8 cm. broad; flowers white with red, pink, or white spots or streaks, in dense axillary, stalked, bracteate spikes 2.5 to 7.5 cm long; bracts conspicuously leafy, 1-flowered; calyx deeply divided into 5 lobes, pubescent; corolla 2-lipped, pubescent outside; upper lip notched, curved, lower lip 3-lobed; capsules 2.5 cm or more long, 0.8 cm broad, clavate, pubescent, 4-seeded; seeds suborbicular, rugose. Flowers and fruit December to April; in some areas flowers May—June also.[278]

Germplasm — Reported from the Indochina-Indonesia Centers of Diversity, Malabar Nut or cvs thereof is reported to tolerate fungus, insects, mycobacteria, and weeds.[82]

Distribution — Common to tropical India from Punjab to southern India, Sri Lanka, N. Burma, Pakistan (Karachi, Sind, Khyber, Wazir, Kurram, Dir); Hong Kong, China, Yunnan, where common.[278]

Ecology — Abundant and gregarious in many areas of China and India, growing in full sun, at edges of forests, in hilly regions often as the co-dominant shrub with *Capparis sepiaria* L. Also grows in full sun on flood plains and in meadows. In Curacao, it grows well on weathered diabase, in south Florida on oolitic limestone. In Sub-Himalayan region ascends to 1,300 m altitude, more frequent at altitudes about 200 to 300 m. Requires a subtropical to tropical climate with moderate precipitation. Though killed to the ground by brief frosts, it recovers rapidly. Ranging from Warm Temperate Dry through Tropical Very Dry Forest Life Zones, Malabar nut is reported to tolerate annual precipitation of 5 to 42 dm (mean of 5 cases = 22), annual temperature of 15 to 27°C (mean of 5 cases = 24), and pH of 4.5 to 7.5 (mean of 4 cases = 6.1).[82,278]

Cultivation — As plants are quite common, often abundant, and gregarious in regions of adaptation and where people use the plant, the plant is cultivated mainly in areas of habitation, as hedges, wind-breaks, and for reclaiming soil. Propagation is by seeds broadcast in areas of need, or in waste areas about areas of cultivation. Any forest edge is a likely place to seed, so that the leaves or branches will be handy for use on other cultivated plants. No particular care is taken, as the plants thrive on any tropical soil that is well-drained and has sufficient precipitation. The plants, also propagated readily from cuttings, are said to coppice well.[278]

Harvesting — Harvesting leaves and branches varies according to the needs of the local farmer, for green manure, covering fruits or protection, etc. As plants are evergreen, leaves are available year-round.[278]

Yields and economics — No data available. However, plants are plentiful, and supply all the leaves and twigs needed by those who use them. An important plant for reclaiming waste land in areas of adaptation, as in India and Sri Lanka. Also used as weedicide, insecticide, and fungicide in tropical areas. Mainly used in tropical Southeast Asia, S. China, India, and Sri Lanka. One ton of leaves can yield 2 kg vasicine equivalent to 2 million human doses.[13]

Energy — I was surprised to see this listed in a book on firewood trees.[232] They note that it has a particularly desirable wood for quick, intense, long-lasting cooking fires, with little or no odor, smoke or sparks. The moderately hard wood has been used to manufacture gunpowder charcoal.[232] If vasicine becomes commercialized, the biomass residues (>99%) following vasicine extraction could conceivably serve as a pesticidal mulch or for conversion to alcohol. Perhaps this should be viewed like the neem tree in the third world, stripping the leaves as a pesticidal mulch, using the woody ''skeleton'' for firewood.

Biotic factors — Fungi reported attacking this plant include the following species: *Aecidium adhatodae, Alternaria tenuissima, Cercospora adhatodar, Chnoospora butleri, Phomopsis acanthi (Phoma acanthi).*[186] Plants are parasitized by *Cuscuta reflexa*. Not browsed by goats or other animals. One source states that this plant ''is never attacked by any insect . . . even the voracious eater, Bihar Hairy Catterpillar (sic) (*Dicresia obliqua*) avoids this plant.''[14]

ALEURITES FORDII Hemsl. (EUPHORBIACEAE) — Tung-Oil Tree

Uses — Tung trees are cultivated for their seeds, the endosperm of which supplies a superior quick-drying oil, utilized in the manufacture of lacquers, varnishes, paints, linoleum, oilcloth, resins, artificial leather, felt-base floor coverings, and greases, brake-linings and in clearing and polishing compounds. Tung oil products are used to coat containers for food, beverages, and medicines; for insulating wires and other metallic surfaces, as in radios, radar, telephone, and telegraph instruments.[111,260]

Folk medicine — Reported to be emetic, hemostat, and poisonous, tung-oil tree is a folk remedy for burns, edema, ejaculation, masturbation, scabies, swelling, and trauma.[91]

Chemistry — The fruit contains 14 to 20%; the kernel, 53 to 60%; and the nut, 30 to 40% oil. The oil contains 75 to 80% alpha-elaeo stearic-, 15% oleic-, ca 4% palmitic-, and ca. 1% stearic acids. Tannins, phytosterols, and a poisonous saponin are also reported.[187]

Description — Trees up to 12 m tall and wide, bark smooth, wood soft; leaves dark green, up to 15 cm wide, heart-shaped, sometimes lobed, appearing usually just after, but sometimes just before flowering; flowers in clusters, whitish, rose-throated, produced in early spring from terminal buds of shoots of the previous season; monoecious, male and female flowers in same inflorescence, usually with the pistillate flowers surrounded by several staminate flowers; fruits spherical, pear-shaped or top shaped, green to purple at maturity, with 4 to 5 carpels each with one seed; seeds usually 4 to 5, but may vary from 1 to 15, 2 to 3.2 cm long, 1.3 to 2.5 cm wide, consisting of a hard outer shell and a kernel from which the oil is obtained. Flowers February to March; fruits late September to early November.[278]

Germplasm — Reported from the China-Japan and North American Centers of Diversity, tung-oil tree, or cvs thereof, is reported to tolerate bacteria, disease, frost, insects, poor soil, and slope.[82] High-yielding cultivars continue to be developed. Some of the best cvs released by the USDA for growing in the southern U.S. are the following:

- 'Folsom': low-heading, high productivity; fruits large, late maturing, turning purplish when mature, containing 21% oil; highest resistance to low temperature in fall.
- 'Gahl': low-heading, productive; fruits large, 20% oil content; matures early, somewhat resistant to cold in fall.
- 'Isabel': low-heading, highly productive; fruits large, maturing early, 22% oil content.
- 'La Crosse': High-heading, exceptional productivity; fruits small, late maturing, tending to break segments if not harvested promptly, 21 to 14% oil content; a very popular cv.
- 'Lampton': out-yields all other varieties; very low-heading; fruits large, early maturing; 22% oil content.

Several other species of *Aleurites* are used to produce tung-oil, usually of low quality: *Aleurites cordata*, Japanese wood-oil tree; *A. moluccana*, Candlenut or lumbang tree; *A. trisperma*, Soft Lumbang tree; none of which can be grown commercially in the U.S. *Aleurites montana*, Mu-tree, is the prevailing commercial species in South China and could be grown in Florida.[82,259,278] (zn = zz.)

Distribution — Native to central and western China, where seedlings have been planted for thousands of years; planted in the southern U.S. from Florida to eastern Texas.[278]

Ecology — Ranging from Warm Temperate Dry to Wet through Tropical Very Dry to Moist Forest Life Zones, tung-oil tree is reported to tolerate annual precipitation of 6.4 to 21.0 dm (mean of 22 cases = 14.0), temperature of 18.7 to 27.0°C (mean of 21 cases = 24.0°C), pH of 5.4 to 7.1 (mean of 5 cases = 6.2).[82] Tung trees are very exacting in climatic and soil requirements. They require long, hot summers with abundant moisture, with usually at least 112 cm of rainfall rather evenly distributed through the year. Trees require 350 to 400 hr in winter with temperatures 7.2°C or lower; without this cold requirement, trees tend to produce suckers from the main branches. Vigorous but not succulent growth is most cold-resistant; trees are susceptible to cold injury when in active growth. Production of tung is best where day and night temperatures are uniformly warm. Much variation reduces tree growth and fruit size. Trees grow best if planted on hilltops or slopes, as good air-drainage reduces losses from spring frosts. Contour-planting on high rolling land escapes frost damage. Tung makes its best growth on virgin land. Soils must be well-drained, deep aerated, and have a high moisture-holding capacity to be easily penetrated by the roots. Green manure crops and fertilizers may be needed. Dolomitic lime may be used to correct excessive acidity; pH 6.0 to 6.5 is best; liming is beneficial to most soils in the Tung Belt, the more acid soils requiring greater amounts of lime.[82,278]

Cultivation — Tung trees may be propagated by seed or by budding. Seedlings generally vary considerably from parent plants in growth and fruiting characters. Seedlings which have been self-pollinated for several generations give rather uniform plants. Only 1 out of 100 selected "mother" tung trees will produce seedlings sufficiently uniform for commercial planting. However, a "mother" tree proven worthy by progeny testing may be propagated by budding. The budded trees, which are genetically identical with the original tree, will provide an adequate supply of seed satisfactory for planting. Seedlings are used for the root system for budded trees. Buds from "mother" trees are inserted in stems of 1-year old seedlings, 5 to 7.5 cm above the surface of the soil. Later, the original seedling top is cut off and a new top grown for the transplanted bud, making the tops of budded trees parts of the parent tree. Usually seedling trees outgrow budded trees, but budded trees produce larger crops and are more uniform in production, oil content, and date of fruit maturity. Tung seed are normally short-lived and must be planted during the season following harvest. Seeds are best hulled before planting, as hulls retard germination. Hulled seed may be planted dry, but soaking in water for 5 to 7 days hastens germination. Stratification, cold treatment or chemical treatment of seeds brings about more rapid and uniform germination. Dry-stored seed should be planted no later than February; stratified seed by mid-March; cold-treated and chemical treated seed by early April. Seed may be planted either by hand or with a modified corn-planter, the seed spaced 15 to 20 cm apart, about 5 cm, in rows 1.6 m apart, depending on the equipment to be used for cultivation and for digging the trees. Seeds germinate in 60 days or more; hence weed and grass control may be a serious problem. As soon as seedlings emerge, a side-dressing of fertilizer (5-10-5) with commercial zinc sulfate should be applied. Fertilizer is applied at rate of 600 kg/ha, in bands along each side of row, 20 cm from seedlings and 5 to 7.5 cm deep. Other fertilizers may be needed, depending on the soil. Most successful budding is done in late August, by the simple shield method, requiring a piece of budstock bark, including a bud, that will fit into a cut in the rootstock bar; a T-shaped cut is made in the bark of the rootstock at a point 5 to 7.5 cm above ground level, the flaps of bark loosened, shield-bud slipped inside flaps, and the flaps tied tightly over the transplanted bud with rubber budding stripe, 12 cm long, 0.6 cm wide, 0.002 thick. After about 7 days, the rubber stripe is cut to prevent binding. As newly set buds are susceptible to cold injury, soil is mounded over them for winter. When growth starts in spring, soil is pulled back and each stock cut back to within 3.5 cm of the dormant bud. Later, care consists of keeping all suckers removed and the trees well-cultivated. Trees are transplanted to the orchard late the following winter. Spring budding is done only as a last resort. Trees may be planted at 125 to 750/ha. When trees are small, close planting in rows greatly increases the bearing surface, but at maturity the bearing surface of a crowded row is about the same as that of a row with trees farther apart. However, it is well to leave enough space between row for orchard operations. In contour-planting, distances between rows and total number of trees per hectare vary; rows 10 to 12 m apart, trees spaced 3.3 to 4 m apart in rows, 250 to 350 trees/ha. Tops of nursery trees must be pruned back to 20 to

25 cm at planting. As growth starts, all buds are rubbed off except the one strongest growing and best placed on the tree. A bud 5 cm or more below the top of the stump is preferred over one closer to top.[278]

Harvesting — Tung trees usually begin bearing fruit the third year after planting, and are usually in commercial production by the fourth or fifth year, attaining maximum production in 10 to 12 years. Average life of trees in the U.S. is 30 years. Fruits mature and drop to ground in late September to early November. At this time they contain about 60% moisture. Fruits must be dried to 15% moisture before processing. Fruits should be left on the ground 3 to 4 weeks until hulls are dead and dry, and the moisture content has dropped below 30%. Fruits are gathered by hand into baskets or sacks. Fruits do not deteriorate on the ground until they germinate in spring.[74,278]

Yields and economics — Trees yield 4.5 to 5 tons/ha. An average picker can gather 60 to 80 bushels of fruits per day, depending on conditions of the orchard. Fruits may be gathered all through the winter season when other crops do not need care. Because all fruits do not fall at the same time, 2 or more harvestings may be desirable to get the maximum yield. Fruits are usually sacked, placed in the crotch of the tree and allowed to dry 2 to 3 weeks before delivery to the mill. Additional drying may be done at the mill, but wet fruits contain less oil percentage-wise and prices will be lower. Prices for tung oil depend on price supports, domestic production, imports, and industrial demands. World production in 1969 was 107,000 MT of tung nuts; in 1970, 143,000; and projected for 1980, 199,000. Wholesale prices were about \$0.276/kg; European import prices, \$0.335/kg. Growers received about \$51.10/ton of fruit of 18.5% oil content to about \$63.10/ton for fruits of 22% oil content. Major producing countries are mainland China and South America (Argentina and Paraguay); the U.S. and Africa produce much less. U.S. Bureau of Census figures 1,587,000 pounds of tung oil were consumed during February of 1982, representing a 1,307,000 pound drop from January. The largest application for the oil is paint and varnish, which accounted for 566,000 pounds of total consumption in February.[82,278,351] Dealers in tung oil include:[351]

Alnore Oil Co., Inc.
P.O. Box 699
Valley Stream, NY 11582

Pacific Anchor Chemical Corp.
6055 E. Washington Boulevard
Los Angeles, CA 90040

Industrial Oil Products Corp.
375 N. Broadway
Jericho, NY 11753

Welch, Holme, & Clark Co., Inc.
1000 S. 4th Street
Harrison, NJ 07029

Kraft Chemical Co.
1975 N. Hawthorene Avenue
Melrose Park, IL 60160

Energy — During World War II, the Chinese used tung oil for motor fuel. It tended to gum up the engines, so they processed it to make it compatible with gasoline. The mixture worked fine,[243] Gaydou et al.[107] reported yields of 4 to 6 MT/ha, converting to 1,800 to 2,700 ℓ oil per ha, equivalent to 17,000 to 25,500 kWh/ha.

Biotic factors — Bees are needed to transfer pollen from anthers to pistil. When staminate and pistillate flowers are on separate trees, 1 staminate tree for 20 pistillate trees should be planted in the orchard. Pollination can occur over several days. Tung trees are relatively free of insects and diseases, only a few causing losses serious enough to justify control measures: e.g., *Botroyosphaeria ribis*, *Clitocybe tabescens*, *Mycosphaerella aleuritidis*, *Pellicularia koleroga*, *Physalospora rhodina* and the bacterium, *Pseudomonas aleuritidis*. Other bacteria and fungi reported on tung trees are *Armillaria mellea*, *Botryodiplodia theo-*

bromae, *Cephaleures virescens*, *Cercospora aleuritidis*, *Colletotrichum gloeosporioides*, *Corticium koleroga*, *Fomes lamaoensis*, *F. lignosus*, *Fusarium heterosporum* forma *aleuritidis*, *F. oxysporum*, *F. scirpi*, *F. solani*, *Ganoderma pseudoferreum*, *Coleosporium aleuriticum*, *Glomerella cingulata*, *Pestalotia dichaeta*, *Phyllosticta microspora*, *Phytomonas syringae*, *Phytophthora omnivora*, *Ph. cinnamomi*, *Poria hypolateritia*, *Pythium aphanidermatum*, *Rhizoctonia solani*, *Septobasidium aleuritidis*, *S. pseudopedicellatum*, *Sphaerostilbe repens*, *Uncinula miyabei* var. *aleuritis*, *Ustilina maxima*, *U. zonata*. Insect pests are not a serious problem, since fruit and leaves of tung trees are toxic to most animal life. Nematodes *Meloidogyne* spp. have been reported.[186,257,278]

ALEURITES MOLUCCANA (L.) Willd. (EUPHORBIACEAE) — Candlenut Oil Tree, Candleberry, Varnish Tree, Indian or Belgium Walnut, Lumbang Oil
Syn.: *Aleurites triloba* Forst., *Croton moluccanus* L.

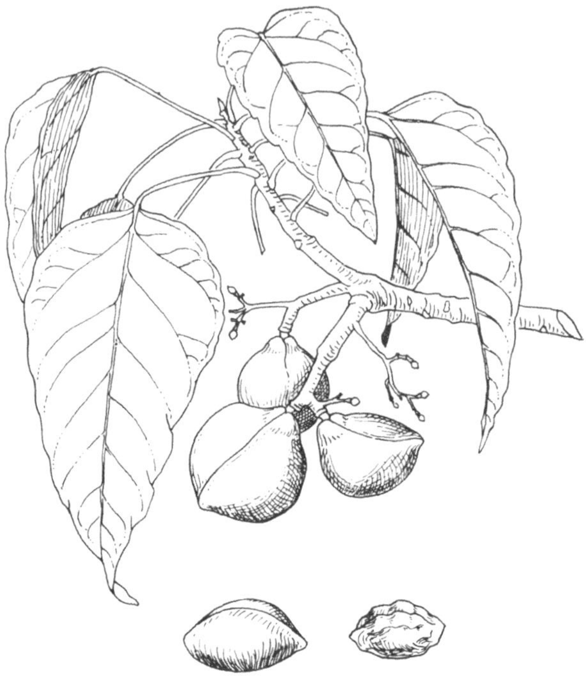

Uses — Seed yields 57 to 80% of inedible, semi-drying oil, liquid at ordinary temperatures, solidifying at $-15°C$, and containing oleostearic acid. The oil is quicker drying than linseed oil, and is used as a wood preservative, for varnishes and paint oil, also as an illuminant, for soap-making, waterproofing paper, in India rubber substitutes and insulating masses. Fruits said to be used as a fish poison. Seeds are moderately poisonous and press cake is used as fertilizer. Kernels, when roasted and cooked are considered edible; may be strung as candlenuts. Oil is painted on bottoms of small craft to protect against marine borers. Tung oil, applied to cotton bolls, stops boll weevils from eating them; also prevents feeding by striped cucumber beetle.[86]

Folk medicine — Bark used on tumors in Japan. Reported to be aperient, aphrodisiac, laxative, poison, purgative, stimulant, sudorific, candlenut oil tree is a folk remedy for asthma, debility, sores, swelling, tumors, unconsciousness, womb ailments, and wounds.[82] The oil is purgative and sometimes used like castor oil. In China, it is applied to sciatica. Kernels are laxative, stimulant, and sudorific. The irritant oil is rubbed on scalp as a hair stimulant. In Sumatra, pounded seeds, burned with charcoal, are applied round the navel for costiveness. Leaves are applied for rheumatism in the Philippines. In Malaya, the pulped kernel enters poultices for headche, fevers, ulcers, and swollen joints. Boiled leaves are applied to headache, scrofula, swollen joints, and ulcers. In Java, the bark is used for bloody diarrhea or dysentery. Bark juice with coconut milk is used for sprue and thrush. Malayans apply boiled leaves to the temples for headache, and to the pubes for gonnorhea.[56] In Yunani medicine, the oil is considered anodyne, aphrodisiac, and cardiotonic, and the fruit is recommended for the brain, bronchitis, bruises, heart, hydrophobia, liver, piles, ringworm, and watery eyes. In Ayurvedic medicine, the fruit is considered apertif, aphrodisiac, antibilious, cardiac, depurative, and refrigerant.[86,165]

Chemistry — The oil cake, containing ca. 46.2% protein, 4.4% P_2O_5, and 2.0% K_2O, is said to be poisonous. A toxalbumin and HCN have been suggested. Bark contains ca. 4 to 6% tannin. Oil also contains glycerides of linolenic, oleic and various linoleic acids. Per 100 g, the seed is reported to contain 626 calories, 7.0 g H_2O, 19.0 g protein, 63.0 g fat, 8.0 g total carbohydrate, 3.0 g ash, 80 mg Ca, 200 mg P, 2.0 mg Fe, 0 mg beta-carotene equivalent, 0.06 mg thiamine, and 0 mg ascorbic acid.[86]

Description — Medium-sized tree, up to 20 m tall, ornamental, with spreading or pendulous branches; leaves simple, variable in shape, young leaves large, up to 30 cm long, palmate, with 3 to 7 acuminate lobes, shining, while leaves on mature trees are ovate, entire, and acuminate, long-petioled, whitish above when young, becoming green with age, with rusty stellate pubescence beneath when young, and persisting on veins and petiole; flowers in rusty-pubescent panicled cymes 10 to 15 cm long; petals 5, dingy white or creamy, oblong, up to 1.3 cm long; ovary 2-celled; fruit an indehiscent drupe, roundish, 5 cm or more in diameter, with thick rough hard shell making up 64 to 68% of fruit, difficult to separate from kernels; containing 1 or 2 seeds. Flowers April to May (Sri Lanka).[196,278]

Germplasm — Reported from the Indochina-Indonesia Center of Diversity, *Aleurites moluccana*, or cvs thereof, is reported to tolerate high pH, low pH, poor soil, and slope.[82] (2n = 44,22).

Distribution — Native to Malaysia, Polynesia, Malay Peninsula, Philippines, and South Seas Islands; now widely distributed in tropics. Naturalized or cultivated in Malagasy, Sri Lanka, southern India, Bangladesh, Brazil, West Indies, and the Gulf Coast of the U.S.[278]

Ecology — Candlenut trees thrive in moist tropical regions, up to 1,200 m altitude. Ranging from Subtropical Dry to Wet through Tropical Very Dry to Wet Forest Life Zones, Aleurites moluccana is reported to tolerate annual precipitation of 6.4 to 42.9 dm (mean of 14 cases = 19.4) annual temperature of 18.7 to 27.4°C (mean of 14 cases = 24.6) and pH of 5.0 to 8.0 (mean of 7 cases = 6.4).[82]

Cultivation — Usually propagated from seed, requiring 3 to 4 months to germinate. Seedlings planted 300/ha. Once established, trees require little to no attention.[196,278]

Harvesting — Bear two heavy crops each year. After harvesting mature fruits, it is difficult to separate kernels from shell, as the kernels adhere to sides of shell.[278]

Yields and economics — As a plantation crop, tree yields are estimated at 5 to 20 tons/ha of nuts, each tree producing 30 to 80 kg. Oil production varies from 15 to 20% of nut weight. Most oil produced in India, Sri Lanka, and other tropical regions is used locally and does not figure into international trade. In the past, oil has sold for 12 to 14 pounds per ton in England. According to the Chemical Marketing Reporter,[351] tung oil prices (then ca. $.65/lb) are likely to rise in the near future if demand remains adequate and Argentinean and Paraguayan suppliers pressure the U.S. market by charging high prices for replacement oil. U.S. imports for the first quarter of 1981 were 58% higher than 1980, despite the absence of Chinese tung from the market.

Energy — Nut yields are estimated at 80 kg/tree, which, spaced at 200 trees per hectare, would suggest 16 MT/ha/yr, about 20% of which (3 MT) would be oil, suitable, with modification, for diesel uses, the residues for conversion to alcohol or pyrolysis. Fruit yields may range from 4 to 20 MT/ha/yr. Commercial production of oil yields 12 to 18% of the weight of the dry unhulled fruits, the fruits being air-dried to ca. 12 to 15% moisture before pressing. The pomace contains 4.5 to 5% oil. This suggests that the "chaff factor" might be ca 0.8. Oil yields as high as 3,100 kg/ha have been reported. As of June 15, 1981, tung oil was $0.65/lb, compared to $0.38 for peanut oil, $1.39 for poppyseed oil, $0.33 for linseed oil, $0.275 for coconut oil, $0.265 for cottonseed oil, $0.232 for corn oil, and $0.21 for soybean oil.[351] At $2.00 per gallon, gasoline is roughly $0.25/lb.

Biotic factors — Following fungi are known to attack candlenut-oil tree: *Cephalosporium* sp., *Clitocybe tabescens*, *Fomes hawaiensis*, *Gloeosporium aleuriticum*, *Phasalospora rhodina*, *Polyporus gilvus*, *Pythium ultimum*, *Sclerotium rolfsii*, *Sphaeronaema reinkingii*, *Trametes corrugata*, *Xylaria curta*, *Ustulina deusta*.[186]

ALEURITES MONTANA (Lour.) Wils. (ANACARDIACEAE) — Wood-Oil Tree, Mu-Oil Tree

Uses — Kernels yield a valuable drying oil, largely used in paints, varnishes, and linoleums. Also used locally for illumination and lacquer-work. Varnish made from this plant possess a high degree of water-resistance, gloss, and durability. There are only slight differences between the oils of *A. montana* and *A. fordii*.[278]

Folk medicine — The oil is applied to furuncles and ulcers.

Chemistry — The oil content of the seed is ca. 50 to 60%. Oil consists chiefly of glycerides of beta-elaeostearic and oleic acids, and probably a little linoleic acid. Oil cake residue is poisonous and is only fit for manuring.

Description — A small tree about 5 m tall, much-branched, partially deciduous, dioecious. Leaves simple, ovate or more or less cordate, apex cuspidate, about 12 cm long, 10 cm broad, sometimes larger and 3-lobed; leaf-blade with 2 large, conspicuous glands at base, petiole up to 24 cm long. Flowers monoecious, petals large, white, up to 3 cm long. Fruits egg-shaped, 3-lobed, wrinkled, about 5 cm in diameter, pointed at summit, flattened at base, generally with 3 or 4 one-seeded segments, the outer surface with wavy transverse ridges, the pericarp thick, hard, and weedy. Flowers and fruits March.[278]

Germplasm — Reported from the China-Japan Center of Diversity, mu-oil tree, or cvs thereof, is reported to tolerate high pH, poor soil, and slope. ($2n = 22$.)[82]

Distribution — Native to South China and some of the S. Shan States (Burma). Introduced and cultivated successfully in Indochina (where it has replaced *A. fordii*), Malawi, and in cooler parts of Florida, and other tropical regions.[278]

Ecology — Ranging from Warm Temperate Moist through Tropical Dry to Moist Forest Life Zones, mu-oil tree is reported to tolerate annual precipitation of 6.7 to 20.2 dm (mean of 8 cases = 13.6), annual temperature of 14.8 to 26.5°C (mean of 8 cases = 21.9°C), and pH of 5.5 to 8.0 (mean of 6 cases = 6.2).[82] Adapted to subtropical regions and high elevations with moderate rainfall. Mainly a hillside species, it can thrive in warmer climates and will withstand heavier rainfall than *A. fordii*, provided the area is well-drained. Maximum temperature 35.5°C, minimum temperature 6°C. It is frost-tender, and does not require a low temperature (below 3°C) as tung-oil trees (*A. fordii*) do, so can be grown in warmer regions. In Assam, grown where rainfall is 175 to 275 cm annually; in Mysore at elevations of 800 to 1,000 m with annual rainfall of 150 cm. Grows well in alluvial soils and is not very exacting in its soil requirements. In richer soils, the growth is more vigorous. A slightly acid soil is preferable.[278]

Cultivation — Trees are propagated from seeds or by budding. In Malawi, propagation is by budding from high-yielding clones. Seeds are usually planted in a nursery and may take from 2 to 3 months to germinate. When seedlings are about 1 year old, they are planted out, spaced 6.6 × 6.6 m or more. Cultural practices are similar to those for *A. fordii*. As soon as the seedlings emerge, a side-dressing of fertilizer (5-10-5) of nitrogen and phosphorus, along with commercial zinc sulfate, should be applied. Fertilizer is applied at rate of 600 kg/ha, in bands along each side of row, 20 cm from seedlings and 5 to 7.5 cm deep. Other fertilizers may be needed, depending on the soil. According to Spurling and Spurling,[312] N is the most important nutrient for tung in Malawi, irrespective of climate or soil. Most successful budding is done in late August, by the simple shield method, requiring a piece of budstick bark, including a bud, that will fit into a cut in the rootstock bark. A T-shaped cut is made in bark of rootstock at a point 5 to 7.5 cm above ground level, the flaps of bark loosened, shield-bud slipped inside flaps, and the flaps tied tightly over the transplanted bud with rubber budding strip 12 cm long and 0.6 cm wide. After about 7 days, the rubber strip is cut to prevent binding. As newly set buds are susceptible to cold injury, soil is mounded over them for winter. When growth starts in spring, soil is pulled back and each stock cut back to within 3.5 cm of the dormant bud. Later care consists of keeping all suckers removed and the trees well-cultivated. Trees may be planted 125 to 750/ha. When trees are small, close planting in rows greatly increases the bearing surface, but at maturity the bearing surface of a crowded row is about the same as for a row with trees further apart. However, it is well to leave enough space between rows for orchard operations. In contour-planting, distances between rows and total number of trees per hectare vary; rows 10 to 12 m apart, trees spaced 3.3 to 4 m apart in rows, 250 to 350 trees/ha. Tops of trees must be pruned back to 20 to 25 cm at planting. As growth starts, all buds are rubbed off except the one strongest growing and best placed on the tree. A bud 5 cm or more below the top of stump is preferred over one closer to the top.[278,312]

Harvesting — Trees begin bearing 2 to 5 years after transplanting with maximum production reached in 8 years and continuing for 40 years. In northern Burma, it has been observed to be more vigorous and disease-resistant than *A. fordii*. In Indochina, it has been successfully planted and its oil is now being produced on a commercial scale, replacing that of *A. fordii*. Fruits mature and drop to ground in late September to early November. They are gathered and dried to 15% moisture before processing. Fruits should be left on the ground 3 to 4 weeks until hulls are dead and dry, and the moisture content has dropped below 30%; fresh they are about 60% moisture. Fruits are gathered by hand into baskets or sacks.[278]

Yields and economics — *A. montana* is reported to give much higher yields of fruits than *A. fordii*. The percentage of kernels in the seeds is about 56%, and of oil in the kernels, about 59.3%. Major producers of the oil from *A. montana* are Burma, Indochina (Vietnam, Cambodia, Laos), Malawi, Congo, East Africa, South Africa, Malagasy Republic, India, and U.S.S.R. It has been considered for introduction in Florida.[278]

Energy — Yields of oil per tree in China is figured to be about 3.2 kg; in Florida, 4.5 to 9 kg. Trees yield about 45 to 68 kg nuts per year, these yielding about 35 to 40% oil. In one Malawi trial, N treatments gave an increase of 519 kg/ha dry seed over a trial mean of 1070 kg/ha. With tung cake and ammonium sulphate, air dry tung seed yields of 12 to 17 year old trees was 2013 to 2367 kg/ha, of 6 to 9 year olds 766 to 1546 kg/ha.[278,312]

Biotic factors — Fungi reported on *A. montana* include the following: *Armillaria mellea*, *Botryodiplodia theobromae*, *Botryosphaeria ribis*, *Cephaleuros mycoidea*, *C. virescens*, *Cercospora aleuritidis*, *Colletotrichum gloeosporioides* var. *aleuritidis*, *Corticium koleroga*, *C. solani (Rhizoctonia solani)*, *Corynespora cassiicola*, *Diplodia theobromae*, *Fusarium arthrosporioides*, *F. lateritium*, *Glomerella cingulata*, *Haplosporella aleurites*, *Mycosphaerella aleuritidis*, *Periconia byssoides*, *Pestalotiopsis disseminata*, *P. glandicola*, *P. japonica*, *P. versicolor*, *Pestalotia dichaeta*, *Phyllosticta microspora*, *Pseudocampton fasciculatum*, *Rhizoctonia lanellifera*, *Schizophyllum commune*, *Thyronectria pseudotrichia*, *Trametes occidentalis*, *Ustulina zonata*.[186]

AMPHICARPAEA BRACTEATA (L.) Fernald (FABACEAE) — Hog Peanut, Wild Peanut

Uses — Ojibwa Indians were said to eat both roots and seeds cooked. (There's not much to the roots.) Meskwaki (Fox) Indians learned that mice gathered the underground nuts and laid them up in stores, which stores the Indians gathered for themselves (Dakota Indians were said to leave corn or other food in exchange). The subterranean seeds are more important as food. They have been likened to garden-bean in flavor, the aerial seeds to soybeans. As late as November in Maryland, the subterranean seeds may be tracked from the dying yellow/brown tops. If eaten raw, seeds might be soaked in warm water or water with hardwood ashes. In October, when both *Amphicarpaea* and *Apios* seeds are available, I find both the aerial and subterranean seeds of the *Amphicarpaea* seeds much more pleasing to the palate raw than the *Apios* seeds. Gallaher and Buhr[405] speculate that the subterranean seeds may "have survival-potential under conditions of intense grazing." I suggest that the subterranean seeds might not set in tightly packed sod. Both aerial and subterranean seeds are eaten by bear, chipmunk, deer, grouse, mice, pheasant, prairie chicken, quail, and wild turkey. Vines are browsed by livestock and probably deer. Once cultivated in southern U.S., hog peanuts have been suggested for planting in poultry forage systems and for intercropping with corn and perhaps ginseng. All members of the genus can be important in soil improvement, as soil cover, and in erosion control.[8,400,401]

Folk medicine — Chippewa drank the root with other roots as a general physic, while, conversely, the Cherokee used it for diarrhea. Cherokee also blew the root tea onto snakebite wounds.[400]

Chemistry — Marshall[402] notes that the aerial seeds, with flavor similar to soybeans, contain ca. 30% protein, 7 to 16% oil. The oil contains 10.3 to 10.4% palmitic-, 1.3 to 1.6% stearic-, 24.9 to 26.7% oleic-, 54.8 to 58.5% linoleic-, and 6.5 to 7.6% linolenic-acids. The cleistogamous, underground seeds, weighing as much as 1 g each, may contain 50% water. Their oil content is lower, and the protein content may be only 14.3%, perhaps[404] reflecting the higher water content.[402] Lectins are also reported. Gallaher and Buhr[405] analyzed Tennessee fodder during early pod-fill stage, reporting for the whole plant ca. 89% organic matter, 26.5 g/kg N, 2.4 g/kg P, 14.2 g/kg K, 17.3 g/kg Ca, 4.1 g/kg Mg, 20 ppm Cu, 40 ppm Zn, 120 ppm Mn, and 360 ppm Fe, averaging slightly lower than pegging peanut forage, but higher in P, Ca, Mn, and Fe. Crude protein in the hog peanut forage was over 16%, slightly below the peanut forage.

Description — Weak, twining, climbing annual (though often cited as perennial) to 2 m long, the stems sparsely appressed short-pubescent to densely villous. Leaves 3-foliolate; leaflets entire, ovate to rhombic-ovate, the laterals often asymmetrical, 2 to 10 cm long, petiolulate, stipellate; usually pubescent. Axillary racemes of 1 to 17 petaliferous flowers, on peduncles 1 to 6 cm long, the ovate bracts 2 to 5 mm long; pedicels 1.5 to 5 mm long; racemes from lower axils slender, elongate, with cleistogamous, apetalous, inconspicuous flowers. Calyx of petaliferous flowers narrowly campanulate; tube 4 to 6 mm long, ca. 2 mm in diameter; upper 2 lobes united, or nearly so, glabrous to densely appressed-pubescent; petals pale purple or lilac to white, 9 to 16 mm long; stamens of the petaliferous flowers diadelphous, 9 and 1; ovary stipitate, style not bearded. Legume from petaliferous flowers flattened, oblong-linear, 1.5 to 4 cm long, 7-10 mm broad, often 3-seeded, valves laterally twisting in dehiscence; fruit from cleistogamous flowers fleshy, often subterranean, usually 1-seeded, indehiscent, cryptocotylar.[276] Duke[403] recognizes four different flower/fruit combinations:

1. Subterranean seed, whose cleistogamous flowers never left the soil (usually one or two); the biggest, juiciest, softest, and most edible (15% protein). For propagation *in situ*.
2. Geotropic seed from cleistogamous flowers at the tip of branches originating in the axils of the first simple aerial leaves. Usually solitary, soft, plump. For propagation nearby.
3. Aerial cleistogamous flowers, whose pods, and usually single hard seeds, develop strictly above ground. For dispersal.
4. Aerial chasmogamous flowers followed by pods with usually three small hard seeds (the smallest, driest, hardest, and least edible, yet 30% protein). For longer distance dispersal. The type 4 flower/fruits are said to occur mostly in sunny situations. If the forest is cleared, the increased sunlight would trigger more dispersal seed, enhancing the chances to move the plant back into the forest.

Germplasm — Reported from the North American Center of Diversity, hog peanut, or cvs thereof, is reported to tolerate alluvium, muck, mulch, sand, shade, slope, and brief waterlogging. *A. bracteata* is said to merge imperceptibly with var. *comosa*, which grows on richer, often calcareous or alluvial soil. Turner and Fearing[406] concluded the genus contained only three species, *A. africana* in the cool high mountains of Africa, *A. edgeworthii* in the Himalayas and eastern Asia, and the American *A. bracteata*, the latter two nearly indistinguishable. (2n = 20,40.)

Distribution — Native to damp shaded woodlands from Quebec to Manitoba and Montana, south to Florida, Louisiana, and Texas.

Ecology — Estimated to range from Warm Temperate Moist to Wet through Cool Temperate Dry to Wet Forest Life Zones, hog peanut is estimated to tolerate annual precipitation of 8 to 20 dm, annual temperature of 8 to 14°C, and pH of 5.5 to 7.5. Although native to damp shaded forest, the plant can be cultivated in sandy, sunny situations. The underground seed must have very different chemistry, ecology, and physiology, destined for immediate survival and not dispersal, as contrasted to the aerial seed, destined for long-term dispersal.

Cultivation — Said to have been cultivated in the South, but few details are available. W. G. Dore[407] sterilizes his soil, plants in the fall, and mulches with such things as sawdust, peat moss, vermiculite, and/or organic muck. Gas-sterilization is all but imperative to control weeds since the clambering habit of the vine precludes cultivation. In fertile soils in full sun, the one-seeded beans grow large and succulent, comparable to peanuts, or even lima beans. Frey[401] suggests intercropping the hog peanut with corn.

Harvesting — The large seeds appear beneath the dead leaves, generally just under the surface of the ground. In weed-free culture, the tangled vines can be raked off preparatory to harvest in fall. In loose sandy soil, the seeds separate out easily with a quarter inch screen. Harvested seed tend to germinate in the refrigerator, if not frozen.

Yields and economics — Unpublished research by W. G. Dore[407] reported yields as high as 1 kg seed per 10 m row. His seed were fall-planted about 10 cm apart in gas-sterilized sandy loam.

Energy — Both biomass (ca. 5 g per plant) and oil yields are low. The biomass raked up before harvesting could conceivably be converted to energy. The nitrogen fixed by the plant could be energetically important, in pastures, forests, and in intercropping scenarios.

Biotic factors — Agriculture Handbook No. 165[4] lists the following as affecting *Amphicarpaea bracteata*: *Cercospora monoica* (leaf spot), and *Erysiphe polygoni* (powdery mildew). Agriculture Handbook No. 165,[4] without reference to a specific species, also lists: *Colletotrichum* sp. (leaf spot), *Parodiella perisporioides* (black mildew), *Puccinia andropogonis* var. *onobrychidis* (rust), and *Synchytrium aecidioides* (false rust, leaf gall). Allen and Allen[8] report that earlier studies showed a relative inability of the hog peanut *Rhizobium* to nodulate legumes from 21 diverse genera. Later plant-infection studies discounted this exclusiveness by showing plant-infection kinships within the cowpea miscellany. Larvae of *Rivella pallida* Lowe, a common and widely distributed species of the dipteran family Platystomatidae (and a potential pest of soybean), attack the N_2-fixing root nodules of *Amphicarpaea*. The nodular contents are completely destroyed, thus eliminating the nodule's ability to fix N_2. Up to 25% of an individual's nodules are damaged in northeastern Ohio. There is one and perhaps a partial second generation per year in northern Ohio, with overwintering occurring as mature larvae in diapause. Eight species of neartic *Rivellia* (including *R. flavimana* Loew and *R. metallica* (Walp)) occur on *Amphicarpaea bracteata* (L.).[408] Chasmogamous flowers are pollinated primarily by *Bombus affinis*.[409]

ANACARDIUM OCCIDENTALE L. (ANACARDIACEAE) — Cashew

Uses — Many parts of the cashew plant are used. The cashew "apple", the enlarged fully ripe fruit, may be eaten raw, or preserved as jams or sweetmeats. The juice is made into a beverage (Brazil cajuado) or fermented into a wine. Seeds of the cashew are consumed whole, roasted, shelled and salted, in Madeira wine, or mixed in chocolates. Shelling the roasted seed yields the cashew nut of commerce. Seeds yield about 45% of a pale yellow, bland, edible oil, resembling almond oil. From the shells or hulls is extracted a black, acrid, powerful vesicant oil, used as a preservative and water-proofing agent in insulating varnishes, in manufacture of typewriter rolls; in oil- and acid-proof cements and tiles, in brake-linings, as an excellent lubricant in magneto armatures in airplanes, and for termite-proofing timbers. Timber is used in furniture making, boat building, packing cases and in the production of charcoal. Bark used in tanning. Stems exude a clear gum, Cashawa gum, used in pharmaceuticals and as substitute for gum arabic. Juice turns black on exposure to air and provides an indelible ink. Along the coast of Orissa, shelter belts and wind breaks, planted to stabilize sand dunes and protect the adjacent fertile agricultural land from drifting sand, have yielded economic cashew crops 5 years after planting.[248,278]

Folk medicine — The fruit bark juice and the nut oil are both said to be folk remedies for calluses, corns, and warts, cancerous ulcers, and even elephantiasis. Anacardol and anacardic acid have shown some activity against Walker carcinosarcoma 256. Decoction of the astringent bark given for severe diarrhea and thrush. Old leaves are applied to skin afflictions and burns (tannin applied to burns is hepatocarcinogenic). Oily substance from pericarp used for cracks on the feet. Cuna Indians used the bark in herb teas for asthma, colds, and congestion. The seed oil is believed to be alexeritic and amebicidal; used to treat gingivitis, malaria, and syphilitic ulcers. Ayurvedic medicine recommends the fruit for anthelmintic, aphrodisiac, ascites, dysentery, fever, inappetence, leucoderma, piles, tumors, and obstinate ulcers.[165] In the Gold Coast, the bark and leaves are used for sore gums and toothache. Juice of the fruit is used for hemoptysis. Sap discutient, fungicidal, repellent. Leaf decoction gargled for sore throat. Cubans use the resin for cold treatments. The plant exhibits hypoglycemic activity. In Malaya, the bark decoction is used for diarrhea. In Indonesia, older leaves are poulticed onto burns and skin diseases. Juice from the apple is used to treat quinsy in Indonesia, dysentery in the Philippines.[165,244,249]

Toxicity — He who cuts the wood or eats cashew nuts or stirs his drink with a cashew swizzle stick is possibly subject to a dermatitis.

Chemistry — Per 100 g, the mature seed is reported to contain 542 calories, 7.6 g H_2O, 17.4 g protein, 43.4 g fat, 29.2 g total carbohydrate, 1.4 g fiber, 2.4 g ash, 76 mg Ca, 578 mg P, 18.0 mg Fe, 0.65 mg thiamine, 0.25 mg riboflavin, 1.6 mg niacin, and 7 mg ascorbic acid. Per 100 g, the mature seed is reported to contain 561 calories, 5.2 g H_2O, 17.2 g protein, 45.7 g fat, 29.3 g total carbohydrate, 1.4 g fiber, 2.6 g ash, 38 mg Ca, 373 mg P, 3.8 mg Fe, 15 mg Na, 464 mg K, 60 mg beta-carotene equivalent, 0.43 mg thiamine, 0.25 mg riboflavin, and 1.8 mg niacin. Per 100 g, the mature seed is reported to contain 533 calories, 2.7 g H_2O, 15.2 g protein, 37.0 g fat, 42.0 g total carbohydrate, 1.4 g fiber, 3.1 g ash, 24 mg Ca, 580 mg P, 1.8 mg Fe, 0.85 mg thiamine, 0.32 mg riboflavin, and 2.1 mg niacin. The "apples" (ca. 30 to 35 kg per tree per annum) yield each 20 to 25 cc juice, which, rich in sugar, was once fermented in India for alcohol production. The apple contains 87.9% water, 0.2% protein, 0.1% fat, 11.6% carbohydrate, 0.2% ash, 0.01% Ca, 0.01% P, .002% Fe, 0.26% vitamin C, and 0.09% carotene. The testa contains alpha-catechin, beta-sitosterol, and 1-epicatechin; also proanthocyanadine leucocyanadine, and leucopelargonidine. The dark color of the nut is due to an iron-polyphenol complex. The shell oil contains about 90% anacardic acid ($C_{22}H_{32}O_3$) and 10% cardol ($C_{32}H_{27}O_4$). It yields glycerides, linoleic, palmitic, stearic, and lignoceric acids, and sitosterol. Examining 24 different cashews, Murthy and Yadava[227] reported that the oil content of the shell ranged from 16.6 to 32.9%, of the kernel from 34.5 to 46.8%. Reducing sugars ranged from 0.9 to 3.2%, nonreducing sugars, 1.3 to 5.8%, total sugars from 2.4 to 8.7%, starch from 4.7 to 11.2%. Gum exudates contain arabinose, galactose, rhamnose, and xylose.[187]

Description — Spreading, evergreen, perennial tree to 12 m tall; leaves simple, alternate, obovate, glabrous, penninerved, to 20 cm long, 15 cm wide, apically rounded or notched, entire, short petiolate; flowers numerous in terminal panicles, 10 to 20 cm long, male or female, green and reddish, radially symmetrical nearly; sepals 5; petals 5; stamens 10; ovary one-locular, one-ovulate, style simple; fruit a reniform achene, about 3 cm long, 2.5 cm wide, attached to the distal end of an enlarged pedicel and hypocarp, called the cashew-apple. The fruit is shiny, red or yellowish, pear-shaped, soft, juicy, 10 to 20 cm long, 4 to 8 cm broad; fruit is reniform, edible, with two large white cotyledons and a small embryo, surrounded by a hard pericarp which is cellular and oily; the oil is poisonous, causing allergenic reactions in some humans. Flowering variable.[278]

Germplasm — Several varieties have been selected, based on yield and nut size. Reported from the South America and Middle America Centers of Diversity, cashew or cvs thereof is reported to tolerate aluminum, drought, fire, insects, laterite, low pH, poor soil, sand, shade, slope, and savanna. ($2n = 42$.)[82]

Distribution — Native to tropical America, from Mexico and West Indies to Brazil and Peru. The cashew tree is pantropical, especially in coastal areas.

Ecology — Ranging from Warm Temperate Moist to Tropical Very Dry to Wet Forest Life Zones, cashew is reported to tolerate annual precipitation of 7 to 42 dm (mean of 32 cases = 19.6), annual temperature of 21 to 28°C (mean of 31 cases = 25.2), and pH of 4.3 to 8.7 (mean of 21 cases = 64). Grows on sterile, very shallow, and impervious savanna soils, on which few other trees or crops will grow, but is less tolerant of saline soil than most coastal plants. Does not tolerate any frost. In Brazil, Johnson[150] summarizes "optimal ecological conditions": annual rainfall 7 to 20 dm, minimum temperature 17°C, maximum temperature 38°C; average annual temperature 24 to 28°C, relative humidity 65 to 80%; insolation 1,500 to 2,000 hr/year, wind velocity 2.25 km/hr, and dry season 2 to 5 months long. It is recommended that cultivation be limited to nearly level areas of red-yellow podzols, quartziferous sands, and red-yellow latosols.[82,278]

Cultivation — Cashew germinates slowly and poorly; several nuts are usually planted to the hole and thinned later. Propagation is generally by seeds, but may be vegetative from grafting, air-layering or inarching. Planting should be done *in situ* as cashew seedlings do not transplant easily. Recommended spacing is 10 × 10 m, thinned to 20 × 20 m after about 10 years, with maximum planting of 250 trees per ha, Once established, the field needs little care. Intercropping may be done the first few years, with cotton, peanut, or yams. Fruits are produced after 3 years, during which lower branches and suckers are removed. Full production is attained by the 10th year, and trees continue to bear until about 30 years old. In dry areas, like Tanzania, flowering occurs in the dry season, and fruits mature in 2 to 3 months. Flowers and fruits in various degrees of development are often present in same panicle.[278]

Harvesting — From flowering stage to ripe fruit requires about 3 months. Mature fruit falls to the ground where the "apple" dries away. In wet weather, they are gathered each day and dried for 1 to 3 days. Mechanical means for shelling have been unsuccessful, so hand labor is required. Cashews are usually roasted in the shell (to make it brittle and oil less blistering), cracked, and nuts removed and vacuum packed. In India, part of the nuts are harvested from wild trees by people who augment their meager income from other crops grown on poor land. Kernels are extracted by people skilled in breaking open the shells with wooden hammers without breaking the kernels. Nuts are separated from the fleshy pedicel and receptacle, seed coat removed by hand, and nuts dried. Fresh green nuts from Africa and the islands off southern India are shipped to processing plants in Western India.[70,278]

Yields and economics — Yields are said to range from 0 to 48 kg per tree per year, with an average yield of 800 to 1,000 kg/ha. Heavy bearing trees often produce nuts considered too small for the trade. Indian field trials showed that fertilizers could increase yields of 15-year-old trees from less than 1 kg to tree to >4 and enabled 6-year-olds to average 5.7. Regular applications of 250 g N, 150 g P_2O_5, and 150 g K_2O per tree resulted in average yield increases of 700 to 1600 kg/ha.[228] In Pernambuco, trees produced 1.5 to 24.0 kg each per year, averaging 10.3 kg per tree.[150] At Pacajus (Ceara, Brazil) trees average 17.4 kg/year with one tree bearing 48 kg/year. Major producers of cashew nuts are India, Tanzania, Mozambique, and Kenya. In 1968 India planted over 224,000 ha in cashews to supply over 200 processing factories operating all year. In 1971 India produced 90,000 MT, the bulk exported to the U.S. and the U.S.S.R. Export price at U.S. ports was $.33/kg. India imports green nuts from the African countries and processes them for resale. Import price in 1971 in India was 1730 rupees/MT. Cashawa Gum is obtained from the West Indies, Portuguese East Africa, Tanzania, and Kenya.

Energy — A perennial species, the cashew has already, in the past, yielded alcohol from the "apple", oil from the nut, and charcoal from the wood. Prunings from the tree and the leaf biomass could also be used as energy sources.

Biotic factors — The cashew tree has few serious diseases or pests. The following are reported disease-causing agents, none of which are considered of economic importance: *Aspergillus chevalieri*, *A. niger*, *Atelosaccharomyces moachoi*, *Balladynastrum anacardii*, *Botryodiplodia theobromae*, *Cassytha filiformis*, *Cephaleuros mycoides*, *Ceratocystis* sp., *Cercospora anacardii*, *Colletotrichum capsici*, *Cytonaema* sp., *Endomyces anacardii*, *Fusarium decemcellulare*, *Gloeosporium* sp., *Glomerella cingulata*, *Meliola anacardii*, *Nematospora corylii*, *Parasaccharomyces giganteus*, *Pestaliopsis disseminata*, *Phyllosticta anacardicola*, *P. mortoni*, *Phytophthora palmivora*, *Pythium spinosum*, *Schizotrichum indicum*, *Sclerotium rolfsii*, *Trichomerium psidii*, *Trichothecium roseum*, *Valsa eugeniae*. *Cuscuta chinensis* attacks the tree. Of insects, *Helopeltis* spp. have been reported in Tanzania. In Brazil, high populations of the nematodes *Criconemoides*, *Scutellonema*, and *Xiphinema* are reported around cashew roots.[185] Four insects are considered major pests: the white fly (*Aleurodicus cocois*), a caterpillar (*Anthistarcha binoculares*), a red beetle (*Crimissa* sp.), and a thrip (*Selenothripes rubrocinctus*). Flowers are visited by flies, ants, and other insects, which may serve as pollinators. Artificial pollination is practiced in some areas.[70,186,278]

APIOS AMERICANA Medik. (FABACEAE) — Groundnut

Uses — An attractively flowered plant, suggestive of Wisteria, *Apios* has been described by the NAS[231] as a "useful, sweet-scented ornamental". I have enjoyed the tubers raw or cooked. During the potato famine of 1845, *Apios* was introduced to Europe (but not for the first time). Its cultivation there as a food crop was abandoned when potato growing again became feasible. The plant was much esteemed by early American settlers, who ate them boiled, fried, or roasted, calling them groundnuts, potato beans, or Indian potatoes. The Pilgrims of New England survived their first few winters thanks to the groundnut. Blackmon[40] presents several groundnut recipes. Erichsen-Brown[96] recounts many of the Indian uses. Menominee preserved the roots by boiling them in maple syrup.[40] Even bread was made from the root. Indians were said to eat the seeds like lentils. I would like to join the ranks of Bill Blackmon,[40] Ed Croom, Janet Seabrook,[299,300] and Noel Vietmeyer, and advocate more studies of the economic potential of this interesting tuber, harvestable all year round. I agree with Blackmon and Reynolds,[40] who, after studying *Apios* intensively stated: "the prognosis for developing *A. americana* as a food crop looks outstanding." Advocates should be aware of its weed potential, at least among uncultivated perennials, e.g., cranberries and azaleas.

Folk medicine — According to Hartwell,[126] the tubers were used in folk remedies for that cancerous condition known as "Proud Flesh" in New England. Nuts were boiled and made into a plaster: "For to eat out the proud flesh they (Indians) take a kind of earth nut boyled and stamped."[143]

Table 1
CHEMICAL COMPOSITION (PERCENT) OF APIOS SPECIES

	Apios americana		*Apios fortunei*		*Apios priceana*	
	Fresh basis	**Dry basis**	**Fresh basis**	**Dry basis**	**Fresh basis**	**Dry basis**
Water	81.00		68.60		61.88	
Fiber	5.20	27.37	1.20	3.82	4.95	12.99
Crude protein	3.12	16.42	4.19	13.34	2.62	6.87
Nonprotein N	0.19	1.00	0.42	1.34	0.15	0.39
Protein N	0.31	1.63	0.25	0.80	0.27	0.69
Crude fat	0.67	3.53	0.19	0.61	0.82	2.15
Ash	0.99	5.21	1.30	4.14	2.67	7.00
Carbohydrate	9.02	47.47	24.52	78.09	27.06	70.97
Starch			18.30	58.28	7.84	20.58
Alcohol-insol. solids					15.08	39.55

From Walter, W. M., Croom, Jr., E. M., Catignani, G. L., and Thresher, W. C., Compositional study of *Apios priceana* tubers, *J. Agric. Food Chem.*, (Jan./Feb.), 39, 1986. Copyright 1986, American Chemical Society. With permission.

Chemistry — Some describe the plant as having a milky juice. Seabrook[299] suggests tha the latex could be used commercially. According to the NAS, the only published analysis[34] records a remarkable protein content of 17.5%. Prompted by the inadequacy of analyses, Duke arranged for new analytical investigations. Sanchez and Duke,[373] based on these analyses provided by Benito de Lumen, report (ZMB): 3.75 crude fat, 5.50% ash, 17.28% crude protein, 28.84% neutral detergent fiber, 44.63% available carbohydrate, and 1.06 nonprotein nitrogen. Per g they report 71.76 mg free amino acids, 1.26 mg nitrate, and 10.36 mg tannin. Subsequently, Walter et al.[374] tabulated the differences in analyses between fresh and dry tubers of *A. americana*, *A. fortunei*, and the endangered *A. priceana* (Table 1). Saponins have been reported in the genus, and the absence of tannins,[109] refuted above. Whether or not the plant exports its fixed nitrogen as ureides (allantoin, allantoic acid) as is typical of many of the subtropical Phaseoleae or as the more soluble amides (asparagine and glutamine) as in such temperate legumes as *Lupinus*, *Pisum*, *Trifolium*, and *Vicia* remains to be seen. Because it is suggested to have a cowpea-type Rhizobium, I predict it will be a ureide exporter. Some calculations suggest it takes ca. 2 1/2 times as much water (remember this is an aquaphyte) to export N as ureides. But the ureides are more economical with a C:N ratio ca. 1:1; cf. 1:1 for asparagine, 5:2 for glutamine.[311] Many legume sprouts are rich in allantoin, widely regarded as a vulnerary medicinal compound. According to the Merck Index, allantoin is a product of purine metabolism in animals, while it is prepared synthetically by the oxidation of uric acid with alkaline potassium permanganate. Medical and veterinary use — "Has been used topically in suppurating wounds, resistant ulcers, and to stimulate growth of healthy tissue (Merck & Co.[210]). Dorland's Illustrated Medical Dictionary[76] puts it differently:

> **allantoin** (ah-lan'to-in). Chemical name: 5-ureidohydantoin. A white crystallizable substance, $C_4H_6N_4O_3$, the diureide of glyoxylic acid, found in allantoic fluid, fetal urine, and many plants, and as a urinary excretion product of purine metabolism in most mammals but not in man or the higher apes. It is produced synthetically by the oxidation of uric acid, and was once used to encourage epithelial formation in wounds and ulcers and in osteomyelitis. It is the active substance in maggot treatment, being secreted by the maggots as a product of purine metabolism.

The direct role of allantoin in gout, if any, should be of great interest to those American males who have gout, especially if they ingest large quantities of legume sprouts or comfrey. *Apios* produces a complex pterocarpan that appears structurally similar to glyceollin III, a phytoalexin of the cultivated soybean.[145]

Description — Twining, herbaceous vine, the stems short-pubescent to glabrate, 1 to 3 m long, the rhizomes moniliform, with numerous fleshy tubers 1 to 8 cm thick. (Some plants have fleshy roots only, others both fleshy roots and tubers, and others only tubers.) In winter, the stems have a distinctive brown color and are locally flattened, enabling the experienced collector to distinguish it from honeysuckle. Leaves once-pinnate, 1 to 2 dm long; leaflets 5 to 7, ovate or ovate-lanceolate to lanceolate, ca. 3 to 6 cm long, glabrous to short-pubescent, obscurely stipellate; petioles mostly 2 to 7 cm long; stipules setaceous, soon deciduous, 4 to 6 mm long. Inflorescence 5 to 15 cm long, nodes swollen, flowers 1 to 2 per node, subtended by linear-subulate bracts 2 to 2.5 mm long; pedicels 1 to 4 mm long with 2 linear-subulate bractlets near apex. Calyx sparsely short-pubescent, broadly campanulate, tube ca. 3 mm long; petals nearly white to brownish purple, the standard obovate or orbicular to obcordate, reflexed, obscurely auricled, 9 to 13 mm long, the wings shorter, slightly auricled, the keel strongly incurved; stamens diadelphous, 1 and 1. Legume linear, 5 to 15 cm long, 4 to 7 mm broad, 2 to 12-seeded, dehiscing by 2 spirally twisted valves.[276] Germination cryptocotylar.[299,375]

Germplasm — Reported from the North American Center of Diversity, groundnut, or cvs thereof, is reported to tolerate acid and bog soils, partial shade, slopes, and waterlogging. In 1982, the Plant Introduction Officer of the USDA suggested to me the possibility of mounting a germplasm expedition to collect germplasm of this species, and its endangered relative, *Apios priceana* Robinson, which produces a single large tuber instead of a string of small tubers. NAS[231] speculates that a bush-like mutant may be found in nature. Seedlings from Tennessee had 22 chromosomes, while plants from the northern part of the range were triploid. Blackmon[40] and Reynolds[375] discuss the variation in germplasm they have already assembled. (2n = 22.)

Distribution — Widely distributed in eastern Canada and the U.S. (often around ancient Indian campsites) (Florida, Texas, to Nova Scotia, Minnesota, and Colorado). Usually in low damp bottomland or riparian woods and thickets. Seems to be associated with *Alnus* in Rocky Gorge Reservoir, Maryland, as well as on the eastern shore of Maryland. Unfortunately, it can become a serious weed in cranberry plots. Uninfested bogs yielded nearly 14 MT/ha cranberries, whereas herbicide plots yielded only ca. 670 to 2,300 kg/ha cranberry.[372] Perhaps the cranberry salesmen could find a market for the groundnuts, since both are Native American food plants.

Ecology — Ranging from Subtropical Dry through Cool Temperate Forest Life Zones, groundnut is reported to tolerate annual precipitation of 9.7 to 11.7 dm (mean of 2 cases = 10.7), annual temperature of 9.9 to 20.3°C (mean of 2 cases = 15.1), and pH of 4.5 to 7.0 (mean of 2 cases = 5.8). Produces well in South Florida and Louisiana. I have successfully germinated fall harvested seed, after soaking in hot water, room temperature water, or frozen water, seeds that sunk and seeds that floated after soaking. These took 4 months from harvest to germination, whereas their unsoaked counterparts had still not germinated. Fall-harvested seed apparently exhibit no dormancy when planted in spring.

Cultivation — According to Vilmorin-Andrieux,[352] since seed do not ripen in France, it is multiplied by division in March and April, or in the latter part of summer. Divisions are planted in good, light, well-drained soil 1 to 1.5 m apart in every direction. Reynolds[375] spaced his seedlings at 2 × 3 feet, tubers at 3 × 3 feet. Stems should be supported by poles or stakes. Ground should be kept free of weeds by an occasional hoeing. Cultivation, if overdone, might discourage the rhizomes and their tubers. Seedlings require at least 2 years growth and a minimum photoperiod of 14 hr to induce flowering.[299] Tuber dormancy can be broken by chilling (several months at 35 to 40°F) or using ethylene.

Harvesting — According to Vilmorin-Andrieux,[352] the tubers are not large enough to be gathered for use until the second or third year after planting. Blackman's results in Louisiana show this is not true where there is a long growing season. Once large enough, they can be dug at any time of the year when the ground is not frozen. If carefully dug, strings of four score tubers can be achieved.

Yields and economics — According to Elliott,[353] Asa Gray once said that if advanced civilization had started in North America instead of the Old World, the groundnut would have been the first tuber to be developed and cultivated. Fernald, Kinsey, and Rollins[354] recount an anecdote indicating the economic value of the groundnuts to the pilgrims, "The great value to the colonists of this ready food is further indicated by a reputed town law, which in 1654 ordered that, if an Indian dug Groundnuts on English land, he was to be set in stocks, and for a second offence, to be whipped." Yields of 30 MT per acre were erroneously reported (should have been 30 MT/ha) for cranberry bog weed populations. Reynolds has attained the equivalent of ca 40 MT/ha from tubers in 1-year studies in Louisiana.[375] Some of his plants yielded more than 3 kg tubers.

Energy — Currently, this looks like a poor prospect for biomass production. However, one should at least consider the possibility of developing the crop for marginal habitat (swamp), the tubers as the main crop; the aerial biomass, as residue, might be used for production of rubber, leaf protein, and power alcohol. The nodulated roots fix nitrogen. Around Rocky Gorge Reservoir, in Maryland, the plant is most commonly intertwined in N-fixing *Alnus* species. Nodules were recorded on *A. americana*, but root-nodule location relative to tuber formation was not specified. Root hairs are said to be lacking on secondary roots of mature plants. Four rhizobial strains isolated from *A. americana* nodules were not tested on the host, but since they produced nodules on cowpea plants, the species was considered a member of the cowpea miscellany. The rhizobia are described as monotrichously flagellated rods with cowpea-type, slow cultural growth.[8] H. Keyser[376] suggests conservatively that *Apios* fixes > 100 kg N per ha. With no idea of the solubility of N fixed by the groundnut, I recommend it be studied as a potential intercrop for marsh and aquatic plants, especially rice and wild rice. It might also be considered for cultivation around the edges of reservoirs used for irrigation, hence adding a small token of nitrogen to the irrigation waters. Because of their tolerance to both acidity and waterlogging, they might be especially advantageous around impoundments in strip-mine reclamations. Certainly the scorings by Roth et al.[355] do not speak well for the energy potential of *Apios*. They give it a score of 14, in a system whereby only species receiving scores of 11 or less were regarded as potential renewable energy sources.

Biotic Factors — Agriculture Handbook No. 165[4] lists the following diseases affecting this species: *Alternaria* sp. (leaf spot), *Cercospora tuberosa* (leaf spot), *Erysiphe polygoni* (powdery mildew), *Microsphaera diffusa*, *Phymatotrichum omnivorum*, and *Puccinia andropogonis* var. *onobrychidis* (rust). Reynolds[375] reported powdery mildew, virus, possibly anthracnose, root-knot nematodes, mealy bugs, spider mites, aphids, white flies, leaf-eating caterpillars, cucumber beetles, grasshoppers, stink bugs, and fire ants. In some cases, the fire ants are responsible for mealy bug infestations. Although most *Erythrinae* are bird pollinated, *Apios* seems to be mostly bee pollinated.[356]

ARECA CATECHU L. (ARECACEAE) — Betel-Nut Palm, Areca, Areca-Nut

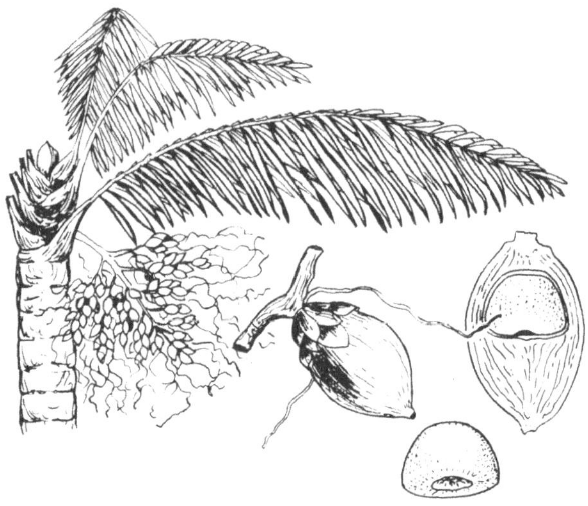

Uses — Chief use of Betel-nut is as a breath sweetening masticatory, enjoyed for centuries by about one-tenth the human population. Often slices of the nut, together with a little lime and other ingredients (cardamom, camphor, cutch, clove, gambier, tobacco) according to taste, are folded in a Betel Pepper leaf (*Piper betel*) and fastened with a clove. Sometimes nuts are ground up with other materials and carried about in a pouch similar to a tobacco pouch. Betel chewing is often considered as an after-dinner or social affair. Chewing colors the saliva red and stains the teeth and gums black, eventually destroying the teeth, at least according to one school of thought. Used in the tanning industry. An extraction of areca-nuts makes black and red dyes. Dried nuts are said to sweeten the breath, strengthen the gums, and improve the appetite and taste. Husks are the most important by-product, being used for insulating wool, boards, and for manufacturing furfural. Innoculated with yeast (*Saccharomyces cervisiae*), leaves used as fermentation stimulant in industrial alcohol production. Large, tough, sheathing parts of leaf-bases, used as substitute for cardboard or strawboard for protecting packages; also used in the Philippines for hats, inner soles for slippers, book-covers, and makes an excellent paper pulp.[86,159]

Folk medicine — The nut, in the form or ghees, powders, bolmes, or enemas, is said to be a folk remedy for abdominal tumors.[126] Reported to be astringent, carminative, deobstruent, dentrifrice, detergent, diaphoretic, diuretic, intoxicant, laxative, masticatory, miotic, panacea, poison, preventative (malaria; mephitis), stomachic, taeniacide, taenifuge, tonic, and vermifuge, betel nut is a folk remedy for ascariasis, beriberi, cancer (esophagus), cholera, circulation problems, colic, diarrhea, dropsy, dysentery, dyspepsia, eruption, fistula, impetigo, malaria, oliguria, rhagades, scabies, smallpox, sores, stomachache, syphilis, and tumors (abdomen).[91] Nuts are astringent, stimulant, and a powerful anthelmintic, especially in veterinary practice. They are also considered digestive, emmenagogue, and are recommended as cardiac, nervine tonic, and as an astringent lotion for eyes, causing dilation of the pupil; once used for glaucoma. Externally, applied to ulcers, bleeding gums, and urinary discharges. Burned and powdered nuts used as a dentifrice in Europe. Once used as antidote to abrin poisoning. Mixed with sugar and coriander, the nuts are given to induce labor in Iran.[138] Unripe fruits are cooling, laxative, and carminative.[86]

Chemistry — Nuts contain the alkaloids, arecoline, arecaine and arecolidine, isoguvacine, guvacine, guvacoline; tannins (18%), fats (1417%), carbohydrates, and proteins, and some Vitamin A.

Toxicity — Per 100 g, the shoot is reported to contain 43 calories, 86.4 g H_2O, 3.3 g protein, 0.3 g fat, 9.0 g total carbohydrate, 1.0 g ash, 6 mg Ca, 89 mg P, and 2.0 mg Fe. Per 100 g, the mature seed is reported to contain 394 calories, 12.3 g H_2O, 6.0 g protein, 10.8 g fat, 69.4 g total carbohydrate, 15.9 g fiber, 1.5 g ash, 542 mg Ca, 63 mg P, 5.7 mg Fe, 76 mg Na, 446 mg K, 0.17 mg thiamine, 0.69 mg riboflavin, 0.6 mg niacin, and a trace of ascorbic acid. Classified by the FDA (*Health Foods Business*, June, 1978) as an Herb of Undefined Safety. Excessive use of betel-nut causes loss of appetite, salivation, and general degeneration of the body. Arecaine is poisonous and affects respiration and the heart, increases peristalsis of intestines, and causes tetanic convulsions.[86,277]

Description — Tall, slender-stemmed palm, up to 30 m, 30 to 45 cm in diameter; stem smooth, whitish, surmounted by crown of pinnate leaves; leaves 0.9 to 1.5 m long, dark-green, with the upper pinnae confluent; lower portion of petiole expanded into a broad, tough, sheath-like structure; inflorescence a spadix encased in a spathe, rachis much-branched bearing male and female flowers; male flowers small and numerous, female ones much larger; fruit a nut, varying in shape from flat to conical or spherical, 5 to 6.5 cm long, 3.7 to 5 cm across, yellow, reddish-yellow to brilliant orange when ripe, size of a nutmeg and with similar internal markings; pericarp hard and fibrous (husk), 65% of fruit mass; kernel (areca-nut), 35% of fruit, grayish-brown, 2.5 to 3.7 cm in diameter, single per fruit, with thin seed-coat and large ruminate endosperm. Flower and fruit seasons variable.

Germplasm — Reported from the Indochina-Indonesia Center of Diversity, the betel palm, or cvs thereof, is reported to tolerate disease, insects, laterite, poor soil, shade, and slope. Varieties are selected on basis of size and shape of fruits and nuts, hardness and astringency of nuts, and various properties of the nuts. Some varieties have large, flat, almost bitter nuts, while others are conical or spherical and so bland in taste as to be called "sweet areca-nuts" (*A. catechu* var. *deliciosa*). *Areca catechu* forma *communis* — fruits orange-red, globose-ovoid, or ovoid-ellipsoid, 4 to 5 cm long, 3 to 4 cm broad; seed subglobose, with a more or less flattish base. *Areca catechu* var. *silvatica* — fruit ovoid-ellipsoid, rather ventricose, smaller than usual, 4 cm long, 3 cm or less broad; seed globose-form from which the commonly cultivated palm has been derived. *Areca catechu* var. *batanensis* — stems shorter and thicker than in forma *communis*, spadix denser, with shorter floriderous branches. *Areca catechu* var. *longicarpa* — fruit narrowly ellipsoid, 5.5 to 7 cm long, 2.5 cm broad; seed ovoid-conical, with blunt apex and flat base, slightly longer than broad. *Areca catechu* var. *semisilvatica*, *A. catechu* var. *alba* and *A. catechu* var. *portoricensis* are other varieties commonly cultivated. (2n = 32.).[82,278]

Distribution — Areca-nut palm is considered native to Malaysia, where it is cultivated extensively. It is also found throughout the East Indies and Philippines. In India, Sri Lanka, Assam, Burma, Madagascar, and East Africa, it is cultivated from the coastal areas up to about 1,000 m. Plants are often spontaneous and occur in second-growth forests, but are rarely found distant from cultivation.

Ecology — Ranging from Subtropical Dry to Wet through Tropical Very Dry to Wet Forest Life Zones, betel nut is reported to tolerate annual precipitation of 6.4 to 42.9 dm (mean of 13 cases = 20.6, annual temperature of 21.3 to 27.5°C (mean of 13 cases = 25.9°C), and pH of 5.0 to 8.0 (mean of 10 cases = 6.4). It requires a moist tropical climate, thriving best at low altitudes, but will tolerate moderate elevations on mountains. Grows in areas with rainfall of 50 cm, if soil is well-drained, but will grow in drier areas with only 5 dm annual rainfall, if suitably irrigated. Uniform distribution of rainfall is very important. Grows in many types of soil varying in texture from laterite to loamy, provided soil has thorough drainage, yet has the ability to retain optimum moisture required by the palm. Light and sandy soils are unsuitable unless copiously irrigated and manured. Maximum temperatures should not exceed 38°C, the optimum temperature for growth being a continuous temperate range from 15.5 to 38°C. These palms are unable to withstand extreme temperatures or a wide variance of daily temperature.[82,278]

Cultivation — Propagation is exclusively from seeds. In southern India and Malaysia, fruits from carefully selected trees are gathered from 25- to 30-year-old trees. In Assam and Bengal, no selection is made. In other areas the middle bunch of fruits is used for seed, and in still other areas the last bunch of the season is preferred. In any case the ripe fruits are gathered in November, dried in the sun for 1 to 2 days, or in shade for 3 to 7 days before being sown. Drying the nuts does not increase germination of seeds. Well-tilled land in a well-drained area in the garden or along an irrigation channel makes a good bed for sowing seed. Seeds sown in rows 15 to 22 cm apart, or in groups of 20 to 50 seeds in pits, or tied up in plantain leaves in rich moist soil to germinate; rarely planted *in situ*. However, seeds may fall from tree and germinate *in situ*. Growth rate of seedling varies, and in about 3 months to 2 years after planting, seedlings are ready to transplant to nursery beds; sometimes up to 4 years may be needed for this stage. Areca-nut is a shade-loving plant and is usually grown as a mixed crop with fruit trees, such as mango, guava, jackfruit, orange, plantain, or coconut. Usually a shade crop, such as bananas, is planted first, spaced about 2.7 m apart in a north-south direction, and allowed to become well-established before transplanting the areca-nut seedlings. Young seedlings are planted in nursery beds 30 × 30 cm, with 3 rows per bed, about 1000 to 1500 trees per ha. After about 20 years, young seedlings are planted between trees and between rows to replace older palms which have become unproductive. After seedlings are planted, the bed is mulched with green or dry leaves, cattle dung, wood ashes, or groundnut cake. Beds are made only in the rainy season and are kept well-irrigated in the summer. Hoeing, weeding and interculture may be practiced. Pepper vines (*Piper betel*) and cardamon may be trained to the trees or grown between them. Farmyard manure, groundnut cake, ammonium sulfate, superphosphate and potassium sulfate have been found to be beneficial. Also leaf manure and green manure may be used.[278]

Harvesting — Palms begin to flower about the 7th year after sowing seed, and reach full production in about 10 to 15 years. With best conditions, trees may begin flowering the 4th year. A plantation may take 30 years to reach maturity. Fruiting life of a tree is between 30 to 60 years after maturity, but trees may live for 60 to 100 years. Economical life span in India is 45 to 70 years. In different regions there are well-defined seasons for flowering and corresponding fruiting seasons. Because of the tall, slender nature of the palm, harvesting the nuts requires skill and dexterity. Primitive methods are often employed. In India certain classes of people who climb palms fast are employed. Sometimes bamboo poles with sickles attached are used to cut the bunches. In Malaysia, trained monkeys are used. Leaves of the palm (usually 4 to 7) begin to drop in December at intervals of 3 weeks, until June. Inflorescences appear in the axils of such leaves, and although as many as five spadices may appear, usually there are only 2 or 3 mature fruits. Spathes open soon after shedding of leaves, and fruits ripen 8 to 11 months later. (Fruits take 6 to 8 months to ripen.) Nuts harvested when bright red. Usually the shedding of a few nuts from a bunch is sufficient indication to harvest the whole bunch. Harvesting season varies with 2 or 3 pickings made in each season: Bombay and Sri Lanka, from August to March; Mysore, from August to January; Bengal, from October to January. In India, areca-nuts are consumed raw or cured; in other areas ripe nuts are masticated during the harvest season. Surplus nuts are stored in pits in soil or water in earthenware jars for 5 to 7 months, and during the off-season are taken out and chewed. Ripe nuts may also be dehusked, cut and dried, or just dried whole in the sun for 6 to 7 weeks, or may be perfumed by smoke or benzoin. Nuts may be processed, a costly and laborious operation on a commercial scale, to improve their color, taste, palatability, and keeping quality. When properly cured and dried, nuts are dark-brown with glossy finish.[278]

Yields and economics — Each tree yields 2 to 3 bunches per year, containing 150 to 250 fruits; varieties with larger fruits may have 50 to 100 fruits per bunch. Fruits weigh from 1.4 to 2.2 kg per 100 fruits. Yield per hectare with 1,000 trees is 440,000 to 750,000 fruits, or about 15 to 25 cwt of dried areca-nuts. Average yield of dried or cured nuts per annum in Mysore is about 17.5 cwt/ha. India and Pakistan are the major producers of areca-nuts, where most of the production is consumed domestically. It is also an item of internal commerce in the Malay Archipelago and the Philippines. Nuts are exported in large quantities from Java, Sumatra, Singapore, and other Malaysian regions to India. Sri Lanka exports to India and the U.S. In 1969 to 1970 Pakistan grew about 1,000,000 acres of betel-nut, producing about 26,500 long tons of nuts. Bavappa et al.[36] suggest that there are 184,000 ha cultivated to Areca, with production of ca. 191,000 MT/year with a value of 2,500 million rupees. Improved cultural practices are leading to higher yields of nuts. Higher-yielding and more disease-resistant plants are being developed through breeding.

Energy — Debris from the plants could serve as a crude energy source. With 2000 to 3000 trees per hectare or more, there might be 8,000 to 21,000 leaves falling between June and December.[70] Fallen spathes and spadices might also be viewed as energy sources. Much energy is consumed in the boiling and drying of this widely used narcotic. On top of this, there might be 1,500 to 2,500 kg/ha dried nuts. In preparing the kernels for market, there is much husk remaining as a by-product, containing nearly 50% cellulose. The wood of cull trees may be used for firewood.

Biotic factors — The two most serious fungal diseases of this palm are *Phytophthora omnivorum* var. *arecae* (Koleroga disease, a fruit rot) and *Ganoderma lucidum* (Foot rot). Other fungal diseases include: *Alternaria tenuis*, *Aspergillus niger arecae* (causing a storage disease), *Botryodiplodia theobromae*, *Brachysporum arecae*, *Ceratostomella paradoza*, *Colletotrichum catechu* (seedling blight), *Coniothyrium arecae*, *Dendryphium catechu*, *Exosporium arecae*, *Gloeosporium catechu*, *Lenzites striata*, *Lichenophoma arecae*, *Melanconium palmarum*, *Montagnellina catechu*, *Mycosphaerella* sp., *Nigrospora sphaerica*, *Phyllosticta arecae*, *Polyporus ostreiformis*, *P. zonalis*, *Stagonospora arecae*, *Thielaviopsis paradoxa* (causes length-wise splitting of stem), *Torula herbarum*, *Ustulina zonata*. Areca-nut is also attacked by the bacterium *Xanthosomas vasculorum*. In Thailand, the following nematodes are known to attack arecanut: *Rotylenchulus* sp., *Tylenchorhynchus dactylurus*, *Tylenchus* sp., and *Xiphinema insigne*. In Mysore and Malaysia, the Rhinoceros beetle (*Orcytes rhinoceros*), leaf-eating caterpillar (*Nephantis serinapa*), borer (*Arceerns fasciculatus*), white ants, and mites cause minor damage.[186,278]

ARENGA PINNATA (Wurmb) Merr. (ARECACEAE) — Sugar Palm, Kaong, Black Sugar Palm

Syn.: *Arenga saccharifera* Labill.

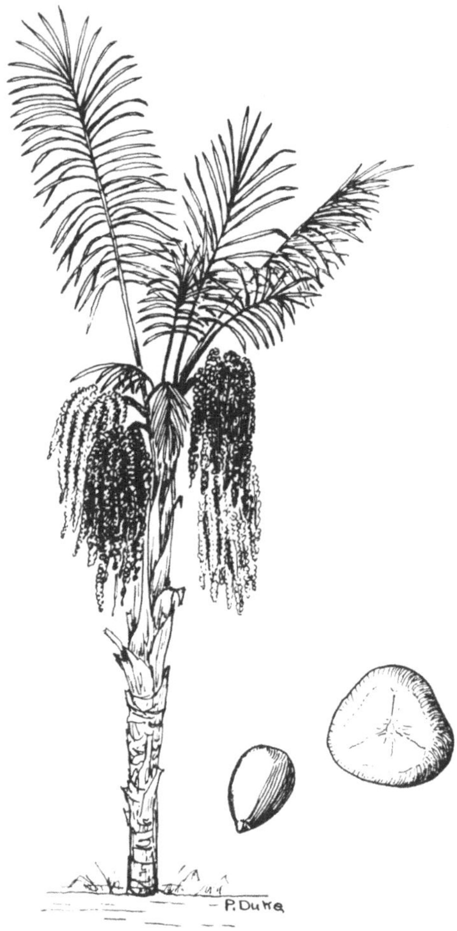

Uses — Sugar palm is grown for its sugar, starch, and fiber. Sap contains 20 to 40% more sucrose than average sugarcane. Juice of the outer covering of fruit is highly corrosive and may cause pain and skin inflammation. Pith of the stem is source of sago starch. Sap may be made into a refreshing fresh drink, or fermented into palm wine, upon distillation yielding Arrack. Alcohol and vinegar may also be made from the sap. Terminal bud or "cabbage" is eaten in salads, raw or cooked. Etiolated leaves, petioles, and pith of young stems eaten in soups or fried, or used as a pickled preserve. Half-ripe fruits are pruned to remove irritating crystals in pericarp; seeds washed and seedcoat removed; endosperm is soaked in lime water for several days and finally boiled in sugary or spicy solutions and eaten as sweetmeats. Young leaf-sheaths produce a valuable fiber used in industrial work. Leaves are used to thatch roofs and are quite durable; leaflets are used for rough brooms and are sometimes woven into baskets. "Wood" is used for water pipes.[34,278] Juice of outer fleshy covering of fruit is used as a fish poison.[278]

Folk medicine — Reported to be intoxicant and piscicide.[91] Sap considered lactogenic in Malaysia. Javanese use a root decoction for kidney stones. Fermented sap taken for tuberculosis in the Philippines and Indonesia; for sprue, dysentery, constipation, and hemorrhoids in Java. The felt-like tomentum at the leaf-base is used as a styptic.[90] Roots used to make a medicine for stone in the bladder in Java. The fresh, sweet toddy used for chronic constipation, phthisis, and dysentery; lactagogue. Applied to wounds as a hemostatic.[56] Diuretic and antithermic; fresh unfermented sap is a purgative and a remedy for sprue in Indonesia. Juice of ripe fruit is poisonous. Roots are a treatment against bronchitis and gravel.[249,278]

Chemistry — Per 100 g, the shoot is reported to contain 19 calories, 94.7 g H_2O, 0.1 g protein, 0.2 g fat, 4.9 g total carbohydrate, 0.5 g fiber, 0.1 g ash, 21 mg Ca, 3 mg P, 00.5 mg Fe, 2 mg Na, 7 mg K, 0.01 mg riboflavin, and 0.1 mg niacin.[89]

Description — Tall, stout palm, 8 to 15 m tall, bole solitary, straight, 40 to 50 cm in diameter; old leaf-bases covering trunk with mat of tough, black fibers and long spines; leaves ascending, pinnate, up to 9.1 m long, 3.1 m wide, with 100 or more pairs of linear leaflets, leaflets whitish or scurfy beneath, dark-green above, 1 to 1.5 m long, 6 to 8 cm (or more) wide, lobed or jagged at apex, auricled at base; petioles 1.5 to 2 m long, very stout, base covered with black fibers and weak spines; plants monoecious, bearing very large pendulous interfoliar inflorescences arising from leaf axils; female inflorescence usually preceding male; male and female inflorescence, which eventually become 1 to 3.3 m long, at first ensheathed in bud by 5 to 7 lanceolate oblong, imbricated, caducous bracts; inflorescence emerging from spathes in 6 to 9 weeks; peduncle large; flowers opening first at base of each branch and successively toward apex; flowers numerous, sessile, either male or female; female flowers usually solitary, male solitary or paired, rarely in threes, occurring in separate inflorescences; in bisexual flowers, stamens usually abortive; male flowers scentless, with 3 green imbricated, persistent sepals, one-fourth length of petals, apex broadly acute, thin-margined; petals 3 to 4, navicular, valvate, 2.5 cm long, red-brown or red-purple on outside, yellow on inside; stamens yellow, numerous, with elongated apiculate anthers, borne on short filaments; no rudimentary ovary; female flowers scentless, 3 unequal green imbricated orbicular sepals, one-third length of petals, persistent; petals coriaceous, 1.5 to 2.5 cm long, light-green,ovate, or triangular, valvate, persistent with sepals as cupule at base of fruit; staminodes absent, or if present, sometimes producing nectar; fruit obovoid to subglobose, smooth, 5 to 6 cm in diameter, with depressed trigonous upper surface; exocarp yellow or yellow-brown, coriaceous; mesocarp fleshy, whitish, gelatinous, very acid due to stinging crystals; endocarp black, smooth, thin, stony; seeds 2 to 3 per fruit, dull-metallic gray-brown, trigonous, oblong, 2.5 to 3.5 cm long, 2 to 2.5 cm wide, with copious endosperm. Flowers and fruits year-round.

Germplasm — Reported from the Indochina-Indonesia and Hindustani Centers of Diversity, sugar palm, or cvs thereof, is reported to tolerate disease, drought, fungus, high pH, insects, poor soil, shade, and slope.[82] Several forms of the sugar palm exist in Malaya, varying mainly in how long is required for plants to begin flowering.[278] (2 n = 26,32.)

Distribution — Native from eastern India and Ceylon, through Bangladesh, Burma, Thailand, southern China, Hainan, Malay Peninsula to New Guinea and Guam. Extensively cultivated in India.[278]

Ecology — Ranging from Subtropical Dry to Moist through Tropical Dry to Wet Forest Life Zones, sugar palm is reported to tolerate annual precipitation of 7 to 40 dm (mean of 8 cases = 19.1), annual temperature of 19 to 27°C (mean of 8 cases = 24.5), and pH of 5.0 to 8.0 (mean of 5 cases = 6.4).[82] More or less a forest tree, but not restricted to jungles; it can be grown on very poor rocky hillsides and in waste places. It flourishes best in humid tropics in a rich moist soil, from sea-level to elevations of 1,200 m, being grown at higher elevations than coconut. It is little subject to drought damage, typhoons, insect pests, or fungal diseases. Trees are hardy, self-sustaining, growing readily in well-drained soil of dark cool valleys, along banks of mountain streams, along forest margins and on partially open hillsides. It develops more slowly in flat, exposed, or sunny habitats.[278]

Cultivation — In forests of Indo-Malaysia, ripe fruits are distributed by various fruit bats, civet cats, and wild swine. Trees are only in semi-cultivation, mainly since trees require many years to begin to be useful. When propagated, seed are used, but it has never been scientifically cultivated. Growing it in plantations for its fiber is too costly.[278]

Harvesting — Various products may be harvested from the sugar palm. Trees reach maturity (flowering stage) in 6 to 12 years and continue to flower for about 15 years before replanting. Flowering is quite irregular. From flowering to ripe fruit takes about 2 years, so the harvest period for the fruit extends over the entire year. Most important industrial product is the black, horsehair-like tough fiber, called gomuta, yunot, or cabo negro, produced at base of petioles in large quantities. It is used in manufacture of a very durable rope used in fresh-and salt-water and for thatching houses; known to last 100 years in the Philippines. Fiber also widely used for filters and for caulking ships. Cost of fiber is high, depending on grade and length of fiber, but is in demand in Europe for industrial purposes. Stiffer fibers are used in Philippines to make floor and hair brushes, and brushes for grooming horses. Thatch-like raincoats are sometimes made from it. Associated with the fibers at basal parts of petiole is a soft, dry, light, punky substance, called barok, varying in color from white to dark shades, used in caulking boats and as a tinder, made by soaking in juice of banana or lye made from ashes of *Vitex negundo* and then dried; 60 to 75 tons of this exported annually from Java to Singapore. Palms commonly tapped for the sweet sap used for producing sugar, vinegar, wine, or alcohol. Trees for sugar production are selected and the young inflorescences beaten with a stick or wooden mallet for a short time each day for 2 to 3 weeks, thus producing wound tissue and stimulating the flow of sap to the injured area. Starch in the trunks is converted into sugar and moves into the inflorescence when it begins to develop. Thus by wounding young inflorescences, the flow of sugar to the wounded tissue can be regulated. The stalk is then cut off at base of the inflorescence and the exuding sap collected. A thin slice is removed from the wounded end of the stalk once or twice a day during sap flow. Flow generally diminishes from 10 to 12 to 2 ℓ/day after 2 1/2 months; some plants yield about 2.8 ℓ/day for about 2 years. Fresh sap is clear with pleasant taste and makes a refreshing drink. Kept awhile, it becomes turbid and acid, and upon fermentation, acquires an intoxicating quality. Flavored with bark of other trees, large quantities of the liquor are consumed. Sap is allowed to ferment, producing "tuba", a palm wine, a popular drink in Philippines; it is supposed to have curative properties. Fermentation begins in the bamboo's tubes in which sap is collected and is usually well-advanced when the product is gathered. Much is converted into a good quality vinegar; alcohol is also distilled from the "tuba". Sugar is made by boiling the sweet, unfermented sap, using a new bamboo joint for the sap each day. To prevent fermentation in the tube, a little crushed ginger or crushed chili-pepper fruit is added to the bamboo joint. Sometimes in Java, bamboo joints are smoked first to reduce fermentation. Sugar is manufactured by boiling thickened juice in an open kettle until the liquid solidifies when dropped on cold surface. Sugar in the Philippines is brown and enters into local commerce in very limited quantities. Yield of sugar is about 20 tons/ha, with 150 to 200 trees/ha. In Java and elsewhere, old trees no longer productive of sugar are felled and cut up into short sections, or the pith is scooped out of trunks cut lengthwise. Fibrous pith is pulverized and washed to remove fibrous material and other impurities. Starch particles in suspension are drawn off and sago starch removed and dried in sun. Starch is light gray-white. A type of tapioca may be prepared from this starch by dropping wet pellets of it on hot plates. Debris, after starch is removed, is boiled and used for hog feed. In Luzon, starch is obtained only from male or sterile trees. Yield of sago meal is about 67.5 kg per tree. Yields of starch vary greatly, with an average yield of 50 to 75 kg per tree.[278]

Yields and economics — Specific yields are stated above for each product. Products of this palm are widely used in areas where it grows, but only the fibers are in international commerce. Sugar and starch, and their by-products are consumed locally, and in very large quantities.[278]

Energy — In *Palms as Energy Sources*, Duke[81] reports that a single sugar palm can yield 2.8 ℓ (sugar content 5 to 8%) toddy per day over a period of about two months. Sugar yields of 20 MT per ha are suggested, all of which could be converted to renewable alcohol. Once flowering, male trees go on producing tappable spadices for 2 to 3 years, until the lowest leaf axil is utilized and the tree is exhausted.[70] A single tree, upon felling, can yield up to 75 kg "sago starch" (true sago may yield 5 times as much). Trees that have been tapped for sugar yield little or no sago). Energy planners cannot then add the sugar and starch, but plan for one or the other. The black reticulate leaf-sheaths have hair-like fibers that are used for tinder.[81]

Biotic factors — Flowers are presumably wild-pollinated. Sugar palm is virtually insect-, pest-, and disease-free, one fungus attacking the palm being *Ganoderma pseudoferreum*. In the East Indies, leaves are damaged by the rhinoceros beetle (*Orcytes rhinoceros*), and dead palms are reported to harbor these beetles, which cause serious damage to coconut palms.[278]

ARTOCARPUS ALTILIS (Parkins.) Fosb. (MORACEAE) — Breadfruit, Breadnut, Pana
Syn.: *Artocarpus communis* Forst.

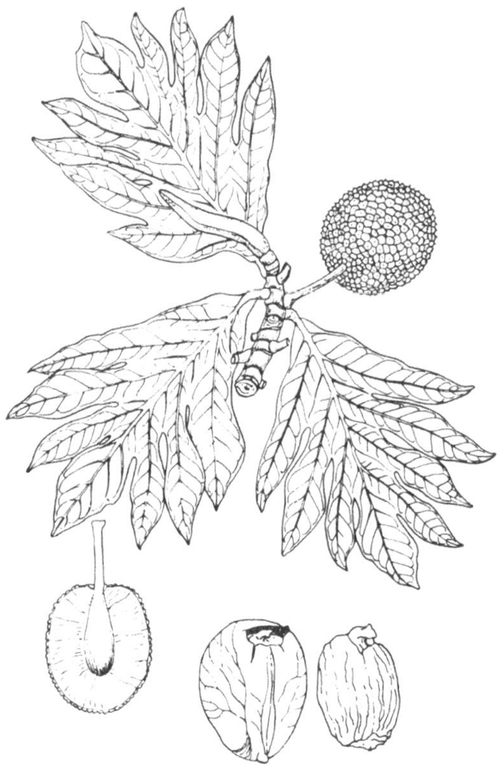

Uses — Cultivated extensively for its fruits and seeds, breadfruit is used as a staple food with Polynesians, Micronesians, and Melanesians. When fruit is not fully ripe, being very starchy, it is peeled, cut into sections, and baked or boiled, and seasoned with salt, pepper, and butter. When fully ripe, the meat is soft and can be baked like sweet potato. Green fruits are roasted, ground into a meal, and used to make breads. Breadfruit can also be used in salads, made into soup, and, when ripe, made into a pulp (coconut milk and sugar being added) and baked as a pudding. Seeds are eaten roasted or boiled. Bark cloth is made from the bark. Leaves furnish fodder for livestock. Wood is used locally for house-building. Latex from the trunk is used in native medicines, as bird lime and to caulk canoes.[278]

Folk medicine — Fruits and leaves used as a cataplasm for tumors in Brazil.[126] Powder of roasted leaves applied for enlarged spleen; ashes of the leaves applied in herpes[56] (Amboina). Reported to be anodyne, laxative, and vermifuge, breadfruit is a folk remedy for backache, blood disorders, boils, burns, diabetes, diarrhea, dysentery, eye ailments, fever, fracture, gout, headache, hypertension, oliguria, rheumatism, sores, stomach-ache, swelling, testicles, worms, and wounds.[91] West Indians have great faith in the leaf decoction for high blood pressure. Colombians cook the fruit with sugar for colic. Virgin Islanders take the plant for coronary ailments, Jamaicans dress liver spots with the latex, Costa Ricans apply it to wounds.[223,224] Chinese use the seeds to aid parturition and to treat typhoid and other fevers. Indonesians use the bark in parturition, poulticing the leaves on splenomegaly. Heated flowers, after cooling, are applied to the gums for toothache, fruits are used for cough, root-bark for diarrhea and dysentery, seeds as an aphrodisiac. Philippinos use the bark decoction for stomach-ache. New Guineans use the latex for dysentery.[249]

Chemistry — Per 100 g, the mature fruit is reported to contain 103 calories, 70.8 g H_2O,

1.7 g protein, 0.3 g fat, 26.2 g total carbohydrate, 1.2 g fiber, 1.0 g ash, 33 mg Ca, 32 mg P, 1.2 mg Fe, 15 mg Na, 439 mg K, 24 mg beta-carotene equivalent, 0.22 mg thiamine, 0.03 mg riboflavin, 0.9 mg niacin, and 29 mg ascorbic acid. Per 100 g, the leaf is reported to contain 75 calories, 75.5 g H_2O, 5.0 g protein, 2.0 g total carbohydrate, 2.0 g ash, 2.0 mg Ca, 170 mg P, 60 mg Fe, 17.5 mg Na, 0.10 mg thiamine, and 70 mg ascorbic acid. Per 100 g, the mature seed is reported to contain 434 calories, 20.2 g H_2O, 15.1 g protein, 29.0 g fat, 34.0 g total carbohydrate, 2.5 g fiber, 1.7 g ash, 66 mg Ca, 320 mg P, 6.7 mg Fe, 41 mg Na, 380 mg K, 280 mg beta-carotene equivalent, 0.88 mg thiamine, 0.55 mg riboflavin, 0.8 mg niacin, and 12 mg ascorbic acid. Quijano and Arango[275] report wetter seeds (56.3% moisture) with (ZMB): 3 to 4 g ash, 12.8 g fat, 16 g soluble carbohydrates, 20 g total protein, and 3.9 g fiber. Of the protein (20%), 6.4 g was nonglobular protein, 13.5 g globular (1.8 g albumins, 3.7 globulins, 3.3 prolamins, and 4.6 g glutelins). The amino acids of the seeds contain 13.04 g/100 g leucine, 12.10 isoleucine, 5.28 g phenylalanine, 15.90 g methionine, 7.24 g tyrosine, 3.62 g proline, 7.68 g alanine, 4.93 g glutamic acid, 3.91 g threonine, 10.43 g serine, 4.78 g glycine, 3.33 g arginine, 4.56 g histidine, and 3.12 g cystine per 100 g protein. Fruits contain papayotin and artocarpin.[187] Leaves contain quercetin and camphorol.[224] Some HCN is reported in the leaves, stem, and root, cerotic acid in the latex.[249]

Description — Handsome tree, 12 to 20 m tall; leaves large, ovate, leathery, rough, glossy, most often lobate or incised, 30 to 90 cm long, 30 to 40 cm broad, dark-green; flowers minute, male and female flowers in separate catkins on the same tree, in axils of newly formed leaves; male inflorescences club-shaped, 15 to 30 cm long, dropping to the ground in a few days; female inflorescences in globose heads about 5 cm in diameter, developing into seedless fruits; some varieties of breadfruit have seeds in profusion; fruit (syncarp) ovoid, spherical, or pear-shaped, 10 to 15 cm long, 10 to 15 cm in diameter, weighing 1 to 4 kg, with white sticky latex, rind yellowish-green or brown, divided into a series of low projections, bearing short spines in some varieties; pulp white or yellowish; in breadnut, rind covered with fleshy spines, with brownish seeds 2.5 cm or more in length and about 2.5 cm in diameter. Flowers and fruits at nearly all stages on the tree at the same time, almost throughout the year.[278]

Germplasm — Reported from the Indochina-Indonesia Center of Diversity, breadfruit, or cvs thereof, is reported to tolerate drought, high pH, heat, laterite, sodium or salt, slope, and virus.[82] Many cvs have developed wherever breadfruit has been long grown. Seeded cvs are of little economic value but are eaten by natives; the seeds, when roasted, taste like chestnuts. Most cvs are seedless. In Ponape, over 50 cvs are known; in Tahiti, about 30; and in the South Pacific area, 165.[278] (2n = 56.)

Distribution — Probably originated in Indonesia and perhaps in New Guinea, where large, spontaneous stands occur. Breadfruit is cultivated throughout the islands of the South Seas. It has been introduced into many tropical areas of the world, including India, West Indies, Mauritius, and southern Florida.[278]

Ecology — Ranging fromWarm Temperate Dry (without frost) through Tropical Dry to Wet Forest Life Zones, breadfruit is reported to tolerate annual precipitation of 7 to 40 dm (mean of 19 cases = 23.0), annual temperature of 17 to 29°C (mean of 19 cases = 24.1), and pH of 5.0 to 8.0 (mean of 14 cases = 6.2). Breadfruit thrives only in humid tropics, where the temperature varies from 16 to 38°C with a humidity of 70 to 80%, and a well-distributed annual rainfall of 250 to 275 cm. Climatic requirements vary according to cv. In Indonesia, some cvs are adapted to moist climates; others endure 6 months of dry weather. It is usually intolerant of climatic extremes in inland regions or high elevations, but grows on high islands (up to 700 m in New Guinea) and on atolls throughout the Pacific. It does not tolerate shade, and irrigation has been unsuccessful. It thrives on alluvial and coastal soils, and can be grown and produced on coral soils of atolls. Some cvs ("Maitarika") grown on atolls such as Gilbert Islands are said to tolerate salinity. Wind easily breaks the branches and may cause many flowers and young fruits to fall.[82,278]

Cultivation — Seeded cvs are propagated from seed; however, seeds lose their viability soon after fruit falls. For seedless varieties, if roots are bruised near soil surface, plants send up root-suckers, which can be removed and planted in a permanent site. Root-cuttings 20 to 25 cm long and 12 to 60 mm in diameter may be laid horizontally in a 12-mm-deep trench and watered daily. Remove and plant sprouts when 20 to 25 cm tall in rainy season. Natural suckers can be air-layered for root initiation planted in a nursery for 2 to 3 months, then transplanted to permanent site. At all times, trees should be spaced about 12 m each way, or about 100 trees per hectare. Trees should be watered for first and second years, given shallow intercultures, and generally no manuring. Some intercropping is practiced.[278]

Harvesting — Trees start bearing in 5 to 6 years, when vegetatively propagated, 8 to 10 years from seed. Under good soil and climatic conditions they will continue to produce well for 50 years. Time of harvest differs in various localities: in Caroline Islands, May to September; Gilbert Islands, May to July; Society Islands, November to April and July to August. For culinary purposes, harvest when still hard. Harvesting is done with a long pole, having a hooked knife and basket at the end so fruits do not fall to ground. Fruits ripen in 4 to 6 days.[278]

Yields and economics — Adult trees yield 50 to 150 fruits, each producing 23 to 45 kg. An 8-year-old tree can bear 800 fruits over the three seasons of a year.[182] A fruit may weigh 1 to 3 kg. Breadfruits are gathered and sold locally where the trees are grown. They rarely enter international trade.[278]

Energy — If an adult tree bears 100 2-kg fruits, and if the fruits contain 70% water, that is still 60 kg dry matter (DM) per tree per year. If one could crowd 100 such productive trees into a hectare that indicates 6 MT DM in fruits alone, a reasonable renewable biomass production. There is a sizable annual accumulation of leaves and limbs as well.

Biotic factors — Hand-pollinated fruits are twice the size of normally developing fruits. The following fungi are known to attack breadfruit: *Capnodium* sp., *Cercospora artocarpi*, *Colletotrichum artocarpi*, *Cephaleuros virescens*, *Corticium salmonicolor*, *Gloeosporium artocarpi*, *G. mangiferae*, *Orbilia epipora*, *Pestalotiopis versicolor*, *Phytophthora palmivora*, *Phyllosticta artocarpi*, *P. artocarpicola*, *Mycosphaerella artocarpi*, *Rhizopus artocarpi*, *Sclerotium rolfsii*, *Uredo artocarpi*, *Zygosporium oscheoides*. Nematodes infesting the tree are *Boledorus* sp., *Helicotylenchus concavus*, *H. cavenessi*, *H. dihystera*, *H. microcephalus*, *H. pseudorobustus*, *Heterodera marioni*, *Meloidogyne incognita acrita*, *Rotylenchulus reniformis*, *Scutellonema calthricaudatum*, *Tylenchorhynchus triglyphus*, and *Xiphinema ifacola*.[186,278]

ARTOCARPUS HETEROPHYLLUS Lam. (MORACEAE) — Jackfruit
Syn.: *Artocarpus integra* (Thunb.) Merr., *Artocarpus integrifolia* L.f.

Uses — Few, if any, tropical fruits can excel the jackfruit in size and usefulness. Cultivated for its multiple fruit, the pulp may be cooked or fried before ripening, or eaten raw when ripe. Fruits I sampled in Brazil were quite adequate, right off the tree. Pulp is sometimes boiled with milk, or made into preserves or curries. Leaves and bark contain a white latex. Leaves are fed to sheep, goats, and cattle as fodder, especially in the dry season. Flower clusters are eaten in Java with syrup and agar-agar or coconut milk. Young fruits may be eaten in soups. When properly fermented, pulp produces a vinegar. Seeds are mealy and are tasty when boiled or roasted. Half-ripe fruits are fed to pigs and used for fattening cattle and sheep. Wood is bright yellow when fresh, darkening on exposure, used for furniture, cabinet work, house-building, doors, window frames, and cart work. The wood chips are distilled in Burma and Sri Lanka to produce the yellow dye used for Buddhist robes. Trees are usually cut for lumber when upwards of 30 years old; wood takes a high polish and is ornamental. Heartwood contains a brilliant yellow dye, similar to fustic. Cyanomaclurin is also present, producing an olive-yellow with chromium, dull yellow with aluminum, and a brighter yellow with tin mordant. Green and red dyes may also be prepared. Sawdust and shavings of wood, when boiled in water, yield a yellow dye used for dying silk. Milky juice is used in some countries as a bird-lime. Bark yields a fiber.[7,278] Shedding nearly 10 MT leaves a year and bearing fruits weighing up to 11 kg each, this species deserves consideration as a shade tree for cardamoms.[70,283]

Folk medicine — According to Hartwell, the plant is used in folk remedies fcr tumors.[126] Reported to be astringent, demulcent, laxative, refrigerant, and tonic, jackfruit is a folk remedy for alcoholism, carbuncles, caries, leprosy, puerperium, smallpox, sores, sterility, stomach problems, toothache, and tumors.[91] Burmese, Chinese, and Filipinos use the sap to treat abscesses and ulcers, and the bark to poultice on such afflictions. Burmese also use the roots for diarrhea and fever. Indochinese use the wood as a sedative in convulsion, the boiled leaves as a lactagogue, the sap for syphilis and worms. Filipinos use the ashes of the leaves to treat ulcers and wounds.[249] Cambodians used the wood to calm the nerves. Munda of India use the leaves for vomiting. Both Ayurvedics and Yunani consider the fruit and seeds aphrodisiac. Ayurvedics use the ripe fruit for biliousness, leprosy, and ulcers. India uses the roots for hydrocoele.[165]

Chemistry — Per 100 g, the leaves contain (ZMB) 18.5 g protein, 5.0 g fat, 66.3 g total carbohydrate, 26.2 g fiber, 10.2 g ash, 2,000 mg Ca, and 110 mg P. Per 100 g, the fruits (ZMB) contain 347 calories, 6.3 g protein, 1.1 g fat, 87.5 g total carbohydrate, 3.3 g fiber, 5.2 g ash, 100 mg Ca, 140 mg P, 2.2 mg Fe, 7.4 mg Na, 1,502 mg K, 867 ug beta-carotene equivalent, 0.33 mg thiamine, 0.41 mg riboflavin, 2.58 mg niacin, and 33 mg ascorbic acid. Per 100 g, the seeds contain 51.6 g H_2O, 6.6 g protein, 0.4 g fat, 38.4 g carbohydrate, 1.5 g fiber, 1.5 g ash, 0.05% Ca, 0.13% P, and 1.2 mg Fe.[90] The latex consists of 65.9 to 76.0% moisture and water solubles and 2.3 to 2.9% caoutchouc. The coagulum contains 6 to 10% caoutchouc, 82.6 to 86.4% resins, and 3.9 to 8.1% insolubles. Dried latex contains the steroketone artostenone $C_{30}H_{50}O$, which has been converted to artosterone, a compound with highly androgenic properties. Seeds, though eaten, contain the hemagglutinin, concavalin A. Hager's Handbook gives structures for six flavones isolated therefrom: artacarpanone, artocarpetin, artocarpin, cyanomaclurin, cycloartocarpin, and morin ($C_{15}H_{10}O_7$).[187] The wood contains a yellow pigment, morin, and cyanomaclurin; the bark has tannin, the latex cerotic acid.[249]

Description — Low or medium-sized evergreen tree, 10 to 25 m high, without buttresses, with dense, rather regular crown. Branchlets terete, with scattered, retrose, crisped hairs, becoming glabrous. Leaves alternate, shortly stalked, oblong or obovate, with cuneate or obtuse base, and obtuse or shortly acuminate apex, entire (lobed only on very young plants), coriaceous, rough, glabrescent, shining dark-green above, pale-green beneath, 10 to 20 cm long, 5 to 10 cm wide, with 5 to 8 pairs of lateral veins, petiole 2 to 4 cm long. Stipules ovate-triangular, acute, hairy on the back, glabrous on the inner side, pale, 1 to 2 cm long, on flowering branches much larger, up to 5 cm. Inflorescences peduncled, solitary in the leaf-axils of short, thick branchlets which are placed on the trunk or on the main branches, unisexual, 4 to 15 cm long; the male ones near the apex, fascicled in the higher axils, oblong-clavate, rounded at both ends; the female ones in the lower axils, solitary or in pairs on longer and thicker peduncles. Flowers very numerous, small, the male ones with a two-lobed perianth and one stamen; the female ones cohering at the base, tubular, style obliquely inserted, stigma clavate. Spurious fruits very large, oblong, glabrous, with short, 3 to 6 angular, conical acute spines.[238]

Germplasm — Reported from the Hindustani Center of Diversity, jackfruit, or cvs thereof, is reported to tolerate aluminum, laterites, limestone, low pH, and shade.[82] Varieties such as "Soft" or "Hard", are selected mainly according to the thickness of the rind.

Distribution — Native to the Indian Archipelago, jackfruit is now widely cultivated throughout the Old and New World tropics, being known in India, Burma, Bangladesh, Sri Lanka, Java, and in South America from the Guianas as far south as Rio de Janeiro, in Brazil, West Indies, and southern Florida.[278]

Ecology — Ranging from Subtropical Dry to Moist through Tropical Very Dry to Wet Forest Life Zones, jackfruit is reported to tolerate annual precipitation of 7 to 42 dm (mean of 14 cases = 22.7), annual temperature of 19 to 29°C (mean of 14 cases = 24.8), and pH of 4.3 to 8.0 (mean of 11 cases = 6.0).[82] As a tropical tree, jackfruit grows well in most soils, but not in moist low places. Cultivated below 1,000 m altitudes, it grows best in deep well-drained soil, but will grow slowly and not so tall in shallow limestone soil. Sensitive to frost in its early stages, it cannot tolerate drought or "wet feet".[278]

Cultivation — Propagated by seeds (viable only 2 to 4 weeks), budding after the modified Forkert method, inarching, air-layering, or grafting. Seeds retain viability for about 30 days at room temperature; however, soaking them in water for 24 hr improves their longevity. Budding with eyes of nonpetioled budwood on stocks from 8 to 11 months old gives best results. The best stock is *Artocarpus champeden*, but *A. rigida* Bl. can also be used. Stock should be slightly shaded. For grafting-tape, dry bark-fibers of *Musa textilis* are used. Budding may be performed throughout the year, provided stocks are old enough. Can be propagated through root shoots. Trees should be spaced 12 to 14 m apart each way. Cattle manure is helpful.[278]

Harvesting — Trees start bearing fruit when 4 to 14 years old; once established, they continue to bear for several decades. Ripe fruits, available almost throughout the year, are much relished. About 8 months is required from time the flowers begin to expand until fruit matures.[278]

Yields and economics — A tree may bear 150 to 250 fruits per year and fruits may weigh 10 to 40 kg. Two hundred 20-kg fruits a tree indicate an incredible 4 tons per tree per year. If 50 trees could bear at this rate, that would be 200 tons fruit per ha. But 75% of this is water. Cultivation in Bangladesh in 1969 to 1970 amounted to 17,760 ha, because of greater demand, producing 212,635 tons of fruit. Because of the fruit, it is marketed locally.

Energy — Grown as a shade tree for cardamom, jackfruit contributed annually 9,375 kg/ha leaf mulch.[357]

Biotic factors — The following fungi are known to attack jackfruit: *Ascochyta* sp., *Botryodiplodia theobromae*, *Capnodium* sp., *Cephaleuros* sp., *Circinotrichum* sp., *Corticium salmonicolor* (pink disease), *Diplodia artocarpi*, *Ganoderma applanatum*, *Gloeosporium artocarpi*, *G. caressae*, *Kernia furcotricha*, *Marasmius scandens*, *Marssonia indica*, *Meliola artocarpi*, *Pestalotia elasticola*, *Phomopsis artocarpi*, *Phyllosticta artocarpi*, *P.*

artocarpina, *Phytophthora palmivora*, *Rhizoctonia solani*, *Rhizopus artocarpi*, *R. stolonifer*, *Rosellinia bunodes*, *Septoria artocarpi*, *Setella coracina*, *Torula herbarum*, *Uredo artocarpi*. Trees are also parasitized by *Dendrophthoe falcata* and *Viscum album*. Among the nematodes known to infest jackfruit trees are *Aphelechus avenae*, *Criconema taylori*, *Criconemoides birchfieldi*, *Helicotylenchus dihystera*, *Heterodera marioni*, *Hoplolaimus seinhorsti*, *Leptonema thornei*, *Meloidogyne* sp., *Oostenbrinkella oostenbrinki*, *Peltamigratus* sp., *Pratylenchus zeae*, *Rotylenchulus reniformis*, *Trichodorus* sp., *Tylenchorhynchus acutus*, *T. martini*, *T. triglyphus*, *Xiphinema americanum*, *X. pratense*, and *X. setariae*.[278,306]

BALANITES AEGYPTIACA (L.) Delile (SIMARUBACEAE) — Desert Date, Soapberry Tree, Jericho Balsam

Uses — Monks of Jericho regarded *Balanites* as the balm of the Biblical verse. An oily gum made from the fruit is sold in tin cases to travelers as the balm of Gilead. Both *Balanites* and *Pistacia* are common in old Palestine, and both are called balm. A desert-loving plant, *Balanites* is also revered by the Mohammedans in western India.[85] The wood is used for axes, cudgels, Mohammedan writing boards, mortars and pestles, walking sticks, and wooden bowls. Since it gives little smoke, it is a favorite firewood for burning indoors. Spiny branches are used to pen up animals. The bark yields a strong fiber. The fruit is fermented to make an intoxicating beverage. In West Africa and Chad, the seed is used for making breadstuffs and soups, while the leaf is used as a vegetable, the pericarp is crushed and eaten.[332] Flowers are eaten in soups in West Africa. The comestible oil, which constitutes 40% of the fruit, is used to make soap. African Arabs use the fruit as a detergent, the bark to poison fish. The active principle, probably a saponin, is lethal to cercaria, fish, miracidia, mollusks,[17] and tadpoles. One fruit weighing 25 g has enough active ingredient to kill the bilharzial mollusks in 30 ℓ water.[332] The Douay Bible of 1609 renders Jeremiah 8:22 to read, "Is there no rosin in Gilead?", resulting in this edition being termed the Rosin Bible. The Bishop's Bible of 1568 reads, "Is there no tryacle in Gilead?", and is termed the Treacle Bible. The tree is recommended for arid zones by UNESCO because of its food value, fixed oil, and protein in the kernel (''nut'') and as a raw material for the steroid industry.[85]

Folk medicine — Fruits are pounded and boiled to extract the medicinal vulnerary oil. The oil was poured over open wounds and apparently acted as an antiseptic and protective covering against secondary infections. One Turkish surgeon regarded this as one of the best stomachics, a most excellent remedy for curing wounds. In Ethiopia, the bark is used as an antiseptic, the leaf to dress wounds, and the fruit as an anthelmintic laxative. In Palestine, the oil is said to be used in folk medicine. Ghanans used the leaves as a vermifuge, whereas Libyans use them to clean malignant wounds. Powdered root bark is used for herpes zoster while the root extracts are suggested for malaria. Ghanans use the bark from the stem in fumigation to heal the wounds of circumcision. Nigerians consider it abortifacient. The oil from the fruits is applied to aching bones and swollen rheumatic joints by the Lebanese. Extracts of the root have proven slightly effective in experimental malaria. The bark has been used in treating syphilis. In Chad, the plant is used as a fumigant in liver disease, the seed as a febrifuge, and the fruit for colds. Ugandans use the oil for sleeping sickness, but the efficacy is questioned. Ayurvedics apply the fruit oil to ulcers, the fruit for other skin ailments and rat bites, regarding the fruit as alexipharmic, alterative, analgesic, anthelmintic, antidysenteric. Unanis use the fruit also for boils and leucoderma.[85,91,165]

Chemistry — A chloroform fraction of the stem bark, chromatographed over a column of silica gel, yielded beta-sitosterol, bergapten, marmesin, and beta-sitosterol glucoside. None of these compounds were active in eight 9KB5 (in vitro) or P0388 (in vivo) systems.[301] Per 100 g, the fruit (ZMB) is reported to contain 339 calories, 6.1 to 11.1 g protein, 0.0 to 1.7 g fat, 79.1 to 88.6 g total carbohydrate, 10.2 g fiber, 5.2 to 8.1 g ash, 130 to 380 mg Ca. 400 mg P, and 39 mg ascorbic acid. Shoots contain (ZMB): 27.5 g protein, 1.5 g fat, 64.4 g total carbohydrate, 23.3 g fiber, 6.6 g ash, 480 mg Ca, and 380 mg P; leaves contain 11.6 g protein 4.2 g fat, 71.5 g total carbohydrate, 13.6 g fiber, and 12.7 g ash. Seeds or ''nuts'' contain (ZMB): 21.9 g protein, 45.7 g fat, and 3.3 g ash (21). The fruit flesh contains 7% saponin, 38 to 40% sugar. The saponin from the pericarp contains glucose and rhamnose; from the seeds, glucose, rhamnose, xylose, and ribose. The seed kernel yields the steroid balanitesin, identical with the sapogenin $C_{27}H_{42}O_3$ called diosgenin. The seed oil (30 to 55%), colored yellow with alpha-carotene has 19% palmitic-, 14% stearic-, 27% oleic-, 40% linoleic-, and traces of arachidonic-acids. Traces of yamogenin, 25-alpha-spirosta-3:5-diene and beta-sitosterol.[187]

Description — Savanna tree, 5 to 7 (to 21) m tall; bark gray to dark-brown, with thick ragged scales and long vertical fissures in which new yellow bark is visible; branchlets green, smooth, armed with green straight forward-directed supra-axillary spines to 8 cm long; leaves gray-green, 2 foliolate; leaflets obovate to orbicular-rhomboid, usually 2.5 to 5 cm long, 1.3 to 3 cm broad, flowers green to yellow-green, small, ca. 1.3 cm in diameter, in supra-axillary clusters or rarely subracemose; fruit a plum-sized drupe, green at first, turning yellow, broadly oblong-ellipsoid, with large, hard, pointed stone surrounded by yellow-brown sticky edible flesh.[278]

Germplasm — Reported from the Mediterranean Center of Diversity, desert date, or cvs thereof, is reported to tolerate drought, high pH, insects, savanna, and waterlogging.[82]

Distribution — Widespread across North Africa, south to Uganda, Ethiopia, Sudan, Chad, Nigeria, Arabia, and Palestine.[278]

Ecology — Ranging from Subtropical Dry to Wet through Tropical Desert (with water) to Dry Forest Life Zones, desert date is reported to tolerate annual precipitation of 1.5 to 17 dm (mean of 9 cases = 10), annual temperature of 18.7 to 27.9°C (mean of 9 cases = 24.3), and pH of 5.0 to 8.3 (mean of 6 cases = 6.9).[82] Commonly found in dry areas occasionally subject to inundation. Sandy well-drained soil with slightly acid pH may be most productive.[82]

Cultivation — Propagates widely by seeds naturally. Seeds germinate readily. Sometimes planted in villages for the fruit and other parts.[278]

Harvesting — Fruits are collected when ripe and spread out, often on roofs, to dry until needed. Other parts of plants collected as needed. Available nearly year round.[278]

Yields and economics — When steroid prices were volatile, this was viewed as an alternative source. World consumption was expected to exceed 1000 MT diosgenin or yamogenin by 1973 and 60 MT hecogenin. Seeds from Nigeria (42.8 to 48.4% oil) yielded 1.11 to 1.74% total sapogenins; from Tanzania (43.1% oil) 0.95% sapogenins; and, from India (50.3%) 0.74% total sapogenins.[122]

Energy — Roots have been used for producing charcoal. The wood, burning with little smoke, is used for fuel wood. The oil could be used for fuel, better transesterified.

Biotic factors: — Desert date trees are attacked by the following fungi: *Phoma balanites*, *Septoria balanites*, *Diplodiella balanites*, *Metasphaeria balanites*, and *Schizophyllum commune*.[278]

BARRINGTONIA PROCERA (Miers) Kunth (MYRTACEAE) — Nua Nut

Uses — While nuts of many species are said to be used as fish poisons (*B. asiatica*, *B. cylindrostachya*, *B. racemosa*), others are used for food (*B. butonica*, *B. careya*, *B. edulis*, *B. excelsa*, *B. magnifica*, *B. niedenzuana*, *B. novae-hiberniae*, *B. procera*). The nua nut is a common component of native meals on Santa Cruz, also eaten in between-meal snacks. Smoked whole fruits can be stored.[209,346]

Folk medicine — No data available.

Chemistry — No data available.

Description — Tree, sparingly branched, to 5 m tall or taller, the broad shiny leaves clustered near the ends of the branches. Flowers in long pendulous cylindrical racemes, yellow. Fruit an ovoid drupe; seed and kernel also ovoid.

Germplasm — Reported from the New Guinea Center of Diversity. The fruit epidermis may be green or purple, the seed coat white or pink. In the Solomon Islands, it is generally believed that the kernels from Santa Cruz are bigger than those elsewhere (see Figures 6 and 7 in Yen[346]). Other edible species known as cut-nuts in the Solomons are similar or closely related.

Distribution — Limited to the Huon Peninsula of New Guinea, the New Guinea Islands, the Solomons, and New Hebrides, grown as a village tree in Fiji.

Ecology — Estimated to range from Subtropical Moist to Rain through Tropical Moist to Wet Forest Life Zones, nua nut is estimated to tolerate annual precipitation of 20 to 60 dm, annual temperature of 23 to 27°C, and pH of 6.0 to 8.4.

Cultivation — Propagated from seed or stem cutting. Seedling trees "are said to reflect the characteristics of the parental tree, as, of course, do cuttings, but the latter tend to grow branched closer to the ground".[346]

Harvesting — Seasons of production are indefinite and nuts are available all year round. The growth rate of fruit after fertilization is fast. There are only 6 weeks between the flowering time of the upper part of the inflorescences and the harvest of such fruits.[346]

Yields and economics — No data available.

Energy — No data available.

Biotic factors — No data available.

BERTHOLLETIA EXCELSA Humb. and Bonpl. (MYRTACEAE) — Brazil Nut, Para Nut, Creme Nut, Castanas, Castanhado Para

Uses — Nutritious Brazil nuts are eaten raw, salted, or roasted. Seeds are consumed in large quantities and are used in international trade. Kernels are the source of Brazil nut oil, used for edible purposes and in the manufacture of soap. The wood is light pinkish-brown, neither very hard nor heavy, and it is limited to cheap work.[177,278]

Folk medicine — There has been a flurry of interest in one certain formula of one Dr. Revici, the formula containing selenium and vegetable oils or natural fatty acids. This combination has been tried with cancer patients and, according to one Washington physician, in AIDS. I am frankly skeptical, but would not hesitate to increase my consumption of Brazil nuts were I suffering AIDS or cancer.

Chemistry — Per 100 g, the mature seed is reported to contain 644 calories, 4.7 g H_2O, 17.4 g protein, 65.0 g fat, 9.6 g total carbohydrate, 3.9 g fiber, 3.3 g ash, 169 mg Ca, 620 mg P, 3.6 mg Fe, 2 mg Na, 5 mg beta-carotene equivalent, 0.20 mg thiamine, 0.69 mg riboflavin, 0.20 mg niacin, and 2 mg ascorbic acid.[89] Hager's Handbook notes ca. 1.8% myristic, 13.5% palmitic, 2.5% stearic, 55.6% oleic, and 21.6% linoleic acid glycerides, and 0.24 to 0.26% barium.[187] Hilditch and Williams[128] tabulate the component fatty acid percentage as 13.8 to 16.2% palmitic, 2.7 to 10.4% stearic, 30.5 to 58.3% oleic, and 22.8 to 44.9% linoleic acids. An analysis by Furr et al.[102] reports the edible portion of the nuts to contain 5.0 ppm Al, 0.02 As, 2.7 B, 1,764 Ba, 87 Br, 1,592 Ca, 0.03 Cd, 1.2 Ce, 246 Cl, 1.9 Co, 0.6 Cr, 1.3 Cs, 18 Cu, 0.1 Eu, 1.7 F, 93 Fe, 0.01 Hg, 0.2 I, 5,405 K, 0.1 La, 0.01 Lu, 3,370 Mg, 8.0 Mn, 7.2 Na, 5.8 Ni, 0.4 Pb, 103 Rb, 0.1 Sb, 0.02 Sc, 11 Se, 1,770 Si, 0.04 Sm, 3.5 Sn, 77 Sr, 0.1 Ta, 6.1 Ti, 0.01 V, 0.1 W, 0.2 Yb, and 41 ppm Zn dry weight. The normal concentration of some of these elements in land plants are 50 ppm B, 14 Ba, 15 Br, 2,000 Cl, 0.5 Co, 0.2 Cs, 14 Cu, 3.200 Mg, 630 Mn, 3 Ni, 20 Rb, 3,400 S, 26 Sr, and 0.2 ppm Se dry weight. They were higher in barium, bromine, cerium, cobalt, cesium, magnesium, nickle, rubidium, scandium, selenium, silicon, strontium, tin, titanium, and ytterbium, and equal to or higher in europium, lanthanum, and tantalum than any of the 12 nut species studied by Furr et al.[102] Of 529 nuts analyzed for Se, 6% contained 100 ppm Se or more. The mean value for all nuts was 29.6 ppm, and the median value was 13.4 ppm. Hexane-extracted high-Se Brazil nut meal in a corn-based diet fed to rats produced toxicity similar to that obtained from seleniferous corn, selenomethionine, or sodium selenite as assessed by weight gain, visually scored liver damage and liver, kidney, and spleen weights. The Se in Brazil nuts may be as biologically potent as that from other sources.[247] Other nuts in this family (Lecythidaceae) contain so much selenium that overingestion can lead to hair loss. Apparently selenium, an anticancer element, is essential in traces, toxic in excess. However, the homeostatic human may cope with moderate excesses. ". . . animals regulate their selenium content through excretion. When the element is in short supply, excretory metabolite production is minimal. When the needs of the organism are being met, excess selenium is eliminated by conversion to the excretory metabolites."[54]

Description — Large forest tree, up to 40 m tall; leaves alternate, short-petioled, leathery, oblong, with wavy margin, 30 to 50 cm long, 7.5 to 15 cm broad; flowers in large erect spike-like racemes, white to cream, sepals united but finally separating into two deciduous sepals; fruit large, brown, woody, globose, 10 to 15 cm in diameter, weighing up to 2 kg, with an aperture at one end which is closed by a woody plug and must be broken open to extract the "nuts" inside; fruit may remain on the trees several months after ripening; seeds 12 to 24 per fruit, triangular, with a brown horny testa.[278]

Germplasm — Reported from the South American Center of Diversity, Brazil nut, or cvs thereof, is reported to tolerate lateritic soils.[82]

Distribution — Native to the Amazon basin of Northern Brazil, Bolivia, Colombia, Peru, Venezuela, and Guianas, mainly along banks of the Amazon and upper Orinoco Rivers and their tributaries. Introduced into Sri Lanka in 1880 and Singapore in 1881.[278]

Ecology — Ranging from Subtropical Moist to Tropical Dry through Wet Forest Life Zones, Brazil nut is reported to tolerate annual precipitation of 13.5 to 41.0 (mean of 7 cases = 29.3), annual temperature of 21.3 to 27.4°C (mean of 7 cases = 25.4), and pH of 4.3 to 8.0 (mean of 6 cases = 5.8).[82] A tropical tree, sometimes gregarious, preferring high land, beyond reach of periodical floods. Thrives best in rich alluvial soil, in a hot moist climate.[283]

Cultivation — Brazil nuts are collected from wild trees and are nowhere cultivated for commercial production. Trees are propagated from seed or by layering. From 10 to 25 years are required for fruiting to begin. Attempts to establish Brazil nut plantations have met with mediocrity, at best.[278]

Harvesting — After fruits have fallen and are gathered, usually during the dry season, the nuts are extracted and shipped to Manaos or Belem do Para, where they are graded and exported to the U.S. and Europe.[278]

Yields and economics — A good tree will yield 300 fruits at a time, ca. 15 months after flowering. An adult tree may yield, in normal years, from 30 to 50 kg of fruits, but yields of more than 2000 kg per tree are reported. Early in the 20th century, with the fall of Brazilian rubber prices in 1910 due to Asian competition, Brazil nuts became a vital export. The first U.S. customs entry recorded was 1873, when more than 1,800 MT unshelled nuts entered at an average price less than $0.15/kg. By 1982, spot prices for unshelled nuts were over $3.00/kg. By 1978, 15,472 MT of in-shell nuts were exported, contrasted to 5,367 shelled nuts. The U.S. is the largest importer, followed by the U.K., West Germany, Italy, France, Australia, and the Netherlands. Brazilian output is predicted to remain steady at around 40 to 60 thousand MT in shell-nuts for both internal and external consumption. The principal producer of Brazil nuts is Brazil. In 1971, the Brazil nut crop in Brazil was 22,500 MT, and in 1970, 40,000 MT. Domestic consumption in Brazil is 1,000 to 2,000 MT per year. Shelled assorted nuts commanded $0.55/lb; unshelled, dehydrated nuts $0.23/lb; and natural unshelled nuts $0.18/lb.[278,283]

Energy — Shells and spoiled kernels supplement firewood in the power plants providing heat for the dryers. Imperfect nuts are used for oil extraction, the press-cake employed as feed for animals, whose manure could be used to extend fuel.

Biotic factors — The following fungi are known to attack this tree: *Actinomyces brasiliensis*, *Aspergillus flavus*, *Cephalosporium bertholletianum*, *Cercospora bertholletiae* (Gray spot), *Cunninghamella bertholletiae*, *Fusarium* sp., *Myxosporium* sp., *Pellionella macrospora*, *Phytophthora heveae*,[6] *Piptocephalus sphaerocephala*, *Phomopsis bertholletianum*, and *Thamnidium elegans*.[186,278] Albuquerque et al.[6] recommend Cuprosan copper oxychloride or difolatan-80-captafol for control of *Phytopthora* leaf blight. The nematode, *Meloidogyne incognita*, has been found causing heavy galling on the roots.[101]

BORASSUS FLABELLIFER L. (ARECACEAE) — Palmyra Palm, Brab Tree, Woman's Coconut
Syn.: *Borassus flabelliformis* Roxb. (?)*Borassus aethiopum* Mart.

Uses — Palmyra palm is grown for the juice or toddy, extracted from the inflorescence from which sugar or jaggery is made. Tender fruits resembling pieces of translucent ice are eaten during hot season. Seeds are eaten as well as fruits. Fleshy scales of young seedling shoots are eaten as a delicacy, especially in northern Sri Lanka, or dried to make a starchy powder (reported to contain a neurotoxin). Salt prepared from leaves. The inflorescence is a source of sugar, wine, and vinegar. Five types of fiber are obtained from different parts of the plant, used for hats, thatching houses, books, writing paper, mats, bags, and all types of utensils for carrying or storing water and food. Timber is black, sometimes with yellow grain, strong, splits easily; said to withstand a greater cross-strain than any other known timber; used for boat making, rafters, water pipes, walking sticks, umbrella handles and rulers. Tree also yields a black gum. A Tamil poem enumerates 801 ways to use this palm. Sometimes planted as a windbreak.[278,397]

Folk medicine — An emollient made from the root is said to be a folk remedy for indurations. Flower or root is a folk remedy for tumors of the uterus (Cambodia).[126] Sprouting seed used as a diuretic and galactagogue. Petiole used as a vermifuge in Cambodia. Root regarded as cooling. Ash of spathe given for enlarged spleens. Juice drunk before breakfast has important medicinal properties, and is stimulant and antiphlegmatic. Juice is diuretic, stimulant, antiphlegmatic, useful in inflammatory affections and dropsy; pulp is demulcent and nutritive.[91,249]

Chemistry — Per 100 g, the mature fruit is reported to contain 43 calories, 87.6 g H_2O, 0.8 g protein, 0.1 g fat, 10.9 g total carbohydrate, 2.0 g fiber, 0.6 g ash, 27 mg Ca, 30 mg P, 1.0 mg Fe, 0.04 mg thiamine, 0.02 mg riboflavin, 0.3 mg niacin, and 5 mg ascorbic acid. Sap contains about 12% sugar. Spontaneous fermentation produces ca. 3% alcohol and 0.1% acids during the first 6 to 8 hr. Beyond this, fermentation goes to 5%, but there is too much butyric acid. A cheap source of vinegar. Accordingto the *Wealth of India*, the nira (fresh sap) contains 85.9% moisture, 0.2% protein, 0.02% fat, 0.29 ash, 13.5% carbohydrates, 12.6% total sugar, and 5.7 mg Vitamin C per 100 g; the gur (boiled-down molasses) contains 8.6% moisture, 1.7% protein, 0.08% fat, 1.8% ash, 88.5% carbohydrate,

(84% total sugar); the seed pulp contains 92.6% moisture, 0.6% protein, 0.1% fat, 0.3% ash, 6.3% carbohydrates, and 13.1 mg/100 g vitamin C.[70,89] The mannocellulose of the endosperm is transformed to glucose via mannose.

Toxicity — Fleshy scale leaves of the germinating seeds, eaten by humans, contain a neurotoxin.

Description — Tall palm, 20 to 30 m high; trunk cylindrical, 30 to 35 cm in diameter, very hard, black, mainly composed of stiff longitudinal fibers, central portion soft and starchy, with crown of 30 to 40 fan-like leaves. Leaves glaucous, palmate, up to 3.3 m wide, stiff, with numerous free pointed tips, petiole 11.3 m long, channeled above, with hard saw-like teeth on margins. Inflorescence stalks among the leaves, long, much-branched; male and female flowers on separate trees; male flowers borne on thick digitate processes, female flowers appearing like small fruits. Fruit a large drupe, 15 to 20 cm in diameter, depressed-globular, brown; exterior smooth, enclosed in a tough matted fiber; interior very fibrous, with 2 to 3 seeds; seeds rounded, but flattish, 3.7 to 5 cm across. Spathes begin to appear in November or December, but flowers in March; fruits July-August.[278]

Germplasm — Reported from the African and, secondarily, the Hindustani Centers of Diversity, palmyra palm, or cvs thereof, is reported to tolerate disease, drought, fire, high pH, salt, sand, slope, savannah, waterlogging, and wind. The genus *Borassus* is believed to contain one or as many as eight species, depending on your taxonomic point of view. Kovoor maintains that the African *B. aethiopum* is distinct from *B. flabellifer*. No dwarf mutants have been reported. (2n = 36.)[82,397]

Distribution — Said to be native to Africa, but also claimed to be indigenous to tropical India and Malaysia, where it is both wild and cultivated, especially in coastal areas. Widely cultivated throughout tropical Asia and Africa (Congo, Gabon, Gambia, Guinea, Guinea-Bissau, Ivory Coast, Malagasy, Mali, Mauritania, Nigeria, Senegal, Sudan, Tanzania, Upper Volta), with huge stands covering thousands of hectares. Grown in comparatively dry parts of Burma, India, Sri Lanka, and Malaya.[278] Kovoor[397] estimates that there are 10,615,000 palmyra in Sri Lanka, 60 million in India, 2,350,000 in Burma, 1,800,000 in Kampuchea, 5 million in east Java.

Ecology — Ranging from Subtropical Dry to Moist through Tropical Very Dry to Wet Forest Life Zones, palmyra palm is reported to tolerate annual precipitation of 6.4 to 42.9 dm (mean of 11 cases = 18.8), annual temperature of 20.6 to 27.5°C (mean of 11 cases = 24.3°C), and pH of 4.5 to 8.0 (mean of 7 cases = 6.4).[82] Palmyra palm is grown in regions with a pronounced dry monsoon, being especially abundant in all sandy tracts near the sea, on embankments, and in mixed coconut and date palm jungles of Bengal.[278] Though drought-tolerant, it suffers little from prolonged flooding. Kovoor[397] suggests that "its natural preference is for rich alluvial soil".

Cultivation — Plants develop from self-sown seed. Seeds germinate, producing a "sinker", which grows downward 1 m before producing growth at top. "Once sprouted, the seedling cannot be transported."[397] Trees are slow-growing, taking 15 to 20 years before showing a stem above ground; in the early stages only the underground portion of the stem increases in thickness. Male and female trees cannot be distinguished until they flower. For food, the seed-bed is prepared and nuts planted as close together as possible about June or July, about 50 seeds to the square meter. In about 3 to 4 months the nuts are dug up, by which time they have germinated, and the sprouts are eaten as a vegetable. Actually, the nut is broken open and the embryo eaten dry or made into a flour, tasting similar to tapioca.[278]

Harvesting — Trees begin to flower when 12 to 15 years old, depending on the region, and continue to flower for about 50 years. Female trees yield about twice as much sap as male trees. Fresh sap, called "sweet toddy" or "nira", containing about 12% sucrose, is obtained by tapping the flower stalk. Juice may be used fresh as a beverage, or, if not treated promptly, begins to ferment into an intoxicating liquor. Fresh juice boiled down into a sugar called jaggery or gur, with about 80% sucrose and 2.5% glucose, is an important sugar in southern India and Burma. Tapping does not injure the tree. However, every 3 years the sap-drawing process is omitted; otherwise the tree would die. A toddy collector climbs the tree, tightly binds the spathes with thongs to prevent further opening, and then thoroughly bruises the embryo flower within to facilitate the exit of juice. This operation is repeated for several days, and on each occasion a thin slice is removed from spathe to facilitate running of sap and to prevent it bursting the bound spathe. In about 8 days, sap begins to exude into an earthen pot placed for that purpose. Pots are emptied twice daily, the pots coated with lime inside to prevent fermentation. In factories, raw gur is heaped on platforms for about 2 months to drain away most of the molasses. Then it is dissolved in water and refined in the usual manner to make crystalline sugar. Molasses obtained during crystallization is used for producing arrack. Five types of fibers may be obtained from the Palmyra palm, each with specific characteristics and uses:

1. Fibers about 60 cm long, separated from leafstalks, called "Bassine", are used for making rope, twine, and sometimes paper.
2. A loose fiber surrounds the base of the leafstalk.
3. "Tar", prepared from the interior of stem without any spinning or twisting, is plaited into fishtraps.
4. A coir is derived from the pericarp.
5. Fibrous materials of the leaves, are torn into strips, prepared, dyed, plaited into braids, and worked up into basketware, fancy boxes, cigar cases and hats. In Bengal, long strips of leaf are employed by children as washable slates.[278]

Kovoor[397] gives good details of various methods for tapping this and other palms.

Yields and economics — Trees yield 4 to 5 quarts of sap daily for 4 to 5 months; one gallon of sap yields about 680 g jaggery sugar, which is about 80% saccharose or sucrose. Joshi and Gopinathan[155] suggest that Asian Indians can more cheaply get nearly twice as much sugar per hectare from palm as sugar cane, i.e., ca. 6,000 kg/ha vs. 3,500 kg/ha. Comparing *Borassus* with other Indian sugar palms, they note that *Borassus* is longest lived (90 to 120 years), and can be tapped more than twice as many years (70 to 95 years) as others, yielding 20 to 70 kg tree, with 1,250 trees per hectare. At one time, one-fourth of the inhabitants of northern Sri Lanka were dependent on this tree for subsistence; in India many also depend on it. Most of the trade in Palmyra goes through the Port of Madras.[278]

Energy — Ironically, the palmyra is better as a fire-breaker in arid regions of West Africa prone to wild fires. Its timber burns very poorly as firewood, and young palms are said to be more fire resistant than old ones. The relatively high yields of sugar could be converted renewably to alcohol for energy purposes. Kovoor[397] notes that low bearers may produce only about 1 ℓ, average ones 6 to 10 ℓ, and exceptional trees 20 ℓ sap per day. Natural fermentation can take these liters to 5 to 6% ethanol.

Biotic factors — The most serious fungus attacking palmyra palm is *Pythium palmivorum* (Bud-rot, which grows into the growing point and ultimately kills the tree). Other fungal diseases include: *Cladosporium borassii*, *Curvularia lunata*, *Graphiola borassi*, *Microxyphium* sp., *Penicillopsis clavariaeformis*, *Pestalotia palmarum*, *Phytophthora palmivora*, *Sphaerodothis borassi*, *Rosellinia cocoes*. Palmyra is attacked by insects which affect coconut palm: Rhinoceros beetle (*Oryctes rhinoceros*); Black headed caterpillar (*Nephantis serinopa*); and Red palm weevil (*Rhynchophorus ferrugineus*).[186,278] Termites and grubs of the Rhinoceros beetle can be very destructive to germinating seeds. In Guinea-Bissau, several insects "commence their destructive careers by turning saprophytic on dead palms". The most predominant of them is *Oryctes gigas*, whereas others like *O. owariensis*, *O. monoceros*, *Rhynchophorus phoenicis*, *Platygenia barbata*, and *Pachnoda marginella* are common. Still, Kovoor[397] concludes that the palm is extraordinarily disease-resistant. One study showed that more than 2% of the trees were infested with scorpions or snakes.

BROSIMUM ALICASTRUM Swartz (MORACEAE) — Breadnut, Ramon, Capomo, Masico

Uses — Branches and leaves used as an important cattle fodder, especially during the drier months in regions where trees are plentiful. Lopped branches (ramon) are relished by cattle; fallen leaves and nuts are also relished by cattle and pigs. Feeding ramon forage is said to augment milk production 1 to 2 ℓ a day in dairy cattle. The milky latex, which flows freely when the trunk is cut, is mixed with chicle or drunk like cow's milk. Sweet pericarp of fruit eaten raw by humans. Fruits boiled and eaten in Costa Rica. The seeds, or breadnuts, with chestnut-like flavor, are eaten raw, boiled, roasted, or reduced to a meal often mixed with corn meal for making tortillas, or baked with green plaintain. They are eaten alone or with plantain, maize, or honey, or boiled in syrup to make a sweetmeat. Seeds used as a coffee substitute. Wood is hard, compact, white, grayish, or tinged with pink, easy to work and used in carpentry, a valuable timber sometimes used in construction, cabinet work, and other purposes in Yucatan.[250,278]

Folk medicine — According to Hartwell,[126] the plant is used in folk remedies for cancer of the uterus. Reported to be lactagague and sedative, ramon is a folk remedy for asthma (latex, leaves), bronchitis, and chest ailments.[91] Guatemalans drink the latex as a pectoral for stomach disorders. Crushed seeds are taken in sweetened water as a lactagogue. The bark shows CNS-depressant activity.[224] Leaf infusions are used in cough and kidney ailments. The diluted latex is used to aid tooth extraction.[91,126,224]

Chemistry — Per 100 g, the leaf is reported to contain 127 calories, 62.0 g H_2O, 3.2 g protein, 1.2 g fat, 30.6 g total carbohydrate, 8.9 g fiber, 3.0 g ash, 530 mg Ca, 68 mg P, 5.4 mg Fe, 820 mg beta-carotene equivalent, 0.24 mg thiamine, 0.51 mg riboflavin, 1.4 mg niacin, and 55 mg ascorbic acid. Per 100 g, the fruit is reported to contain 56 calories, 84.0 g H_2O, 2.5 g protein, 0.5 g fat, 12.1 g total carbohydrate, 1.2 g fiber, 0.9 g ash, 45 mg Ca, 36 mg P, 0.8 mg Fe, 840 mg beta-carotene equivalent, 0.5 mg thiamine, 1.52 mg riboflavin, 0.8 mg niacin, and 28 mg ascorbic acid. Per 100 g, the seed is reported to contain 363 calories, 6.5 g H_2O, 11.4 g protein, 1.6 g fat, 76.1 g total carbohydrate, 6.2 g fiber, 4.4 g ash, 211 mg Ca, 142 mg P, 4.6 mg Fe, 128 mg beta-carotene equivalent, 0.03 mg thiamine, 0.14 mg riboflavin, and 2.1 mg niacin. Another seed analysis shows, per 100 g (oven-dry basis), 361 calories, 40 to 50 g H_2O, 12.8 g protein, 4.6 g fiber, 178 mg Ca, 122 mg P, 3.8 mg Fe, 100 μg beta-carotene equivalent, 0.1 mg thiamine, 0.1 mg riboflavin, 1.6 mg niacin, and 50 mg ascorbic acid. Seed contains an essential oil, resin, wax, mucilage, dextrin, and glucose. The crude protein content of the seeds in higher than corn, the tryptophan content is four times higher, significant among corn-fed Latins.[89,224] Peters and Pardo-Tejeda[250] report the seeds to contain 10.4% leucine, 9.7% valine, 3.3% isoleucine, 4.0% phenylalanine, 2.3% lysine, 2.4% threonine, 2.3% tryptophan, 1.0% hisitidine, 0.7% methionine, 5.1% arginine, 15.3% aspartic acid, 6.7% proline, 9.9% cystine, 2.9% serine, 2.3% glycine, 3.7% tyrosine, and 2.5% alanine.

Description — Evergreen, dioecious, tropical tree, 20 to 35(to 40) m tall, trunk to 1 m in diameter, sometimes with buttresses; latex white to yellow; leafy twigs 1 to 4 mm thick, glabrous or sparsely puberulent; leaves alternate, elliptic to oblong or lanceolate, slightly inequilateral, often broadest above to below the middle, 4 to 28 cm long, 2 to 11 cm broad, chartaceous to coriaceous, acuminate, nearly acute, also acute at the base, or obtuse, truncate or subcordate; margin entire, rarely denticulate; glabrous to sparsely puberulent beneath, and pubescent on the costa, 12 to 21 pairs of secondary veins, with or without some parallel tertiary veins; petioles 2 to 14 mm long; stipules nearly fully amplexicaul, 5 to 15 mm long, glabrous to pubescent; inflorescences solitary, in twos or several together, subglobose to ellipsoid, subsessile or pedunculate, the peduncle up to 1.5 cm long; bracts 0.2 to 2 mm in diameter, puberulent, the basal ones sometimes basally attached; staminate influorescence 3 to 8 mm in diameter, with one central abortive pistillate flower; staminate flowers numerous, perianth absent or minute one, 1 stamen; pistillate inflorescence 2 to 4 mm in diameter, with 1 or 2, occasionally many, abortive flowers, style 1.5 to 8.5 mm long, stigmas 0.2 to 8 mm long; infructescences subglobose, 1.5 to 2 cm in diameter, at maturity yellow, brownish, or orange; seeds small, roundish, yellow or brownish, 1.3 cm or less in diameter, borne singly or in twos, in a thin, paper-like, stout shell, surface of seed smooth or somewhat granular. Flowers throughout the year.[278]

Germplasm — Reported from the Middle and South American Centers of Diversity, ramon, or cvs thereof, is reported to tolerate drought, fungi, insects, limestone, slope, and waterlogging.[82] Besides the typical form with the anthers peltate with fused thecae, found in West Indies and Central America, there is a subsp. *bolivarense* (Pittier) Berg, called Guaimoro (Colombia and Venezuela), and Tillo (Ecuador), in which thecae are free, growing from Panama through the Andes to Guyana and in Brazil to Acre Territory.[82,278]

Distribution — Native from the Pacific Coast of Mexico (Sinaloa) south through Central America to Ecuador, Guyana, and parts of Brazil; also in the West Indies. Introduced and planted in Singapore, Trinidad, and Florida.[278]

Ecology — Ranging from Subtropical Dry to Moist through Tropical Dry to Wet Forest Life Zones, ramon is reported to tolerate annual precipitation of 3 to 40 dm, annual temperature of 19 to 26°C, and pH of 5.0 to 8.0. Nearly pure stands may occur on steep calcareous slopes.[82] In evergreen, semi-evergreen or deciduous forests in tropical climates, from 50 to 800 m (to 1,000 m) altitudes, sometimes in cloud-forests, regionally abundant, but also planted. Trees are extremely tolerant of drought, and grow well in dry habitats as well as in seasonally flooded places as along rivers and in swampy areas. Common on limestone in Jamaica. Thrives on various types of soils in tropical regions.[278]

Cultivation — Propagated by seeds, cuttings, and air-layering. Seeds germinate readily. After trees are established, they grow well without much care. Often form a large portion of the forest tree population in some regions.[278]

Harvesting — Branches are cut by men who climb the trees with machetes, and cut down limbs for stock to browse upon. To increase the yield of fodder, it is suggested that close-planting and regular coppicing may be tested. Nuts collected from the ground by natives are used for food, or for making a black meal for making tortillas and other food-stuffs. Timber is harvested from mature trees and used, especially in the Yucatan.[278]

Yields and economics — Fodder yield of natural and coppiced trees is not known but should be ascertained. There seems to be plenty of fodder material about when it is needed during dry spells.[278] Peters and Pardo-Tejeda[250] put yields at 50 to 75 kg fruit per female tree per year. Based on a rough estimate of the distribution in Vera Cruz, Mexico, it is estimated that 80,000 MT seed could be collected annually with an annual production of 10,000 MT crude protein, leaving the trees standing strong against erosion. Yucatan plantings are producing 10 to 15 MT forage per ha at each lopping. Thus ramon plantations produce almost twice as much (lactogenic) forage as established pasture.[250] Of great value since ancient times in Central America, the West Indies, and northern tropical South America for a fodder food for stock and as a source of seeds for meal, latex for food, and timber for construction and other purposes. Not a commercial trade crop, but very important locally for these many purposes.[278]

Energy — Assuming that the 80,000 MT seed was gathered for the production of 10,000 MT crude protein, there would, of course, be a 70,000 MT biomass available for energy production. I would estimate that litter-fall from this species might approach 5 to 10 MT/ha. Although not a leader among firewoods, the wood could also be renewably gathered for fuel wood. As Gomez-Pampa[112] notes, "With a year-round, food-producing plant, we can liberate a good part of the energy that is currently spent on the production of grain for basic food products in tropical regions," where "weakness of tropical soils for annual crops has always been a limiting factor."

Biotic Factors — Seeds stored when fresh are promptly infested by *Aspergillus,* some of which contain toxic compounds.

BROSIMUM UTILE (H.B.K.) Pitt. (MORACEAE) — Cow Tree, Palo de Vaca
Syn.: *Brosium galactodendron* D. Don in Sweet, *Galactodendron utile* H.B.K.

Uses — Latex from trunk, considered to be highly nutritive, is used by natives as a milk-like beverage, as a cream substitute in coffee, made into a kind of vegetable cheese, and made into a dessert after being chilled, whipped, and flavored. Laborers soak their bread in it. Used as a base for chewing gum. Bark used by Indians for making cloth, blankets, and sails. Plants grown in tropical areas for fruit or nuts and for leaves used for fodder. Fleshy outer layer of fruit eaten by parrots. Humans also eat the fruits, raw or cooked.[80] The soft, white wood, though not durable, has been used for concrete forms, boxes, and sheathing.[9,224,278]

Folk medicine — According to Hartwell,[126] a plaster of the milk is said to be a folk remedy for swelling of the spleen and indolent tumors. Reported to be lactagague and masticatory, cow tree is a folk remedy for asthma, inflammation, and tumors.[91] The latex is taken for asthma in Venezuela, and as an astringent for diarrhea in Costa Rica.[224]

Chemistry — The latex contains 3.8% wax, 0.4% fibrin, 4.7% sugar and gum, and 31.4% resins.[224] Garcia-Barriga[104] states that the latex contains 57.3% water, 0.4% albumen, 31.4% wax of the formula $C_{35}H_{66}O_3$, 5.8% wax of the formula $C_{35}H_{58}O_7$, and 4.7% gum and sugars.

Description — Laticiferous tree, 20 to 25 m tall, with simple trunk 40 to 50 cm in diameter at base, bark thick, grayish, smooth or verrucose, crown elongate; young brancelets subangular, more or less pubescent; leaves coriaceous; petioles 0.5 to 1.5 cm long, thick, canaliculate, sparsely pubescent; blades ovate, elliptic, rounded at base, abruptly acuminate at apex in a drip tip, 10 to 25 cm long, 3.5 to 9.5 cm broad, glabrous on both surfaces, green above, golden-brown beneath, margin entire, venation impressed on upper surface, prominent and slightly pubescent on lower one; primary veins 27 to 30, parallel, straight, almost transverse; stipules about 2 cm long, acute-lanceolate, silky pubescent, canducous, leaving a circular scar at each node; receptacles globose with 1 female flower, solitary in axils of leaves, long-pedunculate, about 7 mm in diameter in flowering stage; bractlet orbicular, thick, sessile, pilose-pubescent; staminal bractlets short (0.5 mm long), broad and ciliate; stamens 0.7 to 1.4 mm long, solitary, with smooth filaments, anthers ovate and 2-celled; ovary inserted 2.5 to 3 mm deep in receptacle; fruit depressed-globose, 2 to 2.5 cm in diameter, epicarp fleshy, 4 to 6 mm thick, yellow at maturity, mesocarp woody, rugose on surface, entirely filled with a single almond-like, white seed. Flowers and fruits September.[278]

Germplasm — Reported from the Middle and South American Centers of Diversity, cow tree, or cvs thereof, is reported to tolerate slope.[82]

Distribution — Native to tropical America from Nicaragua and Costa Rica south into northern South America, Colombia, and Venezuela, sometimes being the common tree in upland forests.[278]

Ecology — Ranging from Tropical Dry to Moist Forest Life Zones, cow tree is reported to tolerate annual precipitation of 3 to 40 dm, annual temperature of 25 to 27°C, and pH of 8.0.[82] Thrives in wet and subtropical climates, especially on hillsides bordering rivers.[278]

Cultivation — Propagation by cuttings over heat. When cuttings are rooted, they are planted in the forest where they soon became established. Plants rarely cultivated as a pure crop, because the trees freely propagate naturally in the forest.[278]

Harvesting — Incisions are made in trunk of tree, after which there is a profuse flow of gluey, thick milk, destitute of acridity and giving off a very agreeable balsamic odor. When exposed to air, the fluid displays on its surface, probably by absorption of atmospheric oxygen, membranes of a highly animal nature, yellowish and thready, like those of cheese. These, when separated from the more watery liquid, are nearly as elastic as those of caoutchouc, but in time they exhibit the same tendency to purify as gelatin. The milk itself, kept in a corked bottle, only deposits a small amount of coagulum and continues to give off the balsamic scent. Large quantities of this vegetable milk are drunk by the natives and it has been noted that workers gain weight during that time of year when the tree produces the most milk.[104,278]

Yields and economics — No yield data available. Widely used in the areas where the tree grows native, namely southern Central America and northern South America. Not known to be of international commercial value.[278]

Energy — The wood can serve as a fuelwood, said to burn green. Resin extracted from the fruits is used to make candles. The latex is mixed with balsa charcoal and wrapped in palm leaves to serve as a torch.[80]

Biotic factors — I find no reports of pests or diseases on this tree.

BRUGUIERA GYMNORRHIZA (L.) Savigny (RHIZOPHORACEAE) — Burma Mangrove
Syn.: *Bruguiera conjugata* Auct.

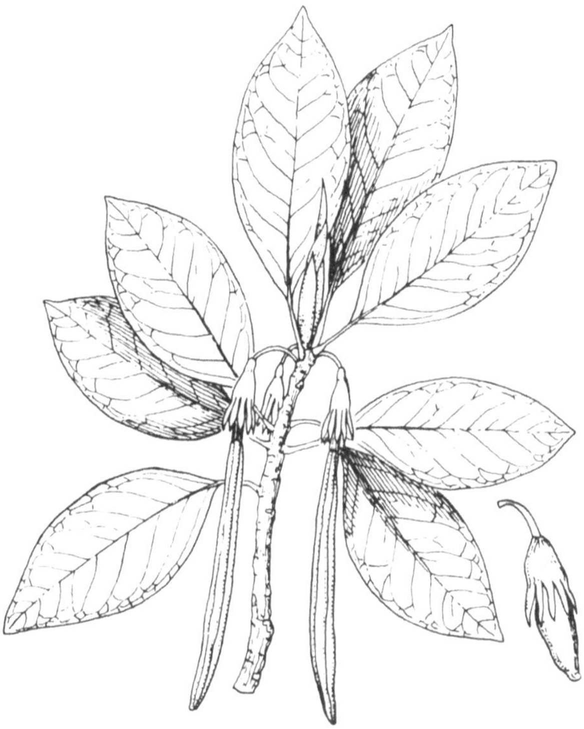

Uses — The heavy wood (sp. grav. 0.87 to 1.08) is durable, but hard to saw and work. It is used for construction, furniture, house-posts, and pilings.[188] Thousands of tons of *Bruguiera* wood chips are exported annually from Indonesia, Sabah, and Sarawak for pulp and for rayon manufacture.[232] Fruits are eaten, but not when anything better is available. More often, they are chewed as astringent with the betel quid. Since it is mostly the seed or embryo of this one-seeded indehiscent fruit that is eaten, this can be called a tropical nut. Embryos of several species are eaten, usually after processing.[209] Chinese in Java make a sweetmeat therefrom. In the South Pacific, fruits are peeled, sliced, and soaked in water for several hours, then steamed or boiled and eaten with coconut cream.[33] Dutch Indians use the bark to flavor raw fish. The leaves and peeled hypocotyls are eaten in the Moluccas after soaking and boiling.[140] In the Loyalty Islands, the embryo is kept for months after sundrying.[33] The phlobaphene coloring matter is used in China and Malaya for black dye.[56] In South Africa, the tree has been planted to stabilize dunes and in fresh-water swamps.

Folk medicine — Reported to be astringent,[91] the bark is used for diarrhea and fever in Indonesia.[249] Cambodians use the astringent bark for malaria.[56]

Chemistry — In Burma, leaves may contain 18.3% H_2O, 13.5% tannin; outer cortex (small trees) 14.6 and 7.9, outer cortex (large trees) 14.2 and 10.8; twig bark 13.1 and 14.8, bole bark (small trees) 16.3 and 31.7; whereas the bole bark of large trees contains 12.5% H_2O, 42.3% tannin. Bark contains from ca.4 to 53.12% tannin, according to Watt and Breyer-Brandwijk[332] and *The Wealth of India*.[70]

Toxicity — Eating too much (bark) is dangerous.[56] The skull and crossbones indicate that Menninger[209] considers the "nuts" to be poisonous.

Description — Evergreen tree 8 to 25(to 35) m high, with straight trunk 40 to 90 cm in diameter, buttressed at base, and with many upright pneumatophores rising to 45 cm from long horizontal roots. Bark gray to blackish, smooth to roughly fissured, thick; inner bark reddish. Leaves opposite, elliptical, 9 to 20 cm long, 5 to 7 cm wide, acute at both ends, entire, without visible veins, thick, leathery, glabrous. Petioles 2 to 4.5 cm long. Flowers single in leaf axils, 3 to 4 cm long, usually drooping on stalk of 1 to 2.5 cm, red to yellowish or cream-colored, with red to pink-red bell-shaped hypanthium. Calyx with 10 to 14 very narrow, leathery lobes. Petals 10 to 14, 13 to 15 mm long, white turning brown, each with 2 narrow lobes ending in 3 to 4 bristles. Stamens 2, nearly hidden at base of each petal. Pistil with inferior 3- to 4-celled ovary, each cell with 2 ovules; style slender; stigma with 3 to 4 short forks. Berry drooping, ovoid or turbinate, 2 to 2.5 cm long. Seed 1, viviparous, finally 1.5 to 2 cm in diameter.[140]

Germplasm — Reported from the Hindustani, African, Australian, and Indonesian-Indochina Centers of Diversity, Burma mangrove, or cvs thereof, is reported to tolerate alkali, disease, high pH, insects, pest, salt, shade, waterlogging.[188,232]

Distribution — Tropical South and East Africa, Madagascar, Seychelles, Sri Lanka, southeastern Asia, Ryukyu; throughout Malaysia to Philippines, Australia, Micronesia, and Polynesia. Introduced into Hawaii.[188]

Ecology — Estimated to range from Tropical Moist to Rain through Subtropical Moist to Rain Forest Life Zones, Burma mangrove is reported to tolerate annual precipitation of 10 to 80 dm, annual temperature of 20 to 26°C, and pH of 6.0 to 8.5. One of the largest trees in the Malayan mangroves, usually on drier well-aerated soils toward the landward side, often dominating with occasional stems >35 m tall. It is probably the longest-lived of the mangroves. It can stand "any amount of shade,"[140] Mostly on brackish or saline silts of depositing shores and marshes.

Cultivation — According to the NAS,[232] planting is usually not needed, because natural regeneration is so successful. In *Avicennia* and *Rhizophora,* direct seeding results in ca.90% survival.

Harvesting — Mostly harvested from natural stands. Species of Rhizophoraceae, growing only from the tips of the branches, are often killed by indiscriminate lopping of branches.[232] After felling, its regeneration is often very scant and there is danger of overgrowth by Acrostichum (but once seedlings have established themselves, the "fern acts rather as a nurse, forcing the seedling up.").[140]

Yields and economics — A good mangrove stand can show annual productivity of 10 to 20(to 25) MT/ha/year, but for firewood purposes, I would reduce that to 10 to 20 (to 25) m^3/ha/year, figuring that as optimal rather than average. Litter-fall may account for 1/3 to 1/2 of above-ground productivity. Because of the heaviness of the wood, a cubic meter of mangrove wood is generally more valuable than the wood of other species.

Energy — Wood widely used for charcoal and fuel.[188] For charcoal, the tree seems to rank with Rhizophora, with an even higher calorific value. According to *The Wealth of India,*[70] the calorific value of moisture-free sapwood is 5,169 cals, heartwood 5,079.

Biotic factors — No data available.

BUCHANANIA LANZAN Spreng. (ANACARDIACEAE) Chirauli Nut, Cuddapah Almond, Cheronjee, Chironjii, Almondette
Syn.: *B. latifolia* Roxb.

Uses — Cuddapah almond is cultivated for the fresh fruit, which has a very agreeable flavor. The delicate nutty-flavored seed is very nutritious, especially when roasted. Seeds are consumed by natives of India and Burma, roasted with milk or as sweetmeats. Seeds are also the source of an excellent oil, which is light yellow, sweet, mild with pleasant aroma, and used as a substitute for olive oil or almond oil in confectionery, and in medicinal preparations — especially applied to glandular swellings of the neck. A gum (Chironji-kigond) is sold at bazaars in India and has adhesive properties. Kernels are used as important articles of trade, in exchange for salt, grain, and cloth. Leaves are used as fodder in Bombay and Punjab. Bark and fruits furnish a natural varnish. A pellucid gum, obtained from wounds on stems, is used in diarrhea. Used to tan leathers of dark reddish-brown color with a somewhat stiff, harsh texture. Wood is light gray to grayish-brown, sometimes with a faint yellow tinge, to dark-brown in heartwood of old trees, rough, very light, straight-grained, coarse-textured, moderately strong, used for boxes, yokes, doors, cheap furniture, posts, and bedsteads. Berar females use the pounded kernels to remove facial spots and blemishes.[70,165,278]

Folk medicine — Reported to be antidotal for fish poisoning and scorpion stings, almondette is a folk remedy for asthma, bronchitis, burns, cholera, consumption, cough, diarrhea, dysuria, fever, gingivitis, phthisis, and snakebite.[91] Describing the genus *Buchanania* as therapeutically inert, Kirtikar and Basu[165] go on to describe the almondette as used in the Ayurvedic and Yunani systems of medicines. Ayurvedics use the roots for biliousness and blood disorders; the fruits for blood diseases, fevers, impotence, thirst, and ulcers; the aphrodisiac cardiotonic seeds for biliousness. Yunani consider the seed aphrodisiac, expectorant, stomachic, and tonic. Useful in fever, gleet, and urinary concretions, it is believed to cause headache. Yunani regard the leaf juice as antibilious, aphrodisiac, depurative, digestive, expectorant, purgative, and refrigerant. The seed oil is applied to glandular swellings on the neck. It is also used for itch, pimples, and prickly heat. In Madras, the gum is given with goat's milk for intercostal pain. Hakims apply the fruit to inflamed or indurate tongue.[91,165]

Chemistry — Seeds contain 51.8% oil, 12.1% starch, 21.6% protein, 5% sugar;[283] bark contains 13.4% tannin.[70] Kernels also contain 152 mg Ca and 499 mg P (per 100 g); deficient in amino acids lysine and methionine.[278] The fatty acid composition of *B. lanzan* seed oil, determined by urea complex formation and GLC, was found to be: myristic, 0.6%, palmitic, 33.4%, stearic, 6.3%, oleic, 53.7%, and linoleic, 6.0%. Triglyceride compositions of the native seed oil were calculated from the fatty acid compositions of the triglycerides and of the corresponding 2-monoglycerides produced by pancreatic lipase hydrolysis. The oil is composed of 3.2, 35.8, 45.5, and 15.5% trisaturated, monounsaturated disaturated, diunsaturated monosaturated, and triunsaturated glycerides, respectively. The special characteristics of *B. lanzan* seed oil is its content of 22.7, 31.0, and 11.3% dipalmitoolein, dioleopalmitin, and triolein, respectively. The percent trisaturated glyceride content of the oil increased from 3.2 to 7.5 by the process of randomization. On directed interesterification, the oil yielded a product with a slip-point of 41.5°C which may be suitable as a coating material for delayed action tablets. The oil also appears to be a promising commercial source of palmitic and oleic acids.[303]

Description — Moderate-sized tree, up to 17 m tall and a girth of 1.3 m; young branches pubescent; leaves alternate, simple, leathery, entire, 12 to 25 × 6 to 12.5 cm, petioled; flowers small, sessile, white, monoecious, in terminal or axillary panicles, crowded; calyx short, persistent, the lobes ciliate; petals 4 to 5, ca. 2.5 mm long, oblong, recurved; stamens 8 to 10, free, inserted at base of disk; fruit black, single-seeded drupe, 1.3 cm in diameter, with scanty flesh; stone crustaceous or bony, 2-valved; seeds (kernels) gibbous, acute at one end, size of small cherries. Flowers spring; fruits summer.[165,278]

Germplasm — Reported for the Hindustani Center of Diversity, almondette, or cvs thereof, is reported to tolerate savanna, slope, and dry deciduous forests.[82]

Distribution — Native to Southeast Asia, mostly India, Burma, and Indochina, especially in mountainous regions, almondette is widely cultivated throughout India, ascending to 1000 m in northwestern India and Nepal, spreading towards Malaya, Thailand, and Yunan.[278]

Ecology — Ranging from Subtropical Moist to Tropical Dry through Wet Forest Life Zones, almondette is reported to tolerate annual precipitation of 7 to 40 dm, annual temperature of 23 to 25°C, and pH of 5.0 to 6.0. Trees are found in dry deciduous forests. Within its natural habitat it is a useful tree for covering dry hillsides.[82,278]

Cultivation — Propagated from seed; not formally cultivated.[278]

Harvesting — Harvested from the wild.[70]

Yields and Economics — In Madras, a tree will yield ca.0.4 kg gum/year. Wood is rather cheap; in 1937, Bombay Rs 25 to 35 per ton, in Orissa, Rs 19 per ton. Fruits are frequently sold at bazaars in India, at about 4 to 6 annas per 1b.[278] It takes 36 kg nuts to yield 10 kg oil as expressed in India.

Energy — In Tropical Dry Forest near Varanasi, *Shorea robusta* may be dominant, followed by *Buchanania lanzan*, with standing biomass of 26.8 and 8.3 MT/ha and annual net production of 2.21 and 0.79 MT/ha respectively. Litter amounts to 1.51 MT and 0.58 MT/ha respectively.[307] A seedling in its first year will produce only 0.19 g biomass, compared to 5.98 g for *Butea monosperma*, 12.43 g for *Areca catechu*.[358]

Biotic factors — Tree attacked by the fungus *Marasmius* sp. and by the parasitic flowering plant, *Dendrophthoe falcata*.[278]

BUTYROSPERMUM PARADOXUM (Gaertn.f.) Hepper (SAPOTACEAE) — Shea Nut, Butterseed
Incl.: ***B. parkii***

Uses — An important oil-producing tree, it is the source of shea butter, an edible fat or vegetable butter extracted from the ripe seeds. Natives use shea butter as cooking fat, an illuminant, a medicinal ointment, dressing for the hair, and for making soap. Shea nut meal used for hog-feed, having 60% carbohydrate and 12% protein. Gutta-shea is a reddish exudation obtained by tapping the tree with removal of pieces of bark with a narrow axe. Latex is removed on the following day, boiled and cleaned of dirt and bark; it is a mixture of resin and gutta, called "balata" or "Red Kano rubber".[146] Wood is dull red, very heavy, termite-proof, difficult to work, but takes a good polish and is very durable. Used for wooden bowls, mortars and pestles; used as firewood, producing great heat and making charcoal. In Sierra Leone, used for ribs of boats and in marine workshops. Ashes from burning of wood commonly used as the lye in indigo dyeing. Flowers provide bee nectar.[71,146]

Folk medicine — Nakanis of West Africa use the bark decoction to bathe children and as a medicine. On the Ivory Coast, it is used in baths and sitz-baths to facilitate delivery. Lobis use the leaf decoction as an eye bath. Young leaves are used in steam vapors to alleviate headache. Oil used as a topical emollient and vehicle for other pharmaceuticals. Medicinally, butter used for rubbing on rheumatic pains or mixed with other medicines to replace other oils. Also used both internally and externally on horses for galls and other sores. Root-bark, boiled and pounded, applied to chronic sores in horses. Crushed bark used as a remedy for leprosy. Latex is not poisonous, but a decoction of the bark is lethal. Root mixed with scourings of tobacco as a poison.[71,278]

Chemistry — Per 100 g, the seed (ZMB) is reported to contain 622 calories, 7.3 g protein, 52.6 g fat, 38.2 g total carbohydrate, 5.6 g fiber, 1.8 g ash, 107 mg Ca, 43 mg P, 3.2 mg

Fe, and 0.56 mg thiamine.[89] The fat contains 45.6% oleic acid, 44.3% stearic acid, 5.5% linoleic acid. Of the 2 monoglycerides, 82.1% was oleates and 14% linoleates.[292] Another report puts it at 5.7% palmitic, 41.0% stearic, 49.0% oleic, and 4.3% linoleic. Allantoin and its intermediary products constitute 24 to 28% of the total N of a water extract of defatted shea kernel meal.[21] Alpha- and beta-amyrin, basseol, parkeol, and lupeol are also reported.[187] According to Roche and Michel,[282] the seed protein contains 8.2% arginine, 1.0% cystine, 9.9% leucine, 2.9% phenylalanine, 1.1% tryptophane, and 1.4% valine.

Description — Stout, much-branched tree to 20 m tall; crown spreading, bark usually gray or blackish, deeply fissured and splitting into squarish or rectangular corky scales; short-shoots with conspicuous angular leaf-base scars; young shoots, petioles and flower buds with rusty pubescence. Leaves oblong to ovate-oblong, 10 to 25 cm long, 4.5 to 14 cm broad, rounded at apex, base acute to broadly cuneate, margin undulate and thickened; the petioles one-third to one-half the length of lamina; both surfaces either pubescent or glabrescent, lateral veins 20 to 30 on each side, regularly and closely spaced, slightly arcuate; leaves reddish when young flowers fragrant, in dense clusters, at tips of branchlets, above leaves of previous year; pedicels up to 3 cm long, puberulous to densely pubescent; outer sepals lanceolate, 9 to 14 mm long, 3.5 to 6 mm broad, pubescent or more or less floccose externally; inner sepals slightly smaller; corolla creamy white, tube 2.5 to 4 mm long, glabrous or pilose externally, lobes broadly ovate, 7 to 11 mm long, 4.5 to 7 mm broad; filaments 7 to 12 mm long, anthers more or less lanceolate, up to 4.5 mm long; staminodes up to 8 mm long; style 8 to 15 mm long. Fruit ellipsoid, greenish, up to 6.5 cm long, 4.5 cm in diameter, subglabrous or with pubescence persistent in patches, containing a sweet pulp surrounding the seed. Seed up to 5 cm long, 3.5 cm in diameter, usually solitary, sometimes up to 3 per fruit, shining dark-brown, with a large white scar on one side. Germination cryptocotylar.[398]

Germplasm — Reported from the African Center of Diversity, shea nut, or cvs thereof, is reported to tolerate drought, fire, grazing, laterite, savanna, and slope.[82] Several morphological and physiological forms differ in shape and size of fruits and seeds, and in chemical analysis of kernels and fruit, thickness of pericarp, and early fruiting period. Besides the common type, there are two recognized varieties or subspecies: subsp. *parkii* (G.Don) Hepper (*Butyrospermum parkii* (G.Don) Kotschy, *Brassia parkii* G.Don) is less dense a plant, with shorter indumentum, smaller flowers, and the style is 8 to 12 mm long; subsp. *niloticum* (Kotschy) Hepper has densely ferrugineous parts, with a corolla tube pilose externally, lobes 9.5 to 11 mm long, 6.5 to 7 mm broad, filaments 10 to 12 mm long, and the style 12 to 15 mm long.[398]

Distribution — Widespread throughout tropical Africa from West Africa (Liberia, Gold Coast, Nigeria, Togo, Dahomey, Senegal, Sierra Leone) to Sudan and Uganda, south to eastern Congo.[278]

Ecology — Ranging from Subtropical Dry to Wet through Tropical Dry to Moist Forest Life Zones, shea nut is estimated to tolerate annual precipitation of 8 to 25 dm, annual temperature of 23 to 27°C, and pH of 4.9 to 6.5.[82] Frequent in savanna regions or as scattered trees in grasslands across central Africa from West Africa to East Africa. Often protected and preserved in cultivated land. Common on dry laterite slopes, but not in alluvial hollows or land subject to flooding. Grows from 950 to 1,500 m elevations.[278]

Cultivation — Seeds germinate readily on the ground under natural conditions. Fresh seed is essential. Seedlings develop a very long taproot, making transplanting hazardous. Trees grow very slowly from seed, bearing fruit in 12 to 15 years, taking up to 30 years to reach maturity. Natural propagation is chiefly from root-suckers.[278]

Harvesting — Fruits mostly harvested at end of July, usually during the rainy season. Shea oil is the native product expressed from kernels in Europe; shea butter is material prepared by native methods. In West Africa, the preparation of shea butter is woman's work.

In Nigeria, nuts are collected by one tribe, sold to another, and the butter bartered back. Preparation of shea butter consists of pounding usually roasted kernels in mortar to a coarse pulp, and then grinding this into a fine oily paste with chocolate aroma. Tannin present makes this form inedible. In some areas, this mass is further worked with a little water in a large pot in the ground, followed by hand-kneading and washing in cold water. From this the butter is extracted by boiling and skimming; then it is boiled again to purify further, after which it is transferred to molds. Locally, nuts are boiled before cracking; extraction is made from the sun-dried kernels. Ordinary oven-drying causes no loss of oil. Clean nuts may be roasted until the latex coagulates and the dry nuts stored. In other areas, fruits are spread in the sun until the pulp separates, or they are fermented by being kept moist for weeks or months in earthenware jars, and then the nuts are subsequently roasted.[278]

Yields and economics — Using native processing approaches, it takes about 4 kg kernels to yield 1 kg butter. Thoroughly dried kernels represent about one-third the weight of the fresh nuts. By native standards, a kerosene tin containing about 12.23 kg (27 lb) kernels yield 3.17 kg (7 lb) shea butter. Kernels contain 45 to 55% by weight of fat, but may be as high as 60%, and 9% proteins. The Giddanchi type of kernel averages 3.2 to 6 cm long, yielding 52.4% of fat. The shea butter tree has economic importance as an oil-seed produced under natural conditions in great abundance in regions where the oil palm does not grow and in areas which are otherwise unproductive. A large volume of shea nuts is exported annually from West Africa, mainly to Holland and Belgium, the chief importers. Belgium imports most of the shea butter. In Uganda, a small local market developed during World War II.[71,146,398]

Energy — As fuel, the wood gives out great heat. Charcoal is prepared from it in some districts.[146]

Biotic factors — Where trees are subjected to annual grass-burning, they are frequently stunted and twisted. The thick corky base gives some protection against fire. Trees are frequently grazed by wild animals and the sugary pulp is eaten by them, but not the nut of the fallen fruit. Unripe fruit exudes latex which remains in the ripe nut but disappears from the ripe pulp. In Senegal, caterpillars of *Cirina butyrospermi* cause defoliation; dried, these caterpillars have long been an article of food in Nigerian markets under the name of mone-mone (Yoruba). Fungi attacking this tree include: *Aspergillus flavus, A. niger, A. tamarii, Botryodiplodia theobromae, Cephaleuros mycoidea, Cercospora butyrospermi, Fusicladium butyrospermi, Meliola butyrospermi, Helminthosporium coffeae, Oothyrium butyrospermi,* and *Pestalotia heterospora*. Parasitic on the tree are *Loranthus dodonaefolius, L. globiferus* var. *salicifolius,* and *L. rufescens;* and *Ficus* may be epiphytic on the tree, causing reduction in fruit yield.[186,278]

CALAMUS ROTANG L. and other species (ARECACEAE) — Rattan Cane, Rotang Cane

Uses — Tender shoots and seed edible. The sweet pulp around the seeds is also edible. Stems provide drinking water, especially in the rainy season.[410] Sturtevant[317] describes the fruit as roundish, large as a hazelnut, and covered with small, shining, imbriate scales. Natives generally suck out the subacid pulp which surrounds the kernels to quench the thirst. Sometimes the fruit is pickled with salt and eaten at tea time. Seeds are eaten by aborigines. Stems and branches form rattan cane of commerce, used as props for crop plants, for manufacture of furniture, baskets, wicker-work, umbrella ribs, cables, and ropes. Rattan ropes are used for dragging heavy weights and for tethering wild animals. Cordage and cables are made by twisting together two or more canes. Canes also are used for building boats, suspension bridges, and as a substitute for whale-bone. Jungle experts make fire by rubbing them backwards and forwards as fast as possible under a branch of dry soft wood in which a hole has been scooped and lined with wooden dust.[324]

Folk medicine — Used for abdominal tumors in India,[126] Root given for chronic fevers, and used as antidote to snake venom. Leaves used in diseases of blood and in biliousness. Wood is a vermifuge.[278] Serrano,[410] without mentioning species, cited asthma, diarrhea, enterosis, rheumatism, and snake-bite as ailments treated with rattan.

Chemistry — Per 100 g, the fruit of the figured species (*C. ornatus*) is reported to contain 79 calories, 79.0 g H_2O, 0.6 g protein, 1.2 g fat, 18.6 g total carbohydrate, 0.5 g fiber, 0.6 g ash, 19 mg Ca, 10 mg P, 1.7 mg Fe, 0.06 mg thiamine, 0.01 mg riboflavin, 0.9 mg niacin, and 5 mg ascorbic acid.[89]

Toxicity — Scrapings from the bark of glossy-coated cane species may contain enough silica to act as an irritant to the mucous membranes.[56]

Description — Stems scandent or climbing, very slender; to as much as 200 m long, leaf-sheaths sparingly armed with short, flat spines, glabrous. Leaves 60 to 90 cm long, on short petioles with small, straight or recurved spines; leaflets numerous, narrowly lanceolate, 20 to 23 cm long, 1.3 to 2 cm broad, median costa unarmed on both surfaces, or armed beneath only, lateral costa unarmed on both surfaces. Male spadix slender, very long, branched, whip-like, sparingly spinous; female flowers scattered along slender branches of spadix; spikelets 1.3 to 2.5 cm long, recurved. Fruit globose to subglobose, very pale, 1.6 to 1.8 cm in diameter; scales many, in vertical rows, straw-colored.

Germplasm — Reported to tolerate slope and shade, the rattan genus is from the Hindustani and Indochina-Indonesia Center of Diversity. It contains ca.300 difficulty distinguishable species of the moister tropics of the Old World (Asia, Africa). Perhaps rattans, climbing spiny palms, represent 600 species in ca.15 genera, more used for furniture and construction than for the nuts. Lapis[411] discusses the 12 major Philippine species, illustrating 3 species of *Calamus*.

Distribution — Native to India, Bengal, Assam, and Sri Lanka[278] *(Calamus rotang)*, with other species, extending to Borneo, the Philippines.

Ecology — Ranging from Subtropical Moist to Tropical Moist through Wet Forest Life Zones, rattan is reported from areas with annual precipitation of 17.3 to 42.9 dm (mean of 4 cases = 32.1 dm), annual temperature of 23.5 to 27.4°C (mean of 4 cases = 25.7°C), and pH of 4.5 to 5 (mean of 2 cases = 4.8). Once common in moist localities, in tropical to subtropical climate, now locally overharvested. Does not tolerate any frost. Apparently fares better in primary than secondary forest.[410] Young plants thrive in soil containing a large quantity of leaf mold. Older plants need soil of a more lasting nature.[25]

Cultivation — A quantity of bonemeal and charcoal in the soil may be advantageous.[25] Young plants thrive in rooting media rich with leaf mold. Older trees need more substantial soil with ground bone, charcoal nutrients, and plenty of water.[379] Loams are best, clay loams okay. For seed extraction, the fruits are peeled and fermented in water for ca.24 hr., then squeezed, after which clean seeds settle to the bottom. These are then removed to dry in the shade. Then they are stratified or mixed with moist sawdust for several days. To prevent fungal infestation, seeds are treated with ca.0.5 lb sodium pentachloropentate, and dissolved in 3 gallons distilled water. Germination starts after 68 to 85 days. Nursery-grown seedlings or earthballed wild plants, as well as young suckers, can be used as planting stock. Seedlings 15 cm tall are ready for planting, if they have 4 to 5 leaves. Two seedlings are placed in each hole, 2m × 2m, at the beginning. Fertilization at 6 g per plant 20:10:5 is recommended.[410] During the first 2 to 3 years, humus mulching encourages growth. At this point, more light is desirable. Some Borneo farmers, in abandoning their temporary forest food plots, plant rattans, letting the forest reclaim the plot, returning 7 to 15 years later to harvest rattan and begin food cropping again.

Harvesting — Some cultivated trees yield usable canes in 6 years. Full production occurs in 15 years. At this age, canes average about 30 m long, 2.5 cm in diameter. Mature rattans can be cut at the base and divided into sections 4 to 5 m long. Thereafter, canes can be cut about every 4 years, from suckers. Canes should be harvested during the dry season, and dried and processed promptly. Canes are scraped to remove the thin silicious coating, bringing out its yellowish luster. Canes should then be dried to less than 20% moisture. Kilns at dry bulb temperatures of ca.65°C, wet bulb temperatures of ca.45°C, will bring moisture contents to 12 to 14% in ca.5 days. A dryer design is discussed by Serrano.[410] Stain fungi may be avoided by treatment of 7 pounds sodium pentachlorophenate in 100 gallons water, applied the same day the canes are cut. Post-powder beetles may be prevented by soaking the poles for 3 min in 0.5% aqueous solutions of Lindane or Dieldrin. The canes may also be steeped in a mixture of diesel oil and coconut or palm oil prior to a final drying.[410]

Yields and Economics — Rattan cane is important in India and elsewhere for the manufacture of cane-bottom chairs, etc. Many species of the large genus are used in various parts of the world for similar purposes.[278] In the Philippines, the rattan industry employs 10,000 workers. One Philippine journal[412] suggested that, already, raw rattan was worth $50 million (U.S.), with the finished manufactured rattan products worth $1.2 billion. In the Philippines, in 1977, there was a report of nearly 66 tons split rattan and nearly 4,000,000 linear meters of unsplit rattan.[413] By 1983, it was closer to 5,000,000 linear meters.[410] Among Tagbanua ethnics in the Philippines, rattan collecting returned ca. $1.00 to $5.00/day whereas agriculture returned closer to $1.00/day.[414] But in the 1950s and 1960s, a worker could collect 200 5-m canes a day, while in 1981, 35 to 50 canes was par, each worth little more than $0.05.

Energy — In Peninsular Malaysia, mean stem lengths of *Calamus manan* was only 1.3 m after 6 years, but the longest stem was ca. 18 m. *Calamus caesius* can grow as much as 5 to 6 m/year for the first 5 years of planting. In Sabah, the number of aerial stems doubled each year for the first 3 years in *C. caesius,* first 4 years for *C. trachylopheus.*[415] Trial cultivation[416] of *Calamus ornatus* in the Philippines yielded canes less than 2 m long, not suggesting much biomass potential. Rejects and prunings might be useful for fuel.

Biotic factors — Rattan plants are attacked by the fungi *Catacaumella calamicola, Doratomyces tenuis,* and *Sphaerodothis coimbatorica,*[186] Undesirable stains are caused by *Ceratocystis and Diplodia.*

CANARIUM INDICUM L. (BURSERACEAE) — Java-Almond, Kanari, Kenari

Syn.: ***Canarium amboinense* Hochr., *Canarium commune* L., *Canarium mehenbethene* Gaertn., *Canarium moluccanum* Bl., *Canarium subtruncatum* Engl., *Canarium shortlandicum* Rech., *Canarium polyphyllum* Krause, *Canarium grandistipulatum* Lauterbach, and *Canarium nungi* Guillaumin.**

Uses — Seeds are highly regarded in Melanesia as a food, a delicacy, and in pastries as a substitute for almonds. Mature fruits, dried over fires, are an important stored food in the Solomon Islands. Nuts are ground and added to grated taro and coconut cream.[346] An emulsion of seeds is used in baby-foods. Oil from the seeds is used as a substitute for coconut oil for cooking and illumination. Resin from the stems (Getah kanari) has the scent of eugenol and is used in printing inks and varnishes. It is the source of a Manila elemi, a resin, used as an incense and fixative in the perfume industry, and for varnishes. Oil derived from the resin is also employed in soap and cosmetics. Old stems are used as fuel and when burning lime. Wood may be used in canoe building and paddles are made from the buttresses. Parts of the plant are used to make cloth and to make moth-repelling bookcases. The tree is planted as a shade-tree in nutmeg plantations and as a road-side tree.[278]

Folk medicine — Resin is applied to indolent ulcers.[70] The fruit is laxative. Medicinally (in Java), it is used as an incense for sick persons to keep the atmosphere clean.[56]

Chemistry — Seeds contain 3.8% moisture, 19.6% protein, 72.8% fat, and 3.8% ash.[89] The oil contains 10.2% stearic, 30.5% palmitic, 39.9% oleic, 18.7% linoleic, and 0.7% linolenic acids. The oleoresin which oozes from the trunk contains 10.4% essential oil, 81.8% resin, 3.7% water solubles, and 2.5% water. The essential oil contains ca. 34% anethole and a small quantity of terpenes.[70,314]

Description — Tree grows up to 40 m, to 1 m in diameter, with buttresses; branchlets 7 to 13 mm thick, glabrescent. Leaves are compound, 3-8 pairs of leaflets, glabrous, with persistent ovate to oblong stipules 1.5 to 2 cm long and 1.2 to 1.4 cm wide, pulverulent to glabrous; leaflets oblong-obovate to oblong-lanceolate, 7 to 35 cm long and 3.5 to 16 cm wide, on long slender petiolules (to 3 cm long); blades herbaceous to coriaceous, the base oblique, rounded to broadly cuneate, the apex gradually to bluntly acuminate, margin entire; inflorescences terminal, many-flowered, 15 to 40 cm long, minutely tomentose. Flowers tomentose, male ones subsessile, about 1 cm long, females short-stalked, up to 1.5 cm long, with a concave receptacle; calyx in male flowers 5 to 7 mm long, in females 7 to 10 mm; stamens glabrous, in male flowers free; in females adnate to disk; pistil in male flowers minute or none, in female glabrous; fruiting clusters with up to 30 fruits; fruits ovoid, round to slightly triangular in cross-section, 3.5 to 6 cm long and 2 to 4 cm in diameter, glabrous; pyrene rounded triangular in cross-section, smooth except the 3 more-or-less acute ribs at base and apex; lids 3 to 4 mm thick; seeds usually 1, the sterile cells slightly reduced. Flowers mainly October to December fruits July to December.[178,179,278]

Germplasm — Reported from the Indochinese-Indonesian Center of Diversity, Java almond, or cvs thereof, is reported to tolerate high pH.[82] Several races are cultivated in Melanesia, varying in form and size of fruits. Two botanical varieties are recognized: (1) var. *indicum*, with branchlets up to 13 mm thick; stipules up to 6 by 5 cm, dentate; leaves up to 7 jugate; leaflets up to 28 by 11 cm, herbaceous; and fruits up to 6 × 3 cm. This is the more widespread variety and the more cultivated form. (2) var. *platycerioideum* Leenhouts, with branchlets up to 2.5 cm thick; stipules sometimes inserted on the bases of the petiole only; leaves 5 to 8 jugate, 80 to 135 cm long; leaflets inequilateral, ovate, 25 to 35 × 13 to 16 cm; fruits 6 by 3.5 to 4 cm. Found only on New Guinea up to altitudes of 30 m.[178,179,278]

Distribution — Native to Moluccas (Ternate, Sula, Ceram, Ambon, Kai), the North Celebes (where it may be naturalized), and Indonesia (New Guinea, New Britain, New Ireland, Solomon Islands, and New Hebrides).[278]

Ecology — Ranging from Subtropical Dry to Moist through Tropical Dry to Moist Forest Life Zones, Java almond is reported to tolerate annual precipitation of 11 to 24 dm (mean of cases = 18), annual temperature of 24 to 27°C (mean of 3 cases = 25.5), and pH of 5.3 to 8.1 (mean of 2 cases = 6.7).[82] Java almond is found in rain-forest at low altitudes, rarely native above 250 m. However, it is planted up to 600 m or more.[278]

Cultivation — Sprouted seeds or larger seedlings are transplanted.

Harvesting — Fruiting peaks at August to October and February to April in Santa Cruz, Solomon Islands, but fruits are available year-round. In the Solomon Islands, where *Canarium* is "probably the most important economic tree species", the plants are usually accepted as wild forest species, exploited by gathering, on the basis of recognized individual ownership.[346]

Yields and economics — Large fruited forms or species on Santa Cruz are said to yield fewer fruits than the smaller fruited forms.[346]

Energy — Seed oil and resin might be viewed for energy potential, over and above the fuel wood. The resin was used for illumination in the Solomon Islands. The abundance of *Canarium* on Ndenia Island may explain why *Agathis* resin was not exploited.[346]

Biotic factors — The following fungi are known to attack Java almond: *Aedicium pulneyensis, Meliola canarii, Oudesmansiella canarii, Skierka canarii,* and *Ustilina zonata.* Seeds dispersed by fruit bats (*Pteropus*).[278]

CANARIUM OVATUM Engl. (BURSERACEAE) — Pili Nut, Philippine Nut

Uses — The pulp is edible when cooked and yields a cooking oil. The nut or kernel is also edible and excellent after roasting. It also yields a good cooking oil.[60] Menninger[209] describes this as the "most important of all the nuts in the world to the millions of people who depend on it for food." Abarquez[1] says pili is second only to cashew as a food nut in the Philippines, where it is considered superior to almonds. The nuts have been used to adulterate chocolate.[209] This species is one source of the commercial resin traded as Manila elemi. Spaniards repaired their ships, in colonial days, with gum elemi. Manila elemi is a yellowish-greenish-white, sticky, soft, opaque, fragrant oil mass which gradually becomes hard when exposed. It is a source of a kind of paper for window-panes as a substitute for glass, and is used in the preparation of medicinal ointment. It is an important ingredient in plastics, printing inks for lithographic works, perfumes, and plasters. This resin gives toughness and elasticity to lacquer, varnish, and paint products. Locally, it is used to caulk boats and as an illuminant for native torches. Recently, the possibility of extracting fuel from resin has proved enticing, suggesting the possibility of driving a car "run by a tree."[1] According to Garcia, pili plantations, in pure stand, can be interplanted with cassava, ginger, papaya, pineapple, coffee, cacao, bananas, and taro.

Folk medicine — The "elemi" was once used as an ointment for healing wounds.[283] Filipinos use the crushed emulsion of the kernels as a substitute for milk for infants. Uncooked nuts are purgative.[249]

Chemistry — Per 100 g, the seed (ZMB) is reported to contain 699 to 714 calories, 12.2 to 15.6 g protein, 73.2 to 75.9 g fat, 6.0 to 10.8 g total carbohydrate, 2.3 to 3.5 g fiber, 3.0 to 3.6 g ash, 130 to 180 mg Ca, 71 to 591 mg P, 2.9 to 4.8 mg Fe, 3.2 to 3.3 mg Na, 521 to 537 mg K, 26 to 35 μg beta-carotene equivalent. 0.75 to 1.04 mg thiamine, 0.07 to 0.13 mg riboflavin, 0.44 to 0.58 mg niacin, and 0 to 25 mg ascorbic acid.[89] Campbell reports the kernel contains 74% fat, 12% protein, and 5% starch.[60] Rosengarten reports 71.1% fat, 11.4% protein, and 8.4% carbohydrates.[283]

Description — Buttressed dioecious trees to 20 m tall, 40 cm DBH, leaves alternate, compound, with 5 to 7 leaflets each 10 to 20 cm long; inflorescences and axillary terminal, many-flowered, flowers yellowish, fragrant, ca. 1 cm long; fruits ellipsoid to oblong, 3 to 7 cm long, with thin, oily pulp, greenish, turning black when ripe; seed solitary, triangular in cross-section with pointed ends, thin, hard shell and a single large kernel.[60,209]

Germplasm — Reported from the Philippine Center of Diversity, the pili nut, or cvs thereof, is reported to tolerate slope and strong winds. Campbell says no cultivars are described,[60] but Menninger says 75 kinds grow in enormous quantities from Africa through India to northern Australia, Malaya, and in the Pacific Islands. Menninger may mean the genus rather than the species.[209]

Distribution — Endemic to the primary forests of Luzon at low and medium altitudes in the Philippines; widely distributed, yet little-known.[60] Introduced successfully into El Zamorano and Lancetilla, Honduras.

Ecology — Estimated to range from Subtropical Moist to Rain through Tropical Moist to Wet Forest Life Zones, pili nut is estimated to tolerate annual precipitation of 20 to 80 dm, annual temperature of 23 to 28°C, and pH of 5.0 to 7.0. Best adapted to the hot, wet, tropics.

Cultivation — Generally grown from seed; superior selections may be grafted. It can be marcotted and budded as well. In the Philippines, it is often planted between rows of coconut. Seedlings, wrapped in banana sheath or bark, are transported carefully to the transplant site at the onset of the rainy season. Spacing is generous, 12 to 15 m apart for densities of only 40 to 50/ha.

Harvesting — Vegetatively propagated pilis may bear at age 7 to 8. Rosengarten[283] says female trees start bearing at age 6, but full production is not reached until 12 to 15. From seed, it takes 7 to 10 years to fruiting, 12 to 13 according to Garcia.[106] Fruits are usually shaken or knocked from the tree. Fresh nuts do not store well, becoming rancid in weeks, if not roasted. Abarquez[1] states, "Integrating resin tapping with nut production, that is, the possibility of getting 2 products without disabling the tree, can be studied. As practiced, it is observed that flogging, girdling, or wounding the bark of trees on the lower trunk part usually increase the production of fruits in some trees like mango . . . Controlling the downward translocation of carbohydrates and other hormones from the canopy to the root system, by wounding the bark on the trunk, would induce the production of flower hormones, and consequently, fruits. Timing the tapping activities so that it complements with the natural budding and fruiting season would give us the desired result."

Yields and economics — Sometimes trees may yield as much as 33 kg nuts. Garcia[106] puts peak yield at 2.5 MT/ha/yr. Other species can yield nearly 50 kg resin per year. Manila exported more than 1000 MT as long ago as 1913. But Abarquez[1] shows only a little more than a ton around 1975. In 1950, the Philippines had more than 8,000 ha planted to pili, reduced to ca. 2,500 by the end of 1976.

Energy — If the seeds were copiously produced, their 75% oil could be viewed as an oil source. Other species of *Canarium* exude valuable resins "which could be a promising alternative for the oil industry."[241] Such species are said to average 45 kg resin per year. But Roecklein and Leung[379] put yields of *C. luzonicum* resin at only 4 to 5 kg/yr. The shells of the nuts are said to be an excellent fuel, a handful enough to cook a simple dish. Garcia[106] describes the wood of the pili as an excellent firewood.

Biotic factors — Campbell[60] states that pests and diseases have not been described.

CARYA ILLINOENSIS (Wangenh.) K. Koch (JUGLANDACEAE) — Pecan
Syn.: ***Carya pecan*** **(Marsh.) Engl. and Graebn.,** ***Carya oliviformis*** **Nutt., and** ***Hicoria pecan*** **Britt.**

Uses — Kernels of nuts eaten raw, roasted, or salted and used in candies, confections, ice cream, mixed nuts, and for flavoring in baking and cookery. Pecan oil, expressed from kernels, is edible and sold for the drug, essential oil, and cosmetic trade. Lumber is hard, brittle, not strong, but is occasionally used for agricultural implements, wagons, and for fuel.[278] More recently, pecan timber has been used for veneer and lumber, flooring, and still for firewood. Smith[310] notes that "the pecan has great possibilities as a shade (and timber) tree in a large area where it cannot be a commerical dependence, but may produce an occasional crop." Doubtless, the deep-rooted pecan can contribute to erosion control. To quote Smith,[310] "During the regime of the tribal leaders in the old Seminole Nation in Seminole County, Oklahoma, they had a law that fined a person five dollars or more for mutilating a pecan tree. Yet some people call the Indian a savage. Whoever calls the Indian a savage should go look at the gullies we white men have made in Oklahoma where the Indian made *none!*" Southerners are intercropping pecans with cattle successfully. Pecan has been described as the number three hardwood in the U.S., behind walnut and black cherry.[16,310,278]

Folk medicine — Reported to be astringent, pecan is a folk remedy for blood ailments, dyspepsia, fever, flu, hepatitis, leucorrhea, malaria, and stomach-ache.[91]

Chemistry — Per 100 g, the seed (ZMB) is reported to contain 711 to 718 calories, 9.5 to 9.7 g protein, 73.7 to 75.3 g fat, 13.4 to 15.1 g total carbohydrate, 2.3 to 2.4 g fiber, 1.6 to 1.7 g ash, 75 to 76 mg Ca, 299 to 334 mg P, 2.5 mg Fe, 0 to 3 mg Na, 624 to 1499 mg K, 20 to 82 μg beta-carotene equivalent, 0.74 to 0.89 mg thiamine, 0.11 to 0.13 mg riboflavin, 0.93 mg niacin, and 2.1 mg ascorbic acid. Leaves and leaf stalks contain a phytosterol similar to squalene, capric-, lauric-, myristic-, palmitic-, stearic-, arachidic-, oleic-, linoleic-, and linolenic-acids. Tannins containing phloroglucin and catechin have been identified; also inositol and 3,4-dihydroxybenzoic acid. The bark contains azaleatin (quercetin-5-methyl ether) ($C_{16}H_{12}O_7 \cdot H_2O$), and caryatin (quercetin-3,5-dimethylether)

($C_{17}H_{14}O_7$).[187] According to Hilditch and Williams,[128] the component acids of the seed fats are 3.3 to 7% palmitic-, 1 to 5.5% stearic-, 51 to 88% oleic-, 14 to 38% linoleic-, and 1 to 2% linolenic-acids.

Toxicity — Langhans, Hedin, and Graves[175] report that leaves and fruits contain juglone, a substance toxic to *Fusicladium effusum* at concentrations as low as 0.1 mg/mℓ (roughly 0.1 ppm). They also report linalool as fungitoxic. Schroeder and Storey[296] report aflatoxins in pecans with sound shells. The mycotoxin zearalenone was extracted from kernels with sound shells after 28 days. For reasons unclear to this author, Hager's Handbook[187] calls it a poisonous plant. The pollen is allergenic.

Description — Deciduous tree, 33 to 60 m tall, with massive trunk to 3.5 m in diameter, buttressed at base, crown round-topped; bark light-brown tinged red, twigs with loose pale-reddish tomentum, becoming glabrous or puberulent, lenticels numerous, oblong, and orange; leaves compound, 30 to 50 cm long, petioles glabrous or pubescent; leaflets lanceolate to oblong-lanceolate, more-or-less falcate, long-pointed, doubly serrate, 10 to 20 cm long, 2.5 to 7.5 cm broad, veins conspicuous; staminate flowers in slender clustered aments 7.5 to 12.5 cm long, from axillary buds of previous year's growth, sessile or nearly so, yellow-green, hirsute on outer surface, bract oblong, narrowed at ends, slightly 4-angled, with yellow pubescence; fruits in clusters of 3 to 11, pointed at apex, rounded at base, 4-winged and angled, 1.5 to 6.5 cm long, up to 2.5 cm in diameter, dark-brown, with yellow scales; husk splitting at maturity to nearly the base, often persistent on tree after nut fallen out; nut ovoid to ellipsoidal, rather cylindrical toward apex, rounded at base, reddish-brown with irregular black markings; shell thin with papery partitions; seed sweet, red-brown, kernel separating rather readily. Flowers early spring; fruits fall.[278]

Germplasm — Reported from the North American Center of Diversity, pecan, or cvs thereof, is reported to tolerate high pH, mycobacteria, salt, slope, smog, and weeds.[82] Many selected cvs have been made, some of the "paper-shell" or "thin-shell" cvs include: 'Curtis', 'Frotscher', 'Moneymaker', 'Pabst', 'Schley', and 'Stuart'. ($2n = 32$.)[278]

Distribution — Native to the valley of the Mississippi River from southern Indiana and Illinois, western Kentucky and Tennessee, to Mississippi and Louisiana, west to Texas; reappearing in mountains of northern Mexico. Largely cultivated in southeastern U.S., most abundant and of its largest size in southern Arkansas and eastern Texas. Improved cvs are widely cultivated.[278]

Ecology — Ranging from Warm Temperate Thorn to Moist through Subtropical Dry to Moist Forest Life Zones, pecan is reported to tolerate annual precipitation of 3 to 13 dm (mean of 11 cases = 8.7), annual temperature of 9 to 21°C (mean of 11 cases = 16.5), and pH of 5.0 to 8.2 (mean of 9 cases = 6.4). Low rich ground along streams is favorite habitat, especially in fertile soil, rich in humus, on land that has been under cultivation for many years. Quite hardy in the north, it has been successfully planted up to the 43rd parallel. While favored by alluvial soils, pecan is by no means restricted thereto.[310] Thrives on a variety of soils, from sandy soils of acid reaction to heavy soil with alkaline reaction, and gradations between these. All soils should be well-drained and pervious to water. Pecan is deep-rooted and requires plenty of water, but will not tolerate water-logged soils.[278] Relative humidity above 80% prevents effective pollination. Pecans require 150 to 210 frost-free days, but have not fared well in the tropics. Madden, Brison, and McDaniel[197] suggest a possible chill requirement of 750 hr below 45°F (7°C). Hardy to Zone 5.[343]

Cultivation — Pecans do not come true from seeds and are difficult to start from cuttings. Therefore, propagation is mainly by budding and grafting, in order to perpetuate desirable varieties. After soil with the proper requirements for pecan production has been selected, young trees are set on 11-m squares; in Texas, squares up to 23 m may be used, especially on river valley soils. In some areas 75 to 100 cm annual rainfall may be sufficient, but usually much more is required. Where the rainfall is abundant, growth of trees is rapid and crowding may be a problem. Trees usually do not begin bearing as early in humid areas because of greater rate of growth. In western pecan orchards, trees are used in the manner of interplanted fruit trees since the trees do not grow so fast and cvs may be selected that are prolific at a relatively young age and size. Western cvs apparently lend themselves to dwarfing by pruning, and can be kept relatively small, preventing crowding. Unlike most trees, pecans do not show a deficiency of moisture by wilting of leaves or shoots. This may be due to the deep taproot which absorbs sufficient moisture from the subsoil to prevent wilting, but when weeds and other plants growing near pecan trees show signs of water deficiency, water should be applied. Trees are often intercropped with cotton, corn, or peaches until trees come into bearing. Planting of trees originally varies from 11 to 33 m, but after a first thinning in 12 to 15 years and a second at the end of 20 to 25 years, trees will be spaced 23 to 66 m apart. Nursery trees usually planted in commercial orchards when rootstocks are 4 to 6 years old and budded or grafted trees are 1 or 2 years old, although older trees are used sometimes. Pecan trees are set in both large and small holes. In heavy soils, holes about 1 m in diameter at top give better results. In lighter soils, post-holes have proven satisfactory. For larger trees, larger holes should be used to accomodate the larger root systems. Trees should be planted about 5 cm deeper than they were in the nursery. After they are set, tops should be cut back and the trunks loosely wrapped for a distance of 30 to 45 cm from ground with burlap of heavy paper, which is tied loosely. Pecan trees are very slow to develop new roots after transplanting and should be supplied with adequate moisture during the first summer to help establish the root system. Young trees must be protected from sunscald and winter injury. Pruning and training trees to proper shape is essential. Young pecan orchards require more frequent cultivation than older orchards, because older trees tend to hold weed growth in check by shading and by competition for moisture and nutrients. Disking or plowing should be frequent enough to prevent rank growth of weeds or grass, the number of cultivations depending on the fertility of the soil. If an orchard is on land subject to overflow and bad erosion, it may be sodded with some suitable grass; during the growing season, it may be mowed, or sheep and cattle may be allowed to graze the land to keep down the vegetation. Where soils are poor, intercropping with legumes and adding fertilizer may be useful.[278]

Harvesting — Pecan nuts are harvested when fully ripe and coming out of hulls with little beating of branches. From bloom to harvest varies from 5 to 6 months. Frequently, the harvest of nuts is facilitated by the use of sheets spread under trees beyond the spread of branches. Such sheets are usually made of heavy cotton sheeting in rectangular pieces ca. 5 × 10 m. Nuts are stored in bags or bins after being cured on trays with hardware cloth bottoms. They trays are placed across supports to allow air circulation. Nuts may also be cured in small burlap bags, provided the bags are arranged so air circulates freely around them. Bags should be turned upside-down occasionally to insure more uniform curing. Much of the work of curing can be eliminated if nuts are allowed to cure in husk before harvesting. Nuts may be stored 2 years without appreciable deterioration, if stored at a temperature of 0°C to 3.5°C. They should be stored as soon as curing is completed, since their quality is impaired at ordinary temperatures, long before rancidity is apparent.[278]

Yields and economics — Trees 8 to 10 years old yield from 2 to 12 (to 350) kg per tree. Improved cvs often yield greater amounts. Trees yielding 1500 to 1600 nuts per tree may have yields of 1,000 to 1,200 kg/ha. As Rosengarten[283] notes, most edible nuts are essentially one-state crops: almonds, pistachios, and walnuts are produced in California; filberts in Oregon, and macadamia nuts in Hawaii. The pecan, on the other hand, is a multi-state crop, stretching across the country from the Southeast to the Southwest throughout some 20 states. U.S. production is tabulated in Table 1. By 1981, a record harvest of nearly 175,000 tons was reported.

Energy — Even native pecans (up to 75% oil) are estimated to yield 750 to 800 kg/ha

Table 1
UTILIZED PECAN PRODUCTION TONS (approx.)

	1978	1979	1980
Alabama	11,000	2,000	10,000
Arkansas	1,600	750	450
Florida	2,100	1,300	3,000
Georgia	67,500	32,500	52,500
Louisiana	4,500	8,000	7,000
Mississippi	5,000	1,250	2,250
New Mexico	7,500	7,350	7,350
North Carolina	2,000	650	850
Oklahoma	7,750	5,000	1,750
South Carolina	3,000	1,000	1,100
Texas	13,000	45,500	5,500
Total U.S. Production	**124,950**	**105,300**	**91,750**

After Rosengarten, Jr., F. *The Book of Edible Nuts*, Walker and Company, New York, 1984, 384.

in Texas.[46] Cultivars may exceed 1000 kg. Prunings and thinnings make very good fuel wood. Perhaps even the leaves could be investigated as sources of lauric acid, juglone (herbicide), leaf protein, with the residues going into ethanol production.

Biotic factors — Squirrels may destroy large quantities of nuts during the season, since they start feeding on them while the nuts are immature and continue until the nuts are harvested. They also gnaw the bark off new shoots so that they die. Squirrel guards on trees may effectively control squirrel damage. Because pecan trees are not sufficiently self-pollinating, various cvs should be interplanted. Orchards should be laid out and cvs planted to allow pollination to occur in the direction of prevailing wind. Although the pollinating cvs need be only about 100 m from female trees, they are often alternated with each other. Phillips et al.[252] give an interesting illustrated account of the insects and diseases of the pecans. The following are known to cause diseases of pecans: *Agrobacterium tumefaciens, Articularia quercina, Aspergillus chevalieri, Botryosphaeria bergeneriana, B. ribis, Caryospora minor, Cephaleuros virescens, Cercospora fusca, Cladosporium effusum, Coniothyrium caryogenum, Elsinoe randii, Eutypa heteracantha, Glomerella cingulata, Gnomonia caryar, G. dispora, G. nerviseda, Helicobasidium purpureum, Microcera coccophila, Microsphaera alni, Microstroma juglandis, Mycosphaerella caryigena, M. dendroides, Myriangium duriaei, M. tuberculans, Nematospora coryli, Pellicularia koleroga, Pestalotia uvicola, Phyllosticta convexula, Phymatotrichum omnivorum, Physalospora fusca, P. rhodina, Phytophthora cactorum, Schizophyllum commune, Septoria caryae,* and *Trichothecium roseum.*[4,186,278] Pecan is attacked by the parasitic flowering plant, mistletoe, *Phoradendron serrulata*. Insect pests attacking pecan include: *Myzocallis fumipennellus* (black pecan aphid), *Chrysomphalus obscurus* (obscure scale), *Curculio varyae* (pecan weevil), *Hyphantria cunea* (fall webworm), *Synanthedon scitulae* (pecan borer), *Acrobasis caryae* (pecan nut-case borer), *Acrobasis palliolella* (pecan leaf-case borer), *Laspeyresia caryana* (hickory shuck worm), *Coleophorae caryaefoliella* (pecan cigar-case borer), *Gretchena bolliana* (pecan bud moth), *Phylloxera devastatrix* (leaf and stem galls), *Strymon melinus* (cotton square borer). Nematodes isolated from pecan trees include: *Caconema radidicola, Ditylenchus intermedius, Dolichodorus heterocephalus, Heterodera marioni, Meloidogyne* spp., *Pratylenchus penetrans, Radopholus similis,* and *Xiphinema americanum.*[4,278]

CARYOCAR AMYGDALIFERUM Mutis (CARYOCARACEAE) — Mani, Achotillo, Cagui, Chalmagra

Uses — Fruits edible, said to taste like almonds. Pulp of fruit is also used as a fish poison. According to the NAS,[229] *Caryocar* kernels are said to be the best edible nuts in the tropics. Oil used for cooking in tropical America.[379]

Folk medicine — Fruits are used as a medicine for leprosy.

Chemistry — Wood, possibly of this species, possibly of *C. brasiliense*, contains 1.5 to 1.8% essential oil.

Description — Trees to 55.0 m tall, the trunk buttressed up to 3.0 m, the young branches sparsely puberulous-glabrescent. Leaves trifoliolate; petioles 2.5 to 11.0 cm long, glabrescent, terete; leaflets shortly petiolulate, the terminal petiolule 5.0 to 7.0 mm long, the lateral petiolules slightly shorter than the terminal one, the petiolules sparsely puberulous, shallowly canaliculate; the laminas elliptic to oblong, slightly asymmetrical, acuminate at apex, the acumen 1.0 to 1.5 cm long, cuneate to subcuneate and often markedly unequal at base, unevenly coarsely serrate at margins, glabrous on both surfaces, the terminal lamina 7.5 to 12.0 cm long, 2.5 to 5.5 cm broad, the lateral laminas slightly smaller than the terminal one; primary veins 10 to 11 pairs, plane to prominulous beneath; venation prominulous beneath; stipels to 5.0 mm long, ellipsoid, inflated, persistent. Peduncles ca. 3.5 to 7.0 cm long, glabrous. Inflorescences clustered racemes, the rachis tomentose, the pedicels elongate, ebracteolate. Calyx cupuliform, ca. 6.0 mm long, glabrous on exterior, the lobes 5, small, rounded, the margins ciliate. Corolla lobes 5, ca. 2.0 to 2.5 cm long, oblong, glabrous, greenish-yellow. Stamens numerous, ca. 200, the filaments shortly united at base in a ring, but into groups, white, sparsely pubescent, the apical portion tuberculate, the innermost filaments much shorter than the rest, the anthers small. Ovary globose, glabrous on exterior, 4-locular. Styles 4, filamentous, shorter than filaments. Fruit globose-ellipsoid, ca. 5.5 cm long, exocarp glabrous, smooth; pericarp thick, fleshy; mesocarp and endocarp enveloping the seed to form an ovoid stone; the exterior of mesocarp not seen, the interior enveloping the endocarp tubercules; endocarp with numerous flattened tubercules ca. 5.0 mm long, and a hard woody interior ca. 1.0 mm thick, glabrous within.[264]

Germplasm — From the South American Center of Diversity, achotillo has been reported to tolerate acid soils.[82] Regrettably, this has been confused in the literature, due to orthographic similarities, with Peruvian *C. amygdaliforme* Don. (almendro blanco). The *mani* has inflated stipels to 5 mm long, the Peruvian species lacks stipels.[264]

Distribution — Native to the forests of the Magdalena River Valley of Colombia.[278] Sturtevant[317] assigns it to Ecuador and says it is the ‘‘almendron’’ of Mariquita.

Ecology — Tropical forest tree, thriving in rich loam in river valleys.[278] Duke[82] reports the species from Tropical Moist to Wet Forest Life Zones, annual precipitation of 23 to 40 dm, annual temperature of 23 to 27°C, and pH of 5.0 to 5.3.

Cultivation — Not known in cultivation.[278]

Harvesting — Fruits collected in season for food and medicinal purposes by natives.[278]

Yields and economics — Of limited use by natives in Colombia.[278]

Energy — Like other tropical tree species, this one probably can produce 25 MT biomass per year. Prunings could be used for energy production.

Biotic factors — No serious pests or diseases reported for this tree.[278] Probably bat pollinated.

CARYOCAR NUCIFERUM L. (CARYOCARACEAE) — Suari Nut, Butternut

Uses — This is probably one of the most popular edible nuts in the genus *Caryocar*. Without voucher material, we can only guess to which species the various data refer. After reading Prance and da Silva's excellent monograph,[264] this author has done his best. Certainly this is the largest, if not the oiliest and tastiest, of the nuts in the genus. The timber of the roots is used for making crooks in boats and for canoes.[56]

Folk medicine — The bark of this or one of the species confused with this is considered diuretic and febrifuge.[91]

Chemistry — Apparently all the species have a high oil content in the pericarp and kernel. The pericarp oil is suggestive of palm oil.[128]

Description — Large tree to 45.0 m tall, young branches glabrous. Leaves trifoliolate, petioles 4.0 to 9.0 (to 15.0) cm long, terete to flattened, glabrous; leaflets petiolulate, terminal petiolule 7.0 to 20.0 mm long, lateral petiolules about equal to the center one; petiolules glabrous, shallowly canaliculate; laminas elliptic, acuminate at apex, acumen 5.0 to 15.0 mm long, entire to weakly crenate at margins, rounded to subcuneate at base, glabrous on both surfaces, terminal lamina 12.0 to 30.0 cm long, 6.0 to 18.0 cm broad, lateral laminas equal or slightly smaller than terminal one, primary veins 8 to 13 pairs, plane above, prominent beneath; venation prominulous beneath; stipels absent. Peduncles 6.0 to 10.0 cm long, glabrous, sparsely lenticellate towards base. Inflorescences of clustered racemes, rachis 1.0 to 4.5 cm long, glabrous; flowering pedicels 4.0 to 6.0 cm long, 5.0 to 8.0 cm thick, glabrous, ebracteolate. Calyx campanulate, ca. 2.0 cm long, glabrous on exterior, lobes 5, rounded. Corolla ca. 6.0 to 7.0 cm long, elliptic, glabrous, deep-red on exterior, paler within. Stamens extremely numerous, over 700, filaments caducous as a unit, united at base up to 2.0 mm, dividing into fused groups before becoming free above, outer ones 7.0 to 8.5 cm long including base, yellow, apical portion tuberculate, with many shorter inner filaments from 3.5 cm long and of all intermediate sizes, inner filaments tuberculate at apex only, anthers small. Ovary globose, 4-locular, glabrous on exterior. Styles 4, filamentous, 8.0 to 9.0 cm long, glabrous. Fruit subglobose to sublobate, to 15.0 cm long, exocarp glabrous, lenticellate; pericarp very thick and fleshy, detaching from mesocarp and endocarp; mesocarp and endocarp enveloping seed to form a large stone ca. 7.0 cm broad, 5.0 cm long, mesocarp becoming lignified and hard, the exterior undulate with short, rounded tubercules; endocarp with tuberculate exterior and hard, thin, woody interior ca. 1.0 mm thick; with 1 to 2 subreniform seeds only developing.[264]

Germplasm — Reported from the South American Center of Diversity. The tuberculate large fruits and large flowers are larger than those of any other *Caryocar*.[264]

Distribution — Native of the primary forests of the Guianas and adjacent Venezuela and Brazil. Recently collected in Panama and Choco, Colombia. Apparently abundant in Choco. Cultivated in the West Indies, and grown in botanical gardens in Nigeria, Singapore, and Sri Lanka.[264]

Ecology — According to MacMillan,[196] the tree grows well in the moist low country of Sri Lanka, especially in rich deep loams or alluvial soils.

Cultivation and Energy — No data available.

Harvesting — According to Burkill,[56] it may fruit at 5 years of age, but usually takes 2 to 3 times as long. Introduced into Singapore in 1899, it did not fruit until 20 years old, but flowered years before. At Peradeniya and Henaratgoda, where it was introduced in 1891, the trees had not fruited when MacMillan[196] went to press, though the Peradeniya specimen started flowering after 19 years. These do not seem to be much more precocious than Brazil nuts. However, MacMillan mentions another specimen from British Guiana which fruited 6 years from planting.

Yields and economics — The nut is exported commercially from the Guianas.

Biotic factors — Probably bat pollinated.

CARYOCAR VILLOSUM (Aubl.) Pers., *CARYOCAR BRASILIENSE* Camb., and *CARYOCAR CORIACEUM* Wittmack (CARYOCARACEAE) — Pequi

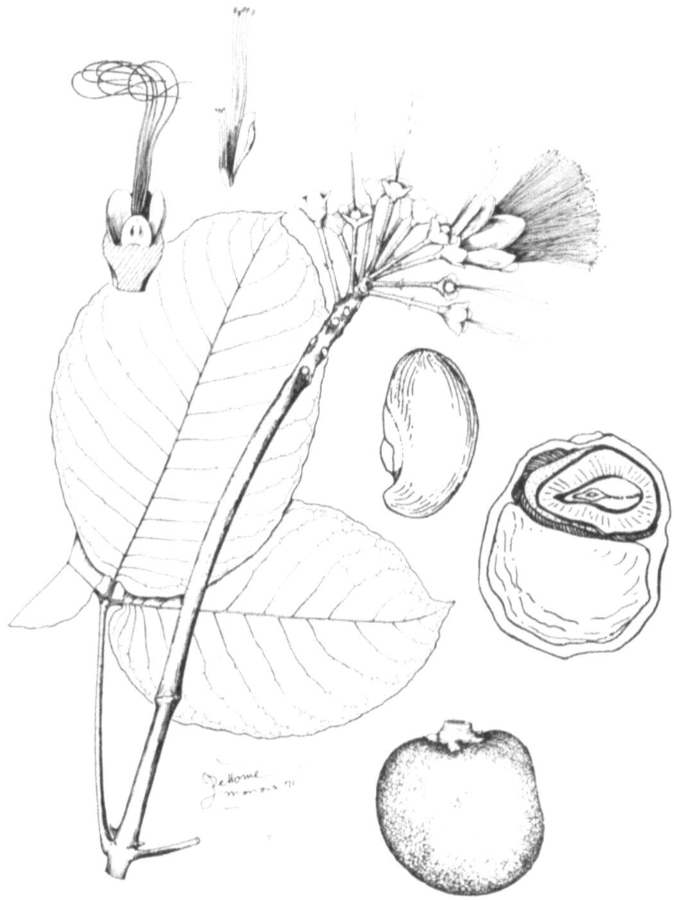

Uses — Several species go under the common name pequi and pequia, said to be one of the best edible nuts in the tropics. But Sturtevant[317] calls it is a sort of chestnut eaten in times of famine. *Caryocar* has several nut-bearing species. These are somewhat more promising because some species are smaller and easier to harvest. The orange-sized fruit contains an oily pulp and kernel that are used for food. So far, they have been employed only in home cooking. The fruit is made into a tasty liqueur, well known in Brazil, especially in the State of Mato Grosso. There is both the fruit oil and the kernel oil. After refining the taxonomy of those species called Pequi and Pequia, Prance and da Silva[264] state that the fruit of *C. villosum* (a huge tree) has an edible pulp and edible cotyledons. The pulp is most often eaten and has a faint smell of rancid butter. It is also used to produce an edible oil. Wood of *C. villosum* is so durable as to be used in boat-building and in heavy construction. Brazilian Indians obtain a yellow dye from *C. brasiliense*.[264]

Folk medicine — The bark of this or one of the species confused with this is considered diuretic and febrifuge.[91]

Chemistry — According to Hager's Handbook,[187] the wood of *C. brasiliense* (or *C. amygdaliferum* or *C. glabrum*) contains 1.5 to 1.8% essential oil. According to a report quoted in Burkill,[56] the inner part of the fruit-wall contains a reddish-orange oil, up to 72.3%. The kernel contains 61.4% (ZMB) or 45% (APB) oil, composed largely of glyceride esters of palmitic and oleic acids. Ripe fruits must be treated as soon as harvested, or enzymes will induce the development of free fatty acids. Lane[174] reported a comparison with Malayan palm oil:

	Palm oil (%)	Pequi pericarp (%)	Pequi kernel (%)
Myristic	2.5	1.5	1.5
Palmitic	40.8	41.2	48.4
Stearic	3.6	0.8	0.9
Oleic	45.2	53.9	46.0
Linoleic	7.9	2.6	3.3

Hilditch and Williams[128] present somewhat different data. The fruit-coat fat of *Caryocar villosum* is interesting, because its fatty acids closely resemble those of palm oils, namely: myristic 1.8, palmitic 47.3, stearic 1.7, oleic 47.3, linoleic 1.9%. It contained only 2% of fully saturated components (tripalmitin), thus differing somewhat from palm oils of similar fatty acid composition. No tristearin was detected in the completely hydrogenated fat and the components of the fat (in addition to 2% tripalmitin) were therefore 42% oleodipalmitins and 56% palmitodioleins — an instance of pronounced "even distribution." Intensive crystallization of the pequia fruit-coat fat yielded five fractions very rich in oleodisaturated glycerides and three more soluble fractions which consisted largely of diunsaturated glycerides. Oleodipalmitin was isolated separately from each of the five fractions and agreed in its transition and melting-points with 2-oleodipalmitin, while the hydrogenated products in each case were 2-stearodipalmitin. The symmetrical form of oleodipalmitin was thus exclusively present. Similar examination of the palmitodistearins obtained by hydrogenation of the palmitodioleins in the three more soluble fractions showed, in contrast, that the latter were present in both the symmetrical and the unsymmetrical configuration, the amounts of each positional isomeride being probably of the same order.[128]

Description — Large tree to 40.0 m tall and up to 2.5 m diameter, the young branches villous-tomentose, becoming glabrous with age. Leaves trifoliolate; petioles 4.0 to 15.0 cm long, villous-tomentose to puberulous, terete to slightly striate; leaflets shortly petiolulate, the terminal petiolule 3.0 to 6.0 mm long, the lateral petiolules 2.0 to 4.0 mm long; petiolules puberulous when young, canaliculate; the laminas elliptic, acuminate at apex, the acumen 3.0 to 10.0 mm long, serrate to crenate at margins, rounded to cordate at base, villous to glabrous above, densely villous-hirsute or with a sparse pubescence on the venation only beneath, the terminal lamina 8.0 to 11.0 cm long, 6.0 to 12.0 cm broad, the lateral laminas slightly smaller; primary veins 12 to 19 pairs, slightly impressed or plane above, prominent beneath; venation extremely prominent beneath; stipels absent. Peduncles 5.0 to 13.0 cm long, tomentellous or puberulous when young, glabrescent, lenticellate. Inflorescences of clustered racemes, the rachis 3.0 to 4.0 cm long, tomentose when young; flowering pedicels 1.8 to 3.5 cm long, puberulous to glabrous, with 2 membraneous subpersistent bracteoles. Calyx campanulate-cupuliform, ca. 1.5 cm long, gray puberulous to glabrous on exterior, the lobes 5, rounded. Corolla ca. 2.5 cm long, the lobes 5, oblong-elliptic, pale yellow. Stamens numerous, ca. 300, the filaments shortly united into a ring at base but not into groups, subpersistent, of two distinct lengths with several of intermediate lengths, the longest ca. 6.5 to 7.0 cm long, yellow, the apical 1.0 to 3.0 mm tuberculate, the shortest ca. 55, 1.0 to 1.5 cm long, with distinct fused portion at base, tuberculate entire length, the anthers small. Ovary globose, 4-locular, glabrous on exterior. Styles 4, filamentous, equalling filaments, glabrous. Fruit oblong-globose, 6.0 to 7.0 cm long, 7.0 to 8.0 cm broad; exocarp glabrous, lenticellate; pericarp thick, fleshy, detaching from mesocarp and endocarp; mesocarp and endocarp enveloping seed to form a reniform stone ca. 5.0 cm broad, the exterior of mesocarp smooth and undulate, the interior enveloping endocarp spines; endocarp with numerous fine spines ca. 3.0 mm long, and a hard wood interior, ca. 1.0 mm thick.[264]

Germplasm — Reported from the South American Center of Diversity, the pequi is sensitive to wind damage. Prance and da Silva[264] key *C. villosum* with acuminate leaflets, *C. brasiliense* as with rounded or acute leaflets. *C. coriaceum* is also found in the complex known to Brazilians as Pequi.

Distribution — French Guiana and Amazonian Brazil (*C. villosum*). Dry woodland of the northern and eastern part of the Planalto of central Brazil (*C. coriaceum*). Brazil and adjacent Bolivia and Paraguay (*C. brasiliense*). Cultivated in Singapore and Sumatra (*C. villosum*).[264]

Ecology — Grows above flood level in the Amazon valley (Burkill, *C. villosum*).

Cultivation — Wickhan, in Lane,[174] figured the trees should be spaced at 100 trees per ha.

Harvesting — Trees have grown to 18 m in 9 years.

Yields and economics — I quote exactly from Wickhan's letter, as quoted by Lane:[174] "Reckoning the fruit as giving some 3/4 lb. of fat, the yield per acre should be from 1300 lbs. to 1/2 tons, it will therefore be at once apparent that this greatly exceeds any existing source of supply — coconut (copra), palm kernels, etc..."

Energy — The husk of the fruit is used, like coconut husks, for fuel, either directly or after conversion to charcoal. Prunings could also be used for fuel.

Biotic factors — *Caryocar villosum* is bat pollinated, with two or three of the many flowers in a given influorescence opening at night, shortly after dark. The pollination process is described in Prance and da Silva.[264]

CARYODENDRON ORINOCENSE Karst. (EUPHORBIACEAE) — Inche, Cacay, Nambi, Arbol de Nuez, Kakari Taccy Nut

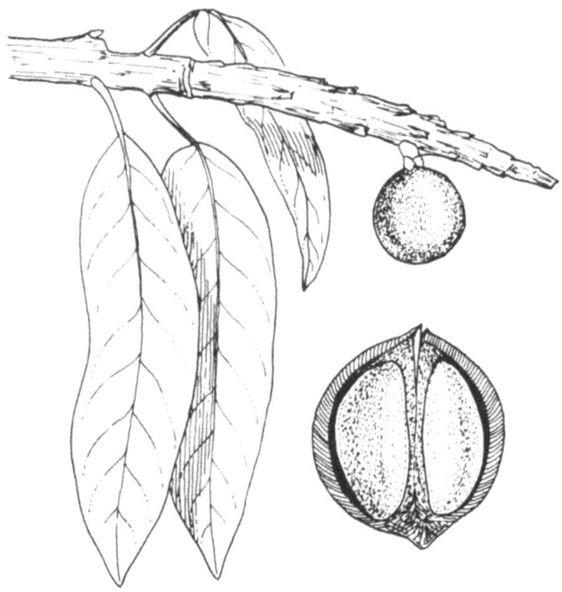

Uses — According to Garcia-Barriga,[105] the oil is used like olive oil, while the toasted seed is very flavorful and nutritious. According to Schultes,[298] the oil is valued for a wide range of uses, from cooking to soap-making and cosmetics. The thin, brown shell surrounding the kernel, is easily broken with the fingers.[283] The tree has been suggested as a plantation crop for Latin America.[105]

Folk medicine — In the Llanos of Colombia, the oil is painted onto skin afflictions. A half-ounce dose is taken as a laxative. It is believed to fortify the lungs.[105]

Chemistry — Seeds contain ca. 50% of a yellowish oil, the husk 17.1%, the pulp 82.9%. The pulp contains 6.6% water.

Description — Tree to 20 m tall, the crown conical; trunk yellowish-ochraceous, striate, with a watery pinkish-yellow latex. Leaves alternate, glabrous, narrowly elliptic or obovate, the margins entire, revolute, 12 to 25 cm long, 4 to 10 cm wide, apically acute, basally cuneate; primary veins ca. 7 to 11; biglandular at the base of the blade. Petiole glabrous, canaliculate above, dilated at both ends, 1 to 5.5 cm long. Flowers unisexual, the male in terminal racemes, with 3 concave tepals, 4 conical glabrous stamens, longidehiscent; disk white. Female flowers with 5 to 6 ovate tepals; ovary trilocular, triovulate, stigma short and trilobate, disk annular, trilobate. Fruit globose-oblong, 5 to 6 cm long, 4 to 5 cm broad, usually 3-seeded. Seeds 3 cm long, 1.7 cm wide.[105,256]

Germplasm — From the South American Center of Diversity, inche is reported to tolerate poor soils in lateritic and savanna situations.

Distribution — According to PIRB,[256] the species is native to the Llanos Orientales and Putumayo of Colombia. I'm told there are plantations in Ecuador, and I have seen plants, apparently thriving, in the humid climate of Talamanca, Costa Rica.

Ecology — I estimate inche ranges from Tropical Dry to Wet through Subtropical Dry to Wet Forest Life Zones, tolerating annual precipitation of 15 to 60 dm, annual temperature of 23 to 29°C, and pH of 4.5 to 7.5. Said to be of the tropical humid zone,[105] ranging from 300 to 1000 m above sea level,[298] where the drier season lasts at least 4 months.

Cultivation — Propagated by seed, the tree has been suggested as an oilseed plantation crop.

Harvesting — Said to start bearing in 4 to 5 years. The determinate height of the tree is said to facilitate harvest.

Yields and economics — According to Garcia-Barriga,[105] each tree produces 280 to 300 kg of fruit, which at the density of 50 trees per ha would calculate to 15 MT fruit. In round figures, 25 fruits would weigh 375 g, or 15 g per fruit. Of that 15 g, there would be about 8 g husk, 1 g testa, and 6 g of kernel. That 6 g kernel should contain about 2 to 3 g oil. This suggests a conversion factor of 20% oil in the fruit. According to PIRB,[256] the cost of establishment and maintenance should be less than that of African Oil Palm. Since the yield is similar, possibly inche could return equal or greater profits.

Energy — If we accept the speculative yield and conversion figures derived above, there could be 3 MT oil, half the expected yield of oilpalm. But, if the same yields were obtained with 100 trees per hectare instead of 50, there could be as good a yield here as with oil palm, with 6 MT oil and possibly 12 MT edible seedcake.

Biotic factors — No data available.

CASTANEA CRENATA Sieb. and Zucc. (FAGACEAE) — Japanese Chestnut, Juri
Syn.: *Castanea stricta* Sieb. and Zucc., *Castanea pubinervis* (Hassk.) C. K. Schneid., and *Castanea japonica* Blume.

Uses — Kernel of nut used as food by Chinese and Japanese, both for humans and for fattening swine.[278] Nut shell extract, bur, and bark used for staining. Male flower used to stain cloth a red-brown color.[343] Wood strong, very hard, heavy, durable in soil, used in Japan for furniture, cabinet work, railroad ties, and in ship-building. Planted in southern Europe for timber. Well adapted for ornamental planting.[278]

Folk medicine — In China and Korea, flowers are used for tuberculosis and scrofula. Decoction of fresh leaves said to allay skin irritation caused by lacquer. Root used for hernia. An ointment for boils made with powdered charcoal from involucres mixed with oil.[249]

Chemistry — Per 100 g, the seed (ZMB) is reported to contain 399 calories, 7.0 g protein, 1.4 g fat, 89.2 g total carbohydrate, 2.3 g fiber, 2.3 g ash, 79.8 mg Ca, 188 mg P, 3.8 mg Fe, 37.6 mg Na, 0.89 mg thiamine, 0.42 mg riboflavin, 3.76 mg niacin, and 68.1 mg ascorbic acid.[89]

Description — Small tree or shrub, often less than 10 m tall, but occasionally much larger, up to 17 m, attaining great girth, with many spreading limbs and slender branches; young shoots at first densely gray-white with short hairs, becoming glabrous or sparsely velutinous; leaves at first densely stellate pubescent all over, retaining on under-surface some pubescence or becoming glabrous, puberulous on veins above, elliptic to oblong-lanceolate, or narrowly oblong, with long acuminate tip and cordate or round at base, margin crenate-serrate or subentire with 10 to 25 bristle teeth on each side, 8 to 16 cm long, 3 to 5 cm broad, thick and heavy, quite crinkly, dark lustrous green above, grayish-green beneath; petiole pubescent, about 2 cm long, stipules soon deciduous, lanceolate, acuminate, gradually broaden at base; winter buds short, ovoid, glabrous, shining crimson; staminate spikes 5 to 20 cm or more long, densely flowered, yellowish-white, erect or suberect; pistillate flowers clustered among the male spikes, occurring in involucres about 5 mm thick, styles exserted, about 3 mm long, densely covered with ascending long gray hairs; bur small in wild types, in cultivated types often to 6 cm in diameter, with long, almost glabrous spines; nuts 2 to 3 per bur, hilum occupying whole basal area.[278]

Germplasm — Reported from the China-Japan Center of Diversity, Japanese chestnut, or cvs thereof, is reported to tolerate disease, frost, and slope.[82] Several Japanese varieties are grown extensively, as 'Alpha', 'Reliance', and 'Parry', the last being a hybrid with *C. dentata,* suitable for planting in California. Other Japanese varieties include: 'Advance', 'Beta', 'Biddle', 'Black', 'Col', 'Eureka', 'Felton', 'Hale', 'Kent', 'Kerr', 'Killan', 'Martin', 'McFarland', 'Prolific', 'Success', and 'Superb'.[278] ($2n = 22,24$.)

Distribution — Native to Japan (Honshu, Shikoku, Kyushu) and Korea. Much planted in Japan for the nuts. Introduced and extensively planted in southern Europe for timber.[228] Introduced to the U.S. in 1876.[208] Hardy as far north as Massachusetts.[278]

Ecology — Ranging from Cool Temperate Moist to Rain through Warm Temperate to Moist Forest Life Zones, Japanese chestnut is reported to tolerate annual precipitation of 9.4 to 23.4 dm (mean of 6 cases = 13.3), annual temperature of 9.9 to 15.8°C (mean of 6 cases = 12.9°C), and pH of 5.0 to 6.8 (mean of 5 cases = 5.9).[82] Trees grow best on well-drained, porous soil, with deep porous subsoil. Withstand temperatures and rainfall of most temperate climates.[278] Hardy to Zone 6.[343]

Cultivation — Trees propagated by whip-grafting to American chestnut (*C. dentata*). American species usually cut down, and the sprouts springing from the remaining roots, when 1.3 to 2 cm in diameter, are grafted with desired varieties of Japanese chestnut. Whip and cleft methods of grafting are used. Trees already grafted with desired varieties may be obtained for the orchard. Seedlings may be top-worked with the permanent kinds after they have become established. Trees set out not less than 10 m apart each way. Trees may be planted closer at first and thinned out for permanent spacing in 10 to 15 years. Meanwhile, trees may be intercropped with vegetables or small tree crops. Two-year old grafts are commonly loaded with burs. It is good practice to keep burs picked from young trees for 3 to 4 years to allow trees to become well-established before crop production is started. If trees are allowed to over-bear, nuts run down in size. Japanese varieties do not abort their burs, and seem to be completely self-fertile.[278]

Harvesting — Trees are very productive and begin to fruit commercially when 6 years old. Nuts are picked from the ground, dried, and stored until marketed or used.[278]

Yields and economics — No yield data available, but all records state that trees are preccocious and very productive. Great quantities of Japanese chestnut are grown and consumed in Japan and China.[278]

Energy — Wood, burs, and husks may be used for fuel or charcoal production.

Biotic factors — The following fungi are known to attack Japanese chestnut: *Actinopelte japonica, Botryosphaeria ribis, Capnodium salicinum, Cronartium quercinum, Cryptodiaporthe castanea, Daedalea quercina, Endothia nitschkei, E. parasitica, E. radicalis, Fomes melanoporus, Fomitopsis castanea, Gloeosporium castanicola, Helicobasidium mompa, Laestadia orientalis, Leszites betulina, Limacinia cheni, Microsphaera alni, Monochaetia desmazierii, M. pachyspora, Ovularia castaneae, Phyllactinia quercus, Phytophthora cambivora, P. cinnamomi, Polyporus cinnabarinus, P. gilvus, P. hirsutus, P. nidulans, P. pargamenus, P. rhodophaeus, P. tulipiferus, P. versicolor, Polystictus hirsutus, P. sanguineus, Puccineastrum castaneae, Pycnoporus coccineus, Schizophyllum commune, Septoria gilletiana, Stereum gausapatum, Trametes dickinsii, T. vittata.* Also *Bacterium castaneae* attacks trees. However, plants are resistant to Eastern Filbert Blight.[186,275] More susceptible to chestnut blight fungus, *Endothia parasitica,* than the Chinese species, *C. mollissima.* Trees may deteriorate slowly or be killed before reaching maturity.[208]

CASTANEA DENTATA (Marsh.) Borkh. (FAGACEAE) — American or Sweet Chestnut
Syn.: *Castanea americana* Raf.

Uses — Native and cultivated trees provide nuts which are sweeter than Old World types. Nuts are gathered and sold in eastern U.S. markets. Reddish-brown wood light, soft, coarse-grained, elastic, moderately strong, easily split, easy to work, tending to warp on drying, resistant to decay. Used in manufacture of cabinet work, caskets, crates, desks, furniture, interior finishes of houses, pianos, railway ties, ship masts, fence posts, telephone poles, rails, mine timbers, siding for barns and other buildings, paper pulp. Tannin in wood used in tanning extracts. Formerly planted in eastern U.S. as an ornamental and for timber, as well as for nuts.[278,324]

Folk medicine — Reported to be astringent, sedative, tonic, and vermifuge, American chestnut is a folk remedy for dysentery and pertussis.[91] Leaves have sedative properties.[278] Indians used the bark to treat worms and dysentery.[168]

Chemistry — Leaves contain 9% tannic acid, which is colored green with ferric salts, and a mucilage insoluble in alcohol.[278] Wood contains from 6 to 11% tannin.[324] Chestnuts are starchy nuts, containing ca. 40 to 45% carbohydrates and less than 1% oil, as compared with pecans with 70% oil and other tree nuts with ca. 60% oil.[278] Nuts contain ca. 1,700 calories/lb.[283] According to Woodroof,[341] chestnuts contain no oil and are very high in carbohydrates, especially starch, making them more easily digestible than other nuts. Ranging from 21 to 25% shells, 4.5 to 6.5% moisture, and 69 to 72% dry matter, native chestnuts are reported to contain 2.66 to 2.72% ash, 12.20 to 12.23% total protein, 2.84 to 3.63% fiber, 65.03 to 66.16% total nitrogen-free extract, and 16.08 to 16.42% fat.[341]

Description — Deciduous tree, up to 50 m tall; trunk straight, columnar, 1 to 1.3 m in diameter; when uncrowned the trunk is shorter and 3.3 to 4 m in diameter; round-topped, with horizontal limbs spreading to 30 m across; branchlets at first yellow-green, tinged with red and puberulous, becoming olive-green and glabrous, eventually becoming dark brown; winter-buds ovoid, about 0.6 cm long, with dark-brown scales, scarious on margins; bark 2.5 to 5 cm thick, dark-brown, deeply ridged with irregular, often interrupted fissures; leaves oblong-lanceolate, apex acute, acuminate, base gradually narrowed and cuneate, 15 to 20 cm long, about 5 cm broad, when young yellow-green and puberulous on upper surface and tomentose beneath, becoming glabrous at maturity, turning yellow late in fall; petioles about 1.3 cm long, slightly angled, puberulous, often reddish; stipules ovate-lanceolate, puberulous, about 1.3 cm long; staminate aments at maturity 15 to 20 cm long, with crowded flower-clusters; androgynous aments slender, puberulous, 6 to 12.5 cm long, with 2 or 3 involucres if pistillate flowers near base; but 5 to 6.5 cm in diameter, covered with glabrous much-branched spines, opening with frost and gradually shedding nut; nut much compressed, 1.3 to 2.5 cm in diameter, broader than long, with thick pale tomentum at apex or nearly to middle, interior of hull lined with rufous tomentum; kernel very sweet. Root system extensive both laterally and vertically. Flowers with strong fragrance, June to July.[278]

Germplasm — Reported from the North American Center of Diversity, American chestnut or cvs thereof, is reported to tolerate drought, frost, heat, poor soil, sand, slope, and weeds.[82] Some varieties grown for nuts are 'Ketchem', 'Watson', and 'Griffin'.[324] Hybrids with blight-resistant Chinese and Japanese species have led to several mixed varieties not in cultivation. Continuous efforts to find immune or resistant strains and repeated attempts to produce resistant hybrids resembling it have failed to give varieties considered safe to plant. 'Clapper' is a hybrid from a cross of Chinese-American hybrid backcrossed to the American chestnut, and is a rapid-growing timber-type.[103,115,278]
($2n = 24$.)

Distribution — Native throughout the eastern U.S. from southern Maine to southern Ontario, south to Delaware, Ohio, southern Indiana, and along the mountains to northern

Georgia and western Florida, from sea-level in Massachusetts to 1,300 m in North Carolina, reaching its greatest height in western North Carolina and eastern Tennessee. Until about 1905, chestnut was important for its durable wood and its nuts. Trees were nearly completely destroyed by the Chestnut Blight, a fungus bark disease (*Endothia parasitica*). Sprouts and shrubby growth from bases of wild trees still appear and sometimes persist long enough to produce fruit.[278]

Ecology — Ranging from Cool Temperate Steppe to Moist through Warm Temperate to Moist Forest Life Zones, American chestnut is reported to tolerate annual precipitation of 4.9 to 11.6 dm (mean of 3 cases = 8.3), annual temperature of 8.4 to 12.5°C (mean of 3 cases = 9.9), and pH of 5.6 to 7.3 (mean of 3 cases = 6.5).[82] Thrives on a variety of soils, from almost pure sand to coarse gravels and shales. Does not grow well on limestone. Prefers dry, well-drained, rocky land of the glacial drift to the richer, more compact alluvial soil of lowlands. Chestnut does not need a rich soil so much as one whose physical structure insures good drainage. Light is essential to the tree, since it is somewhat intolerant to shade. Grows best in a cool climate, but can endure heat and dry sunny situations.[278] Hardy to Zone 4.[343]

Cultivation — Propagation is by direct seedling or by use of nursery-grown seedlings. To prevent drying out and lowering of germination ability, collected seeds should be kept stratified in moist sand until the following spring. The nursery should be located on fresh, well-drained, fertile soil. Thorough cultivation of soil is required. Seed should be planted about 2.5 cm apart in rows about 45 cm apart, at a depth of 2.5 cm. Bushel contains 6,500 to 8,000 nuts, sufficient to plant about 200 m of nursery rows, and to produce about 4,000 plants. While in the nursery, seedlings require careful cultivation and should be kept weed-free. When planting in permanent sites, trees should be set 2 m apart each way. If trees are to be grown directly from seed without transplanting, seed spots should be prepared, spaced as above. Two or three seeds should be planted in each and covered about 2.5 cm deep with fine soil. Only one tree should be allowed to remain in each hill. Little cultivation is necessary after trees become established. Seedlings grow 25 to 37 cm by the end of the first season and 37 to 50 cm per year until the 13th year.[278]

Harvesting — Mature nuts should be gathered every other day during the period of maturity. Frequent and clean collection of nuts is especially important if nuts are likely to be infested with weevils or if weather is hot and dry. Within a week's time, nuts on the ground, or those in opened burs on trees, may become dry or they may mold and spoil. In harvesting nuts from a tree, it is advisable to first knock nuts from opened burs with pole and then gather up nuts from ground. If harvested nuts are infested with weevils, they should be treated by immersing them in water at 49°C for 30 to 45 min, depending on size of nuts. After heating for proper length of time, nuts are removed immediately from hot water and permitted to dry. Nuts, gathered and treated, are spread out in a layer no more than 3 or 4 nuts deep to cure, on a floor or in trays in a well-ventilated building. Nuts should receive no direct sunlight. Time for curing depends upon amount of moisture in air. Usually 1 or 2 days of curing is adequate. Under proper conditions, chestnuts can be stored from time of harvest to late April with minimal spoilage; nuts come out of storage in the same condition as they went in, and they have been found to germinate promptly. Chestnuts for eating may be stored in deep freezers, but they must be cooked promptly after being removed. To freeze the nuts, they should be shelled and the kernels blanched. Nuts must not be roasted slowly, as they will explode in oven or they will not freeze satisfactorily. After shells have been removed, if any of brown skin covering kernel remains, kernels should be blanched. This is done by immersing kernels in boiling water for 1 or 2 min and quickly freezing them in freezer containers. They are then kept −18°C or lower until ready to be cooked, like frozen peas or lima beans. Chestnuts are marketed packaged in cans, woven bags of cotton or jute, or in baskets having tight-fitting lids.[278] Leaves are collected in September and October.[210]

With smaller nuts, the American chestnut, had it survived productive, would be expected to yield somewhat less than the European. Smith reports Italian yields of ca. 1100 kg/ha, French of 1500, and Spanish of ca. 2800 kg/ha.[310]

Yields and economics — Yield varies from 6,500 to 8,000 nuts per bu. At present time, American chestnut is not an item of commercial importance, either for the nuts or for timber, as it was prior to 1907 to 1918, when most of the trees in the eastern U.S. were destroyed. Chestnuts grown at present are hybrids of Chinese chestnuts (*C. mollissima*) or other species, resistant to blight.[278]

Energy — Wood and burs may be used for firewood or for the production of charcoal.

Biotic factors — American chestnut is devastated by Chestnut Blight (*Endothia parasitica*), and since 1918 it has not been a commercial product.[193] Weevils in the nuts are a problem.[278] Browne[53] lists the following as affecting this species: Fungi — *Cryptodiaporthe castanea, Daedalea quercina, Endothia parasitica, Phellinus gilvus, Phytophthora dentata, Polyporus frondosus, P. pargamenus, Urnula craterium;* Acarina — *Oligonychus bicolor*; Coleoptera — *Elaphidion villosum*; Lepidoptera — *Acronicta americana, Alsophila pometaria, Anisota virginiensis, Datana ministra, Disphargia guttivitta, Ennomos magnaria, E. subsignaria, Lymantria dispar, Nematocampa dentata, Symmerista albifrons;* and Mammalia — *Lepus americanus, Odocoileus virginianus.* According to Agriculture Handbook 165,[4] the following are reported as affecting *C. dentata: Actinopelte dryina* (leaf spot), *Aleurodiscus acerinus* (bark patch), *Anthostoma dryophilum, Armillaria mellea* (root and butt rot), *Asconidium castaneae, Botryosphaeria ribis, B. castaneae, Cenangum castaneae, C. albo-atrum, Ceratostomella microspora, Chlorociboria aeruginosa, C. versiformia, Chlorosplenium chlora, Clasterosporium sigmoideum, Clitocybe illudens, C. monadelpha, Colpoma quercinum, Corticium caeruleum, Coryneum* spp., *Cronartium cerebrum* (rust), *Cryptodiaporthe castanea* (twig canker), *Cryptospora cinctula, Cylindrosporium castaneae* (Leaf spot), *Daedalea quercina, D. confragosa, Diaporthe eres, Didymella castanella, Diplodia longispora, Discohainesia oenotherae, Discosia artocreas, Endothia gyrosa, E. radicalis, Favolus alveolaris, Fenestella castanicola, F. phaeospora, Fistulina hepatica, F. pallida, Flammula* sp., *Fomes annosus* (root and butt rot), *F. applanatus* (butt rot or on stumps), *F. everhartii* (white spongy heart rot), *F. ohiensis, F. pinicola* (brown crumbly heart rot), *F. scutellatus, Gnomonia setacea, Hymenochaete rubiginosa, Laestadia castanicola, Lenzites betulina, Leptothyrium castaneae, Marssonina ochroleuca, Melanconis modonia, Melanconium cinctum, Merulius fugax, M. tremellosus, Microsphaerea alni* (powdery mildew), *Monochaetia desmazierii* (leaf spot), *M. pachyspora, Mycosphaerella maculiformis, M. punctiformis, Myxosporium castaneum, Panus rudis, P. stipticus, Pezicula purpurascens, Pholiota adiposa, P. squarrosa, Phoma castanea, Phyllosticta castanea, P. fusispora, Physalospora obtusa, Phytophthora cinnamomi, Pleurotus ostreatus, Polyporus* spp., *Rutstroemia americana, Scolecosporium fagi, Sphaerognomonia carpinea, Steccherinum adustum, Stereum* spp., *Strumella coryneoidea, Trametes sepium* (wood rot), and *Xylaria hypoxylon.*[113,186,278]

CASTANEA MOLLISSIMA Blume (FAGACEAE) — Chinese Hairy Chestnut
Syn.: ***Castanea sativa* var. *formosana* Hayata, *Castanea formosana* (Hayata) Hayata, and *Castanea bungeana* Blume.**

Uses — Nut unexcelled in sweetness and general palatability by any other known chestnut. By far the most common food nut, almost taking the place of potato in parts of the Orient. Eaten raw, boiled, roasted, cooked with meat, made into confections, powdered and mixed with candy, dried whole. Valuable for wildlife where nut production is more important than timber. Recommended for hardiness, blight resistance and large nuts. Wood, leaves, and bark used for their tannin content.[278] Has merit as an ornamental tree.[343]

Folk medicine — Reported to be hemostat, Chinese chestnut is a folk remedy for diarrhea, dysentery, epistaxis, nausea, and thirst.[91] The flower is used for scrofula. Stem-bark used for poisoned wounds; the sap for lacquer poisoning. The fruit pulp is poulticed onto animal bites, rheumatism, and virulent sores; husk astringent and used for dysentery, nausea, and thirst; charred husks applied to boils. The root is used for hernia.[90]

Chemistry — Per 100 g, the seed (ZMB) is reported to contain 403 calories, 11.9 g protein, 2.7 g fat, 83.2 g total carbohydrate, 2.2 g ash, 36 mg Ca, 168 mg P, 3.8 mg Fe, 216 μg beta-carotene equivalent, 0.29 mg thiamine, 0.32 mg riboflavin, 1.44 mg niacin, and 65 mg ascorbic acid.[89]

Description — Deciduous trees, long-lived, 15 to 20 m tall, spreading, round-topped; branches glabrous, branchlets covered with dense pubescence of coarse spreading hairs;

leaves with dense stellate pubescence when young, this persistent on under-surface of mature leaves, alternate, 10 to 20 cm long, 5 to 10 cm broad, with 12 to 20 deep serrations on each side, oblong-lanceolate to elliptic-oblong, base rounded or cordate, apex acute to scarcely acuminate; petioles 2 to 3 cm long, with few long hairs; stipules over 2 cm long, very veiny and rugose, abruptly broadened; staminate spikes axillary or terminal, 20 cm long or more; pistillate flowers in hirsute globose involucres 1 cm thick, situated at base of male spikes or occasionally terminating some spikes; styles about 5 mm long, densely hirsute; bur up to 6 cm thick, with long, very stout, strongly pubescent spines; nut about 2.5 cm thick, with small to rather extensive patch of tawny felt at apex; nut with thin skin which peels readily from kernel. Flowers late May to late June in Maryland, earlier southward.[278]

Germplasm — Reported from the China-Japan Center of Diversity, Chinese chestnut, or cvs thereof, is reported to tolerate disease, frost, and slope.[82] Many cvs have been introduced from China, and several hybrids with Japanese and American chestnuts have been produced in attempts to breed-in blight-resistance. But most of them have failed for one reason or another. Cultivars presently in the trade include: 'Abundance', which produces 110 nuts per kg (84 to 167); 'Kuling', 'Meiling', 'Nanking', and 'Carr'; the latter produces 128 nuts per kg, has good cleaning quality, a sweet, pleasing flavor, and was the first variety grafted in this country, but is not grown at the present time. Most of grafted Chinese chestnuts have shown troublesome stock-scion incompatibility, which causes grafts to fail. Such failures may occur in the first year, but more often after 4 to 6 years of vigorous growth. Failure seems to relate to winter injury and is more frequent northward. Seedlings of selected trees, as 'Hemming' from Maryland and 'Peter Lui' from Georgia, are among the most promising. Seedlings of 'Nanking' come true to type and are planted commercially in the South. Other cvs are hardy northward to Maryland, New Jersey, Pennsylvania, and the warmer areas of New York. Two of the more recent hybrids are 'Sleeping Giant Chestnut' (*C. mollissima* × (*C. crenata* × *C. dentata*)) and 'Kelsy Chestnut' (*C. mollissima* ×?). Also, 'Stoke' is a natural Japanese × Chinese chestnut hybrid.[278]

Distribution — Native to north and west China and Korea. First introduced in the U.S. in 1853 and again in 1903 and 1906. This species has been in cultivation in China for a long time. It is practically the only species of chestnut being planted in the U.S. for commercial purposes.[262,278]

Ecology — Ranging from Warm Temperate Dry to Moist through Subtropical Dry to Moist Forest Life Zones, Chinese chestnut is reported to tolerate annual precipitation of 9.4 to 12.8 dm (mean of 5 cases = 11.6), annual temperature of 10.3 to 17.6°C (mean of 5 cases = 14.2°C), and pH of 5.5 to 6.5 (mean of 5 cases = 5.9).[82]

Chinese chestnut requires much the same conditions of climate, soil, and soil moisture as does peach. Air-drainage must be good, and frost pockets must be avoided. Trees grow naturally on light-textured acid soils, but they show a wide range of tolerance for well-drained soils of different textures. Young trees are sensitive to drought and may be killed. Cultivars and hybrids are about as hardy as peach and may be planted in any areas where peaches do well, most withstanding temperatures to −28°C when fully dormant. Unless leaves are removed soon after turning brown, they are apt to become heavily laden with wet snow or ice and cause severe damage. This situation is particularly common at altitudes of 600 to 700 m, as in West Virginia.[278] Hardy to Zone 4.[343]

Cultivation — Propagated readily from seed, from which selections are made. Nuts lose their viability quickly after harvesting. Seeds may be germinated in nursery beds and the seedlings planted out after two years.[278] As they do not compete well with weeds, young trees should be kept cultivated for the first few years.[343] Trees lend themselves readily to orchard culture, although trees are not particularly vigorous. Trees are self-sterile; in order to produce fruit, two or more cultivars should always be planted near each other for cross-pollination. Spring growth is rapid as long as soil is moist, but root development is shallow during the first few years, and trees must be watered during dry periods. Young trees frequently retain their leaves during much of the winter. Sun-scald on exposed sides of trunks of newly planted trees may be a problem. Usually trees are headed low enough to provide for shading by tops. Trees should be planted as close as 4 × 4 m or 4.6 × 4.6 m each way; such trees should not be pruned. Cutting the lower branches from trunks invites blight infection. Trees do best when left to grow in bush form. Trees planted in this manner must have good cultivation, the same as for apple, peach, or pear trees.[278]

Harvesting — Trees begin to bear when 5 to 6 years old; those for orchard culture with profitable crops begin in 10 to 12 years. Chestnuts should be harvested daily as soon as burs open and nuts fall to ground. Nuts should be placed at once on shelves or in curing containers with wooden or metal bottoms to prevent larvae which may crawl out of nut from reaching the ground. All infected nuts should be promptly burned. For curing the nuts, they should be spread thinly on floors or the like, stirred frequently and held for 5 to 10 days, depending upon condition of nuts and atmospheric conditions at time of harvest. During the curing period, nuts will shrink in weight, and the color will change from lustrous to dull brown. Three weeks is about as long as Chinese chestnuts remain sound without special treatment. Chestnuts should be marketed as promptly as possible to minimize deterioration. Chestnuts in sound condition may be stored in cold storage with temperature just above freezing; this is the simplest method. Stratifying in a wire-mesh container buried deeply in moist, well-drained sand is also very satisfactory. Putting nuts in a tightly closed tin container at refrigerator temperature or in cold storage at 0°C is also acceptable.[218,278]

Yields and economics — Average yields are about 13 to 25 kg per year per tree, with large trees producing from 25 to 126 kg per tree.[278] According to Wyman,[343] trees are said to produce an average crop of 34 to 45 kg edible nuts per tree. Major producers are China and Korea. Very limited cultivation in U.S., with trend increasing.[278]

Energy — Wood and burs may be used to burn, as is, or converted to charcoal.

Biotic factors — The following have been reported as affecting this species: *Cronartium cerebrum* (rust), *Cryptodiaporthe castanea* (canker, dieback), *Cytospora* sp. (twig blight), *Diplodia* sp. (twig blight), *Discohainesia oenotherae*, *Gloeosporium* sp. (blossom-end rot of nuts), *Marssonina ochroleuca* (leaf spot), *Phomopsis* sp. (twig blight), *Septoria gilletiana* (leaf spot), and *Stereum gausapatum* (heart rot).[4] Pollination is carried on by insects. Chinese chestnut is largely, but not wholly, self-sterile. More than a single seedling or grafted cv should be included in any planting. Several seedlings or several cvs would be better. Trees producing inferior fruit should be removed. Chinese chestnut is not immune to the blight (*Endothia parasitica*), but is sufficiently resistant for trees to persist and bear crops. Trees develop bark cankers as a a result of infection, but the lesions usually heal. Nuts are attacked by several diseases, either before or after harvest. Most serious pests are chestnut weevils, often called curculios, which if unchecked, often render whole crops unfit for use.[50] Trees are often planted in poultry yards, in order to decimate the bugs. Japanese beetles are a serious pest on leaves in some areas. June bugs and May beetle also feed on the newest leaves, mainly at night.[278]

Jones et al.[154] report that in commercial and home plantings of Chinese chestnut in 6 southeastern and eastern states, 23 of the trees had main stem cankers incited by *Endothia parasitica*. In general, they found the main stem canker incidence (13 to 93) was higher in plantings of the Appalachian Mountain region than in the Piedmont region (2 to 13 incidence). They found the trees that were damaged most were located in high-wind and cold-winter areas of the Appalachian Mountains.

CASTANEA PUMILA (L.) Mill. (FAGACEAE) — Chinquapin, Allegany Chinkapin

Uses — Kernels of nuts are sweet and edible, but are not consumed by humans very much; they are more of a wildlife food;[278] also used to fatten hogs.[170] Used by Indians in making bread and a drink similar to hot chocolate;[226] boiled and strung to make necklaces.[170] Shrubs useful for planting on dry, rocky slopes, as they are attractive when in flower and again in fall with their light green burs and dark foliage. Often planted as an ornamental. The light, coarse-grained, hard, strong, and dark-brown wood is used for fenceposts, rails, and railway ties.[278]

Folk medicine — Reported to be tonic and astringent, chinquapin is a folk remedy for intermittent fevers.[324]

Chemistry — No data available.

Description — Usually a spreading shrub east of Mississippi River, 2 to 5 m tall, forming thickets, often only 1.3 to 1.6 m tall, westward in range to Arkansas and eastern Texas, becoming a tree to 17 m tall with trunk up to 1 m in diameter, round-topped with spreading branches; branchlets at first bright red-brown, pubescent or nearly glabrous, becoming olive-green or dark-brown; winter-buds reddish, oval to ovoid, about 0.3 cm long, tomentose becoming scurfy pubescent; bark 1.3 to 2.5 cm thick, light-brown tinged red, slightly furrowed; leaves oblong-elliptic to oblong-ovate, 7.5 to 15 cm long, 3 to 5 cm broad, coarsely serrate, acuminate, gradually narrowed, unequal, rounded or cuneate at base, early tomentose on both surfaces, at maturity thick and firm, pubescent beneath, turning dull yellow in fall; petioles pubescent to nearly glabrous, flattened on upper surface, up to 1.3 cm long; stipules pubescent, ovate to ovate-lanceolate to linear at end of branch; staminate catkins single in leaf-axils toward ends of branches, simple with minute calyx and 8 to 20 stamens, yellowish-white, slender, at maturity 10 to 15 cm long, pubescent, flowers in crowded or scattered clusters; pistillate flowers on catkins near very tips of branches, several females near base, numerous males on more distal portion (androgynous), silvery tomentose, 7.5 to 10 cm long; fruits usually several or many in large compact head or spike, each involucre 1-flowered, 2 to 3.5 cm wide; bur 2 to 3.5 cm in diameter, covered with crowded fascicles of slender spines, tomentose at base; nut shining, reddish-brown, ovoid, 1 to 2.5 cm long, about 0.8 cm thick, coated with silvery-white pubescence, shell lined with lustrous tomentum; kernel sweet. Flowers later than leaves, May to early June; fruits September to October.[278]

Germplasm — Reported from the North American Center of Diversity, chinquapin, or cvs thereof, is reported to tolerate frost, poor soil, slope, weeds, and waterlogging.[82] Few selections of chinquapin have been made. More frequently it has been hybridized with chestnuts. *C. pumila* var. *ashei* Sudw. (*C. ashei* Sudw.), the Coastal Chinquapin, grows on sand dunes and in sandy woods along the coast from southeastern Virginia to northern Florida and along the Gulf; small tree to 9 m tall and 26 cm in diameter, or large shrub, leaves smaller, about 7.5 cm long and 3.5 cm broad, and spines on involucre less numerous. Thought by some not to be distinct from species. Trees native to Arkansas and eastern Texas are so much larger than those east of the Mississippi River, that they are considered by some as a distinct species, *Castanea ozarkensis* Ashe; trees to 20 m tall, with narrowly oblong-elliptic leaves often 15 to 20 cm long, distinctly acuminate, coarsely serrate, with triangular acuminate teeth. × *Castanea neglecta* Dode is a natural hybrid with American chestnut (*C. dentata*), with intermediate leaves and involucres containing one large nut; occurring in Blue Ridge areas (Highlands, North Carolina), 'Essate-Jap' is a hybrid between the chinquapin and the Japanese chestnut, which forms a larger tree, with early flowers, and nuts ripening 2 weeks or more before Chinese chestnuts; it grafts better on Japanese stock than on Chinese. ($2n = 24$.)[278]

Distribution — Native in eastern U.S. from southern New Jersey and Pennsylvania to western Florida, through Gulf States to Texas (valley of Nueces River). It is most abundant and attains its largest size in southern Arkansas and eastern Texas.[278]

Ecology — Ranging from Cool Temperate Wet through Subtropical Moist Forest Life Zones, chinquapin is reported to tolerate annual precipitation of 11 to 13 dm, annual temperature of 12 to 19°C, and pH of 5.6 to 5.8.[82] Grows in mixed upland woods, on dry sandy ridges, on hillsides, in sandy wastes, and along borders of ponds and streams, in dry or wet acid soil. Occurs from sea-level to 1,500 m in the Appalachian Mountains. Prefers undisturbed woods with plentiful humus, and a warm temperate climate.[278] Grows in dry woods and thickets.[170] Hardy to Zone 5.[343]

Cultivation — Propagated from seed, or often spreading by stolons. Seeds germinate easily, sometimes sending out hypocotyl before reaching the ground. Although chinquapin is planted, it is not cultivated as a crop. Occasionally, plants are planted for ornamentals, or along edge of woods for wildlife food. Once planted, shrubs require no attention.[278]

Harvesting — Shrubs begin bearing nuts when 3 to 5 years old, and are prolific producers of small, sweet, nutty-flavored nuts. Nuts harvested in fall by man and wildlife.[278]

Yields and economics — According to Rosengarten[283] nuts of *C. pumila,* sweet and very small, yield 400 to 700 nuts per lb (800 to 1540/kg), compared to 75 to 160/lb (165 to 352/kg) for the American chestnut, and 30 to 150/lb (66 to 330/kg) for Chinese chestnut. Nuts are sold in markets in southern and western states. Timber is used west of the Mississippi River. Most valuable as wildlife crop.[278]

Energy — Wood and burs can be used for fuel, as is, or converted to charcoal.

Biotic factors — Although chinquapin is not resistant to the Chestnut Blight (*Endothia parasitica*), shrubs make up for loss of diseased stems by increased growth of remaining stems and by production of new shoots.[278] Agriculture Handbook 165[4] lists the following as affecting *C. pumula: Actinopelte dryina* (leaf spot), *Armillaria mellea* (root and butt rot), *Cronartium cerebrum* (rust), *Cryptospora cinctula, Discohainesia oenotherae, Endothia radicalis, Gnomonia setacea, Lenzites betulina* and *L. trabea* (brown cubical rot of dead trunks and timber), *Marssonina ochroleuca* (brown-bordered leaf spot, eyespot), *Melanconium cinctum* (on twigs), *Microsphaerea alni* (powdery mildew), *Monochaetia desmazierii* (leaf spot), *Phyllosticta castanea* (leaf spot), *Phymatotrichum omnivorum* (root rot), *Phytophthora cinnamomi* (root and collar rot of nursery plants and forest trees), *Polyporus* spp. (various wood rots), and *Stereum* spp. (various wood rots).

CASTANEA SATIVA Mill. (FAGACEAE)—European, Spanish, Italian, or Sweet Chestnut
Syn.: *Castanea vulgaris* Lam., *Castanea vesca* Gaertn., and *Castanea castanea* Karst.

Uses — European chestnuts are grown for the kernels of the nuts, extensively eaten by humans and animals. Nuts used as vegetable, boiled, roasted, steamed, pureed, or in dressing for poultry and meats. In some areas, chestnuts are considered a staple food, two daily meals being made from them.[278] In some European mountainous regions, chestnuts are still the staff of life, taking the place of wheat and potatoes in the form of chestnut flour, chestnut bread, and mashed chestnuts. Flour made of ground chestnuts is said to have provided a staple ration for Roman legions.[283] In Italy, they are prepared like stew with gravy. Dried nuts used for cooking purposes as fresh nuts, or eaten like peanuts. Culled chestnuts used safely for fattening poultry and hogs. Cattle will also eat them.[278] Used as a coffee substitute, for thickening soups, fried in oil; also used in brandy, in confectionary, desserts, and as a source of oil. The relatively hard, durable, fine-grained wood is easy to split, not easy to bend. Used for general carpentry, railroad ties, and the manufacture of cellulose. The bark is used for tanning.[324]

Folk medicine — According to Hartwell,[126] the nuts, when crushed with vinegar and barley flour, have been said to be a folk remedy for indurations of the breasts. Reported to be astringent, sedative and tonic, European chestnut is a folk remedy for circulation problems, cough, extravasation, fever, hematochezia, hernia, hunger, hydrocoele, infection, inflammation, kidney ailments, myalgia, nausea, paroxysm, pertussis, rheumatism, sclerosis, scrofula, sores, stomach ailments, and wounds.[91] Aqueous infusion of leaves used as tonic, astringent, and effective in coughs and irritable conditions of respiratory organs.[278]

Chemistry — Per 100 g, the seed (ZMB) is reported to contain 406 to 408 calories, 6.1 to 7.5 g protein, 2.8 to 3.2 g fat, 87.7 to 88.6 g total carbohydrate, 2.3 to 2.4 g fiber, 2.0 to 2.1 g ash, 30.3 to 56.8 mg Ca, 184 to 185 mg P, 3.4 to 3.6 mg Fe, 12.6 to 32.3 mg Na, 956 to 1705 mg K, 0.46 mg thiamine, 0.46 mg riboflavin, and 1.21 to 1.26 mg niacin.[89] Chemical composition is similar to that of wheat; starch is easily digested after cooking.[278] Woodroof[341] reports Spanish chestnuts to contain 2.87 to 3.03% ash, 9.61 to 10.96% total protein, 2.55 to 2.84% fiber, 73.75 to 77.70% total nitrogen-free extract, and 7.11 to 9.58% fat. In a study on chestnuts from 19 natural stands in southern Yugoslavia, Miric et al.[214] found in most samples the total fat content was between 4 and 5, the highest 5.62. Oleic and linoleic acids predominated, followed by palmitic.

Description — Tree 30 m or more tall, with girth to 10 m; trunk straight, smooth, and blackish or dark-green in youth, finally becoming brownish-gray with deep longitudinal and often spirally curved fissures; branches wide-spreading; young shoots at first minutely downy, becoming glabrous; buds ovoid, obtuse, the terminal one absent; young leaves densely stellately pubescent, becoming glabrous; mature leaves 10 to 25 cm long, 3 to 7 cm broad, oblong-elliptic to oblong-lanceolate, apex long-acuminate, base more or less rounded or cordate, upper surface soft green, lower paler; blades rather thick and stiff, with 6 to 20 bristles on each of the rather deeply serrate margins; petioles minutely velutinous, glandular-lepidote or glabrous, about 2 cm long; stipules lanceolate, long-acute, gradually broadened at base, 1 to 2.5 cm long, not markedly veined; winter buds dull-red, pubescent, long ovoid-conic; staminate spikes 8 to 10 cm long, about 1 cm thick; pistillate flowers at base of male spikes in large globose strigose involucres 1 to 2 cm thick; styles exserted up to 8 mm, sparsely marked with ascending appressed hairs; bur green, 4 to 6 cm in diameter, with numerous slender minutely pubescent spines up to 2 cm long; inside or husk marked by very dense golden felt; nuts shining brown, with paler base, often 2 to 3.5 cm in diameter, at tip thickly pubescent, bearing a short stalked perigynium with its persistent styles; kernel variable from bitter to sweet. Flowers late May to July.[278]

Germplasm — Reported from the Near East Center of Diversity, European chestnut, or cvs thereof, is reported to tolerate bacteria, frost, mycobacteria, and slope.[82] The only disadvantage of the European chestnut is that the skin is astringent, but since most of them are cooked before eating, the skin is removed readily. The skin should not be eaten, as it is indigestible. When European is crossed with American Sweet, this difficulty is modified or eliminated. The following are some of the most generally cultivated cvs. 'Marron Combale', 'Marron Nousillard', and 'Marron Quercy' originated in France; all have the very large light- to dark-brown nuts and are very productive; 'Numbo' and 'Paragon' are the most frequently grown cvs in the U.S.; they have medium-large, roundish nuts of fair quality, and bear regularly. 'Ridgely', originated in Dover, Delaware, has fair-sized nuts, of very good quality and flavor, with 2 or 3 nuts per bur; it is vigorous and productive. 'Rochester' and 'Comfort' are grown to a limited extent. Hybrids with *C. dentata* have leaves with cuneate bases. Some garden forms have variegated leaves or laciniate leaves (var. *asplenifolia* Lodd.). ($2n = 22,24$.)[82,278]

Distribution — Native from southern Europe through Asia Minor to China. Cultivated in many parts of the Himalayas, especially in Punjab and Khasia Hills. Naturalized in central, western, and northern Europe, almost forming forests. Introduced on Pacific Coast of the U.S. Extensively planted for its nuts and timber.[278] Introduced to the U.S. in 1773 by Thomas Jefferson.[283]

Ecology — Ranging from Cool Temperate Steppe to Wet through Warm Temperate Steppe to Dry Forest Life Zones, European chestnut is reported to tolerate precipitation of 3.9 to 13.6 dm (mean of 14 cases = 8.6), annual temperature of 7.4 to 18.0°C (mean of 14 cases = 10.8), and pH of 4.5 to 7.4 (mean of 10 cases = 6.4).[82]

In woods, and often forming forests, on well-drained soils, often on mountain slopes, usually calcifuge. Acclimated to all temperate areas. Trees retain foliage late in fall.[278]

Cultivation — Thorough preparation of soil before planting is essential. For orchard planting, trees are propagated by grafting and budding. Whip grafts on small shoots or stocks about 1.3 to 2 cm in diameter, or cleft grafts on shoots 1.3 to 3 cm in diameter give most successful results. Bark graft on shoots is also successful. In any method of grafting, great care must be used in waxing and then rewaxing in about 2 weeks. Wax should cover cuts made in stock and scion, and should be applied immediately after inserting the latter. Scion should be waxed for its entire length, leaving no bubbles. Cover whole by tying paper bag over top. European types frequently outgrow stocks and cause an enlarged imperfect union. For orchard planting, trees should be spaced 13 × 13 m, 17 × 17 m, or 20 × 20 m; on good soil, the latter is preferable. Row could be set 20 m apart, with trees 7 to 8 m apart in rows. It may be necessary to remove every other tree after they reach a certain size. Good distance between rows provides for better growth of trees, and interplanting with vegetables or small fruits. In any event, do not crowd chestnut trees. Dig holes 75 × 75 cm, breaking down topsoil around rim and allowing it to fall into hole. Always use fine top-soil around roots and firm soil well after planting. Before planting, cut with knife all broken or bruised roots, and clip end of every root. Set trees no deeper than in nursery and in same position, the bark on the north side being greener than that on south side. There will be less loss from sunburn if southern side, hardened by exposure, is again placed to the south. During first year or two, trees should be shaded. Sometimes the trunk is wrapped with paraffined paper or burlap, lightly enough not to interfere with flow of sap. After planting, cut back top to about 1.6 m; if tree is a straight whip, or if branched, cut branches down to 2 or 3 buds from trunk. Staking young trees is desirable, but not necessary. Young trees should be pruned to an open spreading form with 3 to 5 main branches on which top will eventually form, after which trees need little care other than good culture. If trees are allowed to overbear, nuts run down in size. Trees usually develop well without irrigation, but larger yields result when water is applied. While tree is young, regular irrigation is very desirable. Unless intercrops are grown, irrigation may be limited to one application per month during growing season, after trees are bearing. Young trees may require irrigation twice a month. Water should penetrate well into subsoil. Light irrigation induces shallow rooting, which is undesirable. Do not continue irrigation too late in growing season, as it is likely to make nuts crack open or over-develop them. Cracked nuts soon spoil and mold. Cultivation must be thorough, so that free growth is promoted. After maturity, cultivation need not be so intensive. During the first few years, it is advisable to hoe around tree by hand, but after tree is well-established, annual plowing or cultivation after each irrigation is sufficient. Annual cover crops may be used to build up or maintain soil fertility. Two-year-old grafts are commonly loaded with burs, and if such grafted trees show a tendency to bear heavily while young, burs should be thinned out so that very few remain. Otherwise, trees will grow out of shape and be retarded in their development. Sometimes burs are picked from trees for 3 or 4 years until trees become well established, before beginning nut production. With seedlings and grafted trees, a mixture of cvs gives better yield of nuts. If all burs are filled, tree would not stand the weight nor develop nuts to marketable size. Many burs are empty and many have few mature nuts, perhaps a provision of nature, rather than poor pollination. Many trees self-prune (drop) fruits or abort seeds.[278]

Harvesting — Allow burs to mature thoroughly and fall of their own accord. Some cvs stick, so that shaking or jarring the limbs is useful. In other cvs, burs open, and nuts fall to ground. Burs which fall and do not open can be made to shed their nuts by pressure of the feet or by striking with small wooden mallet. Some harvesters use heavy leather gloves and twist nuts out of burs by hand. Nuts should be picked up every morning and stored in sacks, if they are to be shipped at once. If they are to be kept for a while, they should be piled on floor to sweat. Pile should be stirred twice a day for 2 days, and then nuts sacked. Always store nuts so that air can circulate freely. Do not pile up sacks for any length of time, as they will heat and mold. If stacking is necessary, place sticks between sacks for ventilation. In gathering nuts, the collector usually has two pails or containers, one for first-grade perfect nuts, the other for culls.[278]

Yields and Economics — Yields average from 45 to 136 kg per tree.[278] In 60- to 80-year-old stands in Russia, yields average 770 kg/ha, up to 1230 kg/ha in better stands.[191] Italy reports ca. 1100, France ca. 1500 to 2200, and Spain ca. 2800 kg/ha.[310] In the best years, 5,000 kg/ha are reported.[310] Nuts are marketed to a limited degree, but are mostly locally cultivated and used.[278]

Energy — Wood and burs may be used for firewood or for the production of charcoal.

Biotic factors — European chestnut is susceptible to diseases of other chestnuts, especially susceptible to attacks of leaf fungi.[278] Agriculture Handbook 165[4] reports the following as affecting *C. sativa: Actinopelta dryina* (leaf spot), *Cronartium cerebrum* (rust), *Endothia parasitica* (blight), *Exosporium fawcettii* (canker, dieback), *Marssonina ochroleuca* (leaf spot), *Melanconis modonia* (twig blight), *Microsphaera alni* (powdery mildew), *Phyllactinia corylea* (powdery mildew), *Phyllosticta castanea* (leaf spot), *Phymatotrichum omnivorum* (root rot), *Phytophthora cinnamomi* (root and collar rot of seedlings), *Schizophyllum commune* (sapwood rot), and *Stereum versiforme*. Browne[53] adds: Fungi — *Armillaria mellea, Cerrena unicolor, Daedalea quercina, Dematophora* sp., *Diplodina castaneae, Fistulina hepatica, Fomes mastoporus, Ganoderma applanatum, G. lucidum, Ilymenochaete rubiginosa, Inonotus cuticularis, I. dryadeus, Laetiporus sulphureus, Microsphaera alphitoides, Mycosphaerella castanicola, Phyllactinia guttata, Phytophthora cactorum, P. cambivora, P. cinnamomi, P. syringae, Polyporus rubidus, P. squamosus, P. tulipiferae, Rhizinia inflata, Rosellinia radiciperda, Sclerotinia candolleana, Stereum hirsutum, Valsa ambiens, Verticillium alboatrum*; Angiospermae — *Viscum album*; Coleoptera — *Attelabus nitens, Platypus cylindrus, Prionus coriareus, Xyleborus dispar*; Hemiptera — *Lachnus roboris, Myzocallis castanicola, Quadraspidiotus perniciosus*; Lepidoptera — *Carcina quercana, Euproctis scintillans, Lithocolletis messaniella, Pammene fasciana, Suana concolor;* and Mammalia — *Dama dama, Sciurus carolinensis.*

CASTANOSPERMUM AUSTRALE A.Cunn. et Fras. (FABACEAE) — Moreton Bay Chestnut, Black Bean Tree

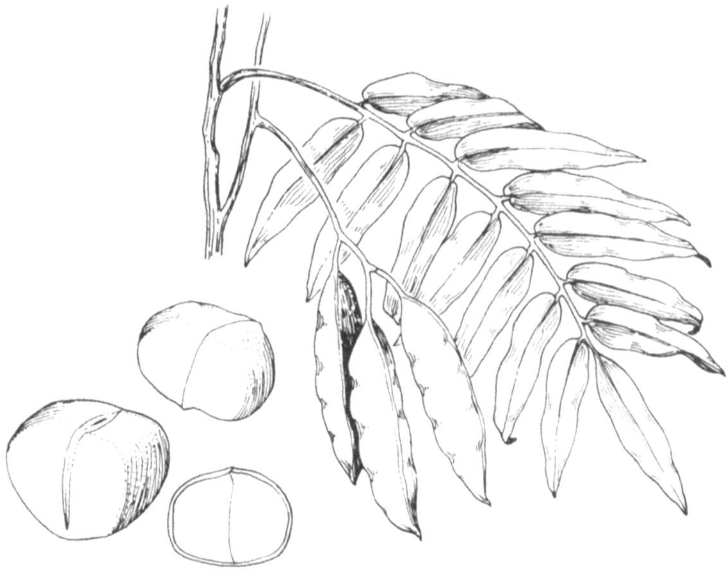

Uses — Australian aborigines processed the seeds for food. Of its edibility, Allen and Allen[8] say, "The edibility of the roasted seed of *C. australe*, often equated with that of the European chestnut, has been overestimated. Some writers rate its edibility about equal to that of acorns, or as acceptable only under dire circumstances of need and hunger . . . The astringency of fresh seeds is reduced or removed by soaking and roasting, although even after such treatment ill effects are known to occur."[8] Commonly cultivated in Australia in home gardens and as a street tree, this species is well known in the timber trade as Black Bean. In view of the shape and configuration of the seeds, I believe "Brown Buns" would be more appropriate. The timber dresses well and is regarded as a heavy cabinet timber. Before synthetics, the wood was used for electrical switchboards, because of its particularly high resistance to the passage of electric current. The wood is also used in inlays, panels, umbrella handles, ceilings, plywood, and carved jewel boxes. In South Africa, it is frequently cultivated for shade and as an oramental, suitable for planting along suburban sidewalks. Around Sydney, Australia, they have become popular as a house plant for short term decoration.[437] The NAS[231] classifies this as a "vanishing timber", used sometimes as a walnut substitute (750 kg/m^3).

Folk medicine — Extracts have given negative antibiotic tests. According to the *Threatened Plants Newsletter*,[440] 100 kg of seed were shipped to the U.S. for cancer and AIDS research, research which is suggesting anti-AIDS activity, in vitro at least. In a letter (1987), Dr. K. M. Snader, of the National Institutes of Health,[439] tells me, "I do not at this moment know if castanospermine will become an AIDS treatment, but it is showing some activity in our screening systems. Indeed, there is enough interest to want to look further at the pharmacology and to explore other products with either similar structures or with the same mechanism of action."

Chemistry — Australian cattle fatalities are reported from grazing the fallen seed during dry periods (mostly October to December). Unfortunately, the cattle may develop a liking for the seed. Also, with the leaves, cattle becoming fond of them may pine away and die if deprived of them.[332] Seeds contain the triterpenoid castanogenin. The structure is outlined in Hager's Handbook.[187] The wood contains bayin ($C_2H_{20}O_9$) and bayogenin. Castanospermine is said to inhibit alpha- and beta-glucosidases, beta-xylosidase, and to inhibit syncytra formation in HIV-infected CD_4 positive cells.[442] According to Saul et al.,[438] castanospermine decreases cytoplasmic glycogen in vivo in rats, showing a dose-dependent decrease in alpha-glucosidase activity in the liver (50% at 250 mg/kg), spleen (50% at 250), kidney (48% at 250), and brain (55% at 50 mg/kg). At doses of 2,000 mg/kg, the rats experienced diarrhea (reversible with diet) with decreased weight gain and liver size. With the HIV, there is a dose-dependent decrease in syncytium formation (H9 human aneuploid neoplastic cells infected with HIV) with complete inhibition at 100 μg/mℓ. Apparently, it affects the envelope protein, not the CD_4 receptor glycoprotein. At 50 μg/mℓ, it inhibits the cell death of infected cells. And there is a dose-dependent decrease in extracellular virus (a million-fold at 200 μg/mℓ).[442]

Toxicity — The unpleasant purgative effects of fresh seeds and their indigestibility are attributed to the 7% saponin content. Later writers question the presence of saponin. The sawdust irritates the nasal mucosa.[8] Brand et al.[47] report an uninspiring 79% water, 3.2% protein, 0.7 g fat, and 0.5 g fiber. Menninger[209] quotes one of his sources, "Recently 14 Air Force personnel were admitted to the hospital after being on a survival mission and eating the seed."

Description — Tall, glabrous, slow-growing, evergreen trees to 45 m tall, 1 to 2 m DBH. Leaves large, imparipinnate; leaflets large, 8 to 17, glossy, short-petioled, elliptic, tapering, leathery; stipels absent. Flowers large, orange-to-reddish yellow, in short, loose racemes in the axils of old branches; bracts minute; braceteoles none; calyx thick, large, colored; teeth broad, very short; standard obovate-orbicular, narrowed into a claw, recurved; wings and keel petals shorter than the standard, free, subsimilar, erect, oblong; stamens 10, free; anthers linear, versatile; ovary long-stalked, many-ovuled; style incurved; stigma terminal, blunt. Pod elongated, 18 to 25 cm long, subfalcate, turgid, leathery to woody, 2-valved, valves hard, thick, spongy inside between the 2 to 6, large, globose, chestnut-brown seeds.[8] Seeds 2 to 4 cm broad. Fruiting February to April in Australia.

Germplasm — Reported from the Australian Center of Diversity, Moreton Bay chestnut, or cvs thereof, should tolerate some salt. It tolerates shade and some drought, but little frost. ($2n = 26$.)

Distribution — Only of local importance in its native Australia and New Caledonia.[204] Native to the tablelands of northeast Australia, Queensland, and New South Wales. Introduced into Sri Lanka ca. 1874. Introduced and surviving as far as 35°S in Australia. Now somewhat common in India and the East Indies. Planted as an ornamental in the warmer and more humid parts of South Africa.

Ecology — Estimated to range from Subtropical Dry to Rain through Tropical Dry to Wet Forest Life Zones, this species is estimated to tolerate annual precipitation of 10 to 60 dm, annual temperature of 20 to 26°C, and pH of 6.0 to 8.5. Apparently damaged by heavy frost (but tolerating 0°C in Sydney). Usually in coastal or riverine forests. Best suited to rich loam, it will succeed on sandy, less-fertile soils.

Cultivation — Seeds should be sown fresh and barely covered (1 to 2 cm) with soil. They should germinate in 10 to 21 days when planted at 20 to 30°C. They can be held 6 to 8 months at 4°C.

Harvesting — For reasons not fully understood, the tree often fails to fruit where it has been introduced as an ornamental. For example, it grows well at Singapore and Manila, apparently without fruiting. Some West Indian introductions have fruited at ca. 20 years of age. The seeds may be steeped in water for 8 to 10 days, then dried in the sun, roasted on hot stones, pounded, and ground into meal.

Yields and economics — Data provided me by the National Cancer Institute (NCI) indicate that one could obtain 100 g pure castanospermine from 1,000 lb seed, suggesting yields of 0.0002203% or ca. 2 ppm. Before the NCI AIDS announcement July 24, 1987, the Sigma

Chemical Company was offering castanospermine at \$22.60 to \$23.80 per mg or \$89.50 to \$94.20 per 5 mg, which translates to \$8 million to \$10 million per pound. There is a newly published synthesis which can produce 100 mg and four isomers for \$10,000. So the price will come down.

Energy — The wood has a density of 800 kg/m^3. If the seed contains only 2 ppm castanospermine, most of the residual biomass could be used for fuel.

Biotic factors — The timber is subject to wood-rotting fungi and to termites. The sapwood is subject to beetle attack. Apparently ornithophilous (pollinated by birds) and distributed by water. Nodulation and rhizobia have not yet been reported.[8]

CEIBA PENTANDRA (L.) Gaertn. (BOMBACACEAE) — Kapok, Silk Cotton Tree
Syn: *Eriodendron anfractuosum* DC. and *Bombax pentandrum* L.

Uses — Valued as a honey plant. Young leaves are sometimes cooked as a potherb. In the Cameroons, even the flowers are eaten. I have used the water from the superficial roots when clean drinking water was unavailable.[80] Silky fiber from pods used for stuffing protective clothing, pillows, lifesavings devices; as insulation material, mainly against heat and cold, because of its low thermal conductivity, and sound, and for caulking various items, as canoes. Fiber contains 61 to 64% cellulose, the rest lignin and other substances, including a toxic substance, making it resistant to vermin and mites. Wrapped around the trunk of a fruit tree, it is supposed to discourage leaf-cutting ants. Fiber is white or yellowish, cylindrical, each a single cell with a bulbous base, resilient, water-resistant, with buoyance superior to that of cork. The floss, irritating to the eyes, is used to stuff life-preservers, mattresses, pillows, saddles, etc., and it also used as tinder. In the U.S., baseballs may be filled with kapok. Mixed with other fibers, like cotton, it is used in the manufacture of carpets, laces, felt hats, "cotton", fireworks, and plushes. Fiber can be bleached or dyed like cotton. Seeds are the source of an oil (20 to 25% in seed, about 40% in kernel), used for illumination, for soap making, or as a lubricant. Seed oil roughly comparable to peanut oil; used for the same purposes as refined cotton-seed oil. West Africans use the seeds, pounded and ground to a meal, in soups, etc. Roasted seeds are eaten like peanuts. Some people sprout the seeds before eating them. The young fruits are a vegetable like okra. Expressed cake serves for fodder. The timber, though little used, is said to be excellent at planing, sanding, and resistance to screw-splitting. Used for boxes, matches, toys, drums, furniture, violins, dugouts (said to float even when capsized), and for tanning leather. Shaping and boring qualities are poor, turning very poor.[55,80,209,278]

Folk medicine — According to Hartwell,[126] *Ceiba* is used in folk remedies for nasal polyps and tumors. Reported to be antidiarrheic, astringent, diuretic, emetic, and emollient, *Ceiba pentandra* is a folk remedy for bowel disorders, foot ailments, female troubles, headache, hydropsy, leprosy, neuralgia, parturition, spasm, sprain, swelling, tumors, and wounds.[91] The Bayano Cuna use the bark in medicine for female troubles. The roots are used in treating leprosy. A bath of a bark infusion is supposed to improve the growth of hair (Colombia). The same infusion is given to cattle after delivery to help shed the placenta.[80] Gum used as tonic, alterative, astringent, or laxative. Young leaves are emollient. Roots used as diuretic and against scorpion stings. Juice of roots used as a cure for diabetes.[278] Ayurvedics used the alexeiteric gum for blood diseases, hepatitis, obesity, pain, splenosis, and tumors. Yunani use the leaves for boils and leprosy, the gum and/or the root for biliousness, blood diseases, dysuria, and gonorrhea, considering them antipyretic, aphrodisiac, diuretic, fattening, and tonic. Others in India use the roots for anasarca, ascites, aphrodisia, diarrhea, and dysentery; the taproot for gonorrhea and dysentery; the gum for menorrhagia, and urinary incontinence in children.[165] Malayans use the bark for asthma and colds. Javanese mix the bark with areca, nutmeg, and sugar candy for bladder stones. Liberians use the infusion as a mouthwash. In Singapore, leaves are mixed with onion and turmeric in water for coughs. Javanese use the leaf infusion for catarrh, cough, hoarseness and urethritis. Cambodians use the leaves to cure migraine and inebriation. In French Guiana, flowers are decocted for constipation. In Reunion, the bark is used as an emetic. Annamese also use the bark as emetic, the flowers for lochiorrhea and plague, the seed oil as an emollient. West Indians use the leaves in baths and poultices for erysipelas, sprained or swollen feet, and to relieve fatigue. The tea is drunk for colic and inflammation. French-speaking West Indians take the root decoction as a diuretic. Latin Americans apply the bark to wounds and indolent ulcers, using the inner bark decoction as antispasmodic, diuretic, emetic, and emmenagogue, and for gonorrhea and hemorrhoids.[224] Colombians use the leaf decoction as a cataplasm or bath for boils, infected insect bites and the like.[105] Nigerians use the seed oil for rheumatism.[55] Bark extracts show curare-like action on anesthetized cat nerves.

Chemistry — Per 100 g, the seed is reported to contain 530 calories, 30.4 to 33.2 g protein, 23.1 to 39.2 g fat, 21.6 to 38.3 g total carbohydrate, 1.6 to 19.6 g fiber, 6.1 to 8.2 g ash, 230 to 470 mg Ca, 970 to 1269 mg P.[89] Contains little or no gossypol, the seeds contain 20 to 25% (kernel, ca. 40%) oil. The percentages of fatty acids in the oil are oleic, 43.0; linoleic, 31.3; palmitic, 9.77; stearic, 8.0; arachidic, 1.2; and lignoceric, 0.23. Analysis of the seed-cake gave the following values: moisture, 13.8; crude protein, 26.2; fat, 7.5; carbohydrate, 23.2; fiber, 23.2; and ash, 6.1%; nutrient ratio, 1:1.5; food units, 107. Analysis of a sample from Indo-China gave: nitrogen, 4.5; phosphoric acid, 1.6; potash, 1.5%. Analysis of the wood gave: moisture, 9.8; ash, 5.9; fats and waxes, 0.62; cellulose, 68.3; and lignone, 25.2%. The yield of bleached pulp was 30%. Destructive distillation of wood from West Africa gave: charcoal, 28.4; crude pyroligneous acid, 43.7; tar, 12.8; and acetic acid, 2.3%.[70] The floss contains pentosans and uronic anhydrides. Root and stem bark contain HCN. Leaves contain quercetin, camphorol, caffeic acid, and resin. Bark contains up to 10.82% tannin.[55]

Toxicity — The air-borne floss can induce allergy and conjunctivitis.

Description — Deciduous, umbraculiform, buttressed, armed or unarmed, medium to large trees to 70 m tall, more often to 33 m tall, spines conical when young; branches horizontal in whorls and prickly when young; leaves alternate, stipulate, long-petiolate, palmately compound with 5 to 11 leaflets, these elliptic or lanceolate, acuminate, entire or toothed, up to 16 cm long, 4 cm broad; flowers nudiflorous, numerous, in axillary dense clusters or fascicles on pedicels 8 cm long, near ends of branches; calyx 5-lobed, 1 to 1.5 cm long, green, bell-shaped, persistent; petals 5, fleshy, forming a short tube and spreading out to form a showy flower 5 to 6 cm in diameter; cream-colored, malodorous; stamens united into a 5-branched column 3 to 5 cm long; ovary 5-celled; fruit a 5-valved capsule, ellipsoid, leathery, 20 to 30 cm long, about 8 cm in diameter, filled with numerous balls of long silky wool, each enclosing a seed; seeds black, obovoid, enveloped in copious, shining silky hairs arising from inner walls of capsule. Flowers December to January; fruits March to April.[278]

Germplasm — Reported from the Indochina-Indonesia, Africa, and Middle America Centers of Diversity, kapok, or cvs thereof, is reported to tolerate drought, high pH, heat, insects, laterite, low pH, slope, and virus.[82] The Indonesian cv 'Reuzenrandoe' (giant kapok) bears some characteristics of the var. *caribaea*. "Some authors believe in an American/African origin of the kapok tree. If America is the sole center of origin, then the African center is secondary. The African kapok tree is divided into the caribaea-forest type and the caribaea-savannah type. The latter type, which has a broadly spreading crown, is planted in market places. It is possible that this type arose from cuttings of plagiotropic branches.[350] Some research has gone into developing whiter floss, indehiscent pods, and spineless trunks.[55] Trees are quite variable in the spininess of the stem, habit of branching, color of flowers, size of fruits, manner of fruit opening, and length, color, and resiliency of fibers of floss. Based on these characteristics, three varieties are recognized: var. *indica*, Indian forms; var. *caribaea*, American forms; and var. *africana*, African forms.[278] (2n = 72,80,82)

Distribution — Probably native to tropical America; widely distributed in hotter parts of western and southern India, Andaman Islands, Burma, Malaysia, Java, Indochina, and southeast Asia, North Borneo; cultivated in Java.[278] According to Zeven and Zhukovsky,[350] it was believed that the kapok tree originated in an area which was later divided by the Atlantic Ocean, so this species is native both to America and Africa. This conclusion is based mainly on the great variability of this plant and on the high frequency of dominant inherited characteristics in these two continents. Another thought is that seeds may have come from America in prehistoric times and that later introduction increased the variability. Because of its chromosome number, a polyploid origin is suggested. If this supposition is correct, the kapok tree can only have arisen in that area where its parents occur. As all other *Ceiba* species are restricted to America, this would also indicate an American origin.[350]

Ecology — Ranging from Subtropical Dry to Wet through Tropical Thorn to Wet Forest Life Zones, kapok is reported to tolerate annual precipitation of 4.8 to 42.9 dm (mean of 134 cases = 15.2), annual temperature of 18.0 to 28.5°C (mean of 129 cases = 25.2°C), and pH of 4.5 to 8.7 (mean of 45 cases = 6.7).[82] Hardy to Zone 10.[343] It thrives best in monsoon climates below 500 m altitude. Where night temperatures are below 20°C, fruits do not set. Trees damaged by high winds and waterlogging. Requires well-drained soil, in areas with annual rainfall of 125 to 150 cm, with abundant rainfall during the growing season and a dry period from time of flowering until pods ripen. In Java, commonly grown around margins of fields and along roadsides.[278]

Cultivation — On plantations, kapok is usually propagated from seeds of high-yielding trees. Planted in nurseries about 30 cm apart, seedlings are moved to the field when about 9 months old, topping them to 125 cm. Field spacings of about 6.5 are recommended. Sometimes trees are propagated from cuttings.[278] In Indonesia, cuttings are set in a nursery for a year and then transplanted at the beginning of the rainy season. The first harvest is usually 3 years later.

Harvesting — Since pods are usually handpicked by climbers, before they dehisce, much hand-labor is involved. Trees begin to fruit when 3 to 6 years old. For kapok, natives harvest the unopened pods with hooked knives on long poles. Since pods do not ripen simultaneously, it is necessary to harvest two or three times a year, before the pods open. Fruits are sun-dried and split open with mallets. The floss is removed with the seeds, and the seeds separated out by beating with a stick. In Java and the Philippine Islands, machines are employed for cleaning the floss. Floss is pressed into bales for export; these are generally packed in gunny cloth, and vary in weight from 80 to 120 lbs and are 8 to 16 ft^3 in volume.[278]

Yields and economics — Trees 4 to 5 years old yield nearly a kilogram of floss, whereas full-grown trees, 15 years old, may yield 3 to 4 kg.[183] Some trees may bear for 60 years or more and may yield 4,500 g kapok per year. It takes 170 to 220 pods to give a kilogram of floss. An adult tree may produce 1000 to 4000 fruits, suggesting a potential yield of 5 to 20 kg floss per tree. If the ratios prevail in kapok that prevail in cotton, we would expect that to correspond to 8 to 30 kg seed, or 2 to 7.5 kg oil per tree. In 1950, Indonesia produced 5,000 MT kapok, 6,500 in 1951, 6,600 in 1952, 7,000 in 1953, exporting ca. 5,000 MT a year.[183] Indonesia has produced as much as 16,000 MT kapok oil per year. Until World War II, Indonesia was the major producer; Ecuador exported over 1.25 million lbs in 1938. Today, Thailand produces about half of the 22 million kg of kapok produced, with the U.S. the largest consumer, using about half. Other exporters include Cambodia, East Africa, India, Indonesia, and Pakistan.[278]

Energy — The seed oil, used for cooking, lamps, lubrication, paints, and soaps, might serve, like the peanut, as a diesel substitute. Six trees could produce a barrel of oil renewably. As firewood, it is of no value, as it only smoulders, but the smouldering is sometimes put to use in fumigation.[55] The specific gravity of the wood is 0.920 to 0.933.[324]

Biotic factors — The following fungi attack kapok trees: *Armillaria mellea, Calonectria rigidiuscula, Camillea bomba, C. sagraena, Cercospora ceibae, C. italica, Chaetothyrium ceibae, Coniothyrium ceibae, Corticium salmonicolor, Corynespora cassiicola, Daldinia angolensis, Fomes applanatus, F. lignosus, F. noxius, Glomerella cingulata, Phyllosticta eriodendri, Physalospora rhodina, Polyporus occidentalis, P. zonalis, Polystictus occidentalis, P. sanguineus, Pycnoporus coccincus, Ramularia eriodendri, Schizophyllum commune, Septoria ceibae, Thanatephorus cucumeris, Ustulina deusta*, and *U. zonata.* The bacterium, *Xanthomonas malvacearum*, also infests trees. The parasite, *Dendropthoe falcata*, also occurs on the tree. The following viruses attack kapok: Cacao virus 1A, 1C, and 1M; Offa Igbo (Nigeria) cacao, Swollen Shoot, and viruses of *Adansonia digitata.* Nematodes isolated by kapok include: *Helicotylenchus concavus, H. multicinctus, H. retusus, H. pseudorobustus, H. dihystera, H. cavenessi, Meloidogyne arenaria, M. javanica, Pratylenchus brachyurus, P. delattrei, Scutellonema brachyurus, S. clathricaudatum, Tylenchorhynchus martini, Xiphinema elongatum*, and *X. ifacolum.*[186,278]

Baker and Harris[27] indicate that the flowers are visited by the fruit bats, *Epomorphorus gambianus, Nanonycteris veldkamp*, and *Eidolon helvum.* Flowers, though bat pollinated, are visited by bees.[55] Logs and lumber very susceptible to insect attack and decay. The wood is nearly always turned blue-gray by sap-staining fungi. This can be prevented by dipping in a fungicide solution shortly after sawing. In addition, Browne[53] lists the following as affecting this species: Coleoptera — *Analeptes trifasciata, Araccerus fasciculatus, Batocera numitor, B. rufomaculata, Chrysochroa bicolor, Hypomeces squamosus, Petrognatha gigas, Phytoscaphus triangularis, Steirastoma breve, Tragiscoschema bertolonii*; Hemiptera — *Delococcus tafoenis, Helopeltis schoutedeni, Icerya nigroarcolata, Planococcoides njalensis, Planococcus citri, P. kenyae, P. lilacinus, Pseudaulacaspis pentagona, Pseudococcus adonidum, Rastrococcus iceryoides, Saissetia nigra;* Lepidoptera — *Anomis leona, Ascotis selenaria, Cryptothelea variegata, Dasychira mendosa, Suana concolor, Sylepta derogata*; Thysanoptera — *Selenothrips rubrocinctus.*

COCOS NUCIFERA L. (ARECACEAE) Coconut

Uses — Coconut is one of the ten most useful trees in the world, providing food for millions of people, especially in the tropics. At any one time a coconut palm may have 12 different crops of nuts on it, from opening flower to ripe nut. At the top of the tree is the growing point, a bundle of tightly packed, yellow-white, cabbage-like leaves, which, if damaged, causes the entire tree to die. If the tree can be spared, this heart makes a tasty treat, a 'millionaire's salad'. Unopened flowers are protected by sheath, often used to fashion shoes, caps, even a kind of pressed helmet for soldiers. Opened flowers provide a good honey for bees. A clump of unopened flowers may be bound tightly together, bent over and its tip bruised. Soon it begins to "weep" a steady dripping of sweet juice, up to a gallon per day, that contains 16 to 30 mg ascorbic acid per 100 g. The cloudy brown liquid is easily boiled down to syrup, called coconut molasses, then crystallized into a dark sugar, almost exactly like maple sugar. Sometimes it is mixed with grated coconut for candy. Left standing, it ferments quickly into a beer with alcohol content up to 8%, called "toddy" in India and Sri Lanka; "tuba" in Philippines and Mexico; and "tuwak" in Indonesia. After a few weeks, it becomes a vinegar. "Arrack" is the product after distilling fermented "toddy" and is a common spiritous liquor consumed in the East. The net has a husk, which is a mass of packed fibers called coir, which can be woven into strong twine or rope, and is used for padding mattresses, upholstery, and life-preservers. Fiber, resistant to sea water, is used for cables and rigging on ships, for making mats, rugs, bags, brooms, brushes, and olive oil filters in Italy and Greece; also used for fires and mosquito smudges. If nut is allowed to germinate, cavity fills with a spongy mass called "bread" which is eaten raw or toasted in the shell over the fire. Sprouting seeds may be eaten like celery. Shell is hard and fine-grained, and may be carved into all kinds of objects, as drinking cups, dippers, scoops, smoking pipe bowls, and collecting cups for rubber latex. Charcoal is used for cooking fires, air filters, in gas masks, submarines, and cigarette tips. Shells burned as fuel for copra kilns or house-fires. Coconut shell flour is used in industry as a filler in plastics. Coconut water is produced by a 5-month-old nut, about 2 cups of crystal-clear, cool sweet (invert sugars and sucrose) liquid, so pure and sterile that during World War II, it was used in emergencies instead of sterile glucose solution, and put directly into a patient's veins. Also contains growth substances, minerals, and vitamins. Boiled toddy, known as jaggery, with lime makes a good cement. Nutmeat of immature coconuts is like a custard in flavor and consistency, and is eaten or scraped and squeezed through cloth to yield a "cream" or "milk" used on various foods. Cooked with rice to make Panama's famous "arroz con coco"; also cooked with taro leaves or game, and used in coffee as cream. Dried, desiccated, and shredded it is used in cakes, pies, candies, and in curries and sweets. When nuts are open and dried, meat becomes copra, which is processed for oil, rich in glycerine and used to make soaps, shampoos, shaving creams, toothpaste lotions, lubricants, hydraulic fluid, paints, synthetic rubber, plastics, margarine, and in ice cream. In India, the Hindus make a vegetarian butter called "ghee" from coconut oil; also used in infant formulas. When copra is heated, the clear oil separates out easily, and is made this way for home use in producing countries where it is used in lamps. Cake residue is used as cattle fodder, as it is rich in proteins and sugars; animals should not have more than 4 to 5 lbs per animal per day, as butter from milk will have a tallow flavor. As the cake is deficient in calcium, it should be fed together with calcium-rich foods. Trunk wood is used for building sheds and other semi-permanent buildings. Outer wood is close-grained, hard, and heavy, and when well seasoned, has an attractive dark-colored grain adaptable for carving, especially ornamentals, under the name of "porcupine wood". Coconut logs should not be used for fences, as decayed wood makes favorable breeding places for beetles. Logs are used to make rafts. Sections of stem, after scooping out pith, are used as flumes or gutters for carrying water. Pith of stem contains starch which may be extracted and used as flour. Pitch from top of tree is sometimes pickled in coconut vinegar. Coconut leaves made into thin strips are woven into clothing, furnishings, screens, and walls of temporary buildings. Stiff midribs make cooking skewers, arrows, brooms, brushes, and used for fish traps. Leaf fiber is used in India to make mats, slippers, and bags. Used to make short-lived torches. Coconut roots provide a dye, a mouthwash, a medicine for dysentery, and frayed out, it makes toothbrushes; scorched, it is used as coffee substitute. Coconut palm is useful as an ornamental; its only drawback being the heavy nuts which may cause injury to man, beast, or rooftop when they hit in falling.[64,80,253,278]

Folk medicine — According to Hartwell,[126] coconuts are used in folk remedies for tumors. Reported to be anthelmintic, antidotal, antiseptic, aperient, aphrodisiac, astringent, bactericidal, depurative, diuretic, hemostat, pediculicide, purgative, refrigerant, stomachic, styptic, suppurative, and vermifuge, coconut — somewhere or other — is a folk remedy for abscesses, alopecia, amenorrhea, asthma, blenorrhagia, bronchitis, bruises, burns, cachexia, calculus, colds, constipation, cough, debility, dropsy, dysentery, dysmenorrhea, earache, erysipelas, fever, flu, gingivitis, gonorrhea, hematemesis, hemoptysis, jaundice, menorrhagia, nausea, phythisis, pregnancy, rash, scabies, scurvy, sore throat, stomach-ache, swelling, syphilis, toothache, tuberculosis, tumors, typhoid, venereal diseases, and wounds.[91]

Chemistry — Per 100 g, the kernel is reported to contain 36.3 g H_2O, 4.5 g protein, 41.6 g fat, 13.0 g total carbohydrate, 3.6 g fiber, 1.0 g ash, 10 mg Ca, 24 mg P, 1.7 mg Fe, and traces of beta-carotene.[70] Per 100 g, the green nut is reported to contain 77 to 200 calories, 68.0 to 84.0 g H_2O, 1.4 to 2.0 g protein, 1.9 to 17.4 g fat, 4.0 to 11.7 g total carbohydrate, 0.4 to 3.7 g fiber, 0.7 to 0.9 g ash, 11 to 42 mg Ca, 42 to 56 mg P, 1.0 to 1.1 mg Fe, 257 mg K, trace of beta-carotene, 0.4 to 0.5 mg thiamine, 0.03 mg riboflavin, 0.8 mg niacin, and 6 to 7 mg ascorbic acid.[89] Coconut oil is one of the least variable among vegetable fats, i.e., 0.2 to 0.5% caproic-, 5.4 to 9.5 caprylic-, 4.5 to 9.7 capric-, 44.1 to 51.3 lauric-, 13.1 to 18.5 myristic, 7.5 to 10.5 palmitic-, 1.0 to 3.2 stearic-, 0 to 1.5 arachidic-, 5.0 to 8.2 oleic-, and 1.0 to 2.6 linoleic-acids.[70] Following oil extraction from copra, the coconut cake (poonac) contains 10.0 to 13.3% moisture, 6.0 to 26.7% oil, 14.3 to 19.8% protein, 32.8 to 45.3% carbohydrates, 8.9 to 12.2% fibers, and 4.0 to 5.7% ash. The so-called coconut water is 95.5% water, 0.1% protein, <0.1% fat, 0.4% ash, 4.0% carbohydrate. Per 100 g water, there is 105 mg Na, 312 K, 29 Ca, 30 Mg, 0.1 Fe, 0.04 Cu, 37 P, 24 S, and 183 mg choline. Leaves contain 8.45% moisture, 4.28% ash, 0.56% K_2O, 0.25 P_2O_5, 0.28 CaO, and 0.57% MgO.[70,89]

Description — Palm to 27 m or more tall, bearing crown of large pinnate leaves; trunk stout, 30 to 45 cm in diameter, straight or slightly curved, rising from a swollen base surrounded by a mass of roots; rarely branched, marked with rings of leaf scars; leaves 2 to 6 m long, pinnatisect, leaflets 0.6 to 1 m long, narrow, tapering; inflorescence in axil of each leaf as spathe enclosing a spadix 1.3 to 2 m long, stout, straw- or orange-colored, simply branched; female flowers numerous, small, sweet-scented, borne toward the top of panicle; fruit ovoid, 3-angled, 15 to 30 cm long, containing a single seed; exocarp a thick, fibrous husk, enclosing a hard, boney endocarp or shell. Adhering to the inside wall of the endocarp is the testa with thick albuminous endosperm, the coconut meat; embryo below one of the three pores at end of fruit, cavity of endosperm filled in unripe fruit with watery fluid, the coconut water, and only partially filled when ripe. Flowers and fruits year-round in the tropics.[278]

Germplasm — Reported from the Indochina-Indonesia and Hindustani centers of origin, coconut has been reported to tolerate high pH, heat, insects, laterites, low pH, poor soil, salt, sand, and slope.[82] Many classifications have been proposed for coconuts; none is wholly satisfactory. Variations are based on height, tall or dwarf; color of plant or fruit; size of nut (some palms have very large fruits, others have large numbers of small fruits); shape of nuts, varying from globular to spindle-shaped or with definite triangular sections; thickness of husk or shell; type of inflorescence; and time required to reach maturity. Many botanical varieties and forms have been recognized and named, using some of the characteristics mentioned above. Cultivars have been developed from various areas. Dwarf palms, occurring in India as introductions from Malaysia, live about 30 to 35 years, thrive in rich soils and wet regions, flower and fruit much earlier than tall varieties, and come into bearing by the fourth year after planting. However, dwarf varieties are not grown commercially, and only on a limited scale, because of their earliness and tender nuts — which yield a fair quantity of coconut water. They are highly susceptible to diseases and are adversly affected by even short periods of drought. Tall coconuts are commonly grown for commercial purposes, living 80 to 90 years. They are hardy, thrive under a variety of soil, climatic, and cultural conditions, and begin to flower when about 8 to 10 years after planting. ($2n = 16$.)[278]

Distribution — Now pan-tropical, especially along tropical shorelines, where floating coconuts may volunteer, the coconut's origin is shrouded in mysteries, vigorously debated. According to Purseglove,[272] the center of origin of cocoid palms most closely related to coconut is in northwestern South America. At the time of the discovery of the New World, coconuts (as we know them today) were confined to limited areas on the Pacific coast of Central America, and absent from the Atlantic shores of the Americas and Africa. Coconuts drifted as far north as Norway are still capable of germination. The wide distribution of coconut has no doubt been aided by man and marine currents as well.

Ecology — Ranging from Subtropical Dry to Wet through Tropical Very Dry to Wet Forest Life Zones, coconut has been reported from stations with an annual precipitation of

7 to 42 dm (mean of 35 cases = 20.5), annual temperature of 21 to 30°C (mean of 35 cases = 25.7°C) with 4 to 12 consecutive frost-free months, each with at least 60 mm rainfall, and pH of 4.3 to 8.0 (mean of 27 cases = 6.0).[82]

Cultivation — Propagated by seedlings raised from fully mature fruits. Seeds selected from high-yielding stock with desirable traits. Seed-nut trees should have a straight trunk and even growth, with closely spaced leaf-scars, short fronds, well oriented on the crown, and short bunch stalks. The inflorescence should bear about 100 female flowers, and the crown should have a large number of fronds and inflorescences. Seed-nuts should be medium-sized and nearly spherical in shape; long nuts usually have too much husk in relation to kernel. Because male parent is unknown and because female parent is itself heterozygous, seed-nuts from high-yielding palms do not necessarily reproduce the same performance in progeny. Records are kept of fruits harvested from each mother palm, such as number of bunches, number of nuts, weight of husked nuts, estimated weight of copra (about one-third weight of husked nuts being considered favorable). After fully mature nuts are picked (not allowed to fall), they are tested by shaking to listen for water within. Under-ripe or spoiled nuts or those with no water, or with insect or disease damage are discarded. Nuts are planted right away in nursery or stored in a cool, dry, well-ventilated shed until they can be planted. Seeds planted in nursery facilitate selection of best to put in field, as only half will produce a high-yielding palm for copra. Also, watering and insect control is much easier to manage in nursery. Soil should be sandy or light loamy, free from waterlogging, but close to source of water, and away from heavy shade. Nursery should have long raised beds 20 to 25 cm high, separated by shallow drains to carry away excessive water. Beds should be dug and loosened to a depth of 30 cm. Loosened soil mixed with dried or rotten leaves and ash from burnt fresh coconut husks at a rate of 25 lbs. of husk-ash per 225 ft.^{2} Nuts spaced in beds ca. 20 × 30 cm, a hectare of nursery accommodating 100,000 seed-nuts. Nuts planted horizontally produce better seedlings than those planted vertically. The germinating eye is placed uppermost in a shallow furrow, about 15 cm deep, and soil mounded up around, but not completely covering them, leaving the eye exposed. Soaking nuts in water for 1 to 2 weeks before planting may benefit germination; longer periods of soaking are progressively disadvantageous. Bright sunlight is best for growing stout sturdy seedlings. Regular watering in the nursery is essential in dry weather. Mulching may preserve moisture and suppress weeds. Paddy straw, woven coconut leaves, and just coconut leaves are used; however, they might encourage termites. Potash fertilizer may help seedlings which probably do not need other fertilizers, the nut providing most of needed nutrition. About 16 weeks after the nut is planted, the shoot appears through the husk, and at about 30 weeks, when 3 seed-leaves have developed, seedlings should be planted out in permanent sites. Rigorous culling of seedlings is essential. All late germinators and very slow growers are discarded. Robust plants, showing normal rapid growth, straight stems, broad, comparatively short, dark-green leaves with prominent veins, spreading outward and not straight upward, and those free of disease symptoms, are selected for planting out. Best spacing depends upon soil and terrain. Usually 9 to 10 m on the square is used, planting 70 to 150 trees per ha; with triangular spacing of 10 m, 115 palms per ha; and for group or bouquet planting, 3 to 6 palms planted 4 to 5 m apart. Holes 1 m wide and deep should be dug 1 to 3 months before seedlings are transplanted. In India and Sri Lanka, 300 to 400 husks are burned in each hole, providing 4 to 5 kg ash per hole. This is mixed with topsoil. Two layers of coconut husks are put into the bottom of the hole before filling with the topsoil-mixed ash. Muriate of potash, 1 kg per hole, is better than ash, but increases cost of planting. The earth settles so that it will be 15 to 30 cm below ground level when seedling is planted. In planting, soil should be well-packed around nut, but should not cover collar of seedling, nor get into leaf axils. As plant develops, trunk may be earthed up, until soil is flush with general ground level. Usually 7 to 8 month old seedlings are used for transplants, best done in the rainy season. In some instances plants up to 5 years old are used, as they are more resistant to termite damage. If older plants are used, care must be taken not to damage roots, as they are slow to recover. In areas with only one rainy season per year, it is simpler to plant nuts in the nursery in one rainy season, and transplant them a year later. Young plantation should be fenced to protect plants from cattle, goats, or other wild animals. Entire areas may be fenced in. In Sri Lanka and southern India, piles of coconut husks are placed around the tree. At the end of the first year after transplanting, vacancies should be filled with plants of the same age held in reserve in nursery. Also any slow-growers, or disease-damaged plants should be replaced. During the first 3 years, seedlings should be watered during drought, at about 16 liters per tree twice a week. Keep trees clear of weeds, especially climbers. Usually a circle 1 to 2 m in radius should be weeded several times a year, the weeds left as mulch. Cover-crops, as *Centrosema pubescens, Calopogonium mucunoides*, or *Pueraria phaseoloides*, are used and turned under before dry season. Catch-crops such as cassava (*Manihot utilissima*), and green gram (*Vigna aureus*) and cowpea (*Vigna unguiculata*), bananas and pineapples, may be used. Sometimes bush crops, in addition to or instead of, ground covers are used as green manures, e.g., *Tephrosia candida, Crotalaria striata, C. uraramoensis, C. anagyroides* — all fast growers. *Gliricidia sepium* and *Erythrina lithosperma* may be grown as hedges or live fences, their loppings used as green manure. Usually the cheapest form of fertilizer materials are used, consisting of 230 to 300 g N, 260 to 460 g P_2O_5, and 300 to 670 g K_2O per palm. Lime is generally not recommended. There is no evidence that salt is beneficial, as sometimes claimed. Coconuts can withstand a degree of salinity, about 0.6%, which is lethal to many other crops. Needing some magnesium, the palms are extremely sensitive to an excess. Cultivation depends on soil type, slope of land, and rainfall distribution. Disk-harrowing at end of moonsoon rains may be all that is necessary to control weeds.[278]

Harvesting — Trees begin to yield fruit in 5 to 6 years on good soils, more likely 7 to 9 years, and reach full bearing in 12 to 13 years. Fruit-set to maturity is 8 to 10 months; 12 months from setting of female flowers. Nuts must be harvested fully ripe for making copra or desiccated coconut. For coir they are picked about one month short of maturity, so that husks will be green. Coconuts are usually picked by human climbers, or cut by knives attached to the end of long bamboo poles. With the pole, a man can pick some 250 palms in a day — by climbing, only 25. In some areas nuts are allowed to fall naturally and collected regularly. Nuts are husked in the field, a good husker handling 2,000 nuts per day. Then the nut is split (up to 10,000 nuts per working day). Copra may be cured by sun-drying, or by kiln-drying, or by a combination of both. Sun-drying requires 6 to 8 consecutive days of good bright sunshine to dry meat without its spoiling. Drying reduces moisture content from 50% to below 7%. Copra is stored in a well-ventilated, dry area. Extraction of oil from copra is one of the oldest seed-crushing industries of the world. Coconut cake is usually retained to feed domestic livestock. When it contains much oil, it is not fed to milk cows, but it used as fertilizer. Desiccated coconut is just the white meat; the brown part is peeled off. It is usually grated, then dried in driers similar to those for tea. Good desiccated coconut should be white in color, crisp, with a fresh nutty flavor, and should contain less than 20% moisture and 68 to 72% oil, the extracted oil containing less than 0.1% of free fatty acid, as lauric. Parings, about 12 to 15% of kernels, are dried and pressed yielding about 55% oil, used locally for soap-making. The resulting residue "poonac" is used for feeding cattle. Coconut flour is made from desiccated coconut with oil removed, and the residue dried and ground. However, it does not keep well. Coir fiber is obtained from slightly green coconut husks by retting in slightly saline water that is changed frequently (requires up to 10 months); then, husks are rinsed with water and fiber separated by beating with wooden mallets. After drying, the fiber is cleaned and graded. The greater part of coir produced in India is spun into yarn, a cottage industry, and then used for rugs and ropes.

In Sri Lanka, most coir consists of mechanically separated mattress and bristle fiber. To produce this, husks are soaked or retted for 1 to 4 weeks, and then crushed between iron rollers before fibers are separated. Bristle fibers are 20 to 30 cm long; anything shorter is sold as superior mattress fiber. In some areas, dry milling of husks, without retting, is carried on and produces only mattress fiber. The separated pith, called bast or dust, is used as fertilizer since the potash is not leached out. Coconuts may be stored at a temperature of 0 to 1.5°C with relative humidity of 75% or less for 1 to 2 months. In storage, they are subject to mold, loss in weight and drying up of the nut milk. They may be held for 2 weeks at room temperature without serious loss.[12]

Yields and economics — For copra, an average of 6,000 nuts are required for 1 ton; 1,000 nuts yield 500 lbs. of copra, which yields 250 lbs. of oil. The average yield of copra per ha is 3 to 4 tons. Under good climatic conditions, a fully productive palm produces 12 to 16 bunches of coconuts per year, each bunch with 8 to 10 nuts, or 60 to 100 nuts per tree. Bunches ripen in about 1 year, and should yield 25 kg or more copra. For coir, 1,000 husks yield about 80 kg per year, giving about 25 kg of bristle fiber and 55 kg of mattress fiber. Efficient pressing will yield from 100 kg of copra, approximately 62.5 kg of coconut oil, and 35 kg coconut cake, which contains 7 to 10% oil. The factor 63% is generally used for converting copra to oil equivalent. Yields of copra as high as 5 MT/ha have been reported, but oil yields of 900 to 1,350 kg/ha have been reported. Pryde and Doty[270] put the average oil yield at 1,050 kg/ha, Telek and Martin,[359] at 600 kg/ha. World production of coconut oil is more than 2 million tons/year, about half of which moves in international trade. Sri Lanka, Philippine Islands, Papua, and New Guinea are the largest producers. Only about 40% of copra produced is exported, the remaining 60% processed into oil in the country of origin. The U.S. annually imports 190 million pounds of coconut oil and more than 650 million pounds of copra; some sources state 300,000 tons copra and over 200,000 tons coconut oil annually.[278]

Energy — The coconut of commerce weighs 0.5 to 1.0 kg. According to Purseglove,[272] the average number of nuts per hectare varies from 2,500 to 7,500, indicating a yield of ca. 1,200 to 7,500 kg/ha. On the one hand, 'Jamaica Talls' fruits average 1.7 kg, nuts 0.7 kg, of which 50% is endosperm; on the other hand, 'Malayan Dwarfs' fruits average 1.1 kg, the nut 0.6 kg, yielding 0.2 kg copra (6,000 nuts per ton copra). Average production yields of copra (3 to 8 nuts per kg copra) range from 200 kg/ha in Polynesia to 1,200 kg/ha in the Philippines, suggesting coconut yields of 1,000 to 8,000 kg/ha. Since about 60% of this constitutes the inedible fruit husk and seed husks, I estimate the chaff factor at 0.6. Coconut oil, cracked at high temperatures, will yield nearly 50% motor fuel and diesel fuel. Coconut destructive distillation is reported to yield 11.5% charcoal, 11% fuel gas, 37.5% copra spirit, 12.5% olein distillate, 1% crude acetate, 0.15% glycerol, and 0.85% acetone plus methanol.[70] As of June 15, 1981, coconut oil was $0.275/lb., compared to $0.38 for peanut oil, $1.39 for poppy seed oil, $0.65 for tung oil, $0.33 for linseed oil, $0.265 for cotton-seed oil, $0.232 for corn oil, and $0.21 for soybean oil.[351] At $2.00 per gallon, gasoline is roughly $0.25/lb. Quick[274] tested linseed oil (Iodine number 180) which cokes up fuel injectors inless than 20 hr: and rapeseed oil (Iodine number ca.100) which logs into the hundreds of hours before the onset of severe injector coking. Coconut oil (Iodine number 10) should be a very good candidate from this viewpoint. This could be very important in developing tropical countries where diesel fuel is scarce and often *more* expensive than coconut oil. One Australian patent suggests that distillation of coconuts at 550° gave 11.5% charcoal, 11% fuel gas, and 37.5% copra spirit, 12.5% olein distillate, 12.5% black oil, 1% crude acetic acid, 0.15% glycerol, and 0.85% (acetone + methanol) which natural fermentation takes to 2.7-5.8% ethanol. Of course, you can't have your coconut toddy and eat or drink or burn it too.[81]

Biotic factors — Coconuts are subject to numerous fungal diseases, bacterial infections, and the most serious virus-like disease, cadang-cadang.[161] Coconut trees are also attacked by numerous nematodes and some insect pests, the most damaging insect being the black beetle or rhinoceros beetle (*Oryctes rhinoceros*), which damages buds, thus reducing nut yield, and breeds in decaying refuse. Diseases and pests of a particular area should be considered and a local agent consulted as to how to deal with them. Agriculture Handbook No. 165[4] lists the following as affecting this species: *Aphelenchoides cocophilus* (red ring disease), *Cephalosporium lecanii, Diplodia epicocos, Endocalyx melanoxanthanus, Endoconidiophora paradoxa* (leaf-bitten disease, leaf scorch, stem-bleeding), *Gloeosporium* sp., *Pellicularia koleroga* (thread blight), *Pestalotia palmarum* (gray leaf spot, leaf-break), *Phomopsis cocoes* (on nuts), *Phyllosticta* sp. (on leaves), *Physalospora fusca* (on leaves), *P. rhodina* (on roots and trunk), *Phytopthora palmivora* (bud rot, leaf drop, wilt), *Pythium* sp. (wilt). Stevenson[380] adds: *Aschersonia cubensis, Aschersonia turbinata, Botryosphaeria quercuum, Cytospora palmicola, Escherichia coli, Flammula earlei, Herpotrichia schiedermayeriana, Hypocrea rufa, Marasmius sacchari, Pestalotia gibberosa, Pestalotia versicola, Polyporus lignosus, Polyporus nivosellus, Polyporus zonalis, Rosellinia saintcruciana, Thielaviopsis paradoxa, Valsa chlorina.*

COLA ACUMINATA (Beauv.) Schott and Endl. (STERCULIACEAE) — Kola Nuts, Cola, Guru
Syn: *Sterculia acuminata* Beauv.

Uses — Widely used as a flavor ingredient in cola beverages, but has also been used in baked goods, candy, frozen dairy deserts, gelatins, and puddings.[180] Kola plays an important role in the social and religious life of Africans. Beverage made by boiling powdered seeds in water, equal in flavor and nutriment to cocoa. Seeds also used as a condiment. Dye utilized from red juice. Wood valuable, light in color, porous, and used in ship-building and general carpentry. Tree often planted as ornamental.[100,278] Cola is said to render putrid water palatable.[86]

Folk medicine — According to Hartwell,[126] the powdered bark is used for malignant tumors and cancer. The tea made from the root is said to alleviate cancer. Reported to be aphrodisiac, cardiotonic, CNS-stimulant, digestive, diuretic, stimulant, and tonic, kola is a folk remedy for cancer, hunger, nerves, and tumors.[91] Nuts used as diuretic, heart tonic and masticatory to resist fatigue, hunger and thirst. A small piece of nut is chewed by Africans before mealtime to improve digestion. On the other hand, it is chewed as a stimulant and appetite depressant, e.g., during religious fasts. Jamaicans take grated seed for diarrhea.[224] Powdered cola is applied to cuts and wounds.[117] Formerly used as a CNS-stimulant and for diarrhea, migraine, and neuralgia.[126] The fresh drug is used, especially in its native country, as a stimulant, social drug, being mildly euphoric.[86,187]

Chemistry — Per 100 g, the fruit (ZMB) is reported to contain 399 calories, 5.9 g protein, 1.1 g fat, 90.8 g total carbohydrate, 3.8 g fiber, 2.2 g ash, 156 mg Ca, 232 mg P, 5.4 mg Fe, 67 μg beta-carotene equivalent, 0.08 mg thiamine, 0.08 mg riboflavin, 1.62 mg niacin, and 146 mg ascorbic acid. The aril (ZMB) is reported to contain 371 calories, 9.0 g protein, 3.6 g fat, 86.2 g total carbohydrate, 4.8 g fiber, 1.2 g ash, 18 mg Ca, 102 mg P, 8.4 mg Fe, 180 μg beta-carotene equivalent, 0.06 mg thiamine, 0.30 mg riboflavin, 4.19 mg niacin, and 60 mg ascorbic acid.[89] Contains 1.28 to 3% of fixed oil.[332] Kola nut important for its caffeine content and flavor; caffeine content 2.4 to 2.6%. Nuts also contain theobromine (<0.1%) and other alkaloids, and narcotic properties. Seeds also contain betaine, starch, tannic acid, catechin, epicatechin, fatty matter, sugar and a fat-decomposing enzyme. From a bromatological point of view, cola fruits contain, per 100 g, 148 calories, 62.9% water, 2.2 % protein, 0.4% fat, 33.7% carbohydrates, 1.4% fiber, 0.8% ash, 58 mg Ca, 25 mg carotene, 0.03 mg thiamine, 0.03 mg niacin, 0.54 mg riboflavin, and 60 mg ascorbic acid. Hager's Handbook suggests 1.5 to 2% caffeine, up to 0.1% theobromine, 0.3 to 0.4% D-catechin, 0.25% betaine, 6.7% protein, 2.9% sugar, 34% starch, 3% gum, 0.5% fat, 29% cellulose, and 12% water.[187]

Toxicity — Caffeine in large doses is reported to be carcinogenic, mutagenic, and teratogenic.[180] Caffeine is also viricidal, suppressing the growth of polio, influenza, herpes simplex, and vaccinia viruses, but not Japanese encephalitis virus, Newcastle disease, virus, and type 2 adenovirus.[344] Tyler[323] produces a chart comparing various caffeine sources to which I have added rounded figures from Palotti.[245]

Source	Caffeine content (mg)
Cup (6 oz.) expresso coffee	310
Cup (6 oz.) boiled coffee	100
Cup (6 oz.) instant coffee	65
Cup (6 oz.) tea	10—50
Cup (6 oz.) cocoa	13
Can (6 oz.) cola	25
Can (6 oz.) Coca Cola	20
Cup (6 oz.) mate	25—50
Can (6 oz.) Pepsi Cola	10
Tablet caffeine	100—200
Tablet (800 mg) Zoom (Paullinia cupana)	60

In humans, caffeine 1,3,7-trimethylxanthine, is demethylated into three primary metabolites: theophylline, theobromine, and paraxanthine. Since the early part of the 20th century, theophylline has been used in therapeutics for bronchodilation, for acute ventricular failure, and for long-term control of bronchial asthma. At 100 mg/kg, theophylline is fetotoxic to rats, but no teratogenic abnormalities were noted. In therapeutics, theobromine has been used as a diuretic, as a cardiac stimulant, and for dilation of arteries. But at 100 mg, theobromine is fetotoxic and teratogenic.[65] Leung[180] reports a fatal dose in man at 10,000 mg, with 1,000 mg or more capable of inducing headache, nausea, insomnia, restlessness, excitement, mild delirium, muscle tremor, tachycardia, and extrasystoles. Leung also adds "caffeine has been reported to have many other activities including mutagenic, teratogenic, and carcinogenic activities; . . . to cause temporary increase in intraocular pressure, to have calming effects on hyperkinetic children . . . to cause chronic recurring headache . . . "[180]

Description — Long-lived evergreen tree, up to 14 m tall, resembling an apple tree; bark smooth, green, thick, fissured in old trees. Leaves alternate, on petioles 2.5 to 7.5 cm long; young leaves pubescent, often once or twice cut near base about half-way to midrib; mature leaves 16 to 20 cm long, 2.5 to 5 cm broad, leathery, obovate, acute and long-acuminate, with prominent veins below, margin entire, dark-green on upper surface. Flowers yellow, numerous, unisexual or bisexual, 15 or more in axillary or terminal panicles, no petals; calyx petaloid, greenish-yellow or white, purple at edges, tube green, limb 5-cleft, lobes ovate-lanceolate; male flower with slender column, shorter than calyx, bearing a ring of 10 2-lobed anthers, the anthers divergent; perfect flowers with subsessile anthers in a ring, ovary 5-lobed, 5-celled, stellate pilose, with 5 linear, re-flexed, superposed styles; ovules anatropous, attached in a double row to the ventral surface of each carpel. Fruit oblong, obtuse, rostrate, warty coriaceous to woody, 5 to 17 cm long, 5 to 7.5 cm thick, brown resembling alligator skin, pericarp thick, fibrous, cells filled with resinous colored matter used as dye. Seeds 5 to 12 per fruit, 2.5 to 5 cm long, 1.3 cm thick, yellow, soft, internally whitish, pinkish or purple, brown when dry; cotyledons often 3, flatly ovate or auriculate, cells containing starch and albuminous material. Flowers December to February, and May to July; fruits May to June, and October to November.[278]

Germplasm — Reported from the African Center of Diversity, kola, or cvs thereof, is reported to tolerate low pH, shade, and slope. (2n = 40.)[82]

Distribution — Native and cultivated along west coast of tropical Africa, now cultivated pantropically from 10°N to 5°S latitude, especially in West Indies, South America, Sri Lanka, and Malaya. Occurs naturally in forests from Togo and southern Nigeria eastward and southward to Gabon, Congo, and Angola. Extensively planted in Nigeria.[278]

Ecology — Ranging from Subtropical Dry to Wet through Tropical Dry to Wet Forest Life Zones, kola is reported to tolerate annual precipitation of 6.4 to 40.3 dm (mean of 12 cases = 19.8), annual temperature of 21.3 to 26.6°C (mean of 12 cases = 25.2), and pH of 4.5 to 8.0 (mean of 7 cases = 5.5).[82] Thrives in tropical areas where mean annual temperatures are uniformly 21 to 27°C, moist, with 200 to 225 cm rainfall, mostly at sea level to 300 m altitude. Frequently forms forests in coastal areas. Requires a rich, well-drained soil, but will grow on deep sandy loams in West Indies, with high organic content.[278]

Cultivation — Propagated from seed, which must be sown perfectly fresh. Seeds planted singly in pots and young trees kept growing until needed for permanent planting. Only light shade, if any, is required after trees are 3 years old. Planting distances about 6 to 8 m each way, equalling about 270 trees per ha. Cultivation very easy. Trees respond to fertilizers, and produce highest yields only when weeds are kept controlled. Propagation also by cuttings of softwood or ripe wood, using bottom heat.[273,278]

Harvesting — Trees begin to flower 5 to 10 years after planting, reaching full production by the 20th year, continuing to bear for 70 to 100 years. In many regions, trees flower and fruit throughout the year, but usually two peak crops are produced in May and June and again in October and November. Fruits require about 4 to 5 months to mature. Harvest when pods turn chocolate-brown and begin to dehisce. Pods are shaken from tree and immediately gathered. Seeds removed from pods and first coat cut off, leaving bare cotyledons. Nuts are then carefully graded. Fresh kola nuts tend to mold and spoil easily. Nuts packed and transported for local consumption is homemade baskets lined with leaves and wrapped in canvas or hide to prevent drying out. Kola nuts imported by the U.S. are split in half, sundried, and shipped in bags. Entire seeds are kola nuts of native consumer; kola nuts of commerce are the separated, dried cotyledons only.[278]

Yields and economics — After 10 years, kola trees may be expected to yield 400 to 500 (to 800) pods annually, this being equivalent to 40 to 50 (to 80) lbs of dried nuts.[196,278] Purseglove[272] reports ca. 575 kg/ha salable nuts. Within the tropics, trade of this nut is immense. In West Tropical Africa, kola nut ranks second to the oil-palm (*Elaeis*), with exports over 16 million lbs per year. Although most kola nuts are harvested from wild trees in West Africa coastal areas, the U.S. imports most of its kola nuts from Jamaica, about 170 tons per year.[278]

Energy — Husks, prunings, and fallen leaves can be used for energy production.

Biotic Factors — Poor yields some years have been attributed to poor pollination. Fungi known to attack kola trees include: *Botryodiplodia theobromae, Calonectria rigidiuscula, Cephaleuros mycoidea, Fomes lignosus, F. noxius, Marasmius byssicola, M. scandens*, and *Pleurotus colae*.[186,278]

COLA NITIDA (Vent.) Schott and Endlicher (STERCULIACEAE) — Gbanja Kola

Uses — Kola possesses the central stimulating principle of caffeine. This species is more valued than *C. acuminata* as it contains more caffeine. Nuts are used in West Africa to sustain people during long journeys or long hours of work. Kola, Cola, or Kola-nuts is the dried cotyledon of *Cola nitida*, or of some other species of *Cola*. In the U.S., the kola-nut is used in the manufacture of nonalcoholic beverages. The tree is valued for its wood, which is whitish, sometimes slightly pinkish when fresh; the heartwood is dull yellowish-brown to reddish-tinged. Wood is suitable for carpentry and some construction work as house-building, furniture, and boat-building. Wooden platters, domestic utensils, and images are often carved from the wood. Sometimes trees are planted for ornamental purposes.[86]

Folk medicine — Reported to be astringent, nervine, poison, restorative, sedative, stimulant, stomachic, and tonic, gbanja kola is a folk remedy for digestion, dysentery, exhaustion, hunger, malaria, nausea, and toothache.[91] Dried cotyledons are nervine, stimulant, tonic, and astringent.[324] The seeds are used by natives as a stimulant; when chewed, nuts increase powers of endurance of the chewer.[86,280]

Chemistry — Speaking generically, Hager's Handbook[187] stated that the nuts contain 1.5 to 2% caffeine, a compound the Germans call colarot (= ?cola red) $C_{14}H_{13}(OH)_5$, and glucose. Colarot splits into phloroglucin and a reddish dye. Also contains up to 0.1% theobromine, 0.3 to 0.4% D-catechin ($C_{15}H_{14}O$) ("colatine"), L-epicatechin, essential and fatty oils, colalipase, colaoxydase, a tannic glycoside, 0.25% betaine, 6.7% protein, 2.9% sugar, 34% starch, 3% gum, 0.5% fat, 29% cellulose, 12% water, and procyanidin ($C_{30}H_{26}O_{12}$).[187] The glucoside kolanin is a heart stimulant.[272]

Description — Trees 13 to 20 m tall, with dense crown, the branches and leaves nearly touching the ground. Leaves alternate, 7.5 cm or more long, broadly lanceolate, sharply acuminate, leathery. Flowers yellowish-white, sometimes with red stripes or blotches; fruits 2 in a cluster, covered with a thick green wrinkled coat, each fruit containing 6 to 10 or more nuts; nuts usually red or pink, sometimes white. Fruits commonly longitudinally rugose and wrinkled, nodular to some degree and dorsally keeled; seed separable into only two cotyledons (*C. baileyi* Cornu, from West Equatorial Africa, has 6 cotyledons with very little caffeine.) Flowers and fruits in spring and autumn, with two harvests. The main cola season in West Africa is from October to February.[278]

Germplasm — Reported from the African Center of Diversity, gbanja kola, or cvs thereof, is reported to tolerate low pH, shade, slope, and virus.[82] Chevalier has divided *C. nitida* into four subspecies: *rubra, alba, mixta*, and *pallida*. *C. nitida* subsp. *rubra* Chev., wild in Ivory Coast and Ashanti, has nuts larger than those of the cultivated plants and is the common cultivated kola of Ashanti; subsp. *rubra* Chev., from the Ivory Coast, is a distinct race based on characters other than those of color of the seeds; subsp. *mixta* Chev., known only in cultivation, has red and white nuts on the same tree, and sometimes on the same follicle; and subsp. *alba* Chev., also only known in cultivation, has only white seeds. There is much variation in other characteristics, as size of fruits and nuts and flavor. (2n = 40.)[82,278]

Distribution — Native to West Africa from Sierra Leone to the Congo. Introduced to East Africa, Sri Lanka, Singapore, Indonesia, Brazil, and West Indies, particularly Jamaica.[205,278]

Ecology — Ranging from Subtropical Moist to Wet through Tropical Dry to Moist Forest Life Zones, gbanja kola is reported to tolerate annual precipitation of 13.6 to 27.8 dm (mean of 6 cases = 22.0), annual temperature of 23.3 to 26.6°C (mean of 6 cases = 25.4°C), and pH of 4.5 to 5.3 (mean of 4 cases = 4.9).[82] Kola trees flourish where the mean annual temperature is between 20 to 26°C and the annual rainfall is 250 cm or more. It is found at low altitudes ranging up to several meters above sea-level. Thrives in deep sandy loam with much humus.[278]

Cultivation — Propagation is by seeds (usual), cuttings, air-layering, or grafting. Seeds are planted in seed-beds in well-prepared soil containing much humus. Seedlings are planted in rows 6.6 m apart each way. Trees respond to fertilizers and produce the highest yields only when weeds are cut back regularly. Best crops obtained on soils that are deep, sandy, and with a high content of organic matter. Plantain or other plant is used as shade for the first year or two. Cassava is a catch-crop for the tree until it gets large enough to bear fruits. Trees may also be propagated vegetatively from cuttings. Terminal cuttings set without any hormones retain their leaves and start callusing within 3 to 4 weeks after setting. The roots usually appear at an acute angle from the callus. New flush growth on the rooted cuttings starts at about the third month after potting and is commonly slow. Most cuttings flower the first year of growth. Cuttings set out in the field grow rapidly and flower and fruit within three years. When propagated by air-layering, about 98% of all branches treated are heavily callused within 3 to 4 weeks; within 6 weeks, most branches have developed roots 5 to 8 cm long. About 95% of all marcots become established satisfactorily in the field. Those obtained from mature, already fruiting trees, flowered in 6 to 7 months after cutting them from the mother plant, or 3 to 4 months after transplanting. Propagation by budding is successful at all times of the year, with the highest bud-take from patch or flute budding techniques obtained between January and April, the lowest between September and December.[19,278]

Harvesting — Kola trees produce two crops per year; in Jamaica, pods ripen in May and June and again in October-November; in West Africa, the main crop is harvested from October to February. The chocolate brown pods, which range in size from 5 to 10 cm long, are shaken from the tree and gathered immediately, or are cut off by tree-climbers with knives on long sticks. Harvesters climbing trees are occasionally attacked by ants. The seeds are removed from the pods and the outer coat is cut off, exposing the bare cotyledons. These are carefully graded inasmuch as only sound cotyledons do not deteriorate quickly. Fresh kola-nuts tend to mold and spoil rather easily. They must be taken to market quickly for local consumption. Kola-nuts of commerce are freed from the white covering, usually after soaking or by fermentation in broad leaves. Occasionally, the nuts are buried to keep them sound for a favorable market; in the equatorial regions, it is done in ant hills. The main trade is in good-sized nuts. Packing is done in baskets along with broad leaves, and with occasional moistening, the nuts can be transported for a month, free from mold. Kola-nuts prepared for shipment to the U.S. are split in half, sun-dried, and shipped in bags. They are usually soaked in water for 2 to 3 hr and the juice thrown off. For export to Europe, peat is recommended as a packing material suitable for all conditions of temperature, and the nuts, which are mainly used for drugs and wine, are shipped in the dry condition.[278]

Yields and economics — Depending on how they are propagated, trees begin to bear fruit in 4 to 5 years and reach full production in 10 to 15 years, or begin in 7 to 9 years and reach maturity in 15 to 20 years. Then they continue to bear good crops of fruits for 50 years or more. Usually after a tree is 10 years old, it may be expected to yield, in two harvests, about 56 kg of dried nuts per year.[278] Speaking generically, Purseglove[272] notes that of nearly 250 trees in Nigeria, ca. 20% gave no yield at all, ca. 60% gave mean annual yields up to 300 nuts, while 20% produced 72% of the total yield of the plot. The average was 210 nuts per tree, the 10 best trees averaging 1,415 nuts, while the best yielded 2,209 nuts per year. With an average 60 nuts per kilogram, that is more than 36 kg for the big yielder. Purseglove concludes there are an average 210 salable nuts per tree, or ca. 575 kg/ha.[272] Although most kola-nuts are harvested from wild trees of the West African coast, the U.S. imports most of its kola-nuts from Jamiaca. In the U.S., most kola-nuts are used for manufacturing nonalcoholic beverages.[278]

Energy — Husks, prunings, and fallen leaves can be used for energy production.

Biotic factors — Self-pollinated trees produce only white fruits (white-colored nuts bring the highest prices); the production of colored (red or pink) nuts may therefore be due to cross-pollination. The following fungi have been reported on this species of kolanut: *Auricularia delicata, Botryodiplodia theobromae, Corticium koleroga, Fomes lignosis, F. noxius, Graphium rhodophaeum, Irenopsis coliicola, Marasmius equicrinus, M. scandens, Nectria delbata, Phaeobotryosphaeria plicatula*; twig blight, root rot, and thread blight. Nematodes isolated from this tree include the following species: *Helicotylenchus cavenessi, H. pseudorobustus, Scutellonema clathricaudatum*, and *Xiphinema* sp. Insect pests include borers, cola weevils (*Balanogastris kolae*), and larvae of the moth *Characoma*. Trees are also attacked by pests found on cocoa, as the caspid *Sahlbergella singularis* and by *Mesohomotoma tessmanii*.[186,278]

COLA VERTICILLATA (Thonn.) Stapf ex A.Chev. (STERCULIACEAE) — Owe Cola, Slippery Cola, Mucilage Cola

Syn.: *Cola johnsonii* Stapf. and *Sterculia verticillata* Thonn.

Uses — Seeds, indistinguishable from true cola in appearance, are edible, though very bitter and considered unfit to eat.[146] Nuts are used to make a beverage. In some districts, the people gather the fruit, or at least chew it where found; in others, they usually regard it as a "monkey kola". Wood of the tree is white and hard, and is used in S. Nigeria to make fetish images.[146,278,287]

Folk medicine — Containing caffeine, this species no doubt shares some pharmacological properties and folk uses with other *Cola* species.

Chemistry — Nuts contain a fair proportion of caffeine.[278]

Description — Trees large, 8 to 10(to 25) m tall; branches sparsely puberulent, rarely cylindrical, brownish dark-red, often weeping. Leaves verticillate in threes or fours, opposite in the lower nodes, simple, entire, subcoriaceous to coriaceous; stipules 5 to 6 mm long, puberulent on lower surface; petiole 2 to 6 mm long, sparsely puberulent; blades obovate-elliptic, oblong or oblanceolate, cuneate at base, attenuate to apex, 12 to 25 cm long, 3 to 9 cm broad, glabrous, subcoriaceous, green on upper surface, puberulent and brownish dark-red beneath; secondary veins in 5 to 8 pairs, ascending. Panicles axillary, isolated in groups of 2 to 3; flowers small, 1 to 3 cm long, puberulent; bracts oval, cuspidate, concave, about 6 mm long, more or less persistent; calyx 5- to 8-lobed, densely puberulent on external surface, sparsely so on inner surface; male flowers on pedicels 3 to 7 mm long, articulate at summit, puberulent, calyx 4 to 5(to 8) mm long with 5 to 6 lobes longer than the tube; androphore 1 mm long, puberulent, corona of stamens in 2 verticels, female flowers and perfect flowers on pedicels 12 mm long, articulate near the summit, with 5 to 7 lobes about 7 mm long; ovary with 5 carpels in 2 tiers of 4 ovules. Fruit on pedicels 3 to 4 cm long; follicles subsessile, oblong, up to 20 cm long, 9 cm broad, with short beak, obtuse, and more or less recurved, glabrous. Seeds 4 to 9, sometimes up to 12 per follicle, ovoid-elliptic, 3 mm long, 2 cm broad, either red or white, with 3 to 4 cotyledons.[278]

Germplasm — Reported from the African Center of Diversity, owe kola, or cvs thereof, is reported to tolerate shade, slope, and virus. ($2n = 40$.)[82]

Distribution — Native to Tropical West Africa from Ivory Coast and Ashanti to Cameroons and Lower Congo; planted in N. Nigeria and elsewhere, but nowhere much cultivated. Some cultivation in Nigeria, Cameroons, Ghana, Dahomey, Gabon, and Cabinda.[278] Often found in planting of *C. nitida*. The only kola found on the Mambilla Plateau in Northern Nigeria.[350]

Ecology — Ranging from Subtropical Moist through Tropical Dry to Moist Forest Life Zones, owe kola is reported to tolerate annual precipitation of 13.6 to 22.3 dm (mean of 3 cases = 17.7), annual temperature of 23.5 to 26.4°C (mean of 3 cases = 17.7°C), and pH of 4.8 to 5.0 (mean of 2 cases = 4.9).[82] Indigenous to damp forests of the tropical zone, especially in swamps and by streams; requiring the jungle-type habitat. Often planted in villages.[278]

Cultivation — Most trees are self-seeded in humid forests of tropical West Africa. Propagated by seed planted in site where desired. No special care given after tree is established.[278]

Harvesting — Fruits are gathered from trees in the wild in some districts. Occasionally trees are planted in villages; fruits are collected when ripe to make beverages.[278]

Yields and economics — A fruit of minor importance in area of adapation, used mostly by natives as a source of caffeine, for a beverage, and for wood.[278]

Energy — Husks, prunings, and fallen leaves can be used for energy production.

Biotic factors — The fungus *Irenopsis aburiensis* has been reported on this tree. No serious pests are reported.[278]

CORDEAUXIA EDULIS Hemsl. (CAESALPINIACEAE) — Yeheb Nut

Uses — Seeds said to be edible raw or cooked, likened by one author to a chestnut, by another to a cashew.[231] Much relished by the Somalians, often preferred to the usual diet of rice and dates. The leaves are infused to make a tea. Leaves, eagerly grazed by livestock, contain a brilliant red dye that will stain the hands, even the bones of goats who eat it. Somalians use the magenta-red coloring matter to stain textiles.[8,212,324]

Folk medicine — No data available.

Chemistry — Per 100 g, the seed (ZMB) contains 448 calories, 12.1 g protein, 13.5 g fat, 71.9 g total carbohydrate, 1.6 g fiber, 2.5 g ash, 36 mg Ca, 208 mg P, and 7.2 mg Fe. The NAS (1979) reports 37% starch, 24% sugar, 13% protein, and 11% fat. The protein contains amino acids in proportions similar to other pulses, deficient in methionine.[231] Miege and Miege[212] report 10.8 g arginine, 3.5 g histamine, 3.9 isoleucine, 6.4 g leucine, 6.8 g lysine, 0.7 g methionine, 3.9 g phenylalanine, 3.6 g threonine, 4.8 g valine, 1.9 g tyrosine, 0.6 g cystine, 9.1 g asparagine, 23.8 g glutamine, 3.9 g serine, 6.6 g prolamine, 4.9 g glycine, and 4.5 g alanine per 100 g protein. The albumins have trypsin inhibitors, the globulins nearly 10 times as much. Phytohemagglutinins, alkaloids, or glucosides are said to be absent.[8] The red stain is due to cordeauxione, the only naphthoquinone found in legumes.

Description — Dwarf multistemmed evergreen shrub to 3 m tall; lower branches dense, straight, broomlike, hard. Leaves paripinnate; leaflets usually oval-oblong, 4-paired, leathery, dotted below with reddish, scale-like glands; stipules none. Flowers few, yellow, in apical corymbs; calyx short; lobes 5, blunt, valvate, glandular; petals 5, subequal, clawed, spoon-shaped; stamens 10, free; filaments hairy below; anthers versatile; ovary short-stalked, 2-ovuled, densely glandular; stigma obtuse. Pod leathery, compressed-ovoid, curved, apex beaked, 2-valved, dehiscent, 1 to 4 seeded, seeds ovoid, endosperm lacking, cotyledons thick.[8] Germination epigeal, the eophylls 1 to 8-foliolate, the first eophylls often opposite.[212]

Germplasm — Reported from the Ethiopian Center of Diversity, yeheb, or cvs thereof, is reported to tolerate drought, high pH, poor red sandy soils, sand, and savanna.

Distribution — Endemic to Somalia, Malawi, and Ethiopia.

Ecology — Estimated to range from Subtropical Desert to Thorn through Tropical Desert to Thorn Forest Life Zones, yeheb is estimated to tolerate annual precipitation of 1 to 8 dm (1 to 2 reported), annual temperature of 23 to 30°C, and pH of 6 to 8.5 (reported 7.8 to 8.4). In its native habitat, yeheb occurs in savannas, elevation 300 to 1,000 m, with poor red sandy soils, two rainy seasons, annual rainfall of 250 to 400 mm, and no frosts.[231]

Cultivation — Only recently brought under cultivation at the Central Agricultural Research Station at Afgoi, Somalia, and at Voi and Galana Ranch, Kenya. Seeds germinate as high as 80%, the seedlings quickly developing a thin but tough tap root, which complicates transplanting. Hence, field seeding is recommended.

Harvesting — Starts fruiting at age 3 or 4 years, fruits said to ripen in only 5 to 6 days.[324]

Yields and economics — Overexploitation, overgrazing, nonflowering in drought, and war in its native habitat, have all jeopardized the very existence of the yeheb. "The plant is in great danger of extinction."[212] Somalis use ca. 200 g pulverized leaves to dye 90,000 cm^2 calico. In the old days of British Somaliland, sacks of the nuts were brought down to the coast for sale.[231]

Energy — No data available.

Biotic factors — Although the shrub itself is essentially free of insect pests, the nuts are attacked by weevils and moth larvae. Rhizobia are not reported, but root nodules are reported on younger roots.[8]

CORYLUS AMERICANA Walt. (BETULACEAE) — American Hazelnut or Filbert

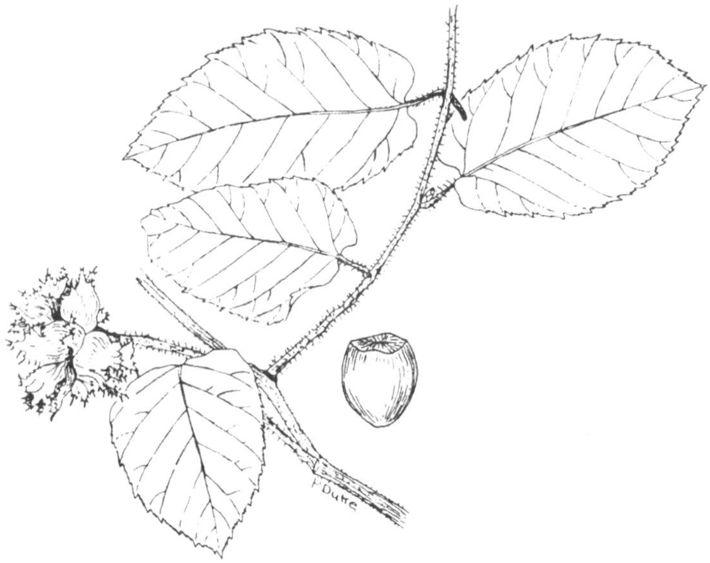

Uses — Cultivated for production of nuts for home use and wildlife, for cover and shelterbelt use, and for an ornamental. Kernels eaten raw or roasted, alone or mixed with other nuts.[278] Nuts may be beaten to a powder and used like flour to make filbert bread.[209]

Folk medicine — According to Hartwell,[126] the bark is used in folk remedies for a poultice for tumors. Reported to be a panacea, American hazelnut is a folk remedy for stomatitis and tumor.[91] Chippewa used the charcoal, pricked into the temples with needles, as analgesic; Ojibwa used a poultice of boiled bark to help close and heal cuts.[216] Said to cause allergic rhinitis, bronchial asthma, and/or hypersensitivity pneumonitis.[184]

Chemistry — Smith[310] reports the nuts to contain 5.4% water, 16.5% protein, 64.0% fat, 11.7% carbohydrates, 2.4% ash, and 3,100 calories per pound.

Description — Deciduous shrub, forming dense thickets, 1 to 3 m tall; branchlets pubescent and glandular bristly. Leaves 7.5 to 15 cm long, slightly cordate or rounded at base, broadly ovate or obovate, irregularly doubly serrate, sparingly pubescent above, paler and finely tomentose beneath. Involucre pubescent but not bristly, compressed, about twice as long as nut, the 2 bracts sometimes connate and usually tightly enclosing it with rather short, triangular, denate lobes, the whole 1.5 to 3 cm long; nut roundish-ovate, compressed, 1 to 1.5 cm long. Flowers March to April; fruits July to October; seed dispersal July to winter.[278]

Germplasm — Reported from the North American Center of Diversity. Of the botanical varieties, the following are sometimes recognized: *C. americana* forma *missouriensis* (DC.) Fern., plants without stipitate glands, and found occasionally throughout the range of the species. *C. americana* var. *indehiscens* Palm. and Steyerm., with fruiting bracts united on one side, found from North Carolina to Missouri. *C. americana* var. *calyculata* Winkl. (*C. calyculata* Dipp.), the involucre with 2 very large bracts at base. The most important cvs of the American filbert are 'Rush' and 'Winkler', both very widely cultivated and the most hardy of all filberts. Four hybrids — 'Bixly', 'Buchanan', 'Reed', and 'Potomac', have been introduced; these have intermediate characteristics between European and American varieties. The cultivars of American hazelnut have smaller nuts than those of European filberts. (2n = 22.)[278]

Distribution — Native from eastern Canada and Maine west to Ontario and Saskatchewan, south to Florida, Georgia, Oklahoma, and the Dakotas. Also usually cultivated within that range.[278]

Ecology — Estimated to range from Warm Temperate Dry to Wet through Cool Temperate Moist to Wet Forest Life Zones, the American hazelnut is estimated to tolerate annual precipitation of 5 to 30 dm, annual temperature of 6 to 14°C, and pH of 5.0 to 8.0.[82] Naturally grows in or along edge of woods and thickets, on both dry and moist soils. However, it grows best on moderately rich, well-drained soils. Filberts should be planted on soils which are deep, fertile, and well-drained. Heavy clay or silt soils as well as coarse, deep sand should be avoided. When planted on poorly drained soils, shrub grows poorly, is subject to winter injury and bears few nuts.[278] Hardy to Zone 4.[343]

Cultivation — Most filberts offered for sale by nurserymen have been propagated by layering and are on their own roots. Trees that have been grown for 1 or 2 years in the nursery after the layers have been removed from parent stock are preferable to older trees. Some nurserymen propagate their trees on Turkish filbert roots that do not produce suckers. This rootstock generally outgrows the scion to some extent. Filbert can be propagated from seed which have been stratified 60 days at 5°C, plus 67 days at 18°C, and 30 days at 5°C. Stratified seed should be sown in spring. Seed should be drilled in fall and protected from rodents. Germination is hypogeous. Horticultural cvs are propagated by suckers, layering, budding, or grafting and cuttings. Filberts of all varieties should be planted 5 to 7 m or more apart. In most cases, trees are planted in late winter or very early spring, after danger of severe freezing is passed. The same general methods of planting should be used as for apple and peach trees. Newly transplanted and young trees should be cultivated sufficiently to destroy all grass and weed growth before the beginning of tree growth in spring and through July. Mulching trees with any type of organic matter is as satisfactory as cultivation, provided that sufficient mulch is applied to a large enough area around each tree to keep grass and weed growth suppressed. In general, the same cultural practices used for peaches are satisfactory for filberts. Filberts generally respond readily to fertilizer applications, although no recommendation would apply to all situations. On most soils, it is not advisable to apply any fertilizer the first year after transplanting. Beginning the second year, about 475 gm (1 lb) per tree of a 5-10-5 or 6-6-5 fertilizer should be broadcast around the tree just before beginning of tree growth. The amount of fertilizer is increased by 475 gm/year until trees are 10 to 12 years old; after that, ca. 5 kg/per tree is adequate. Filberts are pruned to balance top loss with root loss (in planting), or to train young trees to desired form, or to remove dead, broken, or diseased branches, or to stimulate moderate growth of new shoots on old trees. At planting, tree should be cut back to about 60 cm above the ground, leaving 4 to 6 branches to grow. Trees or shrubs should be trained to the central leader form, provided it does not require removal of much wood. The more wood removed from young plants, the later they come into bearing; therefore, only necessary pruning should be done. Pruning should be done after pollen is shed and anthers have fallen. Since American filberts or hazelnuts tend to sucker, the suckers should be removed promptly and the plant trained to a single stem. Suckers should be removed at point on trunk or root where they originate; cutting them off at surface of soil only increases the number that grow. Suckering operations should be done 3 to 4 times a year, as they are easier to remove when young.[278]

Harvesting — Most filbert varieties are self-unfruitful, even though staminate and pistillate catkins are on the same tree or bush. Cross-pollination must be provided for satisfactory fruit-production. In all plantings, 2 or more varieties should be included. The period of pistillate flowering is usually much longer than that of pollen-shedding on a particular variety. Furthermore, pollen of one variety must be shed when pistillate flowers of the other variety are receptive. Nuts, good flavored, should be harvested from bushes in the fall as soon as edges of husks begin to turn brown. As all nuts do not mature at once, 2 to 4 gatherings may be necessary in a season. If nuts drop easily to ground, they should not be allowed to remain there long because of loss to rodents and birds and discoloration and moldiness due to wet weather. Nuts should be promptly dried by spreading them in a thin layer in a dry place having good air circulation. Nuts dried in an unheated building usually require 4 to 6 weeks for drying. During this process, they should be stirred frequently to prevent molding. The temperature of nuts dried by artificial heat should not be higher than 45°C; otherwise they will not store well. After nuts have dried for this time, they are flailed to remove the husk. The nuts, which are the commercial seed, can then be sown, stratified, or stored. Storage in sealed containers at 5°C will retain a large part of viability in *C. americana* for at least 2 years.[278]

Yields and economics — Brinkman[47] reports 491 seed per lb (ca. 1,080/kg). American filberts give good crops every 2 to 3 years, or light crops every year. Yield, size of nut, purity, soundness, and cost of commercial seed vary according to cv.[278] Great quantities of hazelnuts are gathered each year for home use in northeastern U.S. and Canada. Many more are used as food for wildlife.[278]

Energy — Small and erratically bearing, this species does not seem to hold great promise as a firewood or oilseed species. The 64% oil could conceivably serve as an energy source.

Biotic factors — A fungus disease, Eastern filbert blight, may cause severe damage to European filberts in the eastern U.S.; once well-established in a planting, it is very difficult, if not impossible to control. Growers should spot and eradicate early infections. Although this disease is almost always on American filbert plants, it usually does little damage to them. Each spring, trees should be carefully inspected and any diseased branches cut out and burned. Among the insect pests, hazelnut weevil, in severe infestations, may completely destroy the crop of nuts. Leaves are preferred food for Japanese beetles, and plants may be completely defoliated by them. Filbert bug mite and Birch case-borer (*Colephora salmani*) may be pest problems. Stink-bugs and other plant bugs attack developing nuts and cause them to be bitter when mature. As these insects breed on various plants, as legumes and blackberries, control chiefly depends on orchard sanitation and elimination from plantation of host plants on which bugs breed. For control of all pests, consult local state agents.[278] According to Agriculture Handbook No. 165,[4] the following attack this species: *Apioporthe anomala, Cenangium furfuraceum, Cucurbitaria conglobata, Cylindrosporium vermiformis, Diaporthe decedens, Diatrypella frostii, D. missouriensis, Diplodia coryli, Gloeosporium coryli, Gnomoniella coryli, G. gnomon, Hymenochaete cinnamomea, Hypoxylon fuscum, Melanconis flavovirens, Microsphaera alni, Phyllactinia corylea, Phymatotrichum omnivorum, Physalospora obtusa, Polyporus albellus, P. elegans, P. radiatus, P. stereoides, Scorias spongiosa, Septogloeum profusum, Septoria corylina, Sphaeropsis coryli, Taphrina coryli,* and *Valsa ambiens*.

CORYLUS AVELLANA L. (BETULACEAE) — European Filbert, Cobnuts, Hazelnuts, Barcelona Nuts

Uses — Long-cultivated, this is the main source of filberts of commerce. Kernel of nut eaten raw, roasted, or salted, alone or with other nuts; also used in confections and baked goods. Leaves sometimes used for smoking like tobacco. Hazelnut or filbert oil, a clear, yellow, non-drying oil is used in food, for painting, in perfumes, as fuel oil, for manufacture of soaps, and for machinery. Hazelwood or nutwood is soft, elastic, reddish-white with dark lines, and is easy to split, but is not very durable. It is used for handles, sieves, walking sticks, hoops of barrels, hurdles, wattles, and is a source of charcoal made into gunpowder.[278,324]

Folk medicine — According to Hartwell,[126] the paste derived from the bark is said to be a folk remedy for tumors. A salve, derived from the leaves and nuts, in a plaster with honey, is said to be a cure for cancer. Reported to be fumitory and vasoconstrictor, European filberts are a folk remedy for hypotension and parotid tumors.[91] Medicinally, the nuts are tonic, stomachic, and aphrodisiac.[278]

Chemistry — Per 100 g, the seed is reported to contain 620 to 634 calories, 16.4 to 20.0 g protein, 54.3 to 58.5 g fat, 21.4 to 22.9 g total carbohydrates, 3.3 to 5.9 g fiber, 1.8 to 3.7 g ash, 201 mg Ca, 462 mg P, 4.5 mg Fe, 1044 mg K, 10.80 μg beta-carotene equivalent, 0.17 mg thiamine, 0.44 mg riboflavin, 5.40 mg niacin, and 2.2 mg ascorbic acid.[89] *The Wealth of India*[70] reports the kernel to contain 12.7% protein, 17.7% carbohydrate, 60.9% fat, 0.35% P; rich in phosphorus. Kernel contains 50 to 65% of a golden yellow oil. The fatty acid components are 88.1% oleic, 1.9% linoleic, 3.1% palmitic, 1.6% stearic, and 2.2% myristic. The leaves contain myricitroside, a rhamnoside of myricetol and allantoic acid. The bark contains lignoceryl alcohol, betulinol, and sitosterol.[70] Pollen contains guanosine ($C_{10}N_{13}N_5O_5$) and n-triacosan. The wood contains cellulose, galactan, mannan, araban, and xylan. The ripe fruit contains 50 to 60% fat. Corylus oil contains 85% oleic- and 10% palmitic-acid esters; in addition, 0.5% phytosterol, protein, corylin (?), 2 to 5% sucrose, 2 to 5% ash, melibiose ($C_{12}H_{22}O_{11}$), manninotriose ($C_{18}H_{32}O_{16}$), raffinose ($C_{18}H_{32}O_{16}$), and stachyose ($C_{24}H_{42}O_{21}$). Leaves contain taraxerol ($C_{30}H_{50}O$), β-sitosterol, 3α, 7α,22α-trihydroxystigmasterol, n-nonacosan? ($C_{29}H_{60}$), myricitrin ($C_{21}H_{50}O$), sucrose, essential oil, 18% palmitic-acid, 6.6% ash (52.8% CaO, 5.8% SiO_2, 2.6% Fe_2O_3). The bark contains tannic acid, lignocerylalcohol, sitosterol, and betulin ($C_{30}H_{50}O_2$).[187]

Description — Deciduous shrub or small tree, up to 6 m tall, often thicket-forming; dark-brown, smooth, with glandular-hairy twigs; leaves 5 to 12 cm long, orbicular, long-pointed, hairy on both surfaces; margin doubly serrated; catkins appearing before leaves; staminate catkins 2 to 8 cm long, pendulous, in clusters of 1 to 4; pistillate flowers about 5 mm long, bud-like, erect; fruit in clusters of 1 to 4; nuts 1.5 to 2 cm in diameter, brown, invested by deeply lobed irregularly toothed bracts as long as nut. Flowers January to March; fruits fall.[278]

Germplasm — Reported from the Near Eastern and Mediterranean Centers of Diversity, European filbert, or cvs thereof, is reported to tolerate disease, frost, high pH, low pH, and slope.[82] European filberts are varieties or hybrids of *C. avellana* and *C. maxima*, both natives of Old World. In Europe, filberts are those varieties with tubular husks longer than nut, which is usually oblong; cob-nuts are roundish, angular, with husks about length of nut. In America, all varieties of *C. avellana* are filberts, and native species of *Corylus* are hazelnuts. Many hybrids between *C. avellana, C. maxima*, and the American filberts have been produced and many selections have been made. Hybrids with 'Rush' (a selection of *C. americana*) have produced some very hardy and productive plants, as 'Bixly', 'Buchanan', 'Potomac', and 'Reed'. Mixed hybrid seedlings are often sold as 'Jones Hybrids'. 'Barcelona' is the principal variety cultivated in Oregon, with 'Daviana' and 'DuChilly' as pollinizers.

'Cosford', 'Medium Long', and 'Italian Red' are the best of over 100 varieties grown in New York. 'Purple Aveline' is grown for its deep-red foliage in spring. *C. avellana* var. *pontica* (C. Koch) Winkler (Pontine Hazel or Trabzon Filbert) with lacerated, tubular husks, with nuts maturing by end of August, easily propagated by layering or grafting, long cultivated in Asia Minor.[99,278] Three varieties popular for ornamental planting are 'Aurea' (yellow leaves), 'Contorta' (twigs definitely curled and twisted), and 'Pendula' (with pendulous branches).[244,308,343] (2n = 22,28.)

Distribution — Native throughout most of Europe, except some islands and in the extreme north and northeast, east to the Caucasus and Asia, south to North Africa and temperate western Asia. Widely cultivated in temperate zones of Old and New World. Common in gardens on hill country in India, but unsuccessful on plains there; cultivated in Oregon and Washington.[278] Cultivated varieties introduced to the west coast of the U.S. in 1871.[209]

Ecology — Ranging from Boreal Wet through Subtropical Thorn to Dry Forest Life Zones, European filbert is reported to tolerate precipitation of 3.1 to 13.6 dm (mean of 29 cases = 7.0), annual temperature of 5.9 to 18.6°C (mean of 29 cases = 10.3°C), and pH of 4.5 to 8.2 (mean of 21 cases = 6.5).[82] Grows and is cultivated principally in countries where summer temperatures are comparatively cool and winter temperatures uniform and mild. Trees often injured during both mild and severe winters. Low temperatures, following periods of warm weather during latter half of winter generally cause more cold injury to catkins and wood than do abnormally low temperatures earlier in the season. Winters of continuous mild temperatures or those with severe but steady low temperatures (not lower than −5°C) usually result in little injury. Winters of alternating thawing and freezing cause most damage. High summer temperatures, as in Eastern and Central U.S., often cause leaves to scorch and burn and are an important factor in preventing trees from growing and fruiting satisfactorily. Much of this trouble probably results from inadequate soil moisture supply at critical times, as filbert does not have a deep taproot, and the feeding roots are fibrous and shallow. Hence, commercial filbert production in the U.S. is confined to the Northwest where climatic conditions are more favorable.[278] Hardy to Zone 3.[343]

Cultivation — The site for filberts should be selected so as to delay opening of flowers until the time when temperatures lower than −5°C are no longer to be expected. A northern slope or cover is the most satisfactory type of site. Cold, exposed sites, subject to drying effects of winds, should be avoided. Filberts are usually propagated by layering so that new plants are on their own roots. Some varieties sucker profusely, and soil is mounded up around these in spring to depth of several cm. By the following spring, roots have developed at base of sucker. Then, rooted suckers are cut loose, taken up and grown for a year in the nursery before setting them in a permanent site. Filberts may be propagated from seed, but varieties and cultivars do not come true. Seeds require after-ripening for germination. They may be stratified in sand over the winter. In spring, seeds are planted in the nursery and seedlings grown for 2 years. Buds grafted on *C. colurna* seedlings showed 39% successful union. Filbert trees of most varieties should be planted 5 to 7 m or more apart. Small-growing hybrids can be planted 3 to 5 m apart. In most cases, trees should be planted in late winter or very early spring, after danger of severe freezing weather has passed. The same general methods of planting should be used as that used for apple or peach. Newly transplanted and young trees should be cultivated sufficiently to destroy all grass and weed growth before the beginning of tree growth in spring and through July. Mulching trees with organic matter is equally satisfactory, provided that sufficient mulch is applied to a large enough area around each tree to suppress grass and weed growth. In general, the same cultural practices used for peaches are satisfactory for filberts. Filberts generally respond readily to fertilizer applications, although no recommendation would apply to all conditions. On most soils, it is not advisable to apply any fertilizer the first year after transplanting. Beginning the second year, about 475 g (1 lb) per tree of a 5-10-5 or 6-6-5 fertilizer should be broadcast around tree just before beginning of tree growth. The amount of fertilizer should be increased by 475 g (1 lb) per tree per year until trees are 10 to 12 years old; after that ca. 5 kg per tree per year is sufficient. Filberts are pruned to: (1) balance top loss with root loss in planting operations; (2) train young trees to desired form; and (3) remove dead or broken branches and stimulate moderate new shoot growth on older trees. At planting, the tree should be cut back to about 60 cm above ground, and 4 to 6 branches should be allowed to grow. Trees should be trained to the central leader form, provided it does not require removal of much wood. The more wood removed from young trees, the later they come into bearing; therefore, only necessary pruning should be done. Older trees that make short shoot growth should have branches thinned out and slightly cut back to stimulate production of stronger, more vigorous shoots. Pruning should be done after pollen is shed and catkins have fallen. All filberts except Turkish tend to grow as bushes by suckering from roots. All suckers should be promptly removed and the tree trained to a single stem. Suckers should be removed at the point on the trunk or root where they originate; cutting them off at the soil surface only increases the suckers that grow. Suckering operations should be done 3 or 4 times a year, as young suckers are easier to remove.[169,170,171,278]

Harvesting — Shrubs or trees begin bearing in about 4 years and bear well nearly every year. Staminate and pistillate appear on the same tree in different clusters. Depending on the location and winter weather conditions, pollination begins in January to March and lasts about 1 month. Young nuts do not become visible until late June or early July. There is a 3 to 4 month lapse between pollination and fertilization. Although filbert trees flower when freezing temperatures can be expected, they are generally not injured unless the temperature drops to about −10°C during the period of pollination. Most filbert varieties are self-unfruitful, and cross-pollination must be provided for satisfactory fruit-production. In all plantings, trees of 2 or more varieties should be included. The period of pistillate flowering is usually much longer than that of pollen-shedding on a particular variety. Furthermore, pollen of one variety must be shed at a time when pistillate flowers of the other variety are receptive. Pollen of *C. avellana* is effective on pistils of *C. cornuta* and *C. americana*, but a reverse application is usually sterile. *C. americana* × *C. avellana* hybrids have been used successfully to pollinate *C. avellana*. Nuts soon become rancid when stored at room temperature.[278] With good weather and modern equipment, five experienced workers can harvest ca. 200 acres in 10 days.[209]

Yields and economics — No specific yield data available, as nuts are gathered several times.[278] A good orchard can provide ca. 2,000 kg/ha dry in-shell nuts annually. U.S. imports ca. 45% of filberts consumed annually.[209] Filberts include both *C. avellana* and *C. maxima* and their hybrids, and they are not separated in the trade. In 1969—1970, Turkey exported about 81,300 MT of shelled nuts valued at $103 million, and 1,228 MT of unshelled nuts valued at $783,342. In 1970, production was about 240,000 MT unshelled nuts. Filberts range from $125-$150/ton. Major importers are West Germany, USSR, France, Italy, U.K., Switzerland, U.S., Lebanon, East Germany, and Syria. The U.S. produces about 9,000 tons annually in the shell and imports additional quantities.[278]

Energy — Though not usually considered a firewood species, the wood could undoubtedly serve such a purpose. Specific gravity of 0.917. The oil potential of nearly ca. 1 MT/ha would better be utilized for edible than energy purposes.

Biotic factors — Nuts of some varieties drop freely from husk, while others must be removed from husk by hand. Fallen nuts should be gathered 2 to 4 times during the harvest season, as they do not all mature at the same time. Those that drop early should not be allowed to lie on the ground because of loss to rodents and birds and discoloration or moldiness due to wet weather. Nuts should be promptly dried by spreading them in a thin layer in a dry place having good air circulation. Nuts dried in an unheated building usually require 4 to 6 weeks for drying. During this process, they should be stirred frequently to prevent molding. Temperature of nuts dried by artificial heat should not be higher than 45°C — otherwise they will not store well. Kernels of fully dried nuts are firm and brittle and will break with a sharp snap when hit with a hammer or crushed with the fingers. The following fungi are known to cause diseases on European filbert: *Anthostoma dubium, Apioporthe anomala, Armillaria mellea, Cercospora coryli, Chorostate conjuncta, Ciboria amentacea, Coriolus hoehnelii, Cryptospora corylina, Cylindrosporium coryli, Cytospora corylicola, C. fuckelii, Diaporthe decedens, D. eres, Diatrype disciformis, D. stigma, Diatrypella favacea, D. verrucaeformis, Cryptosporiopsis grisea, Diplodia sarmentorum, D. coryli, Fenestella princeps, Fomes annosus, Fumago vagans, Gloeosporium coryli, G. perexiguum, Gnomonia amoena, G. coryli, G. gnomon, Gnomoniella coryli, Helminthosporium macrocarpum, H. velutinum, Helotim fructigenum, Hypoxylon fuscum, H. multiforme, H. unitum, Labrella coryli, Lachnum hedwigiae, Mamiania coryli, Mamianiella coryli, Marasmius foetidus, Melconis sulphurea, Melanomma pulvis-pyrius, Merulius rufus, M. serpens, Monostichella coryli, Nectria coryli, N. ditissima, Nitschkia tristis, Orbilia crenato-marginata, Peniophora cinerea, Pestalozzia coryli, Pezicula corylina, Phellinus punctatus, Phoma suffulta, Phyllactinia corylea, Phyllosticta coryli, Phytophthora cactorum, Radulum oribculae, Rhizopus nodosus, Sclerotinia fructigena, Septoria avellanae, Sillia ferruginea, Stereum hirsutum, S. rugosum, Sphaeropsis coryli, Stictis mollis, Taphrina coryli, Tyromyces semipileatus, Valsa corylina*, and *Vuilleminia comedens*. European filbert trees are attacked by the bacteria, *Agrobacterium tumefaciens* and *Xanthomonas coryli*. Nematodes isolated from filberts include: *Caconema radicicola, Heterodera marioni, Longidorus maximus*, and *Pratylenchus penetrans*. Few insects attack leaves, branches, or nuts; some may cause severe damage unless controlled. Stink bugs and other plant bugs attack developing nuts and cause them to be bitter when mature. As these insects breed on various plants, as legumes, blackberries, and others, control chiefly depends on orchard sanitation and elimination of host plants on which bugs breed.[4,59,186,278]

CORYLUS CHINENSIS Franch. (BETULACEAE) — Chinese Filbert
Syn.: *Corylus colurna* var. *chinensis* Burk.

Uses — Kernels of nuts edible, used for food, eaten raw, roasted, or in cookery, and as flavoring. Plants used for hybridizing, since they are trees relatively resistant to Eastern filbert blight.[278]

Folk medicine — No data available.

Chemistry — No data available.

Description — Deciduous tree up to 40 m tall; leaves 10 to 17 cm tall, ovate to ovate-oblong, cordate or very oblique at base, glabrous above, pubescent along veins beneath, doubly serrate, petioles 0.8 to 2.5 cm long, pubescent and setulose; fruits 4 to 6, clustered; involucre, not spiny, constricted above nut, with recurved and more or less forked lobes, finely pubescent, not glandular; nuts relatively small, hard-shelled but of high quality.[278]

Germplasm — Reported from the China-Japan Center of Diversity, Chinese filbert, or cvs thereof, is reported to tolerate disease, drought, frost, heat, and slope.[82] Some selections are heavy producers. Cultivated, along with its hybrids, in southern Michigan.(2n = 22.)[278]

Distribution — China;[139] cultivated in Michigan.[278]

Ecology — Ranging from Warm Temperate Dry to Moist Forest Life Zones, Chinese filbert is reported to tolerate annual precipitation of 6.6 to 12.3 dm (mean of 2 cases = 9.5), annual temperature of 14.7 to 15.0°C (mean of 2 cases = 14.9°C), and pH of 4.9 to 6.8 (mean of 2 cases = 5.9).[82] Thrives in soils which permit its strong root system to penetrate to great depths. Trees resistant to cold, heat, drought, and other hazardous conditions of the environment.[278]

Cultivation — Propagated by seeds, but seedlings vary greatly in productivity and bearing age. Often hybridized with other species to get larger nuts and more hardy plants. Trees produce few or no suckers.[278]

Harvesting — Trees begin to bear fruit in about 8 years, and then continue for a long time. Nuts harvested in fall as other filbert tree species. Treatment, drying, and storage methods similar to those used for other filberts and hazelnuts.[278]

Yields and economics — Although no exact figures are available for this species, its selections and hybrids are said to be heavy producers. No specific production figures for this species.[278]

Energy — As a tall tree, this produces better firewood than some of the bushy species of *Corylus*.

Biotic factors — No specific data available for this species, but same precautions should be taken as for other filberts. Trees are relatively resistant to Eastern filbert blight.[278]

CORYLUS COLURNA L. (BETULACEAE) — Turkish Filbert or Hazelnut, Constantinople Nut

Uses — Cultivated for the nuts, the edible kernel used for confections, pastries, and for flavoring. Nuts also used roasted or salted, alone or with other nuts. This species is rarely cultivated for nuts in North America, but rather as an ornamental and for nursery understock.[278]

Folk medicine — Nuts used as a tonic.[165]

Chemistry — According to Hager's Handbook,[187] the nuts contain melibiose, manninotriose, raffinose, and stachyose.

Description — Deciduous shrub or small tree, rarely up to 25 m tall, with regular pyramidal head; leaves 7.5 to 12.5 cm long, deeply cordate, rounded, ovate or obovate, slightly lobed, doubly serrate, nearly glabrous above, pubescent beneath; petioles 2.5 cm long, usually glabbrescent, stipules lanceolate and acuminate; catkins up to 12 cm long, pendent; involucre much longer than nut, open at apex, divided almost to base into many long-acuminate or linear serrate lobes, densely covered with glandular hairs; nut globose or roundish-ovate, about 2 cm long, hard. Flowers late winter to early spring; fruits fall.[278]

Germplsm — Reported from the Near East Center of Diversity, Turkish filbert, or cvs thereof, is reported to tolerate drought, frost, poor soil, shade, and slope.[82] *C. colurna* var. *glandulifera* DC. has glandular-setose petioles and peduncles, with the lobes of involucre less acute and more dentate. Some selectins are heavy producers. Many other named botanical varieties. × *C. colurnoides* C. K. Sch. is a hybrid of *C. avellana* × *C. colurna*, grown in Germany. (2n = 28.)[156,157,278]

Distribution — Native to southeastern Europe and southwestern Siberia, south to the western Himalayas from Kashmir to Kumaon, at altitudes from 1,500 to 3,000 m; common in Kashmir forests; also found in Afghanistan, Balkan Peninsula, and Rumania. Extensively cultivated in Turkey.[278]

Ecology — Ranging from Cool Temperate Moist to Wet through Subtropical Dry Forest Life Zones, Turkish filbert is reported to tolerate annual precipitation of 5.2 to 14.7 dm (mean of 8 cases = 8.8), annual temperature of 8.4 to 18.6°C (mean of 8 cases = 12.0°C), and pH of 5.3 to 7.2 (mean of 8 cases = 6.6).[82] A temperate plant, but not quite hardy northward into the U.S. and Europe. Thrives on deep, fertile, well-drained soils, in regions where summer temperatures are comparatively cool and winters uniform and mild. Winters too mild or too severe injure both catkins and wood. Also winters with alternate thawing and freezing are injurious. For best cultivation, winter temperatures should not drop below −10°C.[278]

Cultivation — Turkish filbert is propagated from seeds or graftings on seedling stock. Since it does not sucker or stool, as do most species of *Corylus*, its seedlings are used as understocks for horticultural varieties of the European and American species. Trees should be planted 5 to 7 m or more apart, except for hybrid varieties, which are small-growing and can be planted 3 to 5 m apart. Trees should be planted in late winter or very early spring, after danger of severe freezing has passed. The same general methods of planting should be used as for apple or peach trees. Newly transplanted and young trees should be cultivated sufficiently to destroy all grass and weed growth before the beginning of tree growth in spring and through July. Mulching trees with any type of organic matter is equally as satisfactory as cultivation, provided that sufficient mulch is applied. In general, the same cultural practices used for peaches are satisfactory for filberts. Filberts generally respond readily to fertilizer applications, although no recommendation would apply to all conditions. On most soils, it is not advisable to apply any fertilizer the first year after transplanting. Beginning the second year, about 475 gm (1 lb) per tree of a 5-10-5 or 6-6-5 fertilizer should be broadcast around tree just before beginning of tree growth. Amount of fertilizer increased

475 gm (1 lb) per year until the tenth or twelfth year, and from then on apply about 4.7 kg (10 lbs) per tree per year. Prune trees to desirable shape and remove dead or broken branches. Since Turkish filberts do not sucker, little attention is given to the trees after they are established.[278]

Harvesting — Nuts are harvested in fall. Trees bear every third year, beginning the eighth year. However, in Turkey where they are extensively cultivated for the nuts, trees yield annually from the fourth year onwards up to the twentieth year. Nuts of Turkish filberts are said to be as good in quality as the English hazelnut. Nuts of some varieties drop free from husk while others must be removed from husk by hand. Fallen nuts should be gathered 2 to 4 times during the harvest season as they do not all mature at same time. Those that drop early should not be left on ground because of loss by rodents and birds, and because of discoloration and moldiness due to wet weather. Nuts should be promptly dried by spreading them in a thin layer in a dry place having good air circulation. Nuts dried in an unheated building usually require 4 to 6 weeks for drying. During this process the nuts should be stirred frequently to prevent molding. Temperature of nuts dried by artificial heat should not exceed 46°C (115°F) — otherwise they will not store well. Kernels of fully dried nuts are firm and brittle and will break with a sharp snap when hit with a hammer or crushed with the fingers.[278]

Yields and economics — No specific data on yields separate from that of other filberts cultivated in same areas, as Turkey and southeast Europe. However, some selections are said to be very good producers of nuts. Extensively cultivated in Turkey, and to a lesser degree in southeast Europe and western Asia, south into temperate Himalayas. Although trees are said to yield a good crop, production figures are not separated from production of other European or Asiatic filberts.[278]

Energy — Like other members of the genus *Corylus*, this holds little promise as an energy species, but can provide firewood and seed oils. As a tree species, it can provide higher quality firewood than shrubby species of *Corylus*.

Biotic factors — The following fungi are known to attack Turkish filbert: *Hyposylon multiforme, Lenzites japonica, Microsphaeria alni, Phyllactinia corylea*, and *Pucciniastrum coryli*. The bacterium *Pseudomonas colurnae* has been isolated from this species. Mycorrhiza are necessary in the soil. As staminate and pistillate flowers do not always become fertile on the same tree at the same time, and since most filberts are self-unfruitful, for commercial production, several varieties should be planted near each other for cross pollination, thus assuring good nut production.[186,278]

CORYLUS CORNUTA Marsh (BETULACEAE) — Beaked Filbert
Syn.: *Corylus rostrata* Ait.

Uses — Nuts used for human food and wildlife food; plants used for erosion control and cover and for basket splints.[278]

Folk medicine — Ojibwa Indians used a poultice of boiled bark to help close and heal wounds; Potawatomi used the inner bark as an astringent.[216]

Chemistry — No data available.

Description — Deciduous shrub, 0.6 to 3 m tall, thicket-forming, sometimes a small tree to 10 m tall; bark smooth; branchlets pubescent, villous or glabrous, later glabrescent, not bristly; leaves 6 to 10 cm long, ovate or narrowly oval, acuminate, cordate or obtuse at base, incised-serrate or serrulate on margins, glabrous or with scattered appressed hairs above, sparsely pubescent beneath, at least along veins; petioles glandless, 0.4 to 0.8 cm long; mature involucral of connate bracts 4 to 7 cm long, densely bristly toward base, usually rather abruptly constructed into an elongated beak, cut at summit into narrowly triangular lobes; nut ovoid, brown, compressed, striate, 1.2 to 2.3 cm long. Flowers February to May; fruits July to September; seed dispersal July to winter.[278]

Germplasm — Reported from the North American Center of Diversity, beaked filbert, or cvs thereof, is reported to tolerate frost, slope, smog, and SO_2.[82] Among botanical varieties are the following: *C. cornuta* forma *inermis* Fern., a form in Quebec with non-bristly involucres; *C. Cornunta* var. *californica* (A. DC.) Sharp *(C. californica* (A.DC.) Rose), a variety found on the West Coast. ($2n = 28$.)[278]

Distribution — Native to eastern North America from Newfoundland and Quebec to British Columbia, south to Georgia and Missouri, and on the west coast from California northward. Cultivated elsewhere.[278]

Ecology — Ranging from Boreal Moist through Cool Temperate Steppe to Wet Forest Life Zones, beaked filbert is reported to tolerate annual precipitation of 3.5 to 11.6 dm (mean of 10 cases = 6.8), annual temperature of 5.7 to 12.5°C (mean of 10 cases = 8.1°C), and pH of 5.0 to 7.5 (mean of 9 cases = 6.5).[82] Naturally thrives in moist woods and thickets, on low hillsides, in rich, well-drained soil. When cultivated, shrubs should be planted in soils which are deep, fertile, and well-drained. Heavy clay or silt soils as well as coarse, deep sand should be avoided. When planted on poorly drained soils, the shrub grows poorly, is subject to winter injury, and bears few nuts.[278]

Cultivation — Most filberts offered for sale by nurserymen have been propagated by layering and are on their own roots. Trees or shrubs grown for 1 or 2 years in nursery after the layers have been removed from parent stock are preferable to older plants. Some nurserymen propagated their stock on Turkish filbert roots that do not produce suckers. This rootstock generally outgrows the scion to some extent. Beaked filberts can be propagated from seed which has been stratified for 60 to 90 days at 5°C. Germination is hypogeous. Natural seed dispersal is chiefly by animals. Stratified seed are planted in spring. However, seed may be planted in fall in drills and protected from rodents. Horticultural varieties are propagated by suckers, layering, budding or grafting, and cuttings. Filberts of all varieties should be planted 5 to 7 m or more apart. In most cases, trees or shrubs are planted in late winter or very early spring, after danger of severe freezing is passed. The same general methods of planting should be used as for apple or peach trees. Newly transplanted and young plants should be cultivated sufficiently to destroy all grass and weed growth before the beginning of tree growth in spring and through July. Mulching plants with any type of organic matter is equally as satisfactory as cultivation, provided that sufficient mulch is applied. In general, the same cultural practices used for peaches are satisfactory for filberts. Filberts generally respond favorably to fertilizer applications, although no recommendation would apply to all situations. On most soils it is not advisable to apply any fertilizer the first year after transplanting. Beginning the second year, about 475 g (1 lb) per tree of a 5-10-5 or 6-6-5 fertilizer should be broadcast around the tree just before the beginning of tree growth. The amount of fertilizer is increased 475 g/year until plants are 10 to 12 years old; after that, about 4.7 kg per plant is sufficient. Pruning filberts is done to balance top with loss of roots in planting operations, to train young trees to desired form, to remove dead, broken or diseased branches, or to stimulate moderate growth on new shoots on old trees. At planting, tree should be cut back to about 60 cm above the ground, leaving 4 to 6 branches to grow. Trees or shrubs should be trained to the central leader form, provided it does not mean removal of much wood. The more wood removed from young plants, the later they come into bearing; therefore, only necessary pruning should be done. Pruning should be done after pollen shedding is over and anthers have fallen. Since beaked filberts or hazelnuts tend to sucker, the suckers should be removed promptly and the plant trained to a single stem. Suckers should be removed at the point on the trunk or root where they originate; cutting them off at surface of soil only increases the number that grow. Suckering operations should be done 3 to 4 times a year, as they are easier to remove when young.[278]

Harvesting — Fruits should be gathered by hand from bushes as soon as edges of husks turn brown. Fruits should be spread out in a thin layer to dry for a short time, for about 4 to 6 days. Then husks are removed by flailing. The nuts, which are the commerical seeds, may be sown, stratified, or stored. Storage in sealed containers at 5°C will retain some viability in *C. cornuta* for at least 2 years.[278]

Yields and economics — Beaked filberts yield well every 2 to 5 years, and give a light crop every year. Great quantities of hazelnuts are gathered each year for local home-use in northeastern and northwestern U.S. and Canada. Many more are used as food for wildlife. No exact figures are available on production. Hazelnuts are usually sold as mixed nuts, especially during the winter months and holidays.[278]

Energy — Probably no more promising than other *Corylus* species for energy potential.

Biotic factors — Most filbert varieties are self-unfruitful, even though staminate and pistillate catkins are on the same tree or bush. Cross-pollination must be provided for satisfactory fruit production. In all plantings, two or more varieties should be included. The period of pistillate flowering is usually much longer than that of pollen-shedding on a particular variety. Furthermore, pollen on one variety must be shed at the time when pistillate flowers of the other variety are receptive. The following fungi are known to attack beaked filberts or hazelnut plants: *Apioporthe anomala, Cercospora corylina, Cucurbitaria conglobata, Diaporthe decedens, Diatrypella minutispora, Gloeosporium coryli, G. rostratum, Gnomoniella coryli, Hymenochaete agglutinans, Melanconis flavovirens, Microsphaeria alni, Nectria coryli, Pezicula corylina, Phyllactinia corylea, Phymatotrichum omnivorum, Polysporus albellus, P. elegans, P. radiatus, P. stereoides, Septoria corylina,* and *Sphaeropsis corylii*. Among the insect pests, hazelnut weevil, in severe infestations, may completely destroy the crop of nuts. Leaves are preferred food for Japanese beetles, and plants may be completely defoliated by them. Filbert bud mite may be a pest problem. For control of all pests, consult local State agent.[278]

CORYLUS FEROX Wall. (BETULACEAE) — Himalayan or Tibetan Filbert
Syn.: *Corylus tibetica* Batal. (*thibetica*) and *Corylus ferox* var. *thibetica* Franch.

Uses — Kernel of nut edible, used raw, roasted, or in cookery, and as a flavoring.[278]

Folk medicine — No data available.

Chemistry — No data available.

Description — Deciduous tree to 10 m tall; young branches silky-hairy; leaves 7.5 to 12.5 cm long, oblong, ovate to obovate-oblong, usually rounded at base, acuminate, doubly serrate, glabrous except along veins beneath, 12 to 14 pairs of veins; involucre glabrescent to tomentose, forming a spiny bur about 3 cm across, longer than nut, consisting of 2 distinct bracts; nuts from about twice in diameter as long to twice as long as wide.[278]

Germplasm — Reported from the China-Japan Center of Diversity, Himalayan filbert, or cvs thereof, is reported to tolerate frost and slope. (2n = 22,28.)[82]

Distribution — Native to central and western China to Tibet and central Himalaya, up to 3,300 m altitude.[278]

Ecology — Ranging from Warm Temperate to Moist Forest Life Zones, Himalayan filbert is reported to tolerate annual precipitation of 12.0 dm, annual temperature of 14.8°C, and pH of 5.5.[82] Thrives in temperate forests on well-drained soils.[278]

Cultivation — This filbert is rarely cultivated, but rather, trees are taken care of in the forest. Propagation is by natural distribution of seeds.[278]

Harvesting — Nuts are collected from native trees in the forest in the fall. Drying and storage procedures are about the same as for other filberts.[278]

Yields and economics — No yield data available. Locally in central Asia, these filberts are gathered and sold in local markets. They do not enter international trade.[278]

Energy — Not a promising energy species.

Biotic factors — No data available.

CORYLUS HETEROPHYLLA Fisch. ex Besser (BETULACEAE) — Siberian Filbert

Uses — Kernels of nuts used raw, roasted, cooked, or in confections.[278]

Folk medicine — Reported to be aperitif and digestive.[91]

Chemistry — No data available.

Description — Deciduous shrub or small tree to 4 m tall; branchlets pubescent and glandular-pilose when young; leaves 5 to 12 cm long and about as wide, orbicular-obovate to deltoid-obovate, cordate at base, nearly truncate and abruptly acuminate at apex and with a very short point, margins irregularly toothed or incisely serrate, green on both sides, glabrous above, pubescent on veins beneath, petioles up to 13 cm long, pubescent and glandular-pilose; involucre companulate, 2.5 to 3.5 cm long, somewhat longer than nut, striate, glandular-setose near base, lobes of bracts entire or sparingly dentate, triangular; nuts 1 to 3 in a cluster, at ends of branchlets, on stalks to 3 cm long, subglobose, about 1.5 cm across. Flowers May; fruits August.[42,278]

Germplasm — Reported from the China-Japan and Eurosiberian Centers of Diversity, Siberian filbert, or cvs thereof, is reported to tolerate frost, low pH, and slope.[82] Several botanical varieties are known, and some are cultivated in northern Asia. *C. heterophylla* var. *yezoensis* Koidz. (*C. yezoensis* (Koidz.) Nakai) — leaves obovate-orbicular to broadly obovate, abruptly short acuminate, rarely glandular-pilose; involucres sparsely glandular-pilose; Japan (Hokkaido, Honshu, Kyushu). Other varieties are var. *thunbergii* Blume, var. *crista-galli* Burkill, var. *setchuensis* Franch., and var. *yunnanensis* Franch.[82,278] (2n = 28.)

Distribution — Native to eastern Siberia, eastern Mongolia, Manchuria, northern China (Tschili), Ussuri, Amur, Korea; introduced and cultivated in Japan and France; probably elsewhere.[95,278]

Ecology — Ranging from Cool Temperate to Moist through Warm Temperate Dry to Moist Forest Life Zones, Siberian filbert is reported to tolerate annual precipitation of 12.0 to 14.7 dm (mean of 2 cases = 13.4), annual temperature of 14.8 to 14.8°C (mean of 2 cases = 14.8°C), and pH of 5.3 to 5.5 (mean of 2 cases = 5.4).[82] Naturally found along woods and on mountain slopes, often forming dense thickets. Thrives in cool temperate regions on soil with good drainage.[278]

Cultivation — Modest requirements greatly facilitate cultivation. Propagated from seed, usually distributed naturally in the forest, and by suckers. The most elementary care of wild stands results in considerable improvement in yield and quality of nuts.[278]

Harvesting — Nuts are probably collected in the fall.

Yields and economics — Nuts harvested commercially in Northern Asia, usually from wild plants only. Does not enter international markets; usually marketed locally.[278]

Energy — Not a promising energy species.

Biotic factors — No data available.

CORYLUS MAXIMA Mill. (BETULACEAE) — Giant or Lambert's Filbert
Syn.: *Corylus tubulosa* Willd.

Uses — Widely cultivated for the nuts in Europe; used as roasted or salted nuts, or as flavoring in confections and pastries. Sometimes naturalized, and of some interest as an ornamental, especially the red-leaved form, found in parks in the Caucasus. This species is considered the progenitor in Europe from which most cultivated filberts have been developed: *C. avellana* is more often called the cobnut.[278]

Folk medicine — No data available.

Chemistry — No data available.

Description — Deciduous shrub or small tree, up to 10 m tall; branches somtimes glabrous, mostly stipitate-glandular; leaves 7.5 to 15 cm long, 6 to 10 cm broad, orbicular, cordate at base, short-acuminate, slightly lobed, doubly serrate, very often red, pubescent beneath; petiole 1 to 2.5 cm long; staminate aments to 10 cm long,1 cm in diameter; involucre tubular, contracted above the nut, forming a gradually narrowed elongated deeply laciniate husk, dentate at apex, finely pubescent outside, lower part fleshy, enveloping nut, splitting at maturity; nut ovoid, sometimes subcylindrical, acuminate; kernel with thin red or white skin. Flowers March; fruits September.[278]

Germplasm — Reported from the Central Asia and Near East Centers of Diversity, giant filbert, or cvs thereof, is reported to tolerate frost, low pH, and slope. *C. maxima* var. *purpurea* Rehd. (*C. avellana* (var.) *purpurea* Loud., *C. maxima* var. *atropurpurea* Dochnahl) has dark purple-red leaves. There are many varieties with large nuts. Cultivated forms are partly hybrids with *C. avellana*. (2n = 22,28.)[82,278]

Distribution — Native to southeastern Europe, from Italy and Yugoslavia to Greece, Turkey, and western Asia. Widely cultivated elsewhere in Europe and sometimes naturalized. Cultivated in Crimea and on the Black Sea Coast for more than a century.[278]

Ecology — Ranging from Cool Temperate Wet through Warm Temperate Moist to Wet Forest Life Zones, the giant filbert is reported to tolerate annual precipitation of 6.3 to 16.7 dm (mean of 2 cases = 10.5 dm), annual temperature of 9.7 to 14.8°C (mean of 2 cases = 12.3°C), and pH of 5.3 to 6.8 (mean of 2 cases = 6.1). Thrives in a cool to warm temperate climate under soil and climatic conditions similar to those for *C. avellana*.[82,278]

Cultivation — See *Corylus americana*.

Harvesting — The harvesting of nuts begins in September. The beaked involucre must be removed by hand, and then the nuts are dried for storage until marketed or used. After removing the husk, nuts are spread out to dry in thin layers in a dry place having good air-circulation. Nuts dried in an unheated building usually require 4 to 6 weeks for drying. They should be stirred frequently to prevent molding. The temperature of nuts dried by artificial heat should not be higher than 45°C; otherwise they will not store well.[278]

Yields and economics — The species and its cultivars and hybrids are reported to be good producers. Southeastern Europe and southwestern Asia, especially Crimea and the Black Sea Region, are major producers. However, this filbert is not separated from the Turkish and other filberts grown in the region. Prices vary from $125 to $150/ton for Turkish filberts. About 240,000 MT of nuts are produced annually in Turkey and adjacent areas.[278]

Energy — Although not a promising energy species, this is one of the better species of *Corylus* for energy production.

Biotic factors — Fungi known to attack this filbert include: *Mycosphaerella puntiformis*, *Phyllactinia corylea*, and *Sphaeragnmonia carpinea*. The bacterium, *Xanthomonas coryli*, also attack the plant. In some areas, winter injury may be serious. Pests include: *Lecanium corni* (soft scale) and *Myzocallis coryli* (aphids).[166,278]

COULA EDULIS Baill. (OLACACEAE) — African Walnut, Gabon Nut, Almond Wood

Uses — The fruits, sold in Cameroon markets, have agreeably edible kernels, resembling hazelnuts or chestnuts. They are eaten fresh, boiled in the shell, roasted, boiled, and pounded and made into cakes. Some tribes ferment the fruits underground. The timber is red to reddish-brown, closegrained, hard, heavy, resistant to water, and immune to insects, e.g., termites, through liable to split. Suitable for house posts, railway sleepers, bridge-piles, and charcoal, it has been suggested for heavy carpentry, stair treads, doors, turnery, and boat and carriage construction. Durable under water, the wood can be used for bridges and pilings.[71] The fruit shells make finger-rings in Nigeria.[146]

Folk medicine — The stomachic bark decoction is used for dysentery in Liberia. Powdered bark is used in Equatorial Africa for dressing sores, and in decoctions to stimulate appetite and counteract anemia, or in enemas for dysentery. Liberians believe the fruits eliminate boils.[146]

Chemistry — Per 100 g, the seed (ZMB) is reported to contain 505 calories, 7.9 g protein, 25.7 g fat, 64.3 g total carbohydrate, 2.4 g fiber, 2.1 g ash, 180 mg Ca, and 269 mg P.[89] Dalziel[71] and Irvine[146] suggest that the oil content is closer to 50 than 25%. The seed fat is very high in oleic acid (87 to 95%), with 3% linoleic, and 1.7% palmitic + stearic acids.[128] Menninger cites a source suggesting 87% oleic.[209]

Toxicity — Leaves said to be poisonous.[146] Wood can cause allergy or asthma in woodworkers.[215]

Description — Medium-sized tree to 20 m tall and 2 m girth; crown deep, dense; buttresses slight or none; bark fairly smooth, thin, brownish-green; slash brown or yellow, white and resinous in young trees, darkening to pink; young parts reddish-brown-hairy. Leaves 30 × 8 cm, often rusty, papery, elliptic to oblong-elliptic, glabrous, alternate; tip long-caudate-acuminate; base cuneate; midrib slightly raised above; lateral nerves up to 14 pairs, sub-parallel, sunken above and raised below; petiole 2 to 3 cm long, usually twisted, rusty-puberulous. Flowers (April to May, October to January) in rusty-brown axillary panicles; calyx small, cup-shaped; petals 5, fairly thick; stamens 10. Fruit a drupe, ellipsoid-globose, 4 × 3 cm, nut-shell hard, rough ca. 4 mm thick, breaking into 3 portions when ripe, difficult to break.[146]

Germplasm — Reported from the African Center of Diversity.

Distribution — Sierra Leone to Gabon and Zaire; Liberia, Ivory Coast, Gold Coast, Nigeria, Cameroon.[146]

Ecology — Reported from evergreen and deciduous forests, gabon nut is estimated to range from Subtropical Dry to Wet through Tropical Dry to Moist Forest Life Zones, tolerating annual precipitation of 8 to 35 dm, annual temperature of 23 to 28°C, and pH of 6.0 to 8.0.

Cultivation — Can be grown as a plantation timber crop with the oil or nut as a by-product. [146]

Harvesting — In Angola, north of the Congo River, the nuts mature from December to April.[209] In Nigeria, it flowers January to May, fruiting in August.

Yields and economics — Apparently sold only in Cameroon markets.[146]

Energy — The wood is suitable for charcoal[146] and it is so used in Gabon. The extremely hard wood has a density of 1.073.

Biotic factors — The wood is termite resistant.

CYCAS CIRCINALIS L. (CYCADACEAE) — Cica, Crozier Cycas

Uses — Speaking of Cycads in general, Egolf (in Menninger[209]) says

> Cycad nuts are rather large, many of them an inch across. They are fat and rounded, full of starch, and mostly covered by a brilliant orange or reddish outer coat. They look as if they are meant to be good to eat. The poisonous substance in Cycads is soluble in water. It can be leached from the nuts or from the starchy center of the trunk by water, rendering them fit to eat. It is impossible now to tell what primitive genius first discovered that such tempting nuts could be made free of their poison. Perhaps some tribesman, wits sharpened by hunger, found that Cycad nuts shed into a jungle pool, partially decomposed by water, could be eaten whereas those fresh from the plant could not. Where the nuts are eaten they may be treated whole, with repeated changes of water, and then beaten to a flour for cooking, or the raw nuts may be beaten and the pulp washed in water and strained through a cloth . . . However it happened, in nearly every tropical country where Cycads grow men sooner or later found they could use the nuts for food. They are not an important staple, because nowhere do Cycads grow in dense profusion, but in times of famine, when there is little else to eat, they are as welcome as the finest delicacy.

In Guam, they eat the green husk, fresh or dried, or they cut, soak, and sun-dry it. Indians eat the fruit with sugar. In Java, Sumatra, Sri Lanka, and the Phillipines, the shoots and leaves are used as a potherb. In Fiji they boil the kernels until they are soft. Indochinese pound, soak, settle, and dry the kernels. Africans split the seeds, sun dry them for ca. 4 days, ferment them in a tin with banana leaves for a week, remove the mold, soak another day, pulverize, and use as a porridge.[338] Sap from the kernels has been said to be given to children in the Celebes for "population control." Crushed seed also used to poison fish. A gum can be extracted from breaks in the megasporophylls.[319] Surface fibers from the leaves have been made into cloth.

Folk medicine — Reported to be carminative, narcotic, and poison, *C. circinalis* is a folk remedy for nausea, sores, swellings, and thirst. Terminal buds are crushed in rice-water for adenitis, furuncles, and ulcerous sores. Seeds are applied to malignant and varicose wounds and ulcers. Seeds are squeezed and grafted onto tropical ulcers in Guam. The gum is used for snakebite in India. Filipinos roast and grate the seeds, applying them in coconut oil to boils, itch, and wounds. Indians poultice the female cones onto nephritic pain, using the male bracts as aphrodisiac, anodyne, and narcotic. The gum, which expands many times in water, is said to produce rapid suppuration when applied to malignant ulcers. The gum also has a reputation for treating bugbite and snakebite.[91,249]

Chemistry — Seeds contain ca. 31% starch,-2 toxic glycoside, pakoeine, phytosterin, and a reducing sugar.[70] The pollen is said to be narcotic. Seeds possess antibiotic activity.[187] Sequoyitol is also reported, as is alpha-amino-beta-methylaminopropionic acid.

Caution — FATALITIES are attributed to eating improperly prepared nuts. Many of Captain's Cook's voyagers vomited following the ingestion of cycad nuts. Symptoms of poisoning include headache, violent retching, vertigo, swelling of the stomach and legs, depression, stupor, euphoria, diarrhea, abdominal cramps, tenesmus, muscle paralysis, and rheumatism.

Description — Evergreen ornamental shrub or small tree to 6 m tall, unbranched except by accident, such as cutting of apex. Trunk stout with hard outer layer like bark, light brown-gray, slightly scaly, becoming slightly fissured. Leaves apically crowded with stout axis with 2 rows of short spines replacing leaflets toward base. Leaflets thick, stiff, hairless, mostly opposite, 15 to 30 cm long, 1 to 2 cm broad, straight or curved, long-pointed at apex, with prominent yellowish midvein, but without other visible veins. Male cones large, brown, hard, and woody. Female trees produce a ring of light-brown wooly fertile leaves 6 to 12 inches long. Each leaf bears in notches along the axis 4 to 10 naked elliptic or nut-like seeds, hard with thin outer flesh.[189]

Germplasm — Reported from the African and Indochina-Indonesian Centers of Diversity, cica, or cvs thereof, is reported to tolerate some shade and waterlogging.

Distribution — Old World Tropics, Native from Tropical Africa through southern Asia and Pacific Islands. Pantropically introduced.

Ecology — Estimated to range from Subtropical Dry to Moist through Tropical Dry to Wet Forest Life Zones, cica is estimated to tolerate annual precipitation of 10 to 50 dm, annual temperature of 21 to 26°C, and pH of 6.0 to 8.0. Hardy to Zone 10b.[343]

Cultivation — Rarely cultivated for food, more often cultivated as an ornamental. Easily propagated from suckers or sprouts at the base of parent plants. Grows slowly.

Harvesting — For sago starch, the trunks should be felled before fruiting (usually at about 7 years). Since the felling of the trunk precludes fruiting, it follows that seeds are harvested from older trees.

Yields and economics — A cycas is said to produce annually ca. 550 seeds, yielding about as much starch (ca. 2 kg) as an irreplaceable stem ca. 1 m long. Extraction of starch from the seeds is said to be more economical.

Energy — Since felling these trees is fatal, they are rarely, if ever, used as energy sources.

Biotic factors — No data available.

CYCAS REVOLUTA Thunb. (CYCADACEAE) — Cycad Nut, Sotesu Nut

Uses—(see *Cycas circinalis.)* Exported from Japan as an ornamental, used in Japan for bonsai. According to Thieret,[319] the fleshy testa (sweet and mucilaginous) and the starchy kernels are both eaten. The roasted kernels, like so many other nondescripts, are said to taste like chestnuts. Seeds are eaten by the Annamese of China, though preparation is tough. Japanese use the young leaves as a potherb,[319] and the cycad meal as a food extender and for the preparation of sake, the sake called doku sake, or poisonous sake. A sago starch is extracted from the pith and cortex of the stem before fruiting. It has been said, perhaps exaggerated, that a small portion of the pith can support life for a long time. Gum is extracted from wounds on the megasporophylls.[319] Surface fibers from the leaves have been made into cloth. Leaves are used for funeral decorations.[86]

Folk medicine — Reported to be emmenagogue, expectorant, fattening, and tonic, *C. revoluta* is a folk remedy for hepatoma and tumors.[91] The down from the inflorescence has been used as a styptic, the terminal shoot as astringent, and diuretic. Seeds used as astringent, emmenagogue, expectorant, and tonic, used for rheumatism.[86] "The products extracted from the seeds are useful to inhibit growth of malignant tumors."[249] The gum, which expands many times in water, is said to produce rapid suppuration when applied to malignant ulcers. The gum also has a reputation for treating bugbite and snakebite.

Chemistry — Thieret[319] reports the kernels contain 12 to 14% CP and 66 to 70% starch. Whiting[338] reports that fresh kernels contain 7% protein, 33% starch, dry kernels 12% protein, 60% starch, the pith 7 and 41, the fresh outer husk of the seed 4 and 21, the dry outer husk 10% protein, and 46% starch. Airdry stems contain 44.5% starch and 9.15% CP. Male plants run 27 to 61% starch, averaging over 50% over the year; female stems average only 26%. Root nodules contain about 18% starch. Formaldehyde is reported from the kernels, but cycasin ($C_8H_{16}O_7N_2$) is probably the culprit, in both nuts and pith. Thieret[19] reports that the testa contains ca. 4% oil, the seeds 20 to 23.5% oil (an oil used during crises on Okinawa during World War II). Duke and Ayensu[90] report the seeds (ZMB) contain 13.9 to 15.4 g protein, and 0.9 to 1.0 g fat. Also reported to contain 14% crude protein, 68% soluble non-nitrogenous substances, and 0.16 to 0.22% combined formaldehyde, 90% of which can be washed out with water. Seeds may yield 20.44% fat, the component fatty acids of which are palmitic-, stearic-, oleic-, and a small amount of behenic-acid. Seeds contain 0.2 to 0.3% neocycasin A, neocycasin B, and macrozamin, and cycasin. Trunk contains mucilage with xylose, glucose, and galactose. The wax composition is detailed in Hager's Handbook.[187] Cycasin is carcinogenic to pigs and rats if ingested orally.[86] It also induces chromosomal aberrations in onion root tips.

Caution — FATALITIES are attributed to eating improperly prepared nuts.

Description — Trunk 1.8 m, densely clothed with the old leaf-bases. Leaves 0.6 to 1.8 m long; petiole thick, quadrangular; leaflets narrow, margin revolute. Carpophylls 10 to 23 cm long, blade ovate, laciniate nearly to midrib, stalk longer than blade, with 4 to 6 ovules. Immature seed densely tomentose.[165]

Germplasm — Reported from the Sino-Japanese Center of Diversity, this cycad, or cvs thereof, is reported to tolerate drought, floods, poor soil, slope, and typhoons.[319]

Distribution — China, S. Japan, Formosa, Tonkin. Cultivated in Indian gardens.[165]

Ecology — Estimated to range from Warm Temperate Dry (without frost) to Wet through Tropical Dry to Wet Forest Life zones, *C. revoluta* is estimated to tolerate annual precipitation of 8 to 40 dm, annual temperature of 17 to 25°C, and pH of 6.0 to 8.5. Tolerates the poorer steep soils of the Ryukyu's. Hardy to Zone 9.[343]

Cultivation — Rarely cultivated for food, more often cultivated as an ornamental. Easily propagated from suckers or sprouts at the base of parent plants. Grows slowly.

Harvesting — For sago starch, the trunks should be felled before fruiting (usually at about 7 years). Since the felling of the trunk precludes fruiting, it follows that seeds are harvested from older trees.

Yields and economics — Thieret[319] reported that an estimated 3 million cycas leaves with a gross value of ca. $30,000 were imported annually to the U.S.

Energy — Since felled trees do not coppice, these trees are rarely, if ever, used as energy sources.

Biotic factors — In the Ryukyu Islands, the poisonous *habu* viper nests in the top of this cycad.

CYCAS RUMPHII Miq. (CYCADACEAE) — Pakoo Adji, Pakis Adji, Pahoo Hadji, Akor

Uses — (See *Cyas circinalis.*) A well-known oriental ornamental, this fern-like tree is often planted, e.g., in cemeteries. The young shoots, shortly before unfolding, are cooked as a potherb, often with fish. Eating too much is said to cause rheumatism. The poisonous nuts are rendered edible by various types of elaborate processing. Steeping in water seems to be one of the most common methods of preparation. In the Moluccas, a delicacy is made by cutting the kernels into bars, putting them in a porous bag, and steeping in sea-water for a few days. Then the bars are sun dried, pulverized in a basket, and mixed with brown sugar and coconut. The starchy pith and cortex of the stem may be eaten after cooking. Stems for "sago" starch should be harvested before fruiting. Gums are extracted from wounded megasporophylls.[319] Stems are used in Indonesia to build small houses.

Folk medicine — A folk remedy for colic in Java.[91] The resin is applied to malignant ulcers, exciting suppuration in an incredibly short time. In Cambodia, the leafless bulb is brayed in water, rice-water, or water holding fine particles of clay in suspension, and applied to ulcerated wounds, swollen glands, and boils.[165] The gum also has a reputation for treating bugbite and snakebite.

Chemistry — Probably parallels that of *Cycas revoluta.*

Description — Small, dioecious gummiferous tree, 1 to 6 m high, rarely higher. Trunk terete, armored by the persistent petiole bases. Leaves in a dense terminal whorl, glabrous, shortly petioled, pinnate, with 50 to 150 pairs patent leaflets, glaucous, shining above, 1.5 to 2.5 m long; leaflets linear-lanceolate, usually somewhat recurved; 1-nerved; the central leaflets 20 to 35 cm long, 1 to 2 cm wide, the lower ones gradually shorter and narrower; armed on the edges. Male cone stalked, oblong-ellipsoid, yellowish-brown, 30 to 70 cm long, 12 to 17 cm wide, with numerous spirally arranged stamens; stamens cuneate with upcurved acuminate tips, 3.5 to 6 cm long; the higher ones smaller, anantherous. Female cone terminal, after anthesis producing new leaves at the apex; carpophylls numerous, densely crowded, densely yellowish-brown tomentose along the edges with 2 to 9 big, short ovules, 25 to 50 cm long; tips of the carpophylls oblong, serrate, terminated by a long, entire, upcurved point. Seeds ellipsoid or ovoid-ellipsoid, orange when ripe, 3 to 6 cm long, 2.5 to 5 cm diam.[238]

Germplasm — Reported from the Indochinese-Indonesian and Australian Centers of Diversity.

Distribution — Burma, Malaya, Andamans, Nicobars, Moluccas, New Guinea, and N. Australia, cultivated in India.[165]

Ecology — Estimated to range from Subtropical Dry to Moist through Tropical Dry to Wet Forest Life Zones, *C. rumphii* is estimated to tolerate annual precipitation of 10 to 50 dm, annual temperature of 21 to 26°C, and pH of 6.0 to 8.0.

Cultivation — Rarely cultivated for food, more often cultivated as an ornamental. Easily propagated from suckers or sprouts at the base of parent plants. Grows slowly.

Harvesting — For sago starch, the trunks should be felled before fruiting (usually at about 7 years). Since the felling of the trunk precludes fruiting, it follows that seeds are harvested from older trees.

Yields and Economics — A cycas is said to produce annually ca. 550 seeds, yielding about as much renewable starch (ca. 2 kg) as an irreplaceable stem ca. 1 m long. Extraction of starch from the seeds is hence said to be more economical.

Energy — Rarely, if ever, used as energy sources.

Biotic Factors — No data available.

CYPERUS ESCULENTUS L. (CYPERACEAE) — Tigernut, Yellow Nutsedge, Chufa

Uses — Grown for the edible tubers, eaten when dry, raw, boiled, or roasted. Juice pressed from fresh tubers is consumed in quantities in Europe, especially in Spain, as a beverage, called Horchata de Chufas; sometimes it is chilled or frozen. Nuts used as substitute for coffee; or for almonds in confectionery, or made into a kind of chocolate. In Africa, nuts used in the form of milk pap, made by grinding fresh nuts fine and straining; then boiling with wheat flour and sugar. Roasted nuts are ground and sieved to produce a fine meal, a high caloric value, which is added along with sugar and other ingredients to water as a beverage, or even eaten dry. Oil used for soap-making.[278] Used as a famine food.[332] The haulm is grazed by stock, plaited into rough ropes in Lesotho, and is suitable for making paper pulp.[55] Tubers are relished by hogs, which are used to suppress the plant when it becomes weedy.[278] It has already infested more than 1,000,000 ha in the eastern U.S.[135]

Folk medicine — According to Hartwell,[126] the tubers are used in folk remedies for felons and cancers. Reported to be aphrodisiac, astringent, CNS-sedative, CNS-tonic, diaphoretic, diuretic, emmenagogue, emollient, excitant, lactagog, pectoral, puerperium, refrigerant, sedative, stimulant, stomachic, sweetener, and tonic, tigernut is a folk remedy for abscess, boils, cancer, colds, colic, felons, and flux.[91] Medicinally, tubers are stimulant and aphrodisiac.[278] Decoction of rhizomes (including tubers) taken in Senegal for stomach troubles; leaves poulticed onto forehead for migraine. In Lesotho, heavy consumption said to cause constipation.[55] Young Zulu girls eat porridge mixed with a handful of boiled, mashed root to hasten the inception of menstruation. Root chewed by the Zulu for relief of indigestion, especially when accompanied by halitosis.[332]

Chemistry — Per 100 g, the root (ZMB) is reported to contain 461 to 476 calories, 5.5 to 6.5 g protein, 20.0 to 27.4 g fat, 65.1 to 72.6 g total carbohydrate, 10.5 to 11.7 g fiber, 1.9 to 2.8 g ash, 39.4 to 87.5 mg Ca, 230 to 321 mg P, 3.6 to 12.6 mg Fe, 0.13 to 0.44 mg thiamine, 0.14 mg riboflavin, 2.05 mg niacin, and 4.7 mg ascorbic acid.[89] Tubers contain 20 to 36% of a nondrying, pleasant tasting edible oil, similar to olive oil.[278] Another analysis of tubers reported 14.15% moisture, 25.82% oil, 5.21% albuminoids, 22.72% starch, 24.79% digestible carbohydrates, 5.83% fiber, 1.48% mineral matter. The oil is reported to contain 17.1% saturated acids and 75.8% unsaturated acids. The component fatty acids are: 0.01% myristic, 11.8% palmitic, 5.2% stearic, 0.5% arachidic, 0.3% linoceric, 73.3% oleic, and 5.9% linoleic.[70] Burkill[55] reports the oil to be 73% oleic acid, 12 to 13% palmitic acid, 6 to 8% linoleic acid, 5 to 6% stearic acid. Raw tubers of the genus *Cyperus* have been reported to contain per 100 g, 302 calories, 36.5% moisture, 3.5 g protein, 12.7 g fat, 46.1 g carbohydrate, 7.4 g fiber, 1.2 g ash, 25 mg calcium, 204 mg phosphorus, 8.0 mg iron, 0.28 mg thiamine, 0.09 mg riboflavin, 1.3 mg niacin, and 3 mg ascorbic acid. Dried tubers are reported to contain 452 calories, 11.8% moisture, 4.0 g protein, 25.3 g fat, 56.9 g carbohydrate, 4.7 g fiber, 2.0 g ash, 48 mg calcium, 212 mg phosphorus, 3.2 mg iron, 0.23 mg thiamine, 0.10 mg riboflavin, 1.1 mg niacin, and 6 mg ascorbic acid.[89]

Toxicity — Contains cineole, hydrocyanic acid, and myristic acid.[86]

Description — Perennial herb, forming colonies with creeping thread-like rhizomes 1 to 1.5 mm thick; some forms have tuber-like thickenings on rhizomes, these plants rarely flower. Tubers 1 to 2 cm long, roots fibrous; culms erect, 2 to 9 dm tall, simple, triangular. Leaves several, 3-ranked, pale green, 4 to 9 mm wide, about as long as culm, with closed sheaths mostly basal. Umbel terminal, simple or compound, the longest involucral leaf much exceeding the umbel; spikelets 0.5 to 3 cm long, 1.5 to 3 mm broad, yellowish to golden-brown, strongly flattened, mostly 4-ranked, occasionally 2-ranked, along the wing-angled rachis, blunt, tip acute to round; scales thin, oblong, obtuse, distinctly veined, thin, dry at tip, 2.3 to 3 mm long. Achene yellowish-brown, 3-angled, lustrous, ellipsoid or linear to oblong-cylindric, rounded at summit, 1.2 to 1.5 mm long, granular-streaked. Flowers July to September, fruiting through December in extreme south; various in other parts of the world.[278]

Germplasm — Reported from the Mediterranean Center of Diversity, tigernut, or cvs thereof, is reported to tolerate heavy soil, laterite, salt, sand, virus, weeds, and waterlogging, but not shade.[82] Several botanical varieties are recognized. Two varieties in the U.S. are *C. esculentus* var. *angustispicatus* Britt., with spikelets less than 2 mm wide, tapering to slender points, and *C. esculentus* var. *macrostachys* Boeckl., with spikelets 2 to 3 mm wide, uniformly linear and rounded at apex.[278] ($2n = 18, 108$.)

Distribution — Cosmopolitan, distributed in tropics, subtropics. and warmer temperate regions of world, up to 2,000 m in some areas. Much cultivated in coastal regions of Ghana and in some Mediterranean regions.[278] Listed as a serious weed in Angola, Canada, Kenya, Malagasy, Mozambique, Peru, South Africa, Tanzania, U.S., and Zimbabwe, a principal weed in Australia, Hawaii, India, Mexico, and Switzerland, and a common weed in Argentina, Iran, Portugal.[134]

Ecology — Ranging from Cool Temperate Moist to Wet through Tropical Very Dry to Moist Forest Life Zones, tigernut is reported to tolerate annual precipitation of 1.8 to 27.8 dm (mean of 35 cases = 10.8), annual temperature of 6.9 to 27.5°C (mean of 34 cases =

18.6°C), and pH of 4.5 to 8.0 (mean of 29 cases = 6.3).[82] Common in wet soil, often a weed in cultivated fields and pastures. Often locally abundant and weedy in sandy disturbed, unstable, or loamy soil. Tolerant of nearly any climatic or soil situation, provided there is sufficient water. Often limited to low, poorly drained areas in fields.[82] Hardy to Zone 3.[343] According to Holm et al.,[135] the species grows very well "on all soil types"; including black peat soils, and performs equally well at pH ranges from 5 to 7. More ecological data are reported by Holm et al.[135]

Cultivation — Reproduces by seeds and weak thread-like stolons. Propagated in spring by planting small tubers or chufas, similar to potatoes. Crop requires no cultivation or fertilizers.[278]

Harvesting — Tubers are harvested 5 to 6 months after planting. Two crops can be attained in rainy season. Chufa Oil is obtained by pressing cleaned tubers.[278]

Yields and economics — One tuber weighing 200 mg can produce 36 plants and 332 tubers in 16 weeks, 1,900 plants and 7,000 tubers in one year. Holm et al.[135] report as much as 18 MT/ha tubers in the top 45 cm soil, with perhaps 30,000,000 tubers per hectare. Yields of 800 kg root per hectare in 4 to 6 months have been reported.[82] *Cyperus esculentus* is a serious weed in sugarcane in Hawaii, Puerto Rico, South Africa, and Swaziland; of corn in Angola, South Africa, Tanzania, and the U.S.; of cotton in Mozambique, the U.S., and Zimbabwe; of soybeans in Canada and the U.S.; and of potatoes in Canada, South Africa, and the U.S. More data are presented by Holm et al.[134,135]

Energy — Although 18 MT tubers/ha might sound like good energy potential, it takes a lot of energy to harvest them. Perhaps it is energetically wise to let pigs do the harvesting. Leaving a field fallow 4 years has reduced tuber numbers significantly (912 to 7 per 30 cm^2), the equivalent of 21 to 1.6 MT/ha.[135] Savel'eva et al.[291] have considered this as a possible raw material for industry in Russia.

Biotic factors — Bees visit the flowers in Sierra Leone as a source of pollen.[55] The insect *Bactra verutana* is of interest for biocontrol of the chufa weed. Chufa is an alternative host of the virus which produces lucerne dwarf.[135] The following fungi have been reported on yellow nutsedge: *Aspergillus niger, Puccinia canaliculata, P. conclusa*, and *P. romagnoliana*. Nematodes isolated include: *Caconema radicicola, Heterodera cyperi, Meloidogyne arenaria*, and *M. javanica*.[186,278] In addition, *Ascochyta* sp. and *Phyllachora cyperi* have been reported.[4]

CYPERUS ROTUNDUS L. (CYPERACEAE) Purple nutsedge

Uses — Considered the number one weed in many parts of the world,[135] this sedge has still been suggested as a landscape plant in China, and as a soil binder in India. Tuberous rhizome, eaten in many areas as vegetable or chewed on, may be regarded as a famine food. Plants used as fodder for cattle in West Africa and India. Tubers fed to pigs. Used as bait for catching rats in Tanganyika. The tuber is burnt as a perfume in Tripoli. In Asia and West Africa, the essential oil obtained from tubers, is used as a perfume for clothing and to repel insects, probably due to the camphoraceous odor.[55,278,332]

Folk medicine — According to Hartwell,[126] purple nutsedge is used in folk remedies for phymata, abdominal tumors, glandular tumors, hard tumors, indurations of the stomach, liver, spleen, and uterus, and cervical cancer. Reported to be alterative, analgesic, anodyne, anthelmintic, antihistamine, aphrodisiac, astringent, bactericide, carminative, demulcent, diaphoretic, diuretic, emmenagogue, emollient, fungistatic, lactagogue, stimulant, stomachic, tonic, tranquilizer, vasodilator, vermifuge, and vulnerary, purple nutsedge is a folk remedy for abdominal ailments, amenorrhea, ascites, bladder ailments, bowel ailments, cancer of the cervix, chest ailments, cholera, circulation, colds, congestion, depression, diarrhea, dysentery, dysmenorrhea, dyspepsia, fever, headache, hemicrania, hypertension, impotence, inflammation, metritis, metroxenia, scorpion bites, snake bites, sores, stomach ailments, stomach-ache, toothaches, trauma, tumors of the abdomen, ulcers, and wounds.[91,278] In Mali the tubers are taken as an aphrodisiac. Made into a cough medicine for children. Used in Africa and Asia for urinary troubles, indigestion, childbirth, jaundice, malaria, and many other conditions.[55] Plant used in Vietnam as a diuretic, emmenagogue, headache remedy, and for uterine hemorrhage. The tuber is given to women in childbirth in Indo-China. The fresh tuber, made into a paste or warm plaster, is applied to the breast with galactagogic intent, and, in a dry state to spreading ulcers, in the Indian Peninsula.[332] The tuber, in the form of ghees, powders, bolmes, and enemas is used as a folk remedy for abdominal tumors. In Ghana, an infusion of the plant is given for cattle poisoning due to *Ipomoea repens*.[278]

Chemistry — Per 100 g, the edible tuber should resemble that of *Cyperus esculentus*, which (ZMB) contains 461 to 476 calories, 5.5 to 6.5 g protein, 20.0 to 27.4 g fat, 65.1 to 72.6 g total carbohydrate, 10.5 to 11.7 g fiber, 1.9 to 2.4 g ash, 29 to 88 mg Ca, 230 to 321 mg P, 2.6 to 12.6 mg Fe, 0.13 to 0.44 mg thiamine, 0.14 mg riboflavin, 2.05 mg niacin, and 5 mg ascorbic acid.[89] Tubers of *Cyperus rotundus* include 0.5 to 1.0% essential oil, 0.21 to 0.24% alkaloid, 0.62 to 0.74% cardiac glycosides, 1.25% flavonoids, 1.62% polyphenols, 13.22% saccharides, 9.2% starch, 3.72% pectin, 4.21% resin, and 3.25% total acids (mostly malic), 0.009% vitamin C. In the essential oil, one finds cyperene-1, cyperene-2, patchoulene ($C_{15}H_{22}O$), mutacone ($C_{15}H_{22}O$), beta-seliene, beta-cyperone, cyperenone, 1,8-cineole, limonene, beta-pinene, p-cymol, camphene, isocyperol ($C_{15}H_{24}O$). The fatty oil contains glycerol, linolenic, linoleic, oleic, myristic, and possibly stearic acid. The tuber also contains a substance capable of dissolving several times its weight in lecithin (and other items which cause urinary calculi). Molasses extracted from the tuber contains 41.7% d-glucose, 9.3% d-fructose, and 4% nonreducing sugars.[70,187,332] Salicylic acid may be extracted from leaves and sprouted tubers.[135]

Description — Perennial herb, forming colonies with long, slender, creeping rhizomes, about 1 mm thick, with tuber-like thickenings at intervals, to 1 cm thick; culms slender, 8 to 60 cm tall, simple, smooth, triangular, longer than leaves. Leaves 2 to 6 mm wide, crowded in the basal few centimeters, usually spreading. Inflorescence of simple or slightly compound umbels, 3 to 11 cm long, on 3 to 8 extremely unequal peduncles, each bearing a cluster of 3 to 9 divaricate spikelets; spikelets 0.8 to 2.5 cm long, chestnut-brown to chestnut-purple, acute 12- to 40-flowered; bracts usually 3 or 4, about as long as inflorescence; scales keeled, straight, ovate, closely appressed, nerveless except on keel, 2 to 3.5 mm long, bluntish. Achene linear-oblong, 1.5 mm long, 3-angled, basally and apically obtuse, granular, dull, olive-gray to brown, covered with a network of gray lines. Flowers July to October or December; January to April in southern hemisphere.[278]

Germplasm — Reported from the Euro-Siberian and North American Centers of Diversity, purple nutsedge, or cvs thereof, is reported to tolerate alkali, heat, high pH, insects, laterite, low pH, salt, and weeds. (2n = 108.)[82] Several ecotypes are recognized. Types are described from India with the following variation in glume color: (1) yellowish-white, (2) light-red, (3) coppery-red with metallic luster, and (4) dark-red with blackish tinge.[135]

Distribution — Native to Europe, Japan, and North America; widespread in all tropical, subtropical, and warm temperate regions of the world.[278] *C. rotundus* has been reported from more countries, regions, and localities than any other weed in the world.[135]

Ecology — Ranging from Boreal Moist through Tropical Desert to Wet Forest Life Zones, purple nutsedge is reported to tolerate annual precipitation of 3.0 to 46.1 dm (mean of 192 cases = 16.9), annual temperature of 0.0 to 28.6°C (mean of 156 cases = 20.2), and pH of 4.3 to 9.1 (mean of 75 cases = 6.4).[82] Continuous shading reduces tuber and bulb formation by 10 to 57%. Tubers cannot survive more than 10 days at 45°C or 30 min at 60°C. Tubers held at 50°C more than 48 hr no longer germinate. Exposure to −4°C for 8 hr does not impair viability. Tubers, when dug, contain about 50% moisture. They cannot survive when the level falls below 12 to 15%. Some tubers held in water for 200 days still germinate satisfactorily when removed from water and placed under suitable growing conditions. Thrives in loamy or sandy soil anywhere; in many places up to 2,000 m altitude. In wastelands, gardens, waysides, and in open spots; a troublesome weed in cultivated fields. Requires a warm climate, no colder than the southern U.S., especially the Cotton Belt.[278] According to Holm et al.[135] it seems limited by cold temperatures, but other than this, it grows in almost every soil type, elevation, humidity, soil moisture, and pH, but it cannot stand soils with high salt content. It can survive the highest temperatures known in agriculture. Also found on roadsides, in neglected areas, at the edges of woods, sometimes covering banks of irrigation canals and streams. Nutsedge can take over entire streams or canals as water becomes low. When water supply is low, it may become a problem in paddy rice in which puddling of the soil cannot be done thoroughly.[135]

Cultivation — Propagated by seed and tuber-bearing rhizomes. Because it grows so profusely, it is considered more a weed than a plant to be cultivated.[278]

Harvesting — Plants are harvested from native or naturalized stands. In Africa and Asia it is harvested on a small scale for the oil, but in most areas it is allowed to grow wild.[278] Flowering has been reported as early as 3 weeks in Israel and India, and 4 weeks in Trinidad, with tuber formation occurring at 3 weeks in Hawaii, India, Puerto Rico, Trinidad, and the southern U.S. In Israel, clipping every 2 weeks reduced tuber numbers by 60% and weight by 85%.[135]

Yields and economics — "*Cyperus rotundus* may produce up to 40,000 kilograms of subterranean plant material per hectare." In Mauritius, there may be 30 MT green tops and tubers, withdrawing 815 kg/ha ammonium sulfate, 320 kg/ha muriate of potash, and 200 kg of superphosphate. In Argentina, the weed can reduce sugarcane harvested by 75%, the sugar yield by 65%. Allowed to remain in corn-fields for 10 days in Colombia, it reduces yield by 10%, 30% in 30 days, suggesting a percentage loss for each day it is allowed to remain.[135]

Energy — The 40 tons of underground plant material, convertible to energy, is perhaps most efficiently harvested by grubbing pigs.

Biotic factors — *Cyperus rotundus* is an alternate host of *Fusarium* sp. and *Puccinia canaliculata*, of abaca mosaic virus, and of the nematodes *Meloidogyne* sp. and *Rotylenchus similis*. The nutgrass moth, *Bactra traculenta*, which bores into the stems of *Cyperus* *rotundus*, showed promise for biological control in Hawaii in the early years after its introduction from the Phillippines in 1925. As the populations of *Bactra* increased, so also did those of the insect *Trichogramma minutum*, which parasitizes the eggs of many moths and butterflies. So many of the eggs of *Bactra* were killed that biological control of nutgrass was never attained. The jack bean, *Canavalia ensiformis*, greatly inhibits tuber formation.[135] In addition, *Cintractia minor, Phyllachora cyperi, Puccinia cyperia, Rhizoctonia solani*[4] and *Cintractia peribebugensis, Himatia stellifera*, and *Puccinia cyperi-tergetiformis*[380] are reported.

DETARIUM SENEGALENSE J. F. Gmel. (CAESALPINIACEAE) — Tallow Tree
Syn.: ***D. heudelotianum*** **Baill.**

Uses — The only seeds and the pulp around them are used as food sources in Africa. The pulp can be made into a sweetmeat. The oily kernels, little eaten by humans, are beaten into cattle fodder by the Nupe. Ashes of the fruits are used to prepare a snuff. Seeds are used for necklaces and girdles. An aromatic resin, exuding from the trunk, is used to fumigate African huts and garments. The resin is used as a masticatory and to mend pottery. The wood is used for planks and boat-building in Liberia and sold in England as African Mahogany. Roots are boiled on the Gold Coast to prepare a bird-lime. Seeds are burned to repel mosquitoes.[71,146]

Folk medicine — Senegalese use the wood decoction for anemia and cachexia. In Sierra Leone, young shoots are boiled as a febrifuge. Liberians use the bark decoction for placental retention. In French Guinea, the bark is boiled to make a lotion for itch. Nigerians use the seed for people inflicted with wounds by poisoned arrows. In Ghana, the fruit is used for rubbing chronic backache or tuberculosis of the spine. Fruits are used for chest ailments in West Africa.[71,146]

Chemistry — Per 100 g, the raw fruit is reported to contain 116 calories, 66.9% moisture, 1.9 g protein, 0.4 g fat, 29.6 g carbohydrates, 2.3 g fiber, 1.2 g ash, 27 mg calcium, 48 mg phosphorus, 0.14 mg thiamine, 0.05 mg riboflavin, 0.6 mg niacin, and 1,290 mg ascorbic acid. Dried fruit contains, per 100 g, 299 calories, 14.0% moisture, 3.4 g protein, 0.5 g fat, 78.8 g carbohydrate, 7.1 g fiber, 3.3 g ash, 110 mg calcium, 0.01 mg thiamine, 0.03 mg riboflavin, 3.8 mg niacin, and 3 mg ascorbic acid.[89] Detaric acid has been isolated from the fruits.[146] According to Hager's Handbook,[187] the fruits are among the highest in the world for vitamin C. The figures above suggest that might be true, but the vitamin C is lost in drying. Other sources hint that the seeds or fruits are poisonous.

Description — Tree to nearly 40 m tall, smaller in savanna, with large crown, girth 12 m, bole 12 m; slash pale-salmon, bark bluish, exuding a slightly fragrant gum or gum-resin, twigs rusty. Leaves pinnate, more or less gland-punctuate; leaflets 6 to 12, leathery and rather glaucous or minutely pubescent below, with numerous parallel lateral nerves. Flowers in fragrant creamy axillary panicles, shorter than leaves, flowers small, profuse, sepals 4, white, petals absent, stamens 10, buds glabrous or nearly so, ca. 4 mm long, sepals pubescent within. Fruits round, succulent, like flattened mango, >6 cm in diameter, skin smooth, crustaceous, with intermediate fibrous layer. Flowers May to August; fruits December to January; Ghana.[71,146]

Germplasm — Reported from the African Center of Diversity, tallow tree, or cvs thereof, is reported to tolerate drought and savanna. The savanna form (*senegalense*) is smaller than the closed forest form (*heudelotianum*). Seeds of the latter are more likely to be poisonous.

Distribution — Throughout west Tropical Africa.

Ecology — A tree of the Closed Forest and Fringing Forests of moister savannas.[71,146]

Cultivation — Apparently cultivated only to a limited extent in Senegal.

Harvesting — No data available.

Yieds and economics — No data available.

Energy — The wood burns slowly and is favored as a fuel because of the agreeable odor.[8]

Biotic factors — The heartwood is probably resistant to borers and termites.[71]

ELAEIS GUINEENSIS Jacq. (ARECACEAE [PALMAE]) — African Oil Palm
Syn.: *Elaeis melanococca* J. Gaertn.

Uses — Two kinds of oil are obtained from this palm, palm oil and palm kernel oil. Palm oil is extracted from the fleshy mesocarp of the fruit, which contains 45 to 55% oil which varies from light-yellow to orange-red in color, and melts from 25° to 50°C. For edible fat manufacture, the oil is bleached. Palm oil contains saturated palmitic acid, oleic acid, and linoleic acid, giving it a higher unsaturated acid content than palm kernel or coconut oils. Palm oil is used for manufacture of soaps and candles, and more recently, in manufacture of margarine and cooking fats. Palm oil is used extensively in the tin plate industry, protecting cleaned iron surfaces before the tin is applied. Oil is also used as lubricant in the textile and rubber industries. Palm kernel oil is extracted from the kernel of endosperm, and contains about 50% oil. Similar to coconut oil, with a high content of saturated acids, mainly lauric, it is solid at normal temperatures in temperate area, and is nearly colorless, varying from white to slightly yellow. This nondrying oil is used in edible fats, in making ice cream and mayonnaise, in baked goods and confectioneries, and in the manufacture of soaps and detergents. Press-cake, after extraction of oil from the kernels, is used as livestock feed, and contains 5 to 8% oil. Palm wine is made from the sap obtained by tapping the male inflorescence. The sap contains about 4.3 g/100 mℓ sucrose and 3.4 g/100 mℓ glucose. The sap ferments quickly and is an important source of Vitamin B complex in the diet of people of West Africa. A mean annual yield for 150 palms is 4,000 ℓ/ha, double in value to the oil and kernels from the same number of palms. The central shoot (or cabbage) is edible.

Leaves used for thatching; petioles and rachices for fencing and for protecting the tops of mud walls. Refuse after stripping the bunches is used for mulching and manuring; ash sometimes used in soap-making.[125,278,321]

Folk medicine — According to Hartwell,[126] the oil is used as a liniment for indolent tumors. Reported to be anodyne, antidotal, aphrodisiac, diuretic, and vulnerary, oil palm is a folk remedy for cancer, headaches, and rheumatism.[91]

Chemistry — As the oil is rich in carotene, it can be used in place of cod liver oil for correcting Vitamin A deficiency. Per 100 g, the fruit is reported to contain 540 calories, 26.2 g H_2O, 1.9 g protein, 58.4 g fat, 12.5 g total carbohydrate, 3.2 g fiber, 1.0 g ash, 82 mg Ca, 47 mg P, 4.5 mg Fe, 42,420 mcg beta-carotene equivalent, 0.20 mg thiamine, 0.10 mg riboflavin, 1.4 mg niacin, and 12 mg ascorbic acid. The oil contains, per 100 g, 878 calories, 0.5% H_2O, 0.0% protein, 99.1% fat, 0.4 g total carbohydrate, 7 mg Ca, 8 mg P, 5.5 mg Fe, 27,280 mcg beta-carotene equivalent, 0.03 mg riboflavin, and a trace of thiamine.[89] The fatty composition of the oil is 0.5 to 5.9% myristic, 32.3 to 47.0 palmitic, 1.0 to 8.5 stearic, 39.8 to 52.4 oleic, and 2.0 to 11.3 linoleic. The component glycerides are oleodipalmitins (45%), palmitodioleins (30%), oleopalmitostearins (10%), linoleodioleins (6 to 8%), and fully saturated glycerides, tripalmitin and diapalmitostearin (6 to 8%). Micou[211] notes that vitamin E is a by-product of the process which converts palm oil into a diesel-oil substitute.

Description — Tall palm, 8.3 to 20 m tall, erect, heavy, trunks ringed; monoecious, male and female flowers in separate clusters, but on same tree; trunk to 20 m tall, usually less, 30 cm in diameter. Leaf bases adhere; petioles 1.3 to 2.3 m long, 12.5 to 20 cm wide, saw-toothed, broadened at base, fibrous, green; blade pinnate, 3.3 to 5 m long, with 100 to 150 pairs of leaflets; leaflets 60 to 120 cm long, 3.5 to cm broad; central nerve very strong, especially at base, green on both surfaces. Flower stalks from lower leaf axils, 10 to 30 cm long and broad; male flowers on short furry branches 10 to 15 cm long, set close to trunk on short pedicels; female flowers and consequently fruits in large clusters of 200 to 300, close to trunk on short heavy pedicles. Fruits plum-like ovoid-oblong to 3.5 cm long and about 2 cm wide, black when ripe, red at base, with thick ivory-white flesh and small cavity in center; nuts encased in a fibrous covering which contains the oil. About 5 female inflorescences are produced per year; each inflorescence weighing about 8 kg, the fruits weighing about 3.5 g each.

Germplasm — Reported from the African Center of Diversity, the African oil palm or cvs thereof is reported to tolerate high pH, laterite, low pH, savanna, virus, and water-logging.[82] Ehsanullah[94] reported on oil palm cultivars. African Oil Palm is monoecious and cross-pollinated, and individual palms are very heterozygous. Three varieties are distinguished: those with orange nuts which have the finest oil but small kernels; red or black nut varieties which have less oil, but larger kernels. Sometimes oil palms are classified according to the fruit structure: *Dura*, with shell or endocarp 2 to 8 mm thick, about 25 to 55% of weight of fruit; medium mesocarp of 35 to 55% by weight, but up to 65% in the Deli Palms; kernels large, 7 to 20% of weight of fruit; the most important type in West Africa; the *Macrocarya* form with shells 6 to 8 mm thick forms a large proportion of the crop in western Nigeria and Sierra Leone. *Tenera*, with thin shells, 0.5 to 3 mm thick, 1 to 32% of weight of fruit; medium to high mesocarp 60 to 95% of weight of fruit; kernels 3 to 15% of fruit; larger number of bunches than Dura, but lower mean bunch weight and lower fruit-to-bunch ratio. *Pisifera*, shell-less, with small kernels in fertile fruits, fruits often rotting prematurely; fruit-to-bunch ratio low. Infertile palms show strong vegetative growth, but of little commercial value; however it has now become of greatest importance in breeding commercial palms. *Deli* Palm (Dura type), originated in Sumatra and Malaya, gives high yields in the Far East, but not so good in West Africa. Dumpy Oil Palm, discovered in Malaya among Deli Palms, is low-growing and thick-stemmed. Breeding and selection of oil palms have been aimed at production of maximum quantity of palm oil and kernels per hectare, and resistance to disease. Recently, much attention has been directed at cross-breeding with *E. oleifera* for short-trunk hybrids, thus making harvesting easier. Zeven[349] elucidates the center of diversity, and discusses the interactions of some important oil palm genes.

Distribution — The center of origin of the oil palm is in the tropical rain forest region of West Africa in a region about 200 to 300 km wide along the coastal belt from Liberia to Angola. The palm has spread from 16°N latitude in Senegal to 15°S in Angola and eastwards to the Indian Ocean, Zanzibar, and Malagasy. Now introduced and cultivated throughout the tropics between 16°N and S latitudes. Sometimes grown as an ornamental, as in southern Florida.[278,349]

Ecology — Occurs wild in riverine forests or in fresh-water swamps. It cannot thrive in primeval forests and does not regenerate in high secondary forests. Requires adequate light and soil moisture, can tolerate temporary flooding or a fluctuating water table, as might be found along rivers. It is slightly hardier than coconut. Ranging ecologically from savanna to rain forest, it is native to areas with 1,780 to 2,280 mm rainfall per year. Best developed on lowlands, with 2 to 4 month dry period. Mean maximum temperatures of 30 to 32°C and mean minimum of 21 to 24°C provides suitable range. Seedling growth arrested below 15°C. Grows and thrives on a wide range of tropical soils, provided they have adequate water. Waterlogged, highly lateritic, extremely sandy, stony or peaty soils should be avoided. Coastal marine alluvial clays, soils of volcanic origin, acid sands, and other coastal alluviums are used. Soils with pH of 4 to 6 are most often used. Ranging from Subtropical Dry (without frost) through Tropical Dry to West Forest Life Zones, oil palm is reported to tolerate annual precipitation of 6.4 to 42.6 dm (mean of 27 cases = 22.7), annual temperature of 18.7 to 27.4°C (mean of 27 cases = 24.8), and pH of 4.3 to 8.0 (mean of 22 cases = 5.7).[82]

Cultivation — In wild areas of West Africa the forest is often cleared to let 75 to 150 palms stand per hectare; this yields about 2.5 MT of bunches per hectare per year. Normally, oil palms are propagated by seed. Seed germination and seedling establishment are difficult. A temperature of 35°C stimulates germination in thin shelled cvs. Thick-walled cvs require higher temperatures. Seedlings are outplanted at about 18 months. In some places, seeds are harvested from the wild, but plantation culture is proving much more rewarding. In a plantation, trees are spaced 9 × 9 m; a 410-ha plantation would have about 50,000 trees, each averaging 5 bunches of fruit, each averaging 1 kg oil to yield a total of 250,000 kg oil for the 410 ha. Vegetative propagation is not feasible, as the tree has only one growing point. Because oil palm is monoecious, cross-pollination is general and the value of parent plants is determined by the performance of the progeny produced in such crosses. Bunch-yield and oil and kernel content of the bunches are used as criteria for selecting individual palms for breeding. Controlled pollination must be maintained when breeding from selected plants. Seed to be used for propagation should be harvested ripe. Best germination results by placing seeds about 0.6 cm deep in sand flats and covering them with sawdust. Flats are kept fully exposed to sun and kept moist. In warm climates, 50% of seed will germinate in 8 weeks; in other areas it may take from 64 to 146 days. Sometimes the hard shell is ground down, or seeds are soaked in hot water for 2 weeks, or both, before planting. Plants grow slowly at first, being 6 to 8 years old before the pinnate leaves become normal size. When planting seedlings out in fields or forest, holes are dug, and area about 1 m around them cleared. Young plants should be transplanted at the beginning of rainy season. In areas where there is no distinct dry season, as in Malaya, planting out may be done the year round, but is usually done during months with the highest rainfall. Seedlings or young plants, 12 to 18 months old, should be moved with a substantial ball of earth. Ammonium sulfate and sulfate or muriate of potash at a rate of 227 g per palm should be applied in a ring about the plant at time of planting. Where magnesium may be deficient in the soil, 227 g

Epsom salts or kieserite should be applied also. In many areas oil palms are intercropped with food plants, as maize, yams, bananas, cassava, or cocoyams. In Africa, intercropping for up to 3 years has helped to produce early palm yields. Cover-crops are often planted, as mixtures of *Calopogonium mucunoides*, *Centrosema pubescens*, and *Pueraria phaseoloides*, planted in proportion of 2:2:1 with seed rate of 5.5 kg/ha. Natural covers and planted cover crops can be controlled by slashing. Nitrogen dressings are important in early years. Chlorosis often occurs in nursery beds in the first few years after planting out. Adequate manuring should be applied in these early years. When nitrogen fertilizers, as sulfate of ammonium, are used, 0.22 kg per palm in the planting year and 0.45 kg per palm per year until age 4, should be sufficient. Potassium, magnesium, and trace elements requirements should be determined by soil test and the proper fertilizer applied, according to the region, soil type, and degree of deficiency.[125,278]

Harvesting — First fruit bunches ripen in 3 to 4 years after planting in the field, but these may be small and of poor quality. Often these are eliminated by removal of the early female inflorescences. Bunches ripen 5 to 6 months after pollination. Bunches should be harvested at the correct degree of ripeness, as under-ripe fruits have low oil concentration and over-ripe fruits have high fatty acid content. Harvesting is usually done once a week. In Africa, bunches of semi-wild trees are harvested with a cutlass, and tall palms are climbed by means of ladders and ropes. For the first few years of harvesting, bunches are cut with a steel chisel with a wooden handle about 90 cm long, allowing the peduncles to be cut without injuring the subtending leaf. Usually thereafter, an axe is used, or a curved knife attached to a bamboo pole. A man can harvest 100 to 150 bunches per day. Bunches are carried to transport centers and from there to the mill for oil extraction.[125,278]

Yields and economics — According to the Wealth of India, the oil yield of oil palm is higher than that of any other oilseed crop, producing 2.5 MT oil per ha per year, with 5 MT recorded. Yields of semi-wild palms vary widely, usually ranging from 1.2 to 5 MT of bunches per hectare per year. One MT of bunches yields about 80 kg oil by local soft oil extraction, or 180 kg by hydraulic handpress. Estate yields in Africa vary from 7.5 to 15 MT bunches per hectare per year; in Sumatra and Malaya, 15 to 25 MT, with some fields producing 30 to 38 MT. Estate palm oil extraction yield rates vary accordingly: *Dura*, 15 to 16% oil per bunch; *Deli Dura*, 16 to 18% *Tenera*, 20 to 22%. Kernel extraction yields vary from 3.5 to 5% or more. The U.S. imported nearly 90 million kg in 1966, more than half of it as kernel oil. Recently, palm oil commanded $.31/kg, indicating potential yields of about $1400/ha. In 1968 world producing countries exported about 544,000 long tons of oil and 420,000 long tons of kernels. The main producing countries, in order of production, are Nigeria, Congo, Sierra Leone, Ghana, Indonesia, and Malaysia. The U.K. is the largest importer of oil palm products, importing about 180,000 MT of palm oil and 243,000 MT of palm kernels annually. Japan, and Eastern European and Middle East countries also import considerable quantities of palm oil and kernels. Some palm kernel oil extraction is now being done in the palm oil producing countries. Previously, most of the kernels had been exported, and the oil extracted in the importing countries.[125,272,278]

Energy — Bunch yields may attain 22,000 kg/ha; of which only about 10% is oil, indicating oil yields of only 2,200 kg/ha. Higher yields are attainable. Corley[66] suggests plantation yields of 2 to 6 MT/ha mesocarp oil, experimentally up to 8.5 MT/ha. Hodge,[130] citing oil yields of 2,790 kg/ha, suggests that this is the most efficient oil-making plant species. The seasonal maximum total biomass reported for oil palm is 220 MT wet weight. When replanting occurs, over 40 MT/ha DM (dry matter) of palm trunks are available (conceivably for energy production) after the 70% moisture from the wet material has been expelled.[66] Although annual productivity may approach 37 MT DM/ha, mean productivity during the dry season is 10 g/m^2/day.[334] Averaged over the year, oil palm in Malaysia showed a growth rate of 8 g/m^2/day for an annual phytomass production of 29.4 MT/ha.[41] Fresh fruit bunch yields have been increased elsewhere by 2 MT/ha intercropping with appropriate legumes. Estate yields in Africa are 7 to 15 MT bunches per year, with oil yields of 800 to 1800 kg/ha, and residues of yields of ca. 6 to 13 MT. It is probable that older leaves, leaf stalks, etc., could be harvested with biomass yield of 1 to 5 MT/ha. Based on energetic equivalents of total biomass produced, up to 60 barrels of oil per hectare could be obtained from this species. An energy evaluation of all the wastes from the palm oil fruit was made, and it revealed that this can satisfy ca. 17% of Malaysia's energy requirements. Palm oil could satisfy 20% more.[162] An alcoholic wine can be made from the sap of the male spikes, 150 trees yielding about 4,000 ℓ of palm wine per hectare, per year. Worthy of energetic interest is the suggestion of Gaydou et al.[107] that the oil palm can yield twice as much energetically as sugarcane, at least based on the Malagasy calculations.

Biotic factors — Many fungi attack oil palms, but the most serious ones are the following: Blast (*Pythium splendens*, followed by *Rhizoctonia lamellifera*), Freckle (*Cercospora elaeidis*), Anthracnose (*Botryodiplodia palmarum, Melanconium elaeidis, Glomerella cingulata*), Seedling blight (*Curvularia eragrostidis*), Yellow patch and Vascular wilt (*Fusarium oxysporum*), Basal rot of trunk (*Ceratocystis paradoxa*, imp. stage of *Thielaviopsis paradoxa*), other trunk rots (*Ganoderma* spp., *Armillaria mellea*); Crown disease, rotting of fruit (*Marasmius palmivorus*). Spear rot or bud rot is caused by the bacterium *Erwinia* sp., which has devastated entire areas in S. Congo. *The Agriculture Handbook 165* reports the leaf spot (*Achorella attaleae*) and the Black Mildew (*Meliola melanococcae, M. elaeis*).[4] The following nematodes have been isolated from oil palms: *Aphelenchus avenae, Heterodera marioni, Helicotylenchus pseudorobustus, H. microcephalus cocophilus* (serious in Venezuela), *Scutellonema clathrocaudatus*. The major pests of oil palm in various parts of the world are the following: Palm weevils (*Rhynchophorus phoenicis, R. palmarum, R. ferrugineus, R. schach*), Rhinoceros beetles (*Orcytes rhinoceros, O. boas, O. monoceros, O. owariensis*), Weevils (*Strategus aloeus, Temnoschoita quadripustulata*), Leaf-miners (*Coelaenomenodera elaeidis, Hispolepsis elaeidis, Alurunus humeralis*), Slug caterpillar (*Parasa viridissima*), Nettle caterpillar (*Setoria nitens*), Bagworms (*Cremastophysche pendula, Mahesena corbetti, Metisa plana*). Rodents may cause damage to seedlings and fruiting palms; some birds also cause damage in jungle areas.[186,272,278]

ELAEIS OLEIFERA (HBK) Cortes (ARECACEAE) — American Oil Palm, Corozo
Syn. *Corozo oleifera* (HBK) Bailey; *Elaeis melanococca* Gaertn., emend. Bailey; *Alfonsia oleifera* HBK

Uses — Plants are native and cultivated to a limited extent in South America; the oil is used for soap-making, food, and lamp fuel. Its main value lies in its slow-growing, procumbent trunk and high percent of parthenocarpic fruits, and for its hybridizing potential with *Elaeis guineensis*.[278] American oil palm is better for margarine-making than the African oil palm, because the former has a low level of free fatty acids and a high melting point.[152]

Folk medicine — Reported to be tonic, corozo is a folk remedy for dandruff and other scalp ailments, inflammation, and stomach problems.[91]

Chemistry — The pericarp yields 29 to 50%, the kernel 29 to 45% oil. The pericarp oil contains 48.3% saturated fatty acids (1.0% C_{14}, 32.6% palmitic, 4.7% C_{18}), 47.5% oleic, and 12.0% linoleic, with traces of arachidic acid (0.5%), 0.9% hexadecenoic acid, and 0.8% linolenic acid.[128]

Description — Small palm; trunk procumbent, although an erect habit may be maintained for about 15 years; erect portion 1.6 to 3 m tall, trunks lying on soil up to 8.3 m long; roots formed along entire length of procumbent portion of trunk. Leaves 30 to 37 per plant; leaflets about 6.3 cm broad, all lying in one plane, no basal swellings; spines on petioles short and thick. Male inflorescence with 100 to 200 spikelets 5 to 15 cm long, pressed together until they burst through the spathe just before anthesis, rudimentary gynoecium with 3 marked stigmatic ridges; female inflorescence with spathe persisting after being ruptured by the developing bunch; spikelets ending in a short prong. Flowers numerous, sunk in the body of the spikelet; bunch of fruits surrounded by the fibers of the spathe, with no long spines; bunches round and wide at their center, pointed at top, giving a distinctly conical shape, rarely weighing more than 22.5 kg, usually much smaller, containing a large number of small fruits. Fruits ripen from pale yellow to bright red (a high proportion, up to 90%, parthenocarpic or abortive); perianth persistent as fruit ripens and becomes detached from the bunch; fruits 2.5 to 3.0 cm long, weighing as little as 2 to 3.5 g each with average weights from 8.5 to 12.6 g; nuts with 2 kernels fairly frequent, with 3 occasional.[278]

Germplasm — Reported from the South American Center of Diversity, corozo, or cvs thereof, is reported to tolerate acid soils, drought, savanna, some salt-water, and waterlogging.[82] "Tissue culture has increased interest in the hybrids of *E. guineensis* × *E. oleifera*: the latter produces a high quality unsaturated oil, although the yield of oil is low. The oil yield of the F_1 hybrid is intermediate between both parental species; in back crosses to *E. guineensis*, however, occasional palms are found that combine good yield (from *E. guineensis*) with improved oil quality and reduced height increment (from *E. oleifera*): such palms can now be multiplied clonally."[123] There are some variations in habit of growth and leaf-formation. This species easily hybridizes with the *pisifera* form of *E. gunieensis*, the African Oil Palm, and the fruits are relatively thin-shelled, but have no fiber ring.[236,237,278] (2n = 32.)

Distribution — Native to Central and South America (Brazil, Colombia, Venezuela, Surinam, Panama, and Costa Rica.).[278]

Ecology — Ranging from Subtropical Dry to Moist through Tropical Thorn to Moist Forest Life Zones, corozo is reported to tolerate annual precipitation of 6.4 to 15.2 (to 40) dm (mean of 7 cases = 11.9), annual temperature of 21.0 to 27.8°C (mean of 7 cases = 24.4), and pH of (4 to) 5.0 to 8.0 (mean of 5 cases = 6.5).[82] Actually, I have observed oil palm in much wetter situations than these data indicate. At the Panama-Costa Rica border, where the rainfall is closer to 40 dm, there are abundant strands of corozo, some even said to tolerate brackish water. In Latin America plants grow procumbent in swampy areas, and more upright in drier areas. Best development is in lowland ravines with rainfall between

1,700 and 2,200 mm annually. Mean maximum temperatures of 30 to 32°C and mean minimum temperatures of 21 to 24°C are suitable. Grows and thrives on a wide range of tropical soils, provided they have adequate water; soils with pH 4 to 6 are most often used for cultivation.[278]

Cultivation — When this palm is cultivated, seeds are planted in seedbeds and the seedlings transplanted into the field when about 12 to 18 months old. Fruits are selected from special mother plants, often after pollination with pollen of a selected male palm. Seeds may be germinated in a germinator and the seedlings grown in a pre-nursery, and later in a nursery. Transplants are planted where the bush has been checked. In the nurseries, plants receive water and fertilizer and are shaded to protect them from sunburn. After being planted out, they must receive more fertilizer. Ammonium sulfate and sulfate or muriate of potash at a rate of 227 g per palm should be applied in a ring about the plant at time of planting. Where magnesium may be deficient in the soil, 227 g Epsom salts or kieserite should be applied also. Plants grow slowly at first, being about 7 years old before the typical pinnate leaves form normal size. In many areas, oil palms are intercropped with vegetable and other food crops, as maize, yams, bananas, cassava, or cocoyams. Intercropping for 3 years or so has helped to produce early palm yields. Cover crops are often planted, as mixtures of *Calopogonium mucunoides, Centrosema pubescens*, and *Pueraria phaseoloides,* planted in proportion of 2:2:1 with seed rates of 5.5 kg/ha. Natural covers and planted cover crops can be controlled by slashing. Adequate manure should be applied during the early years to provide nitrogen. When nitrogen fertilizers (e.g., sulfate or ammonium) are used, 0.22 kg per palm in the planting years and 0.45 kg per palm per year until age 4, should be sufficient. Potassium, magnesium, and trace elements requirements should be determined by soil test and the proper fertilizer applied, according to the region, soil type, and degree of deficiency.[278]

Harvesting — Fruits mature from January to June, usually borne only about 1.5 m above the ground, in an averge 5 clusters. Fruits begin to be formed about 4 years after planting in the field. Often the first female inflorescences are cut off to allow better plant development. Bunches ripen about 6 months after pollination. Ripe fruits are harvested about once a week. Bunches are cut with machete or sharp knife, and carried to transport centers, from which they go to the mill for oil extraction.[278]

Yields and economics — Bunches rarely weigh more than 22.5 kg, and generally average 8.5 to 12.67 g each; in Colombia, fruits weigh as little as 2.0 to 3.5 g. Fruit-to-bunch ratio varies from 32 to 44%.[278] Oil yield of *E. oleifera* is much lower than that of *E. guineensis*;[123] a tree can yield annually ca. 25 kg fruit (equalling ca. 12,850 individual fruits).[29,30] Hadcock[120] describes a simple oil palm mill (capacity 250 kg bunches per hr) that would work on either species of oil palm. Bunches are sterilized for 1 hr before stripping. After stripping, the fruit is reheated for 1 hr before it is digested in a rapid digester operated by a 5 h.p engine. Oil is extracted with a hydraulic press. The oil is separated from the crude material by means of a continuous settling clarifier fitted with a heat exchanger to dry the oil. The efficiency of oil recovery is only 75 to 86%. The mill, including the building cost, is U.S. $34,000.[120]

Energy — See African oil palm, which has a somewhat higher energy potential.

Biotic factors — Both bee- and wind-pollinated; but up to 90% of fruits may be parthenocarpic. Bees are common around male inflorescence and may act as pollinating agents. Hermaphroditic inflorescence plants are found in America and in planted trees in the Congo. Most of the pests and diseases of the African oil palm are associated with this palm also, especially where it has been planted with *E. guineensis*.[278]

ELEOCHARIS DULCIS (Burm.f.) Trin. ex Henschel (CYPERACEAE) — Waternut, Chinese Water chestnut, MA TAI, MA HAI

Syn.: ***Andropogon dulce* Burm. f., *Scirpus plantagineus* Tetz., *Scirpus plantaginoides* Rottb., *Scirpus tuberosus* Roxb., *Eleocharis plantaginea* (Retz.) Roem. and Schut., *Eleocharis tuberosa* Schultes**

Uses — Edible tubers or corms are used as a vegetable in many East Indian and Chinese dishes. Sliced, they are esteemed in Chinese soups for their crisp texture and delicious flavor. Sliced water chestnuts are one of the ingredients of chop suey in the U.S. They are recommended, as well, in American salads and soups. Shredded water chestnuts often appear in meat and fish dishes. Raw corms are eaten out of hand in lieu of fresh fruit in China. East Indians cook the tubers, remove the rind, crush the meat with a hammer, sun dry, and fry in coconut oil as a delicacy.[258,278]

Folk medicine — In China,[90] the plant is used for abdominal pain, amenorrhea, anemia, bruises, clots, gas, hernia, inflammation, liver, malnutrition, pinkeye, and swellings. Porterfield[258] notes that Chinese give quantities to children who have ingested coins, in the belief that the water chestnuts will decompose the metal.

Chemistry — Per 100 g, the root (ZMB) contains 360 to 364 calories, 7.4 to 8.5 g protein, 0.7 to 1.6 g fat, 84.8 to 87.6 g total carbohydrate, 3.2 to 3.9 g fiber, 5.1 to 6.0 g ash, 18.4 to 26.5 mg Ca, 299 to 407 mg P, 2.8 to 3.7 mg Fe, 53 to 92 mg Na, 2,304 to 2,545 mg K, 0 μg beta-carotene equivalent, 0.16 to 0.65 mg thiamine, 0.11 to 0.92 mg riboflavin, 4.6 to 5.3 mg niacin, and 18 to 32 mg ascorbic acid.[89] Another analysis suggested 77% carbohydrate (half sugar, half starch) and 8% albuminoids.[90] According to Porterfield,[258] the water chestnut contains ca.77% carbohydrates. The cane-sugar content, in water-free samples, averages about 27.5%, while protein is rather low. The starch content of the fresh corm is ca. 7 to 8%.

Description — Perennial aquatic or paludal rush-like herb, with elongate rhizomes, terminated by a tuber; culms terete, erect, 40 to 80 cm tall, 2.5 mm thick, glaucous-green, smooth, septate-nodose within, arising from short, dark-brown, basal tuber or corm 5 cm or less in diameter. Sheaths 5 to 20 cm long, frequently partially reddish. Spikelets cylindrical, 4 cm long, 3 to 4 mm thick, scales broadly elliptic, 5 to 6 mm long. Achenes obovate-orbicular, 2 mm long, lustrous, tawny, smooth, bristles with short spines at tip, these shorter toward apex, style-base short-deltoid with strongly depressed inconspicuous basal disc. Flowers summer; fruits July to October.[278]

Germplasm — Reported from the China-Japan Center of Diversity, water chestnut, or cvs thereof, is reported to tolerate waterlogging.[82] Two cvs recognized in China are 'Ma Tai', or common water-chestnut, usually black and about 2.5 cm in diameter: 'Mandarin' water chestnut, dark reddish-brown, with slight cover of light-brown skin, and about 3.2 cm in diameter. Shell is a tough hard skin, and the kernel resembles a potato in consistency, color, and composition.[278]

Distribution — Native to the East Indies, China and Japan, Fiji, Philippines, India, and New Caledonia, Chinese water chestnuts are cultivated throughout the Far East, especially in Taiwan, Malaysia, and Ryukyu Islands.[278] Zeven and Zhukovsky[350] mention it for West Africa, as well. Rosengarten[283] suggests its cultivation in the Atlantic and Gulf Coastal States as far north as Virginia. They are suggested also for Puerto Rico and Hawaii.

Ecology — Ranging from Subtropical Dry to Moist through Tropical Dry to Moist Forest Life Zones, water chestnut is reported to tolerate annual precipitation of 8.7 to 24.1 dm (mean of 3 cases = 16.7), annual temperature of 18.7 to 26.6°C (mean of 3 cases = 22.9°C), and pH of 5.3 to 5.5 (mean of 2 cases = 5.4).[82] Hardy only to Zone 9[139] or perhaps to Zone 7,[343] tolerating average annual minimum temperatures of 5 to 10°F (to 15 to 12°C), Chinese water chestnuts grow in shallow water, and are adapted for planting along edges of ponds, in boggy places, or in marshes, remaining green during the fall and winter. In colder areas, plants are grown in pots, tubs, or pools of water.[278]

Cultivation — Planting is done annually in June or July. Old corms are first planted in wet mud and, when sprouted, are planted usually about 15 cm deep in fields of mud covered with some, but not too much, water. Also propagated by offsets from the corms, it spreads by means of horizontal rhizomes. It grows practically throughout the year, at least until ready to replant for the next season. Corms should be planted in rich, fertile soil, one to each 15-cm pot, when grown for local or limited culture. Pots should be submerged so that the soil surface is covered with 15 cm of water. Potted plants may be put out in pools when weather is warm and settled, but should be brought in before frost.[278] DeRigo and Winters[73] recommend 224 kg N (ammonium sulfate), 112 kg P_2O_5 (superphosphate), and 168 kg K_2O (muriate of potash), as the best fertilizer combination for water chestnut growers with soils similar to those in the Savannah station of the U.S. Department of Agriculture.

Harvesting — Corms, mature and ready for use in about 6 months, are harvested as needed. For commerical purposes, toward the end of the season all the tubers in a clump may be harvested. After the tops are removed, tubers may be plowed up and hand picked. At harvest time, the corms, 3 to 5 cm in diameter, may be produced on the rhizomes to a depth of 25 cm. Corms are cleaned, dried, and shipped to markets.[278]

Yields and economics — Plants are very prolific, and a plant may yield 10 to 12 kg of chestnuts per season. Yields as high as 40 MT/ha are reported from China, higher than the 35 MT/ha reported by DeRigo and Winters.[73] The 10 MT/ha reported in *The Wealth of India*[70] may be more realistic. Rosengarten[283] more optimistically suggests 25 to 50 tons per hectare. They are used extensively as food in the East Indies, China, and Japan. Canned Chinese water chestnuts are imported from Hong Kong in large quantities into Europe, Great Britain, and the U.S.[278]

Energy — Accepting tuber yields of 40 MT/ha, the tubers being ca. 75% water, there is a DM yield of 10 MT/ha. This could be used as a food or energy source. Tops, normally discarded, would probably represent even less DM, also available for energy production.

Biotic factors — Attacked by several fungi: *Cladosporium herbarum, Claviceps nigricans, Curvularia lunata, C. maculans, Cylindrosporium eleocharidis, Dermatosorus eleocharidis, Entyloma eleochardis, E. parvum, Epicoccum nigrum, Mucor circinelloides, Pestalotia scripina, Physoderma heleocharidis, Puccinia eleocharidis, P. liberta, Uredo incomposita, Uromyces eleocharidis, Dicaeoma eleocharidis.* The following nematodes have been isolated from Chinese water chestnuts; *Dolichodorus heterocephalus, Hoplolaimus coronatus,* and *Paratylenchus sarissus.*[186,278]

FAGUS GRANDIFOLIA Ehrh. (FAGACEAE) — American Beech
Syn.: *Fagus americana* Sweet, *Fagus ferruginea* Alt., *Fagus atropurpuea Sudw.*

Uses — Nuts eaten raw, dried, or cooked; they usually have a sweet taste. Sometimes roasted and ground for use as a coffee substitute.[278] Beech buds may be eaten in the spring[283] and young leaves cooked as greens in the spring. The inner bark is dried and pulverized for bread flour in times of need and used as emergency food. Beechnuts are used to make cakes and pies.[278] Nuts are a fattening feed for hogs and poultry,[283] and also provide food for wildlife. Trees make excellent ornamentals and provide valuable timber. The wood is heavy, straight-grained, of close texture, hard, but not durable, and hence it is not used as building timber, though extensively used for ordinary lumber ware, furniture, and cooperage stock.[278] Also used for boxes, clothes-pins, crates, cross-ties, flooring, food containers, fuel, general millwork, handles, laundry appliances, pulpwood, spools, toys, veneer, and woodenware. After steaming, the wood is easy to bend and is valuable for the curved parts of chairs. Wood tar (source of creosote) is obtained through destructive distillation of the wood.[324] Early settlers used the wood mainly for fuel wood. Makes excellent charcoal that was used by blacksmiths and in furnaces for smelting iron.[213]

Folk medicine — Reported to be antidote and poison, American beech is a folk remedy for burns, frostbite, rash, and scald,[91] Uphof[324] reports it to be antiseptic, antipyretic, a stimulating expectorant, used for chronic bronchitis, pulmonary tuberculosis, and vomiting seasickness. Guiaicol (from beechwood creosote) is expectorant and intestinal antiseptic.[324] Cherokee Indians chewed the inner bark as a worm treatment. Potawatomi Indians used a decoction of leaves on frostbitten extremities and made a leaf decoction compound for burns.[216] Rappahannock Indians applied it to poison ivy rash three times daily in the form of a wash made by steeping a handful of beech bark, from the north side of the tree, in a pint of water with a little salt.[330]

Chemistry — Per 100 gm, the seed is reported to contain 608 calories, 20.8 g protein, 53.5 fat, 21.7 g total carbohydrate, 4.0 g fiber, and 4.0 g ash.[89] Rosengarten[283] reports beech nuts contain 19.4% protein, 20.3% carbohydrates, and 5,667 calories per kg. Another source lists beech nuts as containing (per 100 g) 568 calories, 19.4 g protein, 50.0 g fat, 20.3 g carbohydrate, and 6.6% water.[226] Smith[310] reports 6.6% water, 21.8% protein, 49.9% fat, 18.0% carbohydrates, 3.7% ash, and 6,028 calories per kg. The wood is a source of methyl alcohol and acetic acid. Guaiacol is derived from beechwood creosote by fractional distillation.[324]

Toxicity — Occasionally nuts cause poisoning in man and domestic animals. There have been reports that indicate gastrointestinal distress, probably caused by a saponin glycoside.[184]

Description — Deciduous tree, to 30 (to 40) m tall and 1 m in diameter, round-topped; bark smooth, gray; winter-buds long, lanceolate, acute; twigs slender, often slightly zigzag. Leaves alternate, short-petioled, simple, ovate-oblong, obovate or elliptical, 6.5 to 12.5 cm long, sharply serrate to denticulate, thin, papery, broadly acute to subcordate at base, straight-veined, densely silky when young, becoming glabrous above and dark bluish-green and usually silky-pubescent beneath, turning yellow in fall. Flowers monoecious, appearing with leaves; staminate flowers in drooping heads, subtended by deciduous bracts, with small calyx, deeply 4 to 8 cleft and 8 to 16 stamens; pistillate flowers in 2 to 4-flowered spikes, usually in pairs at end of short peduncle, subtended and largely concealed by numerous subulate bracts, calyx adnate to ovary with 6 acuminate lobes. Burs prickly, about 2 to 2.5 cm in diameter, dehiscing into 4 valves, partially opening upon maturity; nuts triangular, up to 2 cm long, 2 or 3 in each bur; seed-coat brown, removed from kernel before eating. Root-suckering causes thickets around old trees. Flowers spring; fruits fall.[278]

Germplasm — Reported from the North American Center of Diversity, American beech, or cvs thereof, is reported to tolerate frost, high pH, limestone, low pH, shade, slope, weeds, and waterlogging.[82] Three natural varieties can be distinguished: var. *grandifolia* — prickles of bur 4 to 10 mm long, erect, spreading or recurved, with leaves usually sharply serrate, grows in rich upland soils, from Nova Scotia and New Brunswick to Minnesota, south to Virginia and Kentucky, and in mountains to North Carolina, Illinois, and southeastern Missouri; var. *caroliniana* (Loud.) Fern. and Rehd. — prickles of bur 1 to 3 (to 4) mm long, usually abruptly reflexed from near base, leaves more acuminate and often merely denticulate, found in moist or wet lowland forests, on or near Coastal Plain, Massachusetts to Florida and Texas, and north in the Mississippi Valley to southern Illinois and Ohio; var. *pubescens* Fern. and Rehd. — leaves soft-pubescent below, sometimes only slightly so. Natives in Kentucky and other mountainous areas where both major varieties occur separate them into Red and White Beech, due to color of wood.[278] 'Abrams' and 'Abundance' were introduced into trade in 1926 by Willard Bixby. Both appeared to produce superior nuts. 'Jenner' is said to bear regular crops of exceptionally large nuts.[213] ($2n = 12$.)

Distribution — Generally distributed throughout eastern U.S. and Canada, from Nova Scotia and New Brunswick, south to Florida, west to Minnesota, Wisconsin south to Texas.[278]

Ecology — Ranging from Cool Temperate Moist to Wet through Subtropical Moist Forest Life Zones, American beech is reported to tolerate annual precipitation of 6.7 to 12.8 dm (mean of 9 cases = 10.5), annual temperature of 7.0 to 17.6°C (mean of 9 cases = 10.8°C), and pH of 4.5 to 6.5 (mean of 8 cases = 5.5).[82] Grows well in acid soils on rather dry hillsides, but will grow in lowlands of Coastal Plain. Thrives where soil is protected by mulch of its own leaves. On many rich upland and mountain slopes, this long-lived tree forms nearly pure stands. Southward often found on bottom-lands and along margins of swamps.[278] Hardy to Zone 3.[343]

Cultivation — Propagates readily from seed sown in fall or stratified and kept for sowing in spring. Cover with $^1/_2$ inch of soil; protect from vermin.[278] Fall-sown beds should be mulched until midsummer and kept in half-shade until past mid-summer of first year.[285] Seedlings should be transplanted frequently, for 2 to 3 years, to prevent formation of a long taproot. Horticultural varities are grafted on seedling stock and grown on under glass until planted out.[278] Trees are slow-growing and may live 400 years or more.[283]

Harvesting — Nuts are gathered after heavy frosts have caused them to drop to the ground. Treated like other nuts until used.[278] Fresh nuts will deteriorate within a few weeks if not properly dried. Shells are easily removed with the fingernails.[283] Wood is harvested from trees 60 to 90 cm in diameter.[278]

Yields and economics — Rudolf and Leak[285] report between 2,860 to 5,060 cleaned seeds per kg (1,300 to 2,300/lb). Beech nuts are a very minor product in North America, compared to other nuts. Used more by people with limited supplies of nuts. Lumber is the more important commercial product.[278]

Energy — The heavy wood (sp. grav. 0.65 to 0.75) is used for fuel wood and charcoal. The seeds, though copious at times, are so small that they could hardly be considered an energy source. One could multiply seed yields by 0.5 to get a rough idea of the oil potential.

Biotic factors — Serious bark disease associated with the presence of beech scale, prevalent in Canada and Maine. Dormant oil spray is used to check scale. Nicotine-sulfate can be used when young leaves first appear. Mottle-leaf or scorch disease, resulting in premature leaf-fall, is prevalent on American beech, the exact cause is not yet known.[278] *The Agriculture Handbook 165*[4] reports the following as affecting this species: *Anthostoma turgidum, Armillaria mellea, Botryosphaeria hoffmanni, Ceratostomella echinella, C. microspora, Cercospora* sp., *Coccomyces comitialis, C. coronatus, Coniothyrium fagi, Conopholis americana, Cryptodiaporthe galericulata, Cryptosporella compta, Cytospora* spp., *C. pustulata, Daedalea ambigua, D. confragosa, D. unicolor, Daldinia concentrica, D. vernicosa, Diaporthe fagi, Diatrype* spp., *Dichaena faginea, Discosia artocreas, Endobotrya legans, Endoconidiophora virescens, Endothia gyrosa, Epifagus virginiana, Favolus alveolaris, Fomes applanatus, F. connatus, F. everhartii, F. formentarius, F. igniarius, F. pinicola, Gloeosporium fagi, Graphium album, Hericium coralloides, H. laciniatum, Hymenochaete* spp., *Hypoxylon* spp., *Lasiophaeria pezizula, Libertella faginea, Microsphaera alni, Microstroma* sp., *Mycosphaerella fagi, M. punctiformis, Nectria cinnabarina, N. coccinea, N. galligena, Pholiota* spp., *Phomopsis* sp., *Phoradendron flavescens, Phyllactinia corylea, Phyllosticta faginea, Phytophthora cactorum, Polyporus* spp., *Poria* spp., *Scorias spongiosa, Septobasidium* spp., *Steccherinum ochraceum, S. septentrionale, Stereum* spp., *Strumella coryneoidea, Trametes* spp., *Ustulina deusta, U. linearis, Valsa* spp., *Xylaria corniformis*, and *X. digitata*. Erineum (leaf deformity caused by mites) is also reported. In addition, Browne,[53] lists: Fungi — *Asterosporium hoffmannii, Cerrena unicolor, Ganoderma appalanatum, Gnomonia veneta, Hericium caput-ursi, Hymenochaete tabacina, Hypoxylon blakei, H. cohaerens, Inonotus glomeratus, I. obliquus, Phellinus igniarus, Phyllactinea guttata, Polyporus adustus, P. hirsutus, P. versicolor, Poria laevigata, Sterum fasciatum, S. purpureum, Torula ligniperda, Valsa leucostomoides*. Hemiptera — *Corythucha pallipes, Cryptococcus fagi, Parthenolecanium corni, Phyllaphis fagi, Prociphilus imbricator*. Lepidoptera — *Alsophila pometaria, Cenopis pettitana, Choristoneura fractivittana, Datana integerrima, D. ministra, Disphragia guttivitta, Ennomos magnaria, E. subsignaria, Halisidota maculata, Hemerocampa leucostigma, Lymantria dispar, Nadata gibbosa, Operophtera bruceata, Orgyia antiqua, Pandemis lamprosana, Paraclemensia acerifoliella, Symmerista albifrons, S. leucitys, Tetralopha asperatella*. Mammalia — *Erethizon dorsatum*.

FAGUS SYLVATICA L. (FAGACEAE) — European Beech

Uses — The nuts are sweet and edible when roasted. Roasted nuts can be used as a substitute for coffee. Press-cake from decorticated nuts is used as a feed for cattle, pigs, and poultry.Oil expressed from nuts is used for cooking, illumination, and manufacture of soap. Used as a substitute for butter. Leaves used as a substitute for tobacco. Trees furnish excellent timber. Wood is heavy, hard, straight-grained, close textured, durable, easy to split, strong, resistant to abrasion, and used for flooring, cooperage, furniture, turnery, utensils, wagons, agricultural implements, wooden shoes, spoons, plates, pianos, ship building, railroad ties, brush backs, meat choppers, construction of dams, water-mills, excelsior, wood pulp, and is an excellent fuel. Takes a good polish and can be easily bent when steamed. In Norway and Sweden, boiled beechwood sawdust is baked and then mixed with flour to form the material for bread. Source of creosote, which is used as a preservative treatment of timber. Trees make excellent ornamental plants as leaves remain on tree most of winter.[38,70,209,324]

Folk medicine — Reported to be carminative, poison, analgesic, antidote, antipyretic, antiseptic, apertif, astringent, laxative, parasiticide, refrigerant, and tonic, European beech is a folk remedy for blood disorders and fever.[91] Source of creosote, used as a deodorant dusting powder in cases of gangrene and bed sores when mixed with plaster of paris.[70]

Chemistry — Hager's Handbook[187] reports the leaves to contain pentosane, methylpentosane, idalin, a wax, cerotonic acid, p-hydoxybenzoic acid, vanillic acid, p-coumaric-, ferulic-, caffeic-, chlorogenic-acid, and traces of inositol and sinapic acid; myricetin, leucodelphinidin, quercetin, isoquercitrin, leucocyanidin, and kaempferol; n-nonacosan, beta-sitosterol, alanine, aminobutyric acid, arginine, asparagine, glutamine, hydroxyglutamic acid, glycine, hydroxyproline, leucine, lysine, methionine, phenylalanine, proline, threonine, tyrosine, valine, serine; and a little cystine, tyrosine, and histidine. The seeds contain 25 to 45% oil (3.5% stearic-, ca.5% palmitic-, 40 to 76% oleic-, and ca.10% linoleic-acid); also choline, neurine, trimethylamine, sugar, malic-, citric-, oxalic-, lactic-, and tannic-acids; gums, betaine, sinapic-, caffeic-, and ferulic-acids; saponins, tannins, and the alkaloid fagine. Bark contains 3 to 4% tannin, citric acid, beta-sitosterol, betulin. Arachidylalcohol (arachinalcohol, n-eicosylalcohol $C_{20}H_{42}O$), vanilloside ($C_{14}H_{18}O_8$), docosanol, tetracosanol, hexacosanol; lauric-, myristic-, palmitic-, stearic-, oleic-, and linoleic-acid. Wood contains 0.5% l-arabinose ($C_5H_{10}O_5$), 18% d-xylose ($C_5H_{10}O_5$), l-rhamnose, and d-galactose.[187]

Toxicity — Raw nuts are poisonous, probably due to the presence of a saponin (CSIR, 1948-1976).

Description — Trees deciduous, long-lived, up to 30 m tall, round-topped; trunk smooth, gray; buds slender, fusiform, acute, reddish-brown; branches smooth. Leaves alternate, ovate or elliptic, acute, cuneate or rounded at base, 5 to 10 cm long, glabrous, at least along veins, with 5 to 8 pairs of conspicuous lateral veins, denticulate, shinking dark-green above, turning reddish-brown in fall. Male flowers numerous, in long-stalked aments, perianth divided almost to base; peduncles 5 to 6 cm long. Nut ovate, 12 to 30 mm in diameter, brown; cupule woody, about 2.5 cm wide, deeply divided into 4 valves which are covered outside with awl-shaped spines. April to May.[278]

Germplasm — Reported from the Euro-Siberian Center of Diversity, European beech, or cvs thereof, is reported to tolerate frost, high pH, limestone, low pH, shade, slope, and smog.[82] There are many variations of leaf color and size, and branchlet habit. Some of the horticultural varieties include: var. *albovariegata* — leaves variegated with white; var. *asplenifolia* Lodd. — leaves very narrow, deeply toothed or lobed; var. *atropunicea* Sudw. (var. *atropurpurea* Hort., var. *purpurea* Ait., var. *riversii* Hort., var. *suprea*) — Purple Beech, leaves purple; var. *borneyensis* — intermediate between vars. *pendula* and *tortusa*; leaves coarsely toothed; var. *laciniata* (var. *incisa* Hort., var. *heterophylla* Loud.) — Fernleaf or Cutleaf Beech, leaves deeply toothed or lobed or sometimes entire and linear; var. *latifolia* — leaves to 15 cm long and 10 cm wide; var. *luteovariegata* — leaves variegated with yellow; var. *miltoniensis* — drooping form; var. *pendula* Lodd. — Weeping Beech, branches drooping; var. *purpuero-pendula* Hort. — branches drooping with purple leaves; var. *roseomarginata* — leaves purple edged with pale pink; var. *rotundifolia* — leaves nearly orbicular, 2.5 cm or less long; var. *quercifolia* Schelle (var. *quercoides* Hort.) — leaves deeply toothed and sinuate; var. *quercoides* Pers. — bark dark, rough, oak-like; var. *tortuosa* Dipp. (var. *remillyensis*) — branches twisted and contorted, drooping at tips; var. *tricolor* — leaves nearly white, spotted with green and edged with pink; var. *varigata* — leaves variegated with white or yellow; var. *zlatia* Spaeth — leaves yellow.[278] (2n = 22,24.[82])

Distribution — Central and southern Europe, east to the Caucasus, ascending to 1,700 m in Alps. Introduced to Ireland; widely planted as ornamental. Found as far north as southeastern Norway.[278]

Ecology — Ranging from Cool Temperate Steppe to Wet through Warm Temperate Dry to Moist Forest Life Zones, European beech is reported to tolerate annual precipitation of 3.1 to 13.6 dm (mean of 29 cases = 7.8), annual temperature of 6.5 to 18.0°C (mean of 29 cases = 9.7°C), and pH of 4.5 to 8.2 (mean of 25 cases = 6.3).[82] In woods on well-drained soils, often in mountains and on hillsides. Thrives on northern and eastern exposures, enduring much shade, shunning poor soils and swamps, protecting and improving the soil. Thrives on loamy limestone soil, but will grow on acid soils. Thrives where soils are protected by mulch of its own leaves; growing best in dry sandy loams. Trees are relatively insensitive to unfavorable conditions.[278] Hardy to Zone 4.[343]

Cultivation — Propagation readily attained by seed in fall or stratified and kept for sowing in spring. Protect seeds and seedlings from vermin. Seedlings should be transplanted every second or third year to prevent formation of long taproot. Varieties are grafted on seedling stock under glass. All upright forms may be clipped to form excellent hedges.[278]

Harvesting — Nuts are harvested in fall, usually after they fall to ground. Nuts are also harvested all winter by wildlife. Timber harvested from mature trees.[278]

Yields and economics — Since beech-nuts do not enter markets for human consumption, no data are available. The nuts are not a commercial item, but are especially valuable as food for wildlife. Trees form extensive forests, and the wood is a common hardwood tree in Denmark and Germany, where it is raised as pure growth or as mixed woodland. Nurseries propagate large numbers for ornamentals.[278]

Energy — Cannel[61] presents biomass data showing that trees ca. 100 years old, spaced at 1200 trees per ha, averaged 23.7 m tall, a basal area of 48.2 m^2/ha, and a stem volume of 460 m^3/ha. The stem wood plus the stem bark, on a DM basis, weighed 365 MT/ha, the branches 49, the foliage 5, and the roots were estimated at 50 MT/ha for a total standing biomass of ca. 468 MT/ha. The current annual increment (CAI) of stem wood and bark was 3.6 MT/ha/yr, which total was estimated at 9.3 MT/ha/yr. These data were taken in a brown forest soil in Bulgaria 42 to 43°N, 23 to 25°E, 1400 to 1600 m elevation. On red alluvial soil in Denmark (56°00′N, 12°20′E, elevation 200 m), 200-year-old trees, averaging 26 m tall, had CAIs of only 5.9 m^3/ha/yr compared with 12.7 for 54-year-old trees. Beck and Mittman[38] showed that annual litter fall was close to 5 MT/ha in a pure beech stand in the Black Forest of West Germany (mean annual temperature 8.3°, annual precipitation 10.5 dm; elevation 325 m). In Sweden, Nihlgard and Lindgren[235] cite annual above-ground productivity of 10.4 to 16.7 MT/ha with yearly increments (CAI) of 7.1 to 11.0 MT/ha. Apparently, the annual productivity ranges from 3 to 17 MT/ha. Such biomass could and does serve as a source of energy in temperate forests. The wood is an excellent fuel,[324] and would probably make good charcoal.

Biotic factors — Wooly aphis often covers the surface of leaves of European beech; it is controlled by application of oil spray. Nicotine sulfate also is used when young leaves first appear. Trees are relatively free of fungal and bacterial diseases and are not seriously damaged by insects or other pests.[278] *The Agriculture Handbook 165*[4] lists the following as affecting this species: *Armillaria mellea, Endothia gyrosa, Massaria macrospora, Nectria cinnabarina, Phomopsis* spp., and *Phytophthora cactorum*. Erineum — leaf deformity caused by mites, Leaf Scorch — cause unknown, and Mottle Leaf — cause unknown are also listed. In addition, Browne[53] lists: Fungi — *Armillaria mucida, Asteroporium hoffmannii, Auricularia auricula-judae, Bulgaria inquinans, Cerrena unicolor, Daedalea quercina, Endothia parasitica, Fistulina hepatica, Fomes annosus, F. conchatus, F. fomentarius, F. fraxineus, F. pinicola, Ganoderma applanatum, Gnomonia veneta, Helicobasidium purpureum, Hericium erinaceus, Hydnum cirrhatum, H. diversidens, Hysterographium fraxini, Inonotus cuticularis, I. obliquus, Laetiporus sulphureus, Microsphaera alphitoides, Nectria coccinea, N. coccinea faginata, N. ditissima, N. galligena, Oxyporus populinus, Phellinus igniaris, Pholiota adiposa, Phyllactinia guttata, Phytophthora cinnamomi, P. syringae, Pleurotus ostretus, P. ulmarius, Polyporus adustus, P. giganteus, P. squamosus, P. zonatus, Pythium debaryanum, P. ultimum, Rosellinia quercina, Steccherinum septentrionale, Stereum hirsutum, S. purpureum, S. rugosum, Trametes hispida, Truncatella hartigii, Ustulina deusta, Volvariella bombycina*. Angiospermae — *Viscum album*. Coleoptera — *Agrilus viridis, Apoderus coryli, Byctiscus betulae, Cerambyx cerdo, Leperisinus varius, Melolontha melolontha, Mesosa nebulosa, Phyllobius argentatus, Platypus cylindrus, Prionous coriareus, Rhynchaenus fagi, Rhynchites betulae, Strophosomus coryli, Xyleborus dispar*. Diptera — *Contarinia fagi, Hartigiola annulipes, Mikola fagi, Oligotrophus fagineus, Phegobia tornatella, Phegomyia fagicola*. Hemiptera — *Cryptococcus fagi, Fagocyba cruenta, Phyllaphis fagi*. Hymenoptera — *Caliroa annulipes, Nematus fagi*. Lepidoptera — *Carcina quercana, Cossus cossus, Diurnea fagella, Ectropis crepuscularia, Hepialus humuli, Laspeyresia fagiglandana, Lithocolletis faginella, Lymantria monacha, Nepticula hemargyrella, N. tityrella, Operophtera brumata, Strophedra weirana, Tortrix viridana*. Aves — *Columba palumbus*. Mammalia — *Apodemus sylvaticus, Clethrionomys glarcolus, Dama dama, Microtis agrestis, Sciurus carolinensis, S. vulgaris*.[53]

GINKGO BILOBA L. (GINKGOACEAE) — Ginkgo, Maidenhair Tree

Uses — Valued by the Orientals as a sacred tree, for food, medicine, and ritual. Once the acrid nauseous pulp is removed from around them, the seeds can be boiled or roasted to make a delicacy, the nut, with a flavor likened by one author to mild Swiss cheese. As a delicacy at feasts, the nuts are supposed to aid digestion and alleviate the effects of drinking too much wine. Important in oriental medicine, the ginkgo is now under cultivation as a medicinal plant in the occident. Chinese use the seed to wash clothing. Seed are digested in wine to make a cosmetic detergent.[258] The thick fleshy seed coat is used as an insecticide. The light, yellowish, brittle wood is used for chess-boards and toys. Very valuable in highly polluted air as an ornamental shade tree, along streets and in parks.

Folk medicine — According to Hartwell,[126] the nuts are used in folk remedies for cancer in China, the plant for corns in Japan. In China, macerated in vegetable oil for 100 days, the fruit pulp is traditionally used for asthma, bronchitis, gonorrhea, tuberculosis, and worms.[90] According to Monachino,[417] the nauseous fruit juice becomes antitubercular after immersing in oil for three months. This activity is not lost with sterilization at 100°C for 30 min. Daily administration of 150 gm/kg of the extract of the oil-immersed fruits showed definite activity against *Mycobacterium tuberculosis* in guinea pigs. Pan-fried seeds are used for leucorrhea, polyuria, seminal emissions, and tuberculosis; seeds, seedcoats, or leaves are used for asthma, cough, leucorrhea, spermatorrhea. Seeds are considered antitussive, astringent, sedative. Raw seed is said to be anticancer, antivinous; with a fishy taste; they are consumed, dyed red, at Chinese weddings; said to help bladder ailments, blenorrhea, and uterine fluxes. Used for cardiovascular ailments in Szechuan. Ginkgolic acid is active against the tubercle bacillus. Elsewhere, leaf extracts are used in peripheral arterial circulation problems like arteriosclerotic angiopathy, post-thrombotic syndrome, diabetic vasoconstriction with gangrene and angina, intermittent claudication, Raynaud's disease. Extracts are inhaled for ear, nose, and throat ailments like bronchitis and chronic rhinitis.[90]

Chemistry — Per 100 g, seeds (ZMB) contain 403 calories, 10.2 to 10.5% protein, 3.1 to 3.5% fat, 83.0% total carbohydrate, 1.3 g fiber, 3.1 to 3.8 g ash, 11 mg Ca, 327 mg P, 2.6 mg Fe, 15.3 mg Na, 1139 mg K, 392 mg beta-carotene equivalent, 0.52 mg thiamine, 0.26 mg riboflavin, 6.1 mg niacin, and 54.5 mg ascorbic acid.[89] Dry kernels (ca. 59% of the seed weight) contain: 6% sucrose, 67.9% starch, 13.1% protein, 2.9% fat, 1.6% pentosans, 1% fiber, and 3.4% ash. The globulin of the kernel, accounting for 60% of the total nitrogen, is rich in tryptophane. Fruit pulp, bitter and astringent, contains a volatile oil and a number of fatty acids from formic to caprylic. Press-juice contains: ginnol ($C_{27}H_{56}O$), bilobol ($C_{21}H_{34}O_2$), ginkgol ($C_{24}H_{34}O$), ginkgic acid ($C_{24}H_{42}O_2$), ginkgolic (hydroxy) acid ($C_{22}H_{34}O_3$), ginkgolic (saturated oxy) acid ($C_{21}H_{32}O_3$), ginkgolic acid ($C_{24}H_{48}O_2$), an acid corresponding to the formula $C_{21}H_{42}O_3$, an acidic oil, asparagine, reducing sugars, and phosphoric acid. Autumn leaves contain ginnol, sitosterol ($C_{27}H_{64}$)), ipuranol ($C_{33}H_{56}O_6$), shikimic acid or shikimin ($C_7H_{10}O_5$), linolenic acid, acacetin, apigenin, and substances conforming to the formula $C_{11}H_{14}O_5$ and $C_{11}H_{14}O_6$. Fallen leaves of the plant contain a bright yellow crystalline substance, ginkgetin ($C_{32}H_{22}O_{10}$). Leafy branches contain ceryl alcohol and sterols. Staminate flowers of Paris-grown trees contain 3.27 to 3.57% (ZMB) deoxyribonucleic acid. Male inflorescence may contain raffinose (up to 4% on fresh weight basis). Wood contains raffinose and xylan (2.5%). Bark contains tannin dissolved in a pectinous mucus.[70]

Toxicity — Seeds are reputed to be toxic raw, sometimes resulting in children's deaths.[249] According to Duke and Ayensu,[90] large quantities can induce convulsions, dyspnea, emesis, and pyreticosis. Expressed fruit juice causes erythema, edema, papules, pustules, and intense itching. Some suggest that even old nuts can induce dermatitis. The pollen may cause hay fever.

Description — Deciduous dioecious trees to nearly 35 m tall, often slenderly conical and sparsely branched when young, spreading in age. Leaves on stalks up to 7.5 cm long, fan-shaped, usually 5 to 8 (15 to 20) cm across, with 2 large lobes, usually undulate or notched, but with numerous branching parallel veins. Male and female strobili on different trees. Males appear in early spring as catkins drooping from short shoots (3 to 6 on one shoot), bearing numerous loosely arranged stamens. Female axes arise from short spur shoots in pairs or in threes, each with a long stalk bearing on each side a naked ovule, surrounded at the base by a collar-like rim. Seed with a yellow fleshy outer covering enveloping the woody shell containing the edible kernel.[418] Seeds 400 to 1,150 per kg.

Germplasm — Reported from the China-Japan Center of Diversity, ginkgo, or cvs thereof, is reported to tolerate acid soil, air pollution, disease, frost, insects, and slope. Dallimore and Jackson[418] describe several ornamental cvs, 'Aurea' with leaves yellow even in summer, 'Fastigiata' with the branches almost erect, 'Laciniata' with deeply cut leaves, 'Pendula' with weeping branches, and 'Variegata' with yellow-variegated leaves.

Distribution — Rarely seen wild, even in China and Japan, yet doing well widely in the temperate world as a cultivar. Rosengarten,[283] terming it "unknown in the wild", notes that it has been cultivated as a sacred tree in Chinese Buddhist temple courtyards for over 1,000 years. Introduced into America in 1784, it has generally been successful on good sites in moist temperate areas of the midwestern and eastern U.S., and along the St. Lawrence River in Canada.[5]

Ecology — Estimated to range from Cool Temperate Moist to Wet through Warm Temperate Moist to Wet Forest Life Zones, ginkgo is expected to tolerate annual precipitation of 8 to 12 dm, annual temperature of 9 to 14°C, and pH of 4.5 to 6. Waterlogging, strong winds, hardpan, and alkaline soils are to be avoided. According to Balz,[419] though ginkgo tolerates cold, frost, and snow, it does well with summer temperatures above 25°C and air relative-humidity ca. 50 to 60%. Monthly rainfall in summer should not fall below 40 mm. Deep, light, mellow soils, well-drained and aerated, produce optimal growth. Good growth is reported on soils with 2% coarse sand, 10% fine sand, 37% coarse silt, 40% fine silt, and 11% clay, as well as 5% coarse sand, 45% fine sand, 25% coarse silt, 15% fine silt, and 10% clay. Soils should not contain more than 10 to 15% clay. A pH of 5 to 5.5 is recommended with 100 to 200 ppm P_2O_5, 260 to 400 ppm K_2O, 60 to 120 ppm Mg, 3 to 5% humus, and <1% salts. The soil should warm up early in spring with late autumn leaf fall; i.e., no frost between April 1 and October 31 (7 month or more growing season). Isolation of 1,800 to 2,000 hr/year (250 hr/month midsummer) is considered adequate.

Cultivation — Chinese say that triquetrous seeds produce male trees, lenticular seeds produce females. Seeds germinate readily but grow slowly. Cuttings take as long as 2 years to root. Seed should be cold-stratified 30 to 60 days for seed collected before completion or after ripening. Germinative capacity may vary from about (0 to) 30 to 85%. For amenity plantings, seeds should be sown in furrows in November and covered with 5 to 8 cm soil and a sawdust mulch. Based on limited studies, one Swiss firm, planning to grow the plant in the U.S., suggested sowing the seed under plastic tunnels at a spacing of 25 × 4 cm, equalling ca. 1,000,000 seed per ha. With an 80% germination rate, there were 800,000 plants per ha, held in the tunnel for 2 years, expected to attain 30 cm the first year, 1.2 m the second. In autumn of year 2 or spring of year 3, taproots are shortened to 10 to 15 cm by under-cutting the stems, cut back to 30 cm by mowing. In the spring of the year, plants are outplanted mechanically, at 100 × 30 cm or 33,000 plants per ha.

Harvesting — For the pharmaceutical industry, plants are cut back to 30 cm every year in October. Trees start bearing fruits at ca. age 25[417] (Monachino, 1956) or 30-40 years[5] (Ag. Handbook 450, 1974).

Yields and Economics — In heavy fruiting years, the trees can bear enough fruits to cover 50% of the area circumscribed by the crowns. The Swiss Pharmaceutical firm anticipated 2,400 to 3,200 kg green leaves per ha in the third year (first year outplanted), 6,000 to 8,000 in the second year outplanted, and 20,000 to 25,000 kg in the third year outplanted.[419]

Energy — From a biomass point of view, the ginkgo is not very promising as an energy species. The pulp and seed husks are waste products, when the nuts are gathered. Both could be extracted for chemurgics, then processed into energy products. Extracted leaves could also be useful for biomass fuels.

Biotic factors — According to Monachino,[417] the tree is not attacked by insects and it is resistant to disease. *The Agriculture Handbook 165*[4] reports the following as affecting ginkgo: *Fomes connatus* (sapwood or wound rot), *Glomerella cingulata* (leaf spot, anthracnose), *Meloidogyne* sp. (root knot nematodes), *Phyllosticta ginkgo* (leaf spot), *Phymatotrichum omnivorum, Polyporus* spp. (sapwood rot), and *Xylaria longeana* (seed rot).

GNETUM GNEMON L. (GNETACEAE) — Manindjo, Malindjo, Tangkil

Uses — In India, the seeds are eaten after roasting or cooking. Filipinos use the fruits as a coffee substitute.[331] Fruits are first peeled and then cooked in Java; then the horny testa can be separated; kernels are then pounded and sundried. This mass is then fried in coconut oil and salted to eat with rice. Also sweetened and eaten as a delicacy with tea or coffee. Young leaves are eaten, raw or steamed. Young leaves and inflorescences are cooked with sea food.[238] In Fiji, young leaves are cooked with coconut milk. Bark yields a fiber used for making rope.[56] To obtain the strong fiber, durable in sea water, the branches are peeled and the bark beaten and split into fine filaments. With good tensile and breaking strengths, the fiber is valued for fishing and nets.[70] The wood from old trees is dark, brittle, and not very durable. Younger poles are used for mooring posts for rafts and boats. Branches may be split for cooperage.[332]

Folk medicine — Indochinese use the roots as a general antidote to poison.[249]

Chemistry — Per 100 g, the kernel is reported to contain 30 g H_2O, 10.9 g protein, 1.6 g fat, 52.9 g total carbohydrate, 0.9 g fiber, and 1.7 g ash. Young leaves and stem tips contain 81.9% H_2O, 1.33% ash, 0.24% P_2O_5, 0.11% CaO, and 0.01% Fe_2O_3.[70]

Toxicity — If eaten raw, the young leaves, inflorescences and fruits may irritate the mouth.[249]

Description — Tree (sometimes lianoid) 5 to 22 m high, the crown narrow, conical; trunk straight or somewhat crooked; main branches whorled, often somewhat drooping. Leaves opposite, shortly stalked, oblong-lanceolate or elliptic-oblong; base acute, obtuse or rounded, apex shortly acuminate, acute; entire, thinly coriaceous, above dark-green, shining, beneath light-green, pinnatinerved, 5 to 20 cm long, 3 to 8 cm wide, petiole 0.5 to 1 cm long. Flowers dioecious, sometimes apparently monoecious, in stalked articulate spikes composed of 5 to 8 whorls; whorls supported by an undulate cup. Male spikes single or fascicled, 3 to 5 cm long; female spikes solitary, usually longer than the male ones, to 10 cm long; stalks of the inflorescences 1.2 to 2 cm long. Fruits sessile, ellipsoid, shortly cuspidate, 2 to 2.5 cm long, dark-red when ripe, containing a single large starchy edible seed.[70,238]

Germplasm — Reported from the Indochinese-Indonesian Center of Diversity, manindjo, or cvs thereof, is reported to tolerate alternating dry and wet seasons. Var. *ovalifolium* is considered the wild type, var. *gnemon* the cultivar.

Distribution — Native from Assam to Malaysia and Fiji, introduced to Java, Sumatra, and elsewhere.[70,238]

Ecology — Better adpated to seasonal than to ever-humid tropical forests.[346]

Cultivation — Cultivated in Asian plains, extending easily to an altitude of 1200 m. Sometimes planted in orchards, but mostly in mixed gardens. In the Solomon Islands, seeds may be sown, but more frequently, seedlings are transplanted from beneath established trees. Vegetative propagation is not known in Santa Cruz on the Solomons. Trees sometimes polled to keep them low. They recover readily from pruning.[56,346]

Harvesting — In Santa Cruz, Solomon Islands, fruiting peaks around September to October and March to April.[346] Fibers are said to be best harvested when trees are 5 m tall. Notable for their ability to recover from the near girdling induced by fiber harvest, the older trees, often scarred, may be harvested again.[346]

Yields and economics — Rare in the markets of Malaya, more common in Java.[56]

Energy — In Fiji, at least, the plant is used for firewood.[331]

Biotic factors — No data available.

HELIANTHUS ANNUUS L. (ASTERACEAE) — Sunflower

Uses — Cultivated primarily for the seeds which yield the world's second most important source of edible oil. Sunflower oil is used for cooking, margarine, salad dressings, lubrication, soaps, and illumination. A semi-drying oil, it is used with linseed and other drying oils in paints and varnishes. Decorticated press-cake is used as a high protein food for livestock. Kernels eaten by humans raw, roasted and salted, or made into flour. Poultry and cage birds are fond of raw kernels. Flowers yield a yellow dye. Plants used for fodder, silage and green-manure crop. Hulls provide filler in livestock feeds and bedding.[70,278,283]

Folk medicine — Medicinally, seeds are diuretic, expectorant, and used for colds, coughs, throat, and lung ailments. According to Hartwell,[126] the flowers and seeds are used in folk remedies for cancer in Venezuela, often incorporated in white wine. Reported to be anodyne, antiseptic, aphrodisiac, bactericidal, deobstruent, diuretic, emollient, expectorant, insecticidal, malaria preventive, sunflower is a folk remedy for aftosa, blindness, bronchiectasis, bronchitis, carbuncles, catarrh, cold, colic, cough, diarrhea, dysentery, dysuria, epistaxis, eyes, fever, flu, fractures, inflammations, laryngitis, lungs, malaria, menorrhagia, pleuritis, rheumatism, scorpion stings, snakebite, splenitis, urogenital ailments, whitlow, and wounds.[91]

Chemistry — Per 100 g, the seed is reported to contain 560 calories, 4.8 g H_2O, 24.0 g protein, 47.3 g fat, 19.4 g total carbohydrate, 3.8 g fiber, 4.0 g ash, 120 mg Ca, 837 mg P, 7.1 mg Fe, 30 μg Na, 920 mg K, 30 mg beta-carotene equivalent, 1.96 mg thiamine, 0.23 mg riboflavin, 5.4 mg niacin, and 0 mg ascorbic acid. Seeds contain 25 to 35 % oil, but cultivars have been bred in Russia with up to 50% oil. Oil contains 44 to 72% linoleic acid, and 13 to 20% protein of high biological value and digestibility. Stems and husks are rich in potash.

The forage (ZMB) contains 8.8% protein, 2.9% fat, 77.2% total carbohydrate, 30.3 g fiber, and 11.1 g ash. Young shoots contain: 13.0% protein, 1.9% fat, 70.3% total carbohydrate, 20.4 g fiber, 14.8 g ash, 1,670 mg Ca, and 370 mg P per 100 g. The flowers contain 12.7% protein, 13.7% fat, 64.3% total carbohydrate, 32.9 g fiber, 9.3 g ash, 630 mg Ca, and 80 mg P per 100 g.[70,89] Sunflower oil has a high concentration of linoleic acid, intermediate level of oleic acid, and very low levels of linolenic acid. The saturated acids, palmitic and stearic, rarely exceed 12%, and the minor acids, lauric, arachidic, behenic, lignoceric, eicosenoic, etc. rarely add up to as much as 2%. Tocopherol, or vitamin E, is an important vitamin and natural antioxidant. Sunflower oil is somewhat unique in that the alpha form predominates, with 608, 17, and 11 mg/kg of alpha, beta, and gamma, compared with 116, 34, and 737, respectively, for soybean oil.[77]

Description — Variable, erect, often unbranched, fast-growing, annual herb; stems 0.7 to 3.5 m tall, hirsute. Leaves alternate, ovate, long-petioled, lamina with 3 main veins, 10 to 30 cm long, 5 to 20 cm wide, apex acute or acuminate, lower leaves opposite and cordate. Flowering head terminal on main stem, 10 to 40 cm in diameter, rotating to face the sun, sometimes drooping, heads on lateral branches smaller; outer ray flowers neuter with yellow ligulate corolla, disc florets numerous, spirally arranged, perfect; ovary inferior with single basal ovule. Achenes obovoid, compressed, slightly 4-angled, variable in size and color, seldom less than 1 cm long, usually from 1 to 1.5 cm long, full-colored or striped. Taproot strong, penetrating to depth of 3 m and with large lateral spread of surface roots, Flowers late summer and fall; fruits fall.[278]

Germplasm — Reported from the North American (and secondarily, the Eurosiberian) Center of Diversity, sunflower, or cvs thereof, is reported to tolerate disease, drought, frost, fungi, high pH, laterite, limestone, low pH, mycobacteria, photo-period, poor soil, rust, salt, sand, smog, virus, weeds, and waterlogging.[82] Botanically, the sunflower is treated as the following subspecies: ssp. *lenticularis* is the wild sunflower; ssp. *annuus* is the weedy wild sunflower; and ssp. *macrocarpus* is cultivated for edible seeds. Cultivars are divided into several types: *Giant types*: 1.8 to 4.2 m tall, late maturing, heads 30 to 50 cm diameter, seeds large, white or gray, or with black stripes; oil content rather low; ex. 'Mammoth Russian'. *Semi-dwarf types*: 1.3 to 1.8 m tall, early maturing, heads 17 to 23 cm diameter, seeds smaller, black, gray or striped; oil content higher; ex., 'Pole Star' and 'Jupiter'. *Dwarf types*: 0.6 to 1.4 m tall, early maturing, heads 14 to 16 cm diameter, seeds small, oil content highest; ex., 'Advance' and 'Sunset'. Gene centers are in the Americas, with genuine resources for resistance in southern U.S. and Mexico. Two types of male sterility are known. Although "sunchoke" is the name given to the hybrid with the jerusalem artichoke, much of what is sold as sunchoke in the U.S. is, in fact, straight jerusalem artichoke. ($2n = 34$.)[82,278]

Distribution — Native to western North America, sunflower is one of a few crops to have evolved within the present confines of the U.S. Early introduced to Europe and Russia, the species has now spread to countries both tropical and temperate.[278]

Ecology — Sunflowers are grown from the Equator to 55°N Lat. In the tropics, they grow better at medium to high elevations, but tolerate the drier lowlands. They thrive wherever good crops of corn are grown. Young plants withstand mild freezing. Plants are intolerant of shade. As sunflowers have highly efficient root systems, they can be grown in areas which are too dry for many crops. Plants are quite drought-resistant except during flowering. In South Africa, reasonable yields have been obtained with 25 cm of rainfall by dwarf cvs. Giant types require more moist conditions. Crops may be grown on a wide range of soils, including poor soils, provided they are deep and well-drained. Plants are intolerant of acid or waterlogged soils. Ranging from Boreal Moist through Tropical Thorn to Wet Forest Life Zones, sunflower tolerates annual precipitation of 2 to 40 dm (mean of 195 cases = 11.4), annual temperature of 6 to 28°C (mean of 194 cases = 19.6), and pH of 4.5 to 8.7 (mean of 121 cases = 6.6)[82,278]

Cultivation — Seed, harvested at 12% moisture content and stored, will retain viability for several years. Sunflower production may be adapted to mechanized or unmechanized societies. Propagation is always by seed. Plant with corn or beet planter, 2.5 to 7.5 cm deep, spaced 0.2 m apart in 0.6 to 0.9 m rows; seed rate of 5.6 kg/ha, giving about 62,500 plants per ha. May be planted earlier in spring than corn, since plants are more tolerant to frost. Early weed control is an important factor in yield, so cultivate lightly and early. Sunflowers respond well to a balanced fertilizer based on soil test; usually a 1-2-3 NPK ratio is best, with a need for boron and other trace elements on lighter soils. Application of foliar fertilizers of liquid NPK on plants increases yield 62% with one application and 97% with two applications. Sunflowers should not occur in rotation more than once in every 4 years, and should not be in rotations with potatoes.[278]

Harvesting — Crop matures about 4 months from sowing; some Russian cvs mature in 70 days. Harvest when involucral bracts turn yellow and seeds become loose, but before shedding begins. Harvesting methods are similar to those of corn: heads are gathered, dried, and threshed. For fodder or silage, crop is harvested at the flowering stage. Seed oil is either cold- or hot-pressed. Cold-pressed oil is usually pale-yellow, with a mild taste and pleasant odor, much esteemed as a salad and cooking oil, especially for butter substitutes. Hot-pressed oil is reddish-yellow and is used for technical purposes and as a burning oil. With modern methods, hot-pressed oil may be refined for edible purposes.[278]

Yields and economics — Average yields range from 900 to 1,575 kg/ha of seed; however, yields of over 3,375 kg/ha have been reported. Heads may contain 1,000 to 4,000 florets, with the potential of as many seeds. Yields from dried seeds are 40% oil, 35% protein meal, and 20 to 25% hulls. In 1979, the world low production yield was 308 kg/ha in Algeria, the international production yield was 1,266 kg/ha, and the world high production yield was 2,420 kg/ha in Austria.[70,98] With DM yields ranging from 4 to 9 MT/ha (in 3 months) and seed yields ranging from 300 to more than 3,000 kg/ha, a straw factor of 3 seems appropriate. With an average yield of ca.1,500 kg/ha (North Dakota), a hectare would yield nearly 225 gallons of oil, 75% of which could be extracted on the farm. Twelve to 15 gallons are required to raise a hectare; hence the fuel from one hectare could produce 8 to 11 hectares of crop. In the U.S., the highest average commercial yields occurred in North Dakota and Minnesota, which averaged 1,170 and 1,267 kg/ha respectively, compared with 1,019 kg/ha for Texas. Pryde and Doty[270] suggest average oil yields of 589 kg/ha from 1,469 kg/ha seed. Telek and Martin[359] suggest oil yields of 450 kg/ha. Experimentally, at Davis, California, April plantings yielded 2,592 to 3,181 kg/ha (45.5 to 48.5% oil), May plantings, 2,676 to 3,161 kg/ha (45.5 to 48.4% oil), June plantings 956 to 2,643 kg/ha (40.8 to 43.7% oil), and July plantings 702 to 2,447 kg/ha (40.2 to 42.6% oil). The lowest oil yield was 282 kg/ha, the highest, 1,543 kg/ha.[37] In India, rain-fed sunflower gave seed yields of 1,120 kg/ha in pure stands, 1,050 to 1,070 intercropped with cowpea, and 1,010 to 1,070 kg/ha intercropped with peanuts.[62] Volunteer sunflowers themselves may constitute a weed problem, as few as $3/m^2$ reducing wheat yields by 16%, $23/m^2$ reducing yields by 35%. World production of sunflower seed in 1970 was 9.6 million MT, grown on 8.2 million ha, yielding 1,170 kg/ha. Largest producers are the USSR, Rumania, Bulgaria, Argentina, Yugoslavia, Turkey, and South Africa. In the tropics, Tanzania produces 10,000 to 20,000 MT per year. Cultivars grown in Minnesota contain higher percentages of the desirable linoleic acid than same cultivars in other states. Major importers of sunflower seed were Italy, West Germany, and Japan. Oil prices in the U.S. in 1970 were $331/ton. Production costs in fully mechanized production in the U.S. is about $100/ha with fertilizer, $87 without; hand labor figured at $2/hr. By 1982, sunflower oil was trading at $.59/kg compared to $.50 to .54 for coconut, $.53 for corn oil, $.48 for cottonseed, $.59 for linseed, and $.42 for soybean.[350,360]

Energy — According to the USDA phytomass files,[360] annual productivity ranges from 3 to 15 MT/ha. North Dakota researchers are testing a small auger press, operated on the farm, that can extract ca.75 to 80% of the oil in sunflower seeds, or ca. 55 gallons (barely more than one 42-gallon barrel) from an average yield of 1,400 lb/acre. It takes one acre's production to farm and produce 8 to 11 more acres, our usual 10:1 ratio. In North Carolina, Harwood[127] concluded that sunflower seed was most promising for on-farm production of vegetable oil fuels; soybeans, peanuts, and cottonseed considered not well-suited. Sunflowers yield ca. 2.5 MT/ha, with ca. 40% oil, indicating a potential of 250 gallons of oil per ha if seed were processed in a mill. On-farm processing would produce closer to 200 gallons (ca. 5 barrels) at a cost of more than $2.00 per gallon. Production costs are less than one barrel per hectare. Harwood puts the energetic returns at greater than 5:1 compared to 3:1 for peanuts, 2:1 for soybeans, and 1:1 for cottonseed. Pratt et al.[266] report an endurance test involving engines fueled with various mixtures of sunflower oil (25 to 50%) with diesel oil (75 to 50%). Two motors needed repair, ten were operating with no apparent difficulties, of which two were said to be doing even better. Ohio yields on poor soils (Wood County) were only 260 lb/acre (yielding 9.3 gallons of screw press oil); and on good soils (Champaign County), 1,680 lb/acre (yielding 69.1 gallons oil) cropped after wheat in a double-cropping system. Sunflower oil should be dewaxed before being used as a diesel substitute. In Australia, sunflower was first commercially planted in 1967, has great potential for expansion as a rainfed energy crop. Little water is required for processing oilseeds (unlike ethanol), and the seed coat can provide sufficient energy for heat and steam for oil extraction. Australians figure a net energy gain of 2 ℓ for every 3 ℓ produced. A hundred kg of dry seed will yield about 40 kg oil, 15 to 25 kg hulls, and 40 kg proteinaceous meal. Hulls have been pressed into fuel "logs". Threshed heads are ground and fed to cattle elsewhere. The heads are rich in pectin. Studies have shown that sunflower yields 33.1 MT silage per ha, compared to corn at 19.26 MT/ha. Annual DM productivity ranges from 3 to 15 MT/ha. DM yields averaged closer to 5 MT spaced at 43,000 plants per ha, 8 MT spaced at 172,000 plants per ha near Clarksville, Maryland. In these experiments, the sunflower followed barley.[360] Jake Page's discussion[243] is picturesque: "But I happen to like sunflowers . . . They can be grown almost anywhere in the country and you can grow between 500 and 3,000 pounds of sunflower seeds on an American acre in three months if you're clever. The soil can be lousy, the rainfall terrible . . . if the average American corn farmer put 10 percent of his land into sunflowers, he could become self-sufficient in fuel. It seems that using vegetable oil may be more efficient, in a net energy sense, than growing plants for conversion into alcohol (another nice alternative fuel) because the processing for alcohol is more elaborate, expensive, and energy intensive."

Biotic factors — In the USDA's *Agriculture Research*[361] a new pest of sunflower is reported. A scarab beetle (*Phyllophaga lancolata*) devastated more than 400 ha near Lehman, Texas. *Eucosma womonana*, is also a newly reported sunflower pest in Texas. Seed is set low when selfed, as most cvs are self-incompatible. Florets on one head open over 5 to 6 days and may wait 2 weeks for fertilization. Cross-pollination may be facilitated by 2 to 3 hives of honeybees per ha, the hives spaced in rows 300 to 400 m apart, as they need to be distributed to give coverage to all blooms. Gophers dig up seeds; birds eat tremendous amounts of seeds from the maturing crop. Insects can be destructive to seeds not stored properly. The following fungi are known to cause diseases in sunflowers: *Albugo tragopogonis, Alternaria tenuis, Alternaria zinniae, Armillaria mellea, Ascochyta helianthi, Botrytis cinerea, Cercospora bidentis, Cercospora helianthi, Cercospora helianthicola, Cercospora pachypus, Corticium rolfsii, Cystopus cubicus, Cystopus tragopogonis, Diaporthe arctii, Diplodina helianthi, Entyloma polysporum, Erysiphe chicoracearum, Fusarium acuminatum, Fusarium conglutinans, Fusarium culmorum, Fusarium equiseti, Fusarium javanicum, Fusarium oxysporum, Fusarium sambucinum, Fusarium scirpi, Fusarium semitecum, Fusarium solani, Helminthosporium helianthi, Leptosphaeria helianthi, Leveillula compositarum, Leveillula taurica, Macrophomina phaseoli, Oidium helianthi, Ophiobolus helianthi, Phialea cynthoides, Phoma oleracea, Phymatotrichum omnivorum, Plasmopara halstedii, Puccinia helianthi, Pythium debaryanum, Pythium irregulare, Phythium splendens, Pythium ultimum, Rhabdospora helianthicola, Rhizoctonia rocorum, Rhizoctonia solani, Rhizoctonia bataticola, Rhizopus nodosus, Sclerotinia fuckeliana, Sclerotinia libertiana, Sclerotinia minor, Sclerotinia sclerotiorum, Sclerotium rolfsii, Septoria helianthi, Sphaerotheca fulginea, Sphaerotheca humuli, Uromyces junci, Verticillium albo-atrum, Verticillium dahliae*. Bacteria reported as infecting sunflowers include: *Agrobacterium tumefaciens, Bacterium melleum, Erwinia aroides, Pseudomonas cichorii, Pseudomonas helianthi*, and *Pseudomonas solanacearum*. Virus diseases reported from sunflowers are Apple mosaic, Argentine sunflower, Aster yellows, Brazilian tobacco streak, Cucumber mosaic, Tomato spotted wilt, Peach ringsport, Peach yellow-bud mosaic, Pelargonium leaf-curl, Tobacco necrosis, Tobacco ringspot, and Yellows. Sunflowers are parasitized by the following flowering plants: *Cuscuta pentagona, Cuscuta arvensis, Orobanche aegyptiaca, Orobanche cumana, Orobanche muteli, Orobanche ramosa, Striga hermonthica, Striga asiatica, Striga lutea, Striga senegalensis*. Sunflowers are attacked by many nematodes: *Anguina balsamophila, Aphelenchoides ritzemabosi, Ditylenchus destructor, Ditylenchus dipsaci, Helicotylenchus cavenessi, Helicotylenchus microcephalus, Helicotylenchus microlobus, Helicotylenchus pesudorobustus, Heterodera schachtii, Longidorus maximus, Meloidognye arenaria, Meloidogyne hapla, Meloidogyne incognita acrita, Meloidogyne javanica, Meloidogyne thamesi, Paratylenchus minutus, Pratylenchus penetrans, Rotylenchulus reniformis, Scutellonema clathricaudatum, Trichodorus christiei*, and *Xiphinema ifacolum*.[4,186,278]

HYPHAENE THEBAICA (L.) Mart. (ARECACEAE) Doum Palm

Uses — Unripe kernels are edible, but the ripe kernels are hard as a marble, and even strung together to make a weapon. In Bornu Africa, the nuts are pounded to make a meal sold instead of millet. The rind of the fruits is dry and sweet, edible in some, inedible in others. The part of the germinating seedling just below ground is edible, as is the cabbage or palm heart. Trunks yield a sago starch. Osborn[242] relates how people in Kharga gnaw on the glossy brown fruits. Though fibrous and tough, the fruits have a pleasant flavor suggestive of carob or ginger bread. Beverages are made from the fruits. In parts of the Sahara desert, the spongy internal parts of the fruit are an important dietary element. Mixed with date infusion, the doum nut constitutes a cooling drink much valued medicinally. Stalks of the cotyledons are eaten. Inner leaves are valued for forage, while the outer may be used for fuel. Fronds, usually unexpanded, used in plaited strips to make mats, hats, baskets, fans, bowls, and ropes. A fiber obtained from the root is used for snares and fish nets. Fronds of the palm are used for fuel. The hard fruit, used as vegetable ivory, is also the source of a black dye. Stems are used in house construction. Ashes are used as salt.[70,146,332]

Folk medicine — According to Hartwell[126] the fruits are used in folk remedies for indurations of the limbs. The thick root is used in African folk remedies for hematuria, in some cases due to bilharzia. According to Boulos,[45] the resin from the tree, diaphoretic and diuretic, is recommended both for tapeworm and for the bites of poisonous animals. The fruits are astringent and anthelmintic. Breads made from the fruit have been recommended in fluxes. The beverage made from the fruits is recommended, at least around Kharga, for strenghtening the heart and for gastroenteritis. Mixed with date infusion, the doum is recommended for febrile conditions on the Sahara.[242]

Chemistry — Per 100 g, the dried nut contains 395 calories, 5.7 to 6.2 g H_2O, 2.4 to 5.0 g protein, 4.9 to 8.0 g fat, 6.5 to 11.0 g fiber, 1.9 to 5.4 g ash, 121 to 168 mg Ca, and 170 to 281 mg P. Gohl[110] reports that whole nuts of the doum palm (91.4% DM) contain (ZMB): 4.5% crude protein, 24.7% crude fiber, 3.6% ash, 2.6% fat, and 64.6% nitrogen-free extract. Ground kernels (90.4% DM) from Somalia contain 9.0% CP, 7.3% CF, 2.8% ash, 7.0% EE, and 73.9% NFE. Per 100 g, the seed (ZMB) contains 420 calories, 4.1 g protein, 6.8 g fat, 85.7 g total carbohydrate, 10.0 g fiber, 3.3 g ash, 153 mg Ca, and 240 mg P.[89] According to Watt and Breyer-Brandwijk,[332] the nut contains 50% mannitol, which when hydrolyzed with dilute acid furnishes 56 to 58% reducing sugars, 80% of which is mannose.

Description — Palm to 10 m tall, the trunk branching dichotomously 1 to 2 or more times. Leaves large, flabellate, with linear-lanceolate lobes, and numerous upward-curved hooks on petioles. Flowers small and whitish, monoecious, the male spadices surrounded by pointed male spathe-bracts, branches of female spadix being stouter. Fruits 5 × 5 cm, yellowish-brown, globose-guadrangular, with strong fiber surrounding the hard seeds.[146]

Germplasm — Native to the African and Middle Eastern Centers of Diversity, the doum palm is said to tolerate drought and fire. The closely related *H. indica* Becc., often confused with *H. thebaica*, is probably the only germplasm native to the Middle East.

Distribution — Sometimes gregarious, forming dense stands along rivers in hot dry valleys of tropical Africa, the Middle East, and western India, especially common in the coastal regions of East Africa and in Eritrea.

Ecology — Flourishes in rich sandy loam. Growth, flowering, and fruiting are luxuriant in moist places, but in dry places the fruits become small. With no hard data at hand, I estimate that the palm is most at home in Subtropical to Tropical Thorn to Moist Forest Life Zones, tolerating even drier climates along water-courses. Johnson[151] calls it a promising desert palm for deserts and semideserts up to 600 m. Plants wind-polinated. Fruits disseminated by elephants, baboons, and donkeys, all of which may eat the fruits. Young leaves are eaten by camels.

Cultivation — Cultivated as an ornamental curio, e.g., in India, the palm can be propagated by seed or off-shoots.[70]

Harvesting — Plant parts harvested as needed.

Yields and economics — No data available.

Energy — Around Bornu, Africa, the leaves are used as fuel, especially in boiling down salt. The sap can be used for alcohol production. Since this is a very destructive process, it has been outlawed in Eritrea and Kenya. The pod yields an active charcoal with high decolorizing and absorbing power.

Biotic factors — Nuts and the beads made from them may be attacked by the scolytid beetle, *Coccotrypes dactyliperda* Fabr. Preventive measures are discussed in *The Wealth of India*.[70]

INOCARPUS EDULIS Forst. (FABACEAE) — Tahiti Chestnut, Otaheite Chestnut, Polynesia Chestnut
Syn: *Inocarpus fagiferus* (Parkinson) Fosberg?

Uses — Nuts said to be edible after processing. Menninger[209] says what I have long observed: almost any nut which is difficult to describe is said to taste like a chestnut. So with this one. Seeds are sometimes allowed to ferment in pits in the ground. Natives of Santa Cruz roast the fruits or slowly dry the unhusked fruit over a fire.[346] More often they are boiled or roasted in ashes. Some Samoans make purees from the cooked seeds. Said to be the principle food of the mountaineers of Fiji. Cattle are said to eat the leaves.[56] Seeds are strung as beads.[209] Wood used in furniture.[343]

Folk medicine — Reported to be antidotal to fish poisoning, and useful for blood-shot eyes, diarrhea, and hemorrhage.[91] Mixed with the fern *Drynaria* to treat virulent gonorrhea in Indonesia. Astringent bark is used for intestinal complaints in Malaya.[56] Seed is boiled in coconut milk for parturitional uterine hemorrhage.[331]

Chemistry — Per 100 g, the seed (ZMB) is reported to contain 426 calories, 6.7 g protein, 7.9 g fat, 82.8 g total carbohydrate, 4.4 g fiber, 2.6 g ash, 0.46 mg thiamine, and 4 mg ascorbic acid.[89] Burkill[56] reports the seeds (ZMB) analyze 7% fat, 10% albumens, 2.5% ash, and 80% non-nitrogenous substances, mostly starch.[56]

Description — Tree to 25 m tall, handsome; trunks usually deeply furrowed, commonly fluted or buttressed; crown dense. Leaves simple, entire, large, oblong-lanceolate, short-petioled, pinnately nerved, leathery; stipules very small, soon caducous. Flowers white, cream, or yellow, fragrant, in axillary, simple or branched spikes resembling catkins when young; bracts small, connate with rachis, somewhat pouched; bracteoles small; calyx tubular-campanulate, bilabiate, membranous, irregularly 2- to 5-toothed; petals 4 to 6, usually 5, subequal, imbricate in bud, linear-lanceolate, upper part crinkled; stamens twice the number of petals, alternately long and short, the longer ones briefly joined to the petals; anthers small, uniform, ovary subsessile or short-stalked, 1-, seldom 2-ovuled; style very short; stigma oblique. Pod short-stalked, oblique-obovate, flattened, 2-valved, subdrupaceous, leathery, indehiscent, 1-seeded.[8]

Germplasm — Reported from the Australian and Polynesian Centers of Diversity, the Tahiti chestnut, or cvs thereof, is reported to tolerate swamps, waterlogging, and perhaps some salt. (2n = 20.)[8,350]

Distribution — Native of eastern Malaysia and the Pacific, cultivated in the Malay Peninsula. Cultivated successfully in Peradeniya and Singapore. Allen and Allen describe it as ubiquitous throughout the South Pacific Islands.[8]

Ecology — Estimated to range from Subtropical Dry to Wet to Tropical Dry through Moist Forest Life Zones, Tahiti chestnut is estimated to tolerate annual precipitation of 10 to 50 dm, annual temperature of 22 to 28°C, and pH of 6.0 to 8.0. Rosengarten says, "It prefers a hot, humid, tropical climate at low altitude, with well-distributed rainfall, and thrives along the banks of streams and even in swamps." Often a second-story component of low-lying forest.[283,346]

Cultivation — Rarely cultivated. In Santa Cruz, the seeds are sprouted in the shade and transplanted. It is more gathered than cultivated.[346]

Harvesting — Fruits start bearing at about age 8. In Santa Cruz, Solomon Islands, there are two main harvests per year, and nuts are stored with the fibrous pods intact after cooking in large earth ovens.[346]

Yields and Economics — No data available.

Energy — No data available.

Biotic factors — Nitrogen-fixing nodules were not detected in Philippine specimens.[8]

JATROPHA CURCAS L. (EUPHORBIACEAE) — Physic Nut, Purging Nut

Uses — According to Ochse,[238] "the young leaves may be safely eaten, steamed or stewed." They are favored for cooking with goat meat, said to counteract the peculiar smell. Though purgative, the nuts are sometimes roasted and dangerously eaten. In India, pounded leaves are applied near horses' eyes to repel flies. The oil has been used for illumination, soap, candles, adulteration of olive oil, and making Turkey red oil. Nuts can be strung on grass and burned like candlenuts.[332] Mexicans grow the shrub as a host for the lac insect. Ashes of the burned root are used as a salt substitute.[224] Agaceta, Dumag, and Batolos[3] conclude that it has strong molluscicidal activity. Duke and Wain[91] list it for homicide, piscicide, and raticide as well. The latex was strongly inhibitory to watermelon mosaic virus.[318] Bark used as a fish poison.[332] In South Sudan, the seed as well as the fruit is used as a contraceptive.[187] Sap stains linen and can be used for marking.[215] Little, Woodbury, and Wadsworth list the species as a honey plant.[86,189]

Folk medicine — According to Harwell,[126] the extracts are used in folk remedies for cancer. Reported to be abortifacient, anodyne, antiseptic, cicatrizant, depurative, diuretic, emetic, hemostat, lactagogue, narcotic, purgative, rubefacient, styptic, vermifuge, and vulnerary, physic nut is a folk remedy for alopecia, anasarca, ascites, burns, carbuncles, convulsions, cough, dermatitis, diarrhea, dropsy, dysentery, dyspepsia, eczema, erysipelas, fever, gonorrhea, hernia, incontinence, inflammation, jaundice, neuralgia, paralysis, parturition, pleurisy, pneumonia, rash, rheumatism, scabies, sciatica, sores, stomachache, syphilis, tetanus, thrush, tumors, ulcers, uterosis, whitlows, yaws, and yellow fever.[86,91,187] Latex is applied topically to bee and wasp stings.[332] Mauritians massage ascitic limbs with the oil. Cameroon natives apply the leaf decoction in arthritis.[332] Colombians drink the leaf decoction for venereal disease.[224] Bahamians drink the decoction for heartburn. Costa Ricans poultice leaves onto erysipelas and splenosis. Guatemalans place heated leaves on the breast as a lactagogue. Cubans apply the latex to toothache. Colombians and Costa Ricans apply the latex to burns, hemorrhoids, ringworm, and ulcers. Barbadians use the leaf tea for marasmus, Panamanians for jaundice. Venezuelans take the root decoction for dysentery.[224] Seeds are used also for dropsy, gout, paralysis, and skin ailments.[332] Leaves are regarded as antiparasitic, applied to scabies; rubefacient for paralysis, rheumatism, also applied to hard tumors.[126] Latex used to dress sores and ulcers and inflamed tongues.[249] Seed is viewed as aperient; the seed oil emetic, laxative, purgative, for skin ailments. Root is used in decoction as a mouthwash for bleeding gums and toothache. Otherwise used for eczema, ringworm, and scabies.[90,249] I received a letter from the Medical Research Center of the University of the West Indies shortly after the death of Jamacian singer Robert Morley:

> I just want you to know that this is not because of Bob Morley's illness, why I am revealing this . . . my dream was: this old lady came to me in my sleep with a dish in her hands; she handed the dish to me filled with some nuts. I said to her, "What were those?" She did not answer. I said to her, "PHYSIC NUTS." She said to me, "This is the cure for cancer."

I found this Jamaican dream rather interesting. Four antitumor compounds, including jatropham and jatrophone, are reported from other species of Jatropha.[89] Homeopathically used for cold sweats, colic, collapse, cramps, cyanosis, diarrhea, and leg cramps.[86,187]

Chemistry — Per 100 g, the seed is reported to contain 6.6 g H_2O, 18.2 g protein, 38.0 g fat, 33.5 g total carbohydrate, 15.5 g fiber, and 4.5 g ash.[89] Leaves, which show anti-leukemic activity, contain alpha-amyrin, beta-sitosterol, stigmasterol, and campesterol, 7-keto-beta-sitosterol, stigmast-5-ene-3beta, 7-alpha-diol, and stigmast-5-ene-3beta, 7 beta-diol.[224] Leaves contain isovitexin and vitexin. From the drug (nut?) saccharose, raffinose, stachyose, glucose, fructose, galactose, protein, and an oil, largely of oleic- and linoleic-acids.[187] Poisonous seeds can cause death due to phytotoxin, curcin. Curcasin, arachidic-, linoleic-, myristic-, oleic-, palmitic-, and stearic-acids.[249]

Toxicity — The poisoning is irritant, with acute abdominal pain and nausea about $^1/_2$ hour following ingestion. Diarrhea and nausea continue but are not usually serious. Depression and collapse may occur, especially in children. Two seeds are strong purgative. Four to five seeds are said to have caused death, but the roasted seed is said to be nearly innocuous. Bark, fruit, leaf, root, and wood are all reported to contain HCN.[332] Seeds contain the dangerous toxalbumin curcin.[86]

Description — Shrub or tree to 6 m, with spreading branches and stubby twigs, with a milky or yellowish rufescent exudate. Leaves deciduous, alternate but apically crowded, ovate, acute to acuminate, basally cordate, 3- to 5-lobed in outline, 6 to 40 cm long, 6 to 35 cm broad, the petioles 2.5 to 7.5 cm long. Flowers several to many in greenish cymes, yellowish, bell-shaped; sepals 5, broadly deltoid. Male flowers many with 10 stamens, 5 united at the base only, 5 united into a colum. Female flowers borne singly, with elliptic 3-celled, triovulate ovary with 3 spreading bifurcate stigmata. Capsules 2.5 to 4 cm long, finally drying and splitting into 3 valves, all or two of which commonly have an oblong black seed, these ca. 2 × 1 cm.[189,224]

Germplasm — Reported from the Central and South American Centers of Diversity, physic nut, or cvs thereof, is reported to tolerate slope. There is an endemic species in Madagascar, *J. mahafalensis*, with equal energetic promise.[107]

Distribution — Though native to America, the species is almost pantropical now, widely planted as a medicinal plant which soon tends to establish itself. It is listed, e.g., as a weed in Brazil, Fiji, Honduras, India, Jamaica, Panama, Puerto Rico, and Salvador.[134]

Ecology — Ranging from Tropical Very Dry to Moist through Subtropical Thorn to Wet

Forest Life Zones, physic nut is reported to tolerate annual precipitation of 4.8 to 23.8 dm (mean of 60 cases = 14.3) and annual temperature of 18.0 to 28.5°C (mean of 45 cases = 25.2).[82]

Cultivation — Grows readily from cuttings or seeds. Cuttings strike root so easily that the plant can be used as an energy-producing living fence post.

Harvesting — For medicinal purposes, the seeds are harvested as needed. For energy purposes, seeds might be harvested all at once, the active medicinal compounds might be extracted from the seed, before or after the oil, leaving the oil cake for biomass or manure.

Yields and economics — According to Gaydou et al.,[107] seed yields approach 6 to 8 MT/ha with ca. 37% oil. They calculate that such yields could produce the equivalent of 2,100 to 2,800 ℓ fuel oil per ha (see table under Energy Section). In Madagascar, they have ca. 10,000 ha of purging nut, each producing ca. 2,400 ℓ (or 24 hℓ) oil per ha for a potential production of 240,000 hℓ.[107]

Energy — The clear oil expressed from the seed has been used for illumination and lubrication, and more recently has been suggested for energetic purposes, one ton of nuts yielding 70 kg refined petroleum, 40 kg "gasoil leger" (light fuel oil), 40 kg regular fuel oil, 34 kg dry tar/pitch/rosin, 270 kg coke-like char, and 200 kg ammoniacal water, natural gas, creosote, etc. In their study, Gaydou et al.[107] compare several possible energy species with potential to grow in Malagasy. Oil palm was considered energetically most promising, but this species was considered second most promising.

	Crop production (MT/ha)	Fuel production (ℓ/ha)	Energetic equivalent (kwh/ha)
Elaeis guineenis	18—20	3,600—4,000	33,900—37,700
Jatropha curcas	6—8	2,100—2,800	19,800—26,400
Aleurites fordii	4—6	1,800—2,700	17,000—25,500
Saccharum officinarum	35	2,450	16,000
Ricinus communis	3—5	1,200—2,000	11,300—18,900
Manihot esculenta	6	1,020	6,600

Biotic Factors — *Agriculture Handbook No. 165* lists the following as affecting *Jatropha curcas: Clitocybe tabescens* (root rot), *Colletotrichum gloesporioides* (leaf spot), and *Phakopsora jatrophicola* (rust).[4]

JESSENIA BATAUA (Mart.) Burret. (ARECACEAE) — Seje, Mil Pesos, Jagua, Pataba, Pataua

Syn.: *Jessenia polycarpa* Karst.

Uses — Fruits provide an oil with a taste almost identical to that of the olive. "There is no question about pataua oil being an excellent edible oil."[152] Ripe fruits are harvested and piled up a day or so to encourage further ripening. They are then steamed in water, and the pulp separated from the bony seed with a mortar. Brazilians may simple press out the oil. The seeds are also consumed as food, and the milky residue from oil extraction, the "yucuta", is consumed as a beverage. The oil, used as a cooking or edible oil, is also used in medicine. A chocolate-colored chicha is made by mashing the fruit, straining out the fruits, and adding sugar. Wood is used for both bows and arrow-points.[29,30,32]

Table 1
JESSENIA COMPARISON OF OIL OF *BATAUA* WITH OLIVE OIL

Fatty acid	*Jessenia bataua*[a] samples (%)	Olive oil samples (%)
Palmitic	13.2 ± 2.1	11.2
Palmitoleic	0.6 ± 0.2	1.5
Stearic	3.6 ± 1.1	2.0
Oleic	77.7 ± 3.1	76.0
Linoleic	2.7 ± 1.0	8.5
Linolenic	0.6 ± 0.4	0.5
Other	1.6 (range 0.2 — 4.6)	—

[a] Values given as the mean ± standard deviation of 12 separate samples.

From Balick, M. J. and Gershoff, S. N., *Econ. Bot.*, 35, 261, 1981. Copyright 1981, The New York Botanical Garden. With permission.

Folk medicine — In the Guahibo area, the oil is used for asthma, cough, tuberculosis, and other respiratory problems. Elsewhere it is used for bronchitis, catarrh, consumption, flu, leprosy, and parturition.[30,91] At least four scientists have speculated that natives gain weight, appear healthier with more endurance, and reported fewer respiratory infections during the season of daily consumption of "mil pesos." Colombians consider the oil vermifugal.[104]

Chemistry — I repeat Balick and Gershoff's[32] useful table (Table 1) comparing the oil of bataua with olive oil, because olive oil has recently gotten press as very salubrious. Note that the bataua, like the olive, contains about 80% oleic acid, a feature recently praised in Lubrizol's special high-oleic sunflower. Parenthetically, I add that Johnson[152] reports much lower oleic acid values, 0.48 to 40.67%. He puts the entire fruit's oil content at 7.4%, the mesocarp pulp at 18.2%, and the seed at 3%. If Lubrizol's sunflower is good for the temperate zone, this oil should be great for the tropical zone. I also repeat Balick and Gershoff's[32] Table 3. The data suggest that, though tryptophan and lysine were the limiting amino acids, bataua protein is better than most grain and legume proteins.[32] (see Table 2.) Balick and Gershoff's Table 4[32] compares the "milk" of the seje with human milk, cowmilk, and soybean milk. (see Table 3.)

Description — Unbranched palm to 15 (to 25) m tall, the mature trunk spineless (when young, the trunk is covered with dark brown fibers and spines to 80 cm long). Leaves pinnate, arching, 6 to 8 (to 10) m long, the rachis deep, canaliculate, vaginate at the base. Leaflets alternate, lanceolate, acute, 40 to 75 mm wide. Spathe ca. 1 m long, woody, terminating in an acute process. Spadix with 100 to 225 racemes, flowers cream-colored; petals valvate. Panicles may contain 1,000 fruits, each weighing 10 to 15 g. There may be two panicles per year. Fruits drupaceous, ellipsoid to ovoid, 2.5 to 4 cm long, deep purple when ripe.[29,30,32]

Germplasm — Reported from the South American Center of Diversity, mil peso is reported to tolerate waterlogging. Although taxonomists have tended to recognize at least two species of *Jessenia*, Balick and Gershoff[32] suggest that there is only one. Guajibo Indians distinguish a type with whitish mesocarp and another with purplish or pinkish mesocarp. Further, they recognize a slender variant with a reddish inner skin tissue.[30]

Distribution — Distributed over much of the northern half of South America, including Panama and Trinidad.[29,30,32]

Table 2
AMINO ACID ANALYSIS OF *JESSENIA BATAUA*

Amino acid component	Mg amino acid per g protein (mean ± standard deviation)[a]	Amino acid scoring pattern[b]	Per cent of FAO/WHO scoring pattern
Isoleucine	47 ± 4	40	118
Leucine	78 ± 4	70	111
Lysine	53 ± 3	55	96
Methionine	18 ± 6		
Cystine	26 ± 6		
Methionine ± cystine	44 ± 9	35	126
Phenylalanine	62 ± 3		
Tyrosine	43 ± 5		
Phenylalanine + tyrosine	105 ± 7	60	175
Threonine	69 ± 6	40	173
Valine	68 ± 4	50	136
Tryptophan	9 ± 1	10	90
Aspartic acid	122 ± 8		
Serine	54 ± 3		
Glutamic acid	96 ± 5		
Proline	75 ± 8		
Glycine	69 ± 4		
Alanine	58 ± 4		
Histidine	29 ± 4		
Arginine	56 ± 2		

[a] Values represent mean ± standard deviation for 7 separate samples with the exception of tryptophan, for which only 3 samples were analyzed.

[b] FAO/WHO provisional amino acid scoring pattern. The scoring pattern represents an "ideal protein" containing all the essential amino acids to meet requirements without excess (FAO/WHO, 1973).

From Balick, M. and Gershoff, S. N., *Econ. Bot.*, 35, 261, 1981. Copyright 1981, The New York Botanical Garden. With permission.

Table 3
COMPARISON OF "MILK" OF *JESSENIA BATAUA* AND OTHER MILKS

	Approx. % calories from each component			
	Jessenia bataua milk	Human milk[a]	Cow milk[a]	Soybean milk[a]
Fat	55.3	45.9	49.8	37.6
Protein	7.4	5.6	20.9	37.9
Carbohydrate	37.3	48.5	29.3	24.5

[a] USDA, 1963.

From Balick, M. J. and Gershoff, S. N., *Econ. Bot.*, 35, 261, 1981. Copyright 1981, The New York Botanical Garden. With permission.

Ecology — Estimated to range from Tropical Dry (along river courses) to Rain through Subtropical Dry to Rain Forest Life Zones, the mil pesos is estimated to tolerate annual precipitation of 15 to 100 dm, annual temperature of 21 to 27°C, and pH of 4.5 to 7.5. Once said to have formed solid gallery forests, but also occurring in inland forest up to 1,000 m.

Cultivation — Though not normally cultivated, this palm should be given priority in testing for plantation culture. "It has never been cultivated, the minute amounts of oil that have entered local native markets always having been extracted from wild trees."[298] Seeds apparently take 20 to 40 days to germinate.

Harvesting — Trees may not fruit for 10 to 12 years.[256] Fruits ripen from April to November in Colombia, September to January in Brazil. Natives believed it bears heavier in alternate years like so many of our native fruits.[29] Too often the trees are felled to obtain the fruits. But about two months after felling, the Guajibo also harvest the edible grubs of the palm weevil.[29,30]

Yields and economics — Trees average 14 kg fruit per season.[29] Schultes[298] says the fruit clusters may weigh 30 kg yielding 1.5 to 3 kg oil. The high price of the similar olive oil would suggest introducing this palm into cultivation. An effort towards this end has been initiated by the Centro de Dasarollo Las Gaviotas in the Orinoquia of Colombia.[298] PIRB[256] calculates that the oil can be produced for about $0.20/kg, 1/8 the cost of olive oil. Many Latin Americans, nonetheless, import edible oils. Unfortunately, most of the Brazilian stands are remote from Belem where there are large vegetable oil factories. "The low yield of oil, coupled with a lack of machinery adapted to processing this fruit, have resulted in very limited production."[152]

Energy — "I am not terribly optimistic on *Jessenia* as an oilseed fuel, as the oil is simply too valuable to burn. In the world market, it (is) probably four times the price of palm oil, and thus would be a waste to put in engines."[31] Still, the Colombian natives extract 3 to 4 bottles of oil from a raceme.[104]

Biotic factors — No data available.

JUGLANS AILANTHIFOLIA Carr. (JUGLANDACEAE) — Heartnut, Japanese or Siebold Walnut

Syn.: ***Juglans sieboldiana*** **Maxim.,** ***Juglans mirabunda*** **Koidz.,** ***Juglans lavallei*** **Dode,** ***Juglans sachalinensis*** **(Miyabe et Kudo) Komar.,** ***Juglans allardiana*** **Dode,** ***Juglans coarctata*** **Dode**

Uses — Heartnut is grown primarily for the kernels of the nuts, used in confectioneries and pastries. Wood soft, not strong, of little value as lumber.[278] Wood dark-brown, not easily cracked or warped, used for gunstocks, cabinet work, and various utensils in Japan. Bark and exocarp of fruit used for dying.[324] Good shade tree and often planted as an ornamental.[278]

Folk medicine — Reported to be antitussive and tonic.[91]

Chemistry — Not data available.

Description — Tall erect tree, to 20 m tall, often grown as a low, wide-branching tree; branches grayish-brown, densely glandular-pubescent when young; bark whitish. Leaves large, petiolate, with 9 to 21 leaflets; leaflets ovate-oblong, 8 to 12 cm long, 3 to 4 cm wide, abruptly acute to acuminate, appressed-serrulate, minutely stellate-pubescent above on both surfaces when young, sessile and obliquely truncate at base; petioles and rachis densely glandular. Staminate aments 10 to 30 cm long; pistillate aments 10- to 20-flowered, pedunculate, densely brown pubescent with crisped hairs. Nut pubescent, with hard shell, broadly ovoid to nearly globose, 2.5 to 3.5 cm long, mucronate, rugose, with raised sutures. Very variable. Flowers May; fruits summer to fall.[278]

Germplasm — Reported from the China-Japan Center of Diversity.[82] Nuts vary considerably in size and roughness. Best-known varieties of common Siebold walnut are 'Dardinell' and 'English'. Heartnut (*Juglans ailanthifolia* var. *cordiformis* [Maxim.] Rehd. [Syn.: *J. cordiformis* Maxim; *J. subcordiformis* Dode]) has a cordate or cordate-ovoid, rather depressed shell, with relatively thin shell, is nearly smooth with a shallow groove on each side, and has better shelling quality. 'Fodermaier' and 'Wright' are the best cvs, although a great many selections have been made and named. Most named heartnuts were introduced to the U.S. in the 1920s and 1930s. This cv is extensively cultivated in Japan and the U.S. Hybrids with butternuts (Butterjap or butternut-siebold) resemble the Siebold in branching, leaves, and long racemes of nuts, but resemble the butternut in shape of nut, tree hardiness, and resistance to serious diseases. Leaves larger than in the butternut. In breeding, its high resistance to *Melanconis* fungus is transmitted to its hybrids with butternuts. The small size of the nut has led to selections of clones. Siebold walnut is susceptible to butternut curculio and to witches' broom or bunch disease, the cause of which is unknown, but an insect-transmitted virus is suspected. Hybrid 'Grietz' is better adapted to southern localities than butternut; and 'Helmick' is hardier and very promising. Some cvs are not hardy as far north as New York. *Juglans avellana* Dode and *J. notha* Rehd. are alleged hybrids between *J. ailanthifolia* Carr. and *J. regia* var. *orientis* (Dode) Kitam.[278] A number of cvs and hybrids of heartnut have been developed which should prove useful for cross-breeding. Vigorous hybrids, called "buartnuts" have been produced by crossing heartnuts and butternuts. These hybrids combine the butternut's desirable kernel flavor and superior climatic adaptability with the heartnut's higher yield and better crackability.[283] (2n = 32.)

Distribution — Native to Japan. Introduced to San Jose Valley of California about 1870; now grown more extensively in northeastern U.S. and southern Ontario. Not worth planting in pecan country, and not valued where Persian walnuts (*J. regia*) thrive. Unadapted to extreme temperatures on Northern Plains and Rocky Mountain regions.[278]

Ecology — Ranging from Warm Temperate Dry to Moist Forest Life Zones, heartnut is reported to tolerate annual precipitation of 5.4 to 12.0 dm (mean of 4 cases = 8.3), annual temperature of 14.7 to 25.0°C (mean of 4 cases = 18.1), and pH of 5.5 to 6.8 (mean of

3 cases = 6.4).[82] Thrives on wide range of soils from clay to sand, and even makes rapid and luxuriant growth on rather poor soil.[278] Very common along streams and on wettish plains.[209] Bears early, and endures temperatures to −40°C. However, it is more successfully grown in areas from Nova Scotia, through Wisconsin and Iowa to southern Oregon and British Columbia and south to Virginia, New Mexico, and northern Arizona.[278] Able to withstand winters not too cold for peaches. Grown throughout Atlantic coastal states, Pacific northwest, and more protected northern areas.[226] Foliage is sometimes injured and season's crop destroyed by late spring frosts.[278]

Cultivation — Propagation by grafting, methods being the same as for butternut and black walnut. Siebold grafts easily on its own seedlings and on butternut (*J. cinerea*). It also grafts easily on black walnut, but does not outgrow the stock. Also propagated by layering, by bending low-growing branches to the ground and burying about 10 cm, leaving remainder of branch protruding upright. Limb is cut half through on underside close to trunk, firmly bound with cord to form a girdle, and treated with tree dressing. Bent-down limb should be shaded from trunk to ground to prevent sunscald. Layers require about 2 years to root. Grafted trees or rooted limbs are planted in the orchard about same distances as other walnuts, about 20 m each way.[278]

Harvesting — Fruits are borne in long racemes and in good locations, trees produce prolifically. Nuts fall to ground in late summer and early fall, and should be harvested by picking up the nuts as soon as they fall, to discourage infestation by maggots. Hulls are removed and nuts dried for a few days, and then stored as for other walnuts.[278]

Yields and economics — Heartnuts yield from 106 to 275 nuts per kg, and crack out about one-fourth to one-third kg in kernels.[278] Grown on a noncommercial basis in northeastern U.S. and lower Ontario.[283]

Energy — All walnuts are oilseeds, producing good timber, but their value is greater for ends other than energetic ends. Yielding better than butternut, this might conceivably be a better energy species.

Biotic factors — In some parts of New York State, a beetle burrows in the terminal shoot. Because of Siebold walnut and heartnut's high resistance to *Melanconis* fungus, it is used for hybridizing with butternut, to which it transmits its resistance.[278] Nearly decimated in the U.S. in the early 20th century by walnut bunch disease.[283] The *Agriculture Handbook 165*[4] reports the following as affecting heartnut: *Melanconis juglandis* (canker, dieback), *Meloidogyne* spp. (root knot nematodes), *Xanthomonas juglandis* (bacterial blight). Also listed are brooming disease (virus), rosette (physiogenic, (?) zinc deficiency), and witches' broom (cause unknown).

JUGLANS CINEREA L. (JUGLANDACEAE) — Butternut, White Walnut, Oil Nut

Uses — Butternut grown primarily for its nuts, used fresh, roasted, or salted, in confectioneries, pastries, and for flavoring. Sugar may be made from the sap. Green husks of fruit are used to dye cloth, giving it a yellow-to-orange color.[278] Bark used by pioneers to make a brown dye.[57] Narragansett Indians called the butternut 'wussoquat' and used the nuts to thicken their pottage.[209] Amerindians ate butternuts raw, cooked, or ground into a meal for baking in cakes. Iroquois used seed oil for cooking and as a hair dressing. Nuts were combined with maple sugar in New England to make maple-butternut candy.[283] The early settlers in New England found they could store the nuts for years as insurance against starvation.[129] The wood is coarse-grained, light-brown, turning darker upon exposure, used for boat construction, boxes, buildings that come into contact with the ground, cabinet work, carving, crates, fence posts, furniture, interior finishing of houses, and millwork. Used to make some propellers for early windmills.[169,170]

Folk medicine — According to Hartwell,[126] pills made from the bark and poultices made from the shucks are said to be folk remedies for cancer. Reported to be alterative, cathartic, laxative, stimulant, tonic, and vermifuge, butternut is a folk remedy for cancer, dysentery, epithelioma, liver ailments, mycosis, tapeworms, tumors, and warts.[91] Butternut bark (the inner bark of the root) is used for fevers and as a mild cathartic.[324] Grieve[117] reports the inner bark of the root, collected in May or June, is the best for medicinal use. Has been recommended for syphilis and old ulcers; said to be rubefacient when applied to the skin.

Chemistry — Per 100 g, the seed (ZMB) is reported to contain 654 calories, 24.6 g protein, 63.6 g fat, 8.7 g total carbohydrate, 3.0 g ash, and 7.1 mg Fe.[89] Smith[310] reports the butternut to be 86.40% refuse, 4.5% water, 27.9% protein, 61.2% fat, 3.4% total carbohydrates, 3.0% ash, and 3,370 calories per pound. Butternut bark (the inner bark of the root) contains resinoid juglandin, juglone, juglandic acid, and an essential oil.[278] Roots give off a toxin that poisons many other plants in the root area.[129]

Description — Tree to 35 m, with straight trunk 0.6 to 1 m in diameter, round-topped; bark smooth, light-gray on young branches, becoming light-brown and deeply fissured, to 2.5 cm thick; winter-buds terminal, 1.3 to 2 cm long, flattened, outer scales covered with pale pubescence; axillary buds dark-brown with rusty pubescence, ovoid, flattened, rounded at apex, 0.3 cm long. Leaves 35 to 75 cm long, with stout pubescent petioles, compound with 11 to 17 oblong-lanceolate leaflets, 5 to 7.5 cm long, to 5 cm wide, finely serrate, glandular, sticky, yellow-green and rough above, pale pubescent beneath; leaves turning yellow or brown before falling in fall; hairy fringe present above leaf-scars. Flowers dioecious, staminate flowers in thick aments to 1.2 to 5 cm long, calyx 6-lobed, light-yellow to green, puberulent on outer surface; bract rusty-pubescent, acute at apex, stamens 8 to 12 with nearly sessile dark-brown anthers, slightly lobed connectives; pistillate flowers in 6- to 8-flowered spikes, constricted above the middle, coated with sticky glandular hairs, stigmas red, about 1.3 cm long. Fruits in drooping clusters of 3 to 5, obscurely 2- or 4-ridged, ovoid-oblong, covered with rusty, clammy hairs, 3 to 6 cm long with thick husk; nut elongated, ovoid, deeply ridged with 4 prominent and 4 less-prominent ribs, light-brown, 2-celled at base, 1-celled above the middle; kernel white to cream, sweet, very oily, soon becoming rancid. Flowers April to June; fruits fall.[278]

Germplasm — Reported from the North America Center of Diversity, butternut, or cvs thereof, is reported to tolerate bacteria, fungus, limestone, poor soil, slope, and weeds.[278] Cvs have been selected with excellent shelling qualities, some of them now being grown are 'Kenworthy', 'Kinneyglen', 'Buckley', 'Helmick', 'Craxezy', 'Henick', 'Johnson', 'Sherwood', 'Thrill', and 'Van der Poppen'. × *juglans quadrangulata* Rehd., a natural hybrid between *J. cinerea* and *J. regia*, occurs occasionally in eastern Massachusetts. Hybrids between butternut (*J. cinerea*) and heartnut (*J. ailanthifolia*) have appeared in the U.S.[278]

'Aiken' was the first grafted butternut available.[20] Grafted cv 'Deeming' reported to bear "when it is two feet high".[310] (2n = 32.)

Distribution — Native to eastern North America, from southern New Brunswick to Ontario, Michigan, southern Minnesota, and South Dakota, south to eastern Virginia, central Kansas, and northern Arkansas, and in the mountains to northern Georgia, Alabama, and western Tennessee. Occasionally cultivated elsewhere. Most abundant northward.[278]

Ecology — Ranging from Cool Temperate Moist to Wet through Warm Temperate Dry Forest LIfe Zones, butternut is reported to tolerate annual precipitation of 5.4 to 12.3 dm (mean of 8 cases = 8.6), annual temperature of 8.4 to 18.0°C (mean of 8 cases = 12.1°C), and pH of 4.9 to 7.2 (mean of 7 cases = 6.2).[82] Thrives in rich, moist soils near banks of streams, on low rocky hills, as well as in forests, along fences, and road-sides. However, it cannot be depended upon as an ornamental planting. Succeeds fairly well on poor upland soils, but thrives best on fertile, slightly acid or neutral soils with good drainage. Hardiest of any of the northern nuts, but short-lived under some conditions, apparently due to fungus disease.[278] Hardy to Zone 3.[343]

Cultivation — Trees in the forest and along road-sides develop from natural dispersal of nuts. When cultivated, nuts or small trees can be planted. To assure viability, seeds should not be more than a few years old. Plant where tree is to grow, in spring or fall, burying about 2.5 cm in the ground. Fall-planted nuts should be well protected from nut-hunting squirrels. Spring-planted nuts should be planted as early as possible, so they can be frozen in the ground a few times. Nuts may be stored in freezer a few days before planting to insure sprouting. Mid-summer sprouting seedlings grow rapidly, possibly reaching 1 m by summer's end. Plant 10 to 12 m apart for nut production; 5 m apart for timber production. Generally takes 10 years from planting to first harvest; the first crop should be a big one.[129] Trees are usually grafted either on seedling butternut or black walnut stocks. Black walnut stocks are reported to give earlier bearing trees. Butternut is a rather rapid-growing tree; however, it begins to deteriorate when it reaches medium size. Trunks of older trees are usually hollow. Otherwise, it requires about the same care and cultivation as other nut trees.[278] Ashworth[20] reports that it is difficult to graft, possibly due to high sap pressure and abundant sap flow in the spring.

Harvesting — Nuts are harvested by picking them up from the ground after they have fallen in early to late fall. Husk is removed and nuts are allowed to dry for a few weeks by spreading them one deep on a warm attic floor, a greenhouse bench, a sunny garage floor, etc. Should be stirred up occasionally so they dry thoroughly. Store in a well-ventilated, dry, cool, squirrel-proof place. Kernels are removed by cracking nuts. A hammer and anvil or a block of hard wood seems to be the best cracking method. Another method is to cover the nuts with hot water and soak them until the water cools. They will crack easily and meats come out intact.[129] Kernels may be stored dried, salted, or frozen until used.[278]

Yields and economics — Yield data for this species are usually included with other native and cultivated walnuts. Kernels of butternut are harvested along with other walnuts and sold salted or variously packaged.[278] Two billion board feet of butternut lumber was reported to be cut in 1 year in 1913. Production in 1941 was ca. 920,000 board feet. West Virginia, Wisconsin, Indiana, and Tennessee have been the leading states in production of butternut lumber. West Virginia mills shipped ca. 250,000 board feet to North Carolina furniture plants in 1963. In 1960, the total veneer production was ca. 4 billion square feet; in 1965, ca. 14 billion square feet face veneer was shipped. Butternuts are less important commercially than black walnuts.[297]

Energy — Both timber and seed oils could be used for energy, but they are, at the moment, probably more suitable for other ends. This species is said to yield less than *J. ailanthifolia*.

Biotic factors — The following fungi are known to attack butternut: *Actinothecium* *juglandis, Botryosphaeria ribis, Cercospora juglandis, Cylindrosporium* sp., *Fusarium avenaceum, Gnomonia leptostyla, Marsonia juglandis, Melanconis juglandis, Microstroma brachysporum, M. juglandis,* and *Nectria galligena*. Trees are attacked by Witches' broom, the cause of which is unknown. The nematodes *Caconema radicicola* and *Meloidogyne* sp. have also been isolated from the tree.[186,278]

JUGLANS HINDSII Jeps. ex R.E.Sm. (JUGLANDACEAE) — California or Hind's Black Walnut
Syn.: *Juglans californica* var. *hindsii* Jeps.

Uses — Kernels of nuts edible, of good quality, but small, used for confectioneries, pastries, and roasted or salted nuts. Wood hard, coarse-grained, dark-brown, often mottled, with pale thick sapwood. Often cultivated in California as street and shade tree.[278]

Folk medicine — No data available.

Chemistry — No data available.

Description — Deciduous, round-topped tree 10 to 20 m tall, occasionally to 25 m, with erect, unbranched trunk 3.3 to 13 m, 30 to 60 cm in diameter; bark strong-scented, gray-brown, smoothish, longitudinally fissured into narrow plates; branches pendulous; branchlets villose-pubescent, reddish-brown, lenticels pale. Leaves 22 to 30 cm long, alternate, compound; petioles and rachis villose-pubescent; leaflets 15 to 19, thin, 6 to 10 cm long, 2 to 2.5 cm wide, ovate-lanceolate to lanceolate, long-pointed, often slightly flacate, margin serrate, base rounded cuneate to cordate, upper surface puberulous while young, becoming bright-green and glabrous, lower surface with tufts of hairs and villose-pubescent along midrib and primary veins. Staminate flowers in slender glabrous or villose aments 7.5 to 12.6 cm long, calyx elongated, covered with pubescence, 5- or 6-lobed, stamens 30 to 40, with short connectives bifid at apex; pistillate flowers oblong-ovoid, thickly covered with villose-pubescence about 0.3 cm long. Fruit globose, 3 to 5 cm in diameter, husk thin, dark-colored with soft pubescence; nut nearly globose, somewhat flattened at ends, faintly grooved with remote longitudinal depressions, shell thick; seed small and sweet.[278]

Germplasm — Reported from the North American Center of Diversity, Hind's black walnut, or cvs thereof, is reported to tolerate high pH.[82] In California, natural hybrids are known between this walnut and *Juglans nigra;* also a hybrid 'Paradoxa' (*J. hindsii* × *J. regia*) has been produced artificially. *J. hindsii* var. *quercina* Sarg. (*J. californica* (var.) *quercina* Babcock) has leaves with 1 to 5 leaflets, usually 3, short-stalked or sessile, broadly ovate to oblong, obtuse or emarginate, serrate or entire, 1.3 to 5 cm long. ($2n = 24$.)[278]

Distribution — Native to Coastal region of central California. Sometimes cultivated in California, eastern U.S., and Europe.[278]

Ecology — Ranging from Warm Temperate Thorn to Dry Forest Life Zones, Hind's black walnut is reported to tolerate annual precipitation of 3.1 to 6.6 dm (mean of 2 cases = 4.9), annual temperature of 12.7 to 14.7°C (mean of 2 cases = 13.7), and pH of 6.8 to 8.2 (mean of 2 cases = 7.5).[82] In natural habitats, trees are found along streams and rivers. Trees not suitable for lawn-planting because rootstock is very susceptible to crown rot (*Phytophthora cactorum*), especially if given frequent summer irrigation.[278]

Cultivation — Trees used as stock for Persian walnut (*J. regia*), top-worked high to provide butt logs for walnut timber.[278]

Harvesting — Fruit gathered when ripe in fall. Treated like other walnuts.[278]

Yields and economics — Valued mostly as a shade or street tree in California, and as stock on which to graft varieties of Persian walnut (*J. regia*). Butt logs 45 cm in diameter bring about $200 each.[278]

Energy — Endangered or threatened species are not recommended as energy species. However, if abundant in cultivation, this species could serve as a high-priced oilseed and firewood, though the fruit and timber could find better uses.

Biotic Factors — Trees are resistant to oak root fungus, but particularly susceptible to crown rot (*Phytophthora cactorum*). The following are also reported as affecting this species: *Cacopaurus epacris, Cylindrosporium juglandis* (leaf spot), *Microstroma juglandis, Xanthomonas juglandis* (bacterial blight). Also reported are Black-line (girdle-graft incompatibility) and Little leaf (zinc deficiency).[4]

JUGLANS NIGRA L. (JUGLANDACEAE) — Eastern Black Walnut

Uses — Black walnut is one of most valuable natural forest trees in the U.S. The nuts furnish a food product, used mainly for flavoring baked goods, pastries, and confectioneries. The wood has good texture, strength, and is coarse-grained, very durable, of a rich dark-brown color with light sapwood; used in cabinet-making, gun-stocks, interior finishes of houses, furniture, air-planes, ship-building. Wood is also easy to work, resistant to destructive fungi and insect pests. Woody shells on fruits are used to make jewelry. Green fruit husks are boiled to provide a yellow dye. Trees are used for shade and ornamentals.[144,206]

Folk medicine — The bark and leaves are considered alterative, astringent, detergent, laxative, and purgative. They are used for eczema, herpes, indolent ulcers, scrofula. The unripe fruit is sudorific and vermifugal, and used for ague and quinsy, and is rubbed onto cracked palms and ringworm. Oil from the ripe seeds is used externally for gangrene, leprosy, and wounds. Burnt kernels, taken in red wine, are said to prevent falling hair. Green husks are supposed to ease the pain of toothache. Indians used the root bark as a vermifuge. Macerated in warm water, the husks and/or leaves, are said to destroy insects and worms, without destroying the grass. Insects are said to avoid the walnut; hence it is often used as a poor man's insect repellent. Rubbed on faces of cattle and horses, walnut leaves are said to repel flies. The roots and/or leaves exude substances which are known to inhibit germination and/or growth of many plant species. All parts of the plant contain juglone, which inhibits other plant species. Juglone has antihemorrhagic activity.[91,168,278]

Chemistry — The genus Juglans is reported to contain the following toxins: folic acid, furfural, inositol, juglone, nicotine, and tryptophane.[86] Juglone has an oral LD_{50} of 2500 μg in mice. Chloroform is said to constitute a large part of the essential oil of the leaves. Per 100 g, black walnut contains 3.1% water, 628 calories, 20.5 g protein, 59.3 g fat, 14.8 g total carbohydrate (1.7 g fiber), 2.3 g ash, a trace of Ca, 570 mg P, 6 mg Fe, 3 mg Na, 460 mg K, 300 IU Vitamin A, 0.22 mg thiamine, 0.11 mg riboflavin, and 0.7 mg niacin.[89]

Description — Tree up to 33 m tall, occasionally to 50 m, and often 100 years old; trunk straight, often unbranched for 20 m, 1.3 to 2 m in diameter; branches forming a round-topped crown, mostly upright and rigid; branchlets covered at first with pale or rusty matted hairs, and raised conspicuous orange lenticels; bark 5 to 7.5 cm thick, dark-brown tinged red, deeply furrowed with broad rounded ridges; twigs light-brown with channeled pith; terminal bud as broad as long; no hairy fringe above leaf-scar; leaves compound, deciduous, 30 to 60 cm long, petioles pubescent, with 13 to 23 leaflets; leaflets 7.5 to 8 cm long, 2.5 to 3 cm wide, long-pointed, sharply serrate, slightly rounded at base, yellow-green, thin, glabrous above, soft-pubescent beneath, turning bright-yellow in fall before falling; staminate aments thick, 7.5 to 12.5 cm long, compact, not-stalked, single; calyx 6-lobed, lobes concave, nearly orbicular, pubescent on outer surface, its bract nearly triangular with rusty brown tomentum; stamens 20 to 30, in many series, connectives purple, truncate, nearly sessile; pistillate aments in 2 to 5-flowered spikes, bracts with pale glandular hairs, green, puberulous, calyx-lobes ovate, acute, puberulent on outer surface, glabrous or pilose within; fruit solitary or in pairs, globose, oblong or pointed at apex; husk yellow-green or green, smooth or roughened with clusters of short pale articulate hairs, 3 to 5 cm in diameter, indehiscent; nut oval, oblong or round, rough or sculptured, 3 to 3.5 cm in diameter, dark-brown tinged red, 4-celled at base, slightly 2-celled at apex; kernel sweet, soon becoming rancid. ($2n = 32$.) Flowers April to May; fruits at frost in fall.[82,278]

Germplasm — At present, nearly 100 varieties of black walnuts have been selected and named. Many can be propagated to order, or scions may be obtained for grafting upon established stocks. Varieties or cultivars differ in hardiness, response to length of growing season, summer heat, resistance to diseases and susceptibility to insect damage. 'Thomas' is the most cultivated variety in New York; 'Synder' and 'Cornell' have good cracking quality for northern areas; 'Wiard', for Michigan; 'Huber' and 'Cochrane', for Minnesota; 'Sparrow', 'Stambaugh', and 'Elmer Myers' are all good in parts of the South; 'Ohio' and 'Myers' are good in north central areas. Natural hybrid, × *Juglans intermedia* Carr (*J. nigra* × *J. regia*) has been recorded in the U.S. and Europe. In California, 'Royal' (*J. nigra* × *J. hindsii*) has been artificially produced. Reported from the North American Center of Diversity, walnut is reported to be relatively tolerant to disease, drought, fire, frost, fungi, high pH, heat, insects, limestone, slopes, smog, and weeds.[82,278]

Distribution — Grows naturally in 32 states and in southern Ontario, Canada; most abundant in Allegheny Mountains to North Carolina and Tennessee. Occasionally cultivated as an ornamental in eastern U.S., western and central Europe. Planted in Europe for timber.[278]

Ecology — Wind pollinated, walnut may play a small role in hay fever. Suited to rich bottomlands and fertile hillsides from lower Hudson Valley southward, walnut will grow a few hundred miles outside its natural range, but may not bear nuts. Seedling trees mature fruit rather generally throughout area with a growing season of about 150 days and an average summer temperature of 16.5°C. Best suited to deep, rich, slightly acid or neutral soil, with good drainage, but will not succeed on infertile upland soil or on soils with poor drainage. Reliable indicators for suitable land are good stands of white oak and tulip popular, or where corn grows well. Because trees have a deep tap-root, they are drought-resistant. Black walnut is reported from areas with annual precipitation from 3 to 13 dm (mean of 19 cases = 9), annual temperature from 7 to 19°C (mean of 19 cases = 11), and pH from 4.9 to 8.2 (mean of 15 cases = 6.3).[82,278]

Cultivation — Improved varieties do not come true from seed, hence, propagation is by grafting scions (twigs) from trees of desired varieties onto main stems of 2- to 3-year old native seedlings. Scions develop crowns that bear nuts of their own variety. As there is little information available to indicate the best varieties for different localities, local nurseries should be consulted as to the best for a given locality. Trees are self-fertile, but the sequence of male and female blooming, called dichogamy, can and often does minimize chances of a tree shedding pollen on its own pistils. In different trees pollen may be shed before the receptivity period of female flowers, or at same time, or after pistil receptivity. For greatest possible nut production, plant trees of 2 or more varieties, as different varieties have overlapping pollen-receptivity periods and can pollinate each other. Young plants are best transplanted in early spring, at which time new roots will grow rapidly to replace those lost in transplanting. In the South, young trees may be planted in fall or winter. For nut production, trees are spaced 20 m apart. For trees up to 2.3 m tall, dig hole 0.6 m deep and 1 m wide. Place tree at same depth in hole as it stood in nursery and spread out roots well. Fill hole with topsoil and firm down soil. Form a basin around edge of hole and soak soil immediately. Black walnuts require large quantities of nitrogen and phosphorus. Apply mixed fertilizer (5-10-5 or 10-10-10) each year under tree branches when buds begin to swell in early spring. Use rates of 450 g/year of 5-10-5 fertilizer, or 230 g/year for 10-10-10, per tree. Do not use during first year, because of danger of injuring roots. In strongly acid soils, apply lime to change pH to 6 or 6.5. Do not over-lime, as this makes zinc in soil unavailable to tree. Soils east of Mississippi River are often deficient in magnesium, so crushed dolomite limestone is used to correct this condition and reduce acidity of soil. Prune any suckers that come from below graft on trunk. In orchards, trees over 15 years old may be interseeded with grasses and legumes, and animals may be turned in to pasture, as they will not damage older trees. All black walnuts tend to bear heavy nut crops every second year. No cultural practices have been developed to offset this type of alternating. Some trees bear every year, while others bear every third year. Others mainly react to climatic conditions with no pattern. In the U.S. growing seasons are divided into 3 zones: North of Mason-Dixon Line, 140 to 180 days; south to North Carolina, northern Georgia, Alabama, Mississippi, Arkansas, and Oklahoma, 180 to 200 days; south of that, 220 to 260 days. Varieties are selected for each area. When trees bearing fruits of exceptional quality are found, they are propagated and cultivated for nut production in that area.[278]

Harvesting — Nuts are harvested from native trees as well as from improved selections and cultivars. Fruit ripens in one season, usually by late September or early October. Most production is from wild trees growing on non-crop land, and these represent the main commercial source of kernels for today's market. Nuts should be harvested as soon as they fall, in order to get light-colored kernels with mild flavor. Leaving them on ground causes some discoloration of kernel. Hulls of native trees are thick and heavy, whereas those of 'Thomas' and 'Ohio' have thinner hull, those of 'Myers' being thinnest of all. Hull may be mashed and removed by hand, or by mechanical devices. After removing the hulls, nuts should be washed thoroughly and spread out to dry in direct sunlight. Drying takes 2 to 3 weeks; nuts can then be stored in a cool, dry place until needed. Nuts are cracked and kernels removed for use.[278]

Yields and economics — Although Duke[82] reported yields of 7.5 MT seeds, this is probably highly optimistic. Elsewhere it is said that 95% of the wild black walnut seeds are empty or aborted. Perhaps yields could be as high as 2.5 MT/ha under intensive management, which is attainable in the commercial walnut, ***Juglans regia***. Selections are made based on weight of nuts. Trees may bear at rates of 7,500 seed per ha. Nuts from wild trees weigh about 17 g (27 nuts per lb); for selected varieties, weights vary from 15 to 30 g; those 20 g or over are: 'Michigan' (20); 'Grundy', 'Monterey', 'Schreiber' and 'Thomas' (21); 'Victoria' (22); 'Hare' (23); 'Pinecrest' (25); and 'Vandersloot' (30). 'Thomas', 'Ohio' and

'Myers' begin bearing nuts in second or third year after planting, while native trees usually do not begin to bear until about 10 years after planting. In 5 to 6 years, these three varieties bear about one-fourth bushel of nuts; at 15 to 20 years of age, the first two bear 2 bu of nuts, 'Myers' about 1 bu, and native trees about $^1/_4$ bu. Lumber trees yield about 1150 board feet at 76 years old. Nut shelling industry is centered in and around Arkansas, Kansas, Kentucky, Missouri, Oklahoma, Tennessee, West Virginia, and Virginia. Because of the scarcity of trees and the long growing period required to get wood, walnut lumber is not in great demand as it used to be. More frequently grown in Europe for lumber. Walnuts are grown in the U.S. for nuts and ornamentation. In the U.S. the following are said to deal in walnut oil: Hain Pure Food Company (13660 S. Figueroa, Los Angeles, California) and Tunley Division, Welch, Home and Clark Co. (1000 S. 4th Street, Harrison, New Jersey) Well-formed trees will yield lumber worth thousands of dollars.

Energy — Oil contents of the seeds run about 60%, suggesting that if the walnut yields of 7.5 MT/ha were attained, there might be as much as 4.5 MT oil there. Hulls and exocarp might be used to fuel the processing, as the value of the timber improves with age (one tree commanded $35,000 at an Ohio auction). Prunings and culls, as well as fallen and dead limbs, might about to 5 MT/ha/year.

Biotic factors — Walnut anthracnose is most serious disease to native trees. 'Ohio' is resistant to this disease; 'Myers' is less resistant. Disease over-winters in fallen leaves and reinfects new leaflets in mid-May until mid-June, often defoliating entire trees. Many nuts are empty or contain blackened, shriveled kernels. Bunch disease, of which the cause and means of spread are unknown, stunts growth of the tree and lowers nut production. The most serious insect pests are walnut lace bug, curculios, walnut husk maggot, walnut caterpillar and fall web-worm. Serious damage may also be caused by leaf-eating caterpillars, scales, aphids and twig girdlers. County agricultural agents should be consulted for measures to control these in a particular area.[278] Nematodes include *Meloidogyne* sp., *Pratylenchus coffeae, P. pratensis*, and *P. vulnus*.[382] The following are reported in *Agriculture Handbook 165*[4] as affecting *Juglans nigra: Botryosphaeria ribis, Cercospora juglandis* (leaf spot), *Cladosporium* sp. (? scab), *C. pericarpium, Cylindrosporium juglandis* (leaf spot), *Cytospora* sp. (canker), *C. albiceps, Fomes igniarius, Gnomonia leptostyla* (anthracnose, leaf spot, leaf blotch), *Meloidogyne* spp. (root knot nematodes), *Microstroma juglandis* (downy spot, white mold), *Nectria ditissima, Phleospora multimaculans* (leaf spot), *Phorandendron flavescens* (mistletoe), *Phymatotrichum omnivorum* (root rot), *Phytophthora cinnamomi* (collar rot of seedlings), *Pratylenchus musicola, Rhabdospora juglandis, Sclerotium rolfsii* (seedling blight), *Sphaeropsis druparum, Stereum fasciatum*, and *Xanthomonas juglandis* (bacterial blight).[4]

JUGLANS REGIA L. (JUGLANDACEAE) — English Walnut, Carpathian or Persian Walnut

Uses — Principally valued as an orchard tree for commercial nut production. Nuts are consumed fresh, roasted, or salted, used in confectioneries, pastries, and for flavoring. The shells may be used as antiskid agents for tires, blasting grit, and in the preparation of activated carbon. Ground nut shells are used as an adulterant of spices. Crushed leaves, or a decoction are used as insect repellant and as a tea. Outer fleshy part of fruit, very rich in Vitamin C, produces a yellow dye. Fruit, when dry pressed, yields a valuable oil used in paints and in soap-making; when cold pressed yields a light-yellow edible oil used in foods as flavoring. Young fruits made into pickles, also used as fish poison. Twigs and leaves lopped for fodder in India. Decoction of leaves, bark, and husks used with alum for staining wool brown. Wood hard, durable, close-grained, heavy, used for furniture and gun-stocks. Tree often grown as ornamental.[35,278]

Folk medicine — According to Hartwell,[126] English walnuts are used in folk remedies for aegilops, cancer, carbuncles, carcinoma, condylomata acuminata, corns, excrescences, growths, indurations, tumors, warts, and whitlows, especially cancerous conditions of the breast, epithelium, fauces, groin, gullet, intestine, kidneys, lip, liver, mammae, mouth, stomach, throat, and uterus. Reported to be alterative, anodyne, anthelmintic, astringent, bactericide, cholagogue, depurative, detergent, digestive, diuretic, hemostat, insecticidal, laxative, lithontryptic, stimulant, tonic, and vermifuge, the English walnut is a folk remedy for anthrax, asthma, backache, caligo, chancre, colic, conjunctivitis, cough, dysentery, eczema, ejaculation, favus, heartburn, impotence, inflammation, intellect, intestine, intoxication, kidney, legs, leucorrhea, lungs, rheumatism, scrofula, sore, syphilis, and worms.[91]

Chemistry — Per 100 g, the seed is reported to contain 647 to 657 calories, 2.5 to 4.2 g H_2O, 13.7 to 18.2 g protein, 63.6 to 67.2 g fat, 12.6 to 15.8 g total carbohydrate, 1.6 to 2.1 g fiber, 1.7 to 2.0 g ash, 92 to 106 mg Ca, 326 to 380 mg P, 3.0 to 3.3 mg Fe, 2 to 3 mg Na, 450 to 536 mg K, 0.50 μg beta-carotene equivalent, 0.27 to 0.50 mg thiamine, 0.08 to 0.51 mg riboflavin, 0.7 to 3.0 mg niacin, and 0 to 5 mg ascorbic acid. *Wealth of India*[70] also reports, per 100 g, 2.7 mg Na, 687 mg K, 61 mg Ca, 131 mg Mg, 2.4 mg Fe, 0.3 mg Cu, 510 mg P, 104 mg S, and 23 mg Cl, and 2.8 μg I (as well as Ar, Zn, Co, and Mn). About 42% of the total phosphorus is in phytic acid; lecithin is also present. The immature fruit is one of the richest sources of ascorbic acid, the skin with 1,090 mg/100 g, the pulp with 2,330 mg. The leaves, also rich in ascorbic acid (almost 1% of the weight), are rich in carotene (ca. 0.3% wet weight). Juglone is the active compound in the leaves; also quercetin, cyanadin, kaempferol, caffeic acid, and traces of p-coumaric acid, hyperin (0.2%), quercitrin, kaempferol-3-arabinoside, quercetin-3-arabinoside. The seed oil contains 3 to 7% palmitic, 0.5 to 3% stearic, 9 to 30% oleic, 57 to 76% linoleic, and 2 to 16% linolenic acids. The oil cake, with 86.6% dry matter (DM), contains 35.0% protein, 12.2% fatty oil, 27.6% carbohydrates, 6.7% fiber, 5.1% ash (digestible nutrients: 31.5% crude protein, 11.6% fatty oil, 23.5% carbohydrates, and 1.7% fiber). The shells contain 92.3% DM, 1.7% protein, 0.7% fatty oil, 31.9% carbohydrates, 56.6% fiber, and 1.4% ash.[70,89,187]

Description — Deciduous, monoecious trees, 12 to 15 m tall (Payne vars.), 17 to 20 m tall ('Eureka', 'Placentia', 'Mayette', 'Franquette'), or rarely up to 60 m tall; bark brown or gray, smooth, fissured; leaf-scars without prominent pubescent band on upper edge. Leaves alternate, foetid, pinnate, without stipules; leaflets 15 to 24, opposite, 6 to 15 cm long, ovate-oblong to ovate-lanceolate, acuminate; margin irregularly serrate, glabrescent above, pubescent and glandular beneath. Flowers developing from dormant bud of previous season's growth; staminate flowers in axillary, pendulous aments 5 to 15 cm long, developing 1 to 4 million pollen grains each; flowers in axils of scales, with 2 bracteoles, perianth-segments 1 to 4, stamens 3 to 40; pistillate flowers in clusters of 3 to 9, developing as many nuts; in selected varieties not only terminal bud produces fruit, but all lateral buds on previous years growth also produce; perianth 4-lobed. Fruit 3.5 to 5 cm in diameter, globose or slightly obovoid, pubescent; nut ovoid, acute, strongly ridged, not splitting.[278]

Germplasm — Reported from the Eurosiberian and Central Asian Centers of Diversity, English walnut, or cvs thereof, is reported to tolerate frost, high pH, heat, insects, low pH, and slope. (2n = 32,36.)[82] Varieties are selected on basis of high heat tolerance, resistance to walnut blight (*Xanthomonas juglandis*), tolerance for winter cold, and yield and quality of kernels. Most promising cvs are of Carpathian origin and have been introduced from Poland; they withstand temperatures below those recorded in the the fruit belt of New York. Recent superior cvs include: 'Broadview', 'Schafer', 'Littlepage', 'McKinster', 'Metcalfe', 'Jacobs', and 'Colby'. Other varieties widely grown in the world include: 'Marmot', 'Meylanaise', 'Corne', 'Gourlande', 'Mayette', 'Brantome', 'Ashley', 'Glackner', 'Nugget', 'Poe', 'Franquette', 'Concord', 'Ehrhardt', 'Payne', and 'Waterloo'. Persian walnuts have been hybridized with butternuts, black walnuts, and other European and Oriental walnuts. *Juglans regia* var. *orientis* (Dode) Kitam. *J. orientis* Dode; *J. regia* var. *sinensis* sensu auct. Japan (non DC.) is a widely cultivated Chinese tree, with glabrous leaves and branchlets, leaflets 3 to 9, obtuse, entire, except in young trees, and nuts relatively thin-shelled.[82,278]

Distribution — Native to the Carpathian Mountains of eastern Europe, but often found growing wild eastward to Himalayas and China. Widely cultivated throughout this region and elsewhere in temperate zone of the Old and New World. Thrives in temperate Himalayas from 1,000 to 3,000 m altitude. In North America, thrives as far north as New York State. Introduced from Spain by way of Chile to California about 1867. In 1873 'Kaghazi' was introduced in northern California and a seedling 'Eureka' has become the important source of our commercial cvs.[278]

Ecology — Ranging from Cool Temperate Steppe to Wet through Subtropical Thorn to Moist Forest Life Zones, English walnut is reported to tolerate annual precipitation of 3.1 to 14.7 dm (mean of 25 cases = 8.4), annual temperature of 7.0 to 21.1°C (mean of 25 cases = 12.0), and pH of 4.5 to 8.2 (mean of 21 cases = 6.4). Thrives on rich, sandy loam, well-drained, slightly acid or neutral. Responds well to cultivation and fertilization. In areas where hardiness is a problem, trees should not be forced into excessive vegetative growth. Minimum temperature should not go below −29°C. One fault of Carpathian walnut is that it begins growth early in spring with the result that crop and foliage may be damaged by late frosts. When fully dormant, trees can withstand temperatures from −24°C to −27°C without serious damage. French cvs may be more winter hardy. 'Eureka' is less hardy than newer cvs being produced for northern California, Oregon, and higher altitudes. High summer temperatures damage kernels, slightly at 38°C, severely at 40.5 to 43.5°C. Quite variable resistance to heat among varieties. Reported from areas with pH 4.5 to 8.3, annual rainfall 3 to 15 dm, and annual temperature 7 to 19°C. Rains in late spring and summer increase walnut blight infections.[82,278]

Cultivation — Since trees are deep-rooted, soil should be fertile, well-drained, alluvial, 2 m or more deep, of medium loam to sandy or silt loam texture, and free of alkali salts, especially excessive boron. Seedling trees show great variation as to hardiness, type of fruit and fruitfulness. 'Paradox' hybrids, 'Royal' hybrids and *Juglans hindsii* are used as rootstocks for grafting Persian and Carpathian walnuts. Rootstock of *Juglans regia* may be used if oak root fungus (*Armillaria mellea*) is absent in area. Persian walnuts have been grafted to Chinese wingnut (*Pterocarya stenoptera*).[336] Selected varieties are best whip- or bark-grafted or patch-budded on seedling trees, or top-worked on existing trees. Persian walnuts are planted in the orchard from 10 to 20 m each way; however, many spacings are in use depending on cv and cultivation methods. Intercropping young walnuts may be useful for the first 5 to 10 years. Intercropping may be difficult because of irrigation, spraying, and use of equipment for cultivation of the intercrop. Holes should be dug amply wide to accommodate roots, planting no deeper than they were in the nursery. Roots should never be allowed to dry out. Topsoil should be used to fill hole, firmly tamped around roots. Do not transplant when soil is wet. Nut trees must have tops reduced or cut back, either before or after planting, usually to about 1.5 to 2 m from ground level. Lower buds should be suppressed so the upper ones will be forced to grow and make the framework of the tree. Newly planted trees should be staked, either with a single stake driven close to the tree and tying it to the stake, or driving three stakes equidistant, fastening tree to each with stout cord so as not to injure bark. After trees are planted, they should be watched, and watered during dry spells until established. When irrigated, a total of $2^1/_2$ to 5 acre feet of water per acre should be applied throughout the year, including normal rainfall. The modified central leader system of training young walnuts is recommended for western orchards, in which 4 or 5 main framework branches spaced both vertically and horizontally are developed; the first branch should be started no lower than 2 m from the ground. The trend is toward heavier and more consistent pruning both in young and old trees; very fruitful new varieties respond more readily than some of the older varieties. Standard method for applying zinc to walnut trees is to drive zinc-bearing metal pieces or glazing points into outer sapwood of trees. Other mineral deficiencies which must be corrected are iron, manganese, boron, potassium, magnesium, phosphorus, and copper.[278,376]

Harvesting — Pollination is often a problem, as Persian walnuts are monoecious, with separate staminate and pistillate flowers in different parts of the same tree. Staminate catkins are 10-15 cm long and produce 1-4 million pollen grains each. Sometimes freshly picked catkins are put on paper in room at 21°C and the shed pollen stored in desiccator at 0°C. Then pollen is blown on trees by fan mounted on truck. Helicopters are sometimes used to blow pollen over orchard. Under favorable conditions, the husks of nuts crack open and adhere temporarily to twigs, allowing nuts to fall to the ground, usually between September 1 and November 7. Nut fall may be hastened by shaking the trees with long poles or a boom shaker. During harvest period, nuts are picked up 3 or 4 times before the total crop has matured and dropped. Nuts should not be allowed to remain on ground too long. Nuts are washed, if dirty, and spread out in shallow trays with bottom slats spaced 1.5 to 2 cm apart. Nuts should not be exposed to sun for entire day. Trays are piled up so as to permit ventilation after nuts have become warm. Too-fast drying causes shell to crack and open. In large orchards, a drying house is constructed for curing process. After curing and bleaching, nuts are graded and packed for shipment.[278]

Yields and economics — Newer cvs begin producing nuts in 5 to 6 years; by 7 to 8 years, they produce about 2.5 tons of nuts per hectare. Orchards on relatively poor, unirrigated mountain soil report 1.5 to 2.25 MT/ha; orchards in well-cultivated valleys, 6.5 to 7.5 MT/ha. A grown individual can yield about 185 kg, but 37 kg is more likely.[70] In the U.S., California is the major producing area, with about 129,400 acres producing 77,000 tons nuts per year; Oregon is second with about 3,500 tons annually; the total valued at about $32.3 million. About 60% of Persian walnuts are sold shelled. Lumber from large trees may bring up to $1,500 per tree.[35,278,283]

Energy — If the walnut yields of 7,500 kg/ha[82] yielded all their 65% (63 to 67%) oil, there is a potential oil yield of nearly 5 MT per year, a very worthwhile target, if attainable. The green hulls have recoverable ascorbic acid content (2.5 to 5% of dry weight). Hulls contain 12.2% tannin, bark contain 7.5%, leaf blades contain 9 to 11%. After extraction of the vitamin C and tannin, the residues might be used for fuel or ethanol. Prunings from the trees might contribute another 5 MT biomass per year.

Biotic factors — Seedlings are very susceptible to mushroom root rot, and Walnut girdle disease 'Blackline' is thought to occur when certain horticultural varieties of *Juglans regia* are grafted on rootstocks of *Juglans hindsii* and its hybrids, associated with graft incompatibility.[304,305] Fungi known to attack Persian walnuts include: *Alternaria nucis, Armillaria mellea, Ascochyta juglandis, Aspergillus flavus, Auricularia auricula-judae, Auricularia* *mesenterica, Cerrena* unicolor, *Cladosporium herbarum, Coniophora cerebella, Coprinus micaceus, Coriolus tephroleucus, Cribaria violaceae, Cryptovalsa extorris, Cylindrosporium juglandis, C. uljanishchevii, Cytospora juglandina, Cytosporina juglandina, C. juglandicola, Diplodia juglandis, Dothiorella gregaria, Erysiphe polygoni, Eutypa ludibunda, Exosporina fawcetti, Fomes fomentarius, F. igniarius, F. ulmarius, Fusarium avenaceum, F. lateritium, Ganoderma applanatum, Glomerella cingulata, Gnomonia ceratostyla, G. juglandis, G. leptostyla, Hemitricia leiotyichia, Hypoxylon mediterraneum, Inonotus hispidus, Laetiporus sulphureus, Lentinus cyathiformis, Licea tenera, Marsonia juglandis, Melanconis carthusiana, M. juglandis, Melanconium juglandis, M. oblongum, Melanopus squamosus, Microsphaera alni, M. juglandis, Microstroma juglandis, Mycosphaerella saccardoana, M. woronowi, Nectria applanata, N. cinnabarina, N. ditissima, Oxyporus populinus, Phelliunus cryptarum, Phleospora multimaculans, Phoma juglandis, Phomopsis juglandis, Phoma juglandis, Phyllactinia guttata, Phyllosticta juglandina, P. juglandis, Phymatotrichum omnivorum, Phytophthora cactorum, P. cinnamomi, P. citrophthora, Pleospora vulgaris, Pleurotus ostreatus, Polyporus hispidus, P. picipes, P. squamosus, Polystictus versicolor, Rhizopus nigricans, Stereum hirsutum, Trametes suaveolens, Tubercularia juglandis, T. vulgaris, Verticillium albo-atrum.* Bacteria attacking Persian walnut include: *Agrobacterium tumefaciens, Bacillus mesentericus, Bacterium juglandis, Pseudomonas juglandis, Xanthomomas juglandis. Cuscuta pentagona* also parasitizes the tree. The following nematodes have been isolated from Persian walnut: *Cacopaurus pestis, Diplogaster striatus, Diplogaster coronata, Ditylenchus intermedius, Meloidogyne arenaria, M. javanica, M.* sp., *Pratylenchus coffeae, P. pratensis, P. vulnus, Rhabditis debilicauda, R. spiculigera, Tylolaimophorus rotundicauda.* Among the insect pests of this walnut are the following: Walnut Blister mite (*Eriophytes tristriatus*), Walnut aphid (*Chromaphis juglandicola*), Italian pear scale (*Diaspis piricola*), Calico scale (*Eulecanium cerasorum*), Frosted scale (*Parthenolecanium pruinosum*), Walnut scale (*Quadraspidiotus juglansregiae*), Codling moth (*Cydia pomonella*), Fruit tree leaf-roller (*Archips argyrospila*), Indian meal moth (*Plodia interpunctella*), Walnut caterpillar (*Datana integerrima*), Red-humped caterpillar (*Schizura concinna*), Walnut span worm (*Phigalia plumigeraria*), and Walnut husk fly (*Rhagolestis completa*).[53,186,278]

LECYTHIS MINOR Jacq. (LECYTHIDACEAE) — Coco de Mono
Syn: *Lecythis elliptica* H.B.K.

Uses — These trees are cultivated for the nuts, which have a delicious flavor and possess a high oil content. Small trees are highly ornamental.

Folk medicine — Duke and Wain[91] cite the species as antiasthmatic, depilatory, and poisonous.

Chemistry — The seeds have been reported to be somewhat toxic, especially if eaten in large quantities. Ingesting the nuts is known to cause loss of hair and nails, at least in seleniferous areas. Though seeds taste agreeable, injestion may induce nausea, anxiety, and giddiness. Dickson[75] attributes the temporary loss of hair and fingernails that he experienced after eating 300 to 600 seeds of *L. minor* to toxic elements in the seeds. Throughout northern Colomiba, *L. minor* is thought to be poisonous. Castaneda,[364] however, feels they are nontoxic. The toxicity of the seeds may depend upon the soils. Some evidence suggest that toxic seeds come from plants found on soil high in selenium.[365] Mori[365] suggests that the data suggesting toxicity in *L. ollaria* may in fact refer to this species. Without voucher specimens, we'll never know.

Toxicity — Identified as a selenium-containing analog of the sulfur amino acid, cystathionine, the active compound has the following formula: $HOOC\text{-}Ch(NH_2)\text{-}CH_2\text{-}Se\text{-}CH_2\text{-}CH(NH_2)COOH$.[365]

Description — Small to medium-sized trees, often branched from base when in open habitats, 5 to 25 m tall, to 70 cm DBH, the crown dilated. Twigs gray, glabrous to pubescent. Bark gray, relatively smooth when young, with deep vertical fissures when older. Leaf blades ovate, elliptic, or oblong, 8.5 to 24.5 × 4.5 to 10 cm, glabrous, coriaceous, with 12 to 19 pairs of lateral veins; apex mucronate to acuminate, infrequently acute; base obtuse to rounded, infrequently truncate, narrowly decurrent; margins usually crenulate to serrate, infrequently entire; petiole 5 to 20 mm long, usually puberulous, infrequently glabrous. Inflorescences racemose, unbranched, or once-branched, terminal or in axils of uppermost leaves, the principal rachis 10 to 35 cm long, with 10 to 75 flowers, all rachises pubescent, the pedicels jointed, 1 to 3 mm long below articulation, subtended by an ovate, caducous bract 2 to 4 × 2 to 3 mm, with 2 broadly ovate, caducous bracteoles 3 to 6 × 3 to 4 mm inserted just below articulation. Flowers 5 to 7 cm diameter; calyx with 6 widely to very widely ovate, green lobes 6 to 11 × 6 to 9 mm; petals 6, widely obovate or less frequently widely oblong to oblong, 27 to 42 × 14 to 25 mm, green in bud, usually white, less frequently light-yellow at anthesis; hood of androecium dorsiventrally expanded, 20 to 23 × 19 to 25 mm, with well-developed, inwardly curved, antherless appendages, the outside of hood white or light-yellow, the appendages always light-yellow; staminal ring with 300 to 410 stamens, the filaments 2 mm long, dilated at apex, light-yellow, the anthers 0.5 to 0.7 mm long, yellow; hypanthium usually pubescent, infrequently glabrous; ovary 4-locular, with 3 to 6 ovules in each locule, the ovules inserted on floor of locule at juncture with septum, the summit of ovary umbonate, the style not well differentiated, 2 to 4.5 mm long. Fruits cup-like, globose or turbinate, 5 to 7 × 7 to 9 cm, the pericarp 7.5 to 11 mm thick. Seeds fusiform, 2.4 to 3 × 1.3 to 2 cm, reddish-brown, with 4 to 6 light-brown longitudinal veins when dried, the testa smooth, with cord-like funicle surrounded by fleshy white aril at base.[219,264,265]

Germplasm — Reported from the South American Center of Diversity, coco de mono, or cvs thereof, is reported to tolerate low pH.[82] Very closely related to another coco de mono, *Lecythis ollaria*, found east of the Andes and also suspected to exhibit seed toxicity. (X = 17.)

Distribution — Introduced at Mayaguez, P.R.; La Lima, Honduras: Summit, Panama; and Soledad, Cuba. Ranges from the Maracaibo lowlands of Venezuela to the northern coast of Colombia from where it ascends the Magdalena and Cauca Valleys. This species most often occurs in dry, open, somewhat disturbed habitats, where it grows as a small, much-branched tree. However, it is also found in moister forests, especially along watercourses, where it forms a handsome, single-trunked tree to 25 meters.[219]

Ecology — Ranging from Tropical Very Dry to Wet Forest Life Zones, coco de mono is reported to tolerate annual precipitation of 9.1 to 22.8 dm (mean of 3 cases = 15.1 dm), annual temperature of 24.4 to 26.5°C (mean of 3 cases = 25.3°C), and pH of 5.0 to 8.0 (mean of 3 cases = 6.6).[82] Thrives along rivers in tropical forests.[278]

Cultivation — Trees are easily propagated from seeds, but never systematically cultivated.[278]

Harvesting — Flowers most profusely from April to December and produces mature fruit from December to February throughout its native range. At Summit Gardens, Panama, where it is cultivated as an ornamental, this species flowers during the wet season from April to November.[219] Like Brazil nuts, these nuts are collected from native trees when ripe. Trees may begin to bear fruit when only 2 m tall.[278]

Yields and economics — Formally, before 1968, nuts were distributed regularly, at least locally in Honduras.[278]

Energy — These relatively slow-growing trees and their prunings could serve as energy sources. Annual leaf litter from another species of *Lecythis* was nearly 2 MT/ha/year.[362]

Biotic factors — Probably pollinated by bees and disseminated by fruit bats as in *Lecythis pisonis*.

LECYTHIS OLLARIA L. (LECYTHIDACEAE) — Monkey Pod, Monkey Pot, Olla de Mona

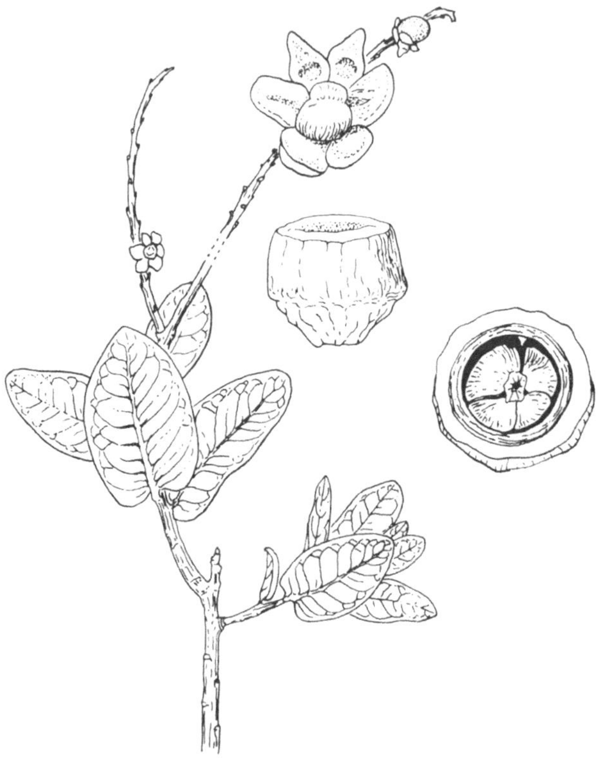

Uses — The Monkey Pot is grown and/or collected for the seeds, which are edible and are the source of an oil used for illumination and for making soap. Sap may be mixed into an agreeable drink. Wood is easy to split, strong, and polishes well. Resistant to insects, termites, and barnacles, it is used for wharves, piles, sluices, and house-framing. Bark is recommended for tanning.[278]

Folk medicine — Oil extracted from the seeds is considered a powerful hemostat.[163] Latex of the pericarp is used by South American Indians as a depilatory.[278]

Chemistry — Ingestion of the seed has associated with alopecia and selenium poisoning, as manifested by acute intoxication, fever, diarrhea, and various neurological symptoms, the active principle being the selenium analog of the sulfur amino acid cystathionine.[164] While I might try the seeds were I suffering cancer or AIDS, there are enough toxicology data to make me avoid the seeds as part of my regular diet. After prolonged exposure to active extracts or the seeds, sacrificed guinea pigs exhibited hair growth inhibition, atrophy and disappearnce of the sebaceous glands, marked atrophy of the epidermis, edema, and intraalveolar hemorrhage of the lungs, necrotic foci of the liver and spleen, and intense sinusoidal congestion of the adrenals.[163] Such symptoms might also result from experimental self-medication.

Description — Small-to-medium tree with warty branches; bark reddish-yellow, hard and heavy; wood reddish-yellow to dark-brown, very strong. Leaves sessile or subsessile, alternate, chartaceous, ovate to oblong-ovate, apex acute to obtuse, base rounded to subcordate, subserrate, the reticulate venation not prominent, 5.2 to 9 cm long, 2.5 to 5 cm wide. Inflorescence in terminal spikes, with ovate deciduous bracts. Flowers variable; sepals 6, oblong, uneven with rounded margins, concave, persistent; petals 6, larger than the sepals, spathulate, subequal, oblong, to subrounded, concave, with a reflex margin, white. Capsule pot-shaped, brown, rounded, 3.5 to 6 × 5.6 to 8.2 cm, with a 6-lobed ring-shaped, obtuse calycine ring; pericarp woody; seeds with brown covering and a yellowish oily meat.[164,265]

Germplasm — Reported from the South American Center of Diversity, monkey pot, or cvs thereof, is reported to tolerate limestone and low pH.[82] Some authors think this is conspecific with *L. minor*. Prance and Mori,[265] tabulating the differences, maintain them as distinct.

Distribution — North-central Venezuela, east of the easternmost branch of the Andes, west of the Paria Peninsula, and north of the Rio Orinoco.[265]

Ecology — Ranging from Subtropical Moist through Tropical Very Dry to Moist Forest Life Zones, monkey pot is reported to tolerate precipitation of 9.1 to 22.8 dm (mean of 4 cases = 13.3), annual temperature of 23.7 to 26.2°C (mean of 4 cases = 24.8°C), and pH of 5.0 to 7.1 (mean of 4 cases = 6.4).[82] Usually a small tree in savanna-like environments, sometimes to 20 m tall in more favorable environments.[219]

Cultivation — Trees are easily propagated from seed, in nature probably disseminated by bats.

Harvesting — Seeds harvested from wild trees as available. Extraction of oil said to be carried out by local populations.

Yields and economics — Seed collected locally and used for oil or as a food, especially by natives of northern South America.[278]

Energy — I can only speculate about these tropical trees with no real yield data. With breeding for dwarfing and improved reliability and quantity of yield, I think these trees could yield 1 to 3 MT oil per ha. Prunings, fruit husks, and leaf litter could also be captured for energy conversion.

Biotic factors — No pests or diseases reported on this plant.[278]

LECYTHIS PISONIS Cambess. (LECYTHIDACEAE) — Sapucaia

Uses — Sapucaia nuts and paradise nuts are almost contradictory terms, paradise implying a good exotic flavor, and sapucaia because, according to one interpretation, the nuts were fed to chickens by Amazonian Indians.[283] Mori is of the opinon that sapucaia is the Tupi-guorani name given to the fruit because of the wailing sound of the wind blowing across the empty open fruits.[219] Some connoisseurs consider them the finest of nuts. The kernels, eaten raw or roasted, are occasionally used to make candies or cakes. An edible oil expressed from the kernels is also used to produce soap and illumination. Since monkeys are fond of the seeds, the empty pods, with lids removed, are baited with corn to trap monkeys who can get their open hands in but have trouble getting their closed hands out. The trees could be widely planted, as they furnish fuel, food, timber, and are ornamental.[209] Still they have their detractions. Falling empty pods are dangerous to pedestrians. Trees are deciduous, so leaves must be raked after they have fallen. The fleshy flowers are also messy.[219]

Folk medicine — The oil is regarded as antipodriagic and cardiotonic. Water preserved in the fruits for 24 hr is said to remove skin blemishes.[363] While I find no anticancer data for this species, I would not hesitate to eat the seeds of the seleniferous varieties if I had cancer or AIDS. I might suffer from nausea and alopecia, side effects common with synthetic chemotherapy. Some people trek to New York to visit with an M.D. (I. Revici) who has "anti-AIDS medications" based on synthetic combinations of selenium and fatty acids or vegetable oils. I urge further testing of seleniferous Lecythidaceous fruits in the U.S. cancer screening program.

Chemistry — Rosengarten[283] suggests that the kernels contain ca. 62% fat and 20% protein. Pereira[363] says fruits contain 9% oil. Finding no more data on these *Lecythis* species, I suggest that they might be comparable in composition to Brazil nuts in component fatty acid percentages, i.e., ca. 15% palmitic-, ca. 5% stearic-, ca. 45% oleic-, and ca. 35% linoleic. Selenium content might be predicted to vary with provenance. Nuts are said to get rancid within a week or two.[209]

Description — Tree to 40 m tall, $1^1/_2$ m DBH, deciduous near the end of the dry season. Leaves simple, alternate, entire, penninerved. Flowers large, attractive, yellow to lilac or lavender or blue, sepals 6, petals 6, stamens numerous, ovary 4-locular. Pods 30 to 40 seeded, operculate. Seeds wrinkled, irregularly oblong, ca. 5 cm long, more rounded than Brazil nuts, lighter brown and with thinner shell. Kernels ivory white, with a creamy texture. Mori and Prance,[221] keenly aware of the taxonomic complexities of the group, list 10 characteristics, the combination of which uniquely identifies the "sapucaia group":

1. Large trees (at maturity they are emergents)
2. Brownish bark with pronounced vertical fissures
3. Laminated outer bark
4. Deciduous leaves which are flushed shortly before or at the same time as the flowers
5. Leaves, flowers, and fruits which possess an unidentified compound that oxidizes bluish-green when the parts are bruised
6. Hood of androecium flat with the proximal appendages anther-bearing and the distal ones antherless
7. Pollen of the hood anthers turning from yellow or white to black after 24 hr
8. Short styles with an annular expansion towards the apex
9. Large, dehiscent, woody fruits
10. Seeds with a long cord-like funicle which is surrounded by a large fleshy aril

Germplasm — Reported from the South American Center of Diversity, sapucaia and closely related species, show a rather general lack of tolerance to environmental extremes.

Such narrow tolerances seem to be characteristics of rainforest species. As defined by Mori and Prance,[221] the sapucaia group consists of three species, in addition to *L. pisonis* (incl. *L. usitata*): *L. ampla (incl. L. costaricensis*) from Nicaragua to Colombia, *L. lanceolata*, from Rio de Janeiro to Bahia, and *L. zabucaja* (incl. *L. tumefacta*), from Venezuela and the Guianas disjunctly to Central Amazonia. Many of the data in the literature on sapucaia may refer to one or the other of these.

Distribution — Common in the coastal forests of eastern Brazil and Amazonia.[221]

Ecology — Estimated to range from Tropical Moist to Wet through Subtropical Dry to Wet Forest Life Zones, sapucaia is estimated to tolerate annual precipitation of 12 to 42 dm, annual temperature of 23 to 27°C, and pH of 4 to 8. Said to occur in samll groups near hilltops in forests. The 'sapucaia' group of *Lecythis* is not found at elevations over 800 m or in the dryer savanna or *caatinga* habitats. They inhabit forests with around 2000 mm or more rainfall per year and in some areas tolerate moderate dry seasons of up to 6 months. Nevertheless, this is a typical moist-forest group which provides a good example of the effects of climatic and geological changes on the distribution and evolution of neotropical lowland trees.[221] The annual leaf litterfall of a 10-year-old stand was estimated at 1,849 kg/ha at Pau-Brasil Ecological Station, on oxisols (haplorthoxs) pH 4.5 to 5.5, annual precipitation 13 to 14 dm, annual temperature 24 to 25°C with annual amplitude of 7 to 8°C.[362] Data on the phenology and floral biology are treated by Mori et al.[220] Over 6 years in Bahia, leaf fall was mostly from September to December, flowering in October and November, and fruiting 7 months later in March and April (southern hemisphere).

Cultivation — Seeds could be planted *in situ* or in pots for transplant later.

Harvesting — Said to start bearing when 8 to 10 years old, the seeds are largely harvested from the wild, often by animals other than man. Bats are the dispersal agent. It is very difficult for man to get the seeds before bats get them.[219]

Yields and economics — Rosengarten[283] quotes estimates of 70 kg nuts per tree. Nonetheless, there are no large plantations, only a few small plantings in Brazil, the Guianas, the West Indies, and Malaysia. The fact that the fruit is dehiscent, exposing the delicious nuts to the nut-eating animals and birds, makes this much less attractive than its relative, the Brazil nut, for commercial exploitation. Dwarfed cvs, which might be bagged for protection from predators, might make the sapucaia a more attractive commercial possiblity.

Energy — Assuming 50 kg nuts per tree and 100 trees per ha (they may bear quite precociously) and 60% oil, there is an incredible 3 MT oil per hectare, if you could capture it all. This edible oil could be used for fuel, if fuel were more valuable than food, and the press-cake, if non-seleniferous, could be used for food or feed. Prunings from the trees, as well as the husks, might be used for fuel. As Periera[363] notes, the dry fruits serve as fuel. Leaf litter alone approaches 2 MT/ha/hr.

Biotic factors — Mori and Prance[221] found that the carpenter bee, *Xylocopa frontalis*, is a regular visitor to the flowers. It transports two types of pollen from the flower, viable pollen from the staminal ring and nonviable pollen (the reward) from the hood of the flowers. Nonviable pollen is collected and placed in pollen baskets; viable pollen, deposited on head and back, causes fertilization.[219] Bats, monkeys, parrots, and peccaries probably obtain most of the production.[209,219,283] In Trinidad, bats (*Phyllostomus hastatus*) remove the seeds, dropping them after eating the aril, either in flight or under their roosts. Bats are the main dispersers.[219]

LICANIA RIGIDA Benth. (ROSACEAE) — Oiticica
Syn.: *Pleragina umbresissima* Arruda Camara

Uses — Seeds of this tree are the source of Oiticica Oil, a drying oil used in place of tung oil for varnishes and protective coatings. Trees are sometimes grown as shade trees in villages where the plants are native. Timber sometimes used in construction.[11,132,192]

Folk medicine — No data available.

Chemistry — Hilditch and Williams[128] indicate that the seed fat contains 61% alpha-licanic acid (4-keto-9,11,13-octadecatrienic acid) ($C_{18}H_{28}O_3$) and 17% alpha- elaeostearic acids. Licanic acid is unique among natural fatty acids in containing a ketonic group. According to Vaughan,[325] the oil most closely resembles tung oil in chemical and physical properties. The oil cake contains 9% protein, but contains so much tannin and residual oil as to be unsuitable for animal feed. Hager's Handbook[187] puts the oil content of the whole fruit at 33 to 45%, the kernels at 49 to 65%. Of this, 70 to 82% is alpha-licanic acid, 4 to 12% oleic-, up to 4% linoleic-, 10 to 11% palmitic-, and stearic- and isolicanic-acid. Myricetin is also reported. Here we have no exception to disprove the rule. In general, tropical oilseeds have higher percentages of saturated fatty acids, compared to their temperate counterparts. In the Rosaceae, seed fats of tropical genera have about 10% saturated fatty acids, temperate genera about 5%. The tropical oils hence become rancid more rapidly.[128,131]

Description — Small tree to 15 m tall, with spreading crown, the young branches lanate to tomentellous, soon becoming glabrous and lenticellate. Leaves oblong to elliptic, 6.0 to 13.0 (to 16.0) cm long, 2.8 to 6.5 cm broad, coriaceous, rounded to cordate at base, glabrous and shining on upper surface, the lower surface with deeply reticulate venation quite or nearly describing stomatal cavities, with lanate pubescence among but not on veins; midrib prominulous above, puberulous toward base when young; primary veins 11 to 16 pairs, prominent on lower surface, prominulous above; 5.0 to 8.0 cm long, tomentose when young, becoming glabrescent with age, terete, with two sessile glands. Stipules linear, to 10.0 mm long, membranous, caducous. Inflorescenes racemose panicles, the rachis and branches gray-tomentose. Flowers 2.5 to 3.5 mm long, in small groups, sessile on primary branches of inflorescence. Bracts and bracteoles 1.5 to 2.5 mm long, ovate, tomentose on exterior, persistent, entire to serrulate, eglandular. Receptacle campanulate, gray-tomentose on exterior, tomentose within; pedicels to 0.5 mm long. Calyx lobes acute, tomentose on exterior, tomentellous within. Petals 5, densely pubescent. Stamens ca. 14; filaments equalling calyx lobes, connate to about half-way from base, densely pubescent. Ovary attached to base of receptacle, villous. Style equalling calyx lobes, villous nearly to apex. Fruit elliptic, 4.0 to 5.5 cm long; epicarp smooth, drying green or black; mesocarp thin, fleshy; endocarp thin, fibrous, fragile, fibers arranged longitudinally promoting longitudinal dehiscence, sparsely pubescent within. Germination hypogeal.[263]

Germplasm — Reported from the South American Center of Diversity, oiticica, or cvs thereof, is reported to tolerate drought. Some efforts have been made to develop high-yielding strains which can be propagated vegetatively. The number of native trees is limited by their habitat requirements and cannot be increased to meet increasing demands for oil.[82,278]

Distribution — Dry forests and gallery forests of northeastern Brazil. According to Prance,[263] this species is cultivated outside its natural range, e.g., in Trinidad, "but is not used commercially outside Brazil." This tree is confined primarily to the arid regions of northeastern Brazil, including the states of Ceara, Rio Grande de Norte, Bahia, Piaui, Maranhao, Paraiba, and northern Pernambuco. Introduced to Trinidad and a few other regions with similar ecological conditions.[278]

Ecology — Oiticica trees thrive on dry tropical lowlands where there is a dry season from July to December and where the annual rainfall varies from ca. 9 to 14 dm.[82] It is often found in dry open grasslands bordering rivers. Plantations should be put on well-drained, alluvial, fertile soils, rich in potash, with a pH of about 7.0. The average temperature should be 31.7 to 32.9°C.[278] Markley[200] suggests that it is especially common along the banks of rivers, said to form dense groves in rich alluvial soils.[325]

Cultivation — Propagation is by seed, grafting, and budding. Seeds lose their viability soon after ripening, seeds 6 months old having lost most of their viability. Best growth is obtained when the seeds are sown in well-watered, good alluvial soils, in a nursery. Seedlings are about 17 cm tall in 60 days. The nursery should be irrigated and deeply cultivated. Transplants are set 0.5 m apart in rows 1 m apart and irrigated every 10 to 15 days during the dry season. Four months after transplanting (when the seedling is about 6 months old) seedlings average nearly a meter tall. Stocks are grafted when 5 to 7 months old. Several methods of grafting, including inarching and budding, have been tried, with budding being most practicable, because of the difficulty in transporting stocks when inarching. Buds sprouted in 25 to 80 days after grafting, mostly in 24 to 40 days. The period between sowing of seed and final setting of the grafted tree in the orchard is about 22 months, depending upon the time of the rainy season.[278]

Harvesting — Usually 3 years after the beginning of nursery work or 2 years and 3 months after grafting, about 12% of the trees were found to flower and set fruit. Then the trees continue to bear for many years, some estimate as long as 75 years. Ripened fruits fall to the ground or are knocked off by shaking the trees. They are collected by men, women, and children, and delivered to local warehouses. Extracting companies maintain collection stations at the end of or along the few available roads or railroads in the regions where the nuts grow native. After the refining companies receive the fruits, they ship them to larger warehouses or the extracting companies where the fruits are cleaned and prepared for processing. Seed (kernel) is easily removed from the husk and the oil obtained by pressure alone, or by pressure plus action of solvents. Because of its unpleasant odor and semisolid state, its uses will be greatly restricted until means are found for refining it and keeping it in a liquid state. After pressing, the oil is transported to the refinery. Harvesting is from December through April.[116,278] As Vaughan[325] puts it, "From December to March, the fruits fall to the ground and are collected."

Yields and economics — Having seen no published yield data on this tree, I estimate that in good years a tree may drop 2 to 3 MT fruits per hectare, suggesting potential kernel yields of 1,200 to 3,000 kg, and oil yields of 700 to 1,800 kg/ha. Concerning the oil yields, the following data may be helpful: average weight per nut = 2.27 to 4.7 g; average percent of kernel per nut = 58 to 70%; average percent of oil per kernel = 52.9 to 60%.[278] Felling the tree and exporting seed are prohibited. Brazil has the monopoly on production of Oiticica Oil, producing annually ca. 20,000 MT, this amount fluctuating from year to year. Vaughan[325] suggested an annual seed production of 54,000 MT. In 1941, Brazil produced 18 to 19 MT, exporting more than 16 MT.[263] Oiticica oil must compete with tung, dehydrated castor oil, and in some cases, with linseed oil. Around 1957, the industry was centered in Ceara, where 14 of the 20 processing mills were located. The largest mill, Brazil Oiticica S.A., had a reported crushing capacity of 3,500 tons per month, mostly oiticica and cashew.[200]

Energy — Prunings and falling biomass from large trees like this could easily add up to 5 to 10 MT/ha. Seed yields should be higher than those of temperate tree members of the Rosaceae, e.g., almond. The press-cake, because of a relatively toxic reputation, might be better for fuel than for food.

Biotic factors — Fertilization of the flowers is by means of insects, but a large number of buds drop before opening or without setting fruit. It has been estimated that for a tree to set 150,000 seeds (458 kg), it would have to bear 12 million buds.[278]

MACADAMIA INTEGRIFOLIA Maiden & Betche, *MACADAMIA TETRAPHYLLA* L. Johnson (PROTEACEAE) — Macadamia Nuts, Australian Nuts

Uses — Macadamia nuts are eaten raw or, after cooking in oil, are roasted and salted; also used to make an edible bland salad oil. Rumsey[286] recommends it also as a timber tree and ornamental. Years ago a coffee-like beverage known as "almond coffee" was marketed from the seeds.[278,316]

Folk medicine — No data available.

Chemistry — Per 100 g, the nut is reported to contain 691 calories, 3.0 to 3.1 g H_2O, 7.8 to 8.7 g protein, 71.4 to 71.6 g fat, 15.1 to 15.9 g total carbohydrate, 2.5 g fiber, 1.7 g ash, 48 mg Ca, 161 mg P, 20 mg Fe, 264 mg K, 0 mg β-carotene equivalent, 0.34 mg thiamine, 0.11 mg riboflavin, 1.3 mg niacin, and 0 mg ascorbic acid.[89] According to MacFarlane and Harris,[195] the oil is high in monounsaturates (79%) and palmitoleic acid (16 to 25%). The composition ranges from 0.1 to 1.4% lauric, 0.7 to 0.8 myristic, 8.0 to 9.2 palmitic, 15.6 to 24.6 palmitoleic, 3.3 to 3.4 stearic, 54.8 to 64.2 oleic, 1.5 to 1.9 linoleic, 2.4 to 2.7 arachidic, 2.1 to 3.1 eicosenoic, and 0.3 to 0.7% behenic acids. The oil-cake contains 8.1% moisture, 12.6% oil, 2.6% crude fiber, 33.4% crude protein, and 43.3% N-free extract.

Description — *Macadamia integrifolia:* trees up to 20 m tall, with spread of 13 m. Leaves opposite in seedlings, later in whorls of 3, pale-green or bronze when young, 10 to 30 cm long, margins with few or no spines, petioles about 1.3 cm long. Flowers creamy white, petalless, borne in groups of 3 or 4 along a long axis in racemes, much like grapes. Fruit consisting of a fleshy green husk enclosing a spherical seed; nuts round or nearly so, surface smooth or nearly so, 1.3 to 2.5 cm in diameter; shell tough, fibrous, difficult to crack; kernel white, of uniform quality, shrinking only slightly after harvesting. Flowers June through to March, some strains almost ever-bearing, flowering while fruiting.

Macadamia tetraphylla: trees up to 20 m tall, with spread of 13 m. Leaves opposite in seedlings, commonly in fours rarely in threes or fives, purple or reddish when young, margins serrate with many spines, up to 50 cm long, sessile or on very short petioles. Flowers pink, in large racemes. Fruit consisting of a fleshy green husk enclosing one seed; nuts usually elliptical or spindle-shaped, surface pebbled; kernel grayish; variable in quality and shrinking some after harvest. Flowers between August and October, producing one main crop. Between these two distinct types are numerous intermediate forms varying in spininess of leaves, color of flower, size of nut and thickness of shell.[278]

Germplasm — Reported from the Australian Center of Diversity, macadamias or cvs thereof are reported to tolerate drought, slope and wind.[82] Since 1956, *Macadamia integrifolia* (smooth-shelled type) and *Macadamia tetraphylla L.* (rough-shelled type) are the names properly applied to the cultivated Macadamia nuts. Prior to this time they had been generally referred to *Macadamia ternifolia.* F. Muell. is a distinct species, bearing small, bitter, cyanogenic seeds less than 1.3 cm in diameter, inedible and never cultivated. Many cultivars have been developed, and grafted trees of promising selections have been made. Three cvs of *M. integrifolia,* 'Kakea', 'Ikaika' and 'Keauhou', have been planted extensively in Hawaii, all giving satisfactory production under favorable conditions. 'Keaau' has been more recently recommended for commercial planting in Hawaii, since it is highly resistant to wind and yields 5 to 10% more than previous cvs, the entire crop maturing and dropping before the end of November. Most of the Australian crop is based upon *M. tetraphylla,* with some orchards of grafted *M. integrifolia* cvs. Among the medium- to thick-shelled selections, used mainly for processing, are 'Richard', 'Tinana', 'Our Choice' and 'Hinde'. Rough-shelled types, mostly grown for table purposes, are 'Collard', 'Howard', 'Sewell' and 'Ebony'. Cvs showing hybrid characteristics are 'Oakhurst' and 'Nutty Glen'. 'Teddington' is a hybrid with thin shell.[278]

Distribution — Native to coastal rain-forests of central east Australia (New South Wales and Queensland). Introduced in other parts of tropics, e.g., Ceylon, and commercially grown in Costa Rica, Hawaii, and France, at medium elevation.[278]

Ecology — Ranging from Warm Temperate Dry (without frost) through Tropical Moist Forest Life Zones, macadamias are reported to tolerate annual precipitation of 7 to 26 dm, annual temperature of 15 to 25°C, and pH of 4.5 to 8.0.[82] Macadamia grows best in rain-forest areas, along coasts with high humidity and heavy rainfall. However, it is tolerant of adverse conditions when once established. Inland crops are usually lighter than coastal crops. Trees produce a deep taproot and relatively few lateral roots; therefore, they may need windbreaks in exposed areas. Under orchard conditions, trees are shapely, robust, and more heavily foliaged than they are in rain-forest. Grows well on a wide range of soils, but fails on infertile coastal sands, heavy clays, or gravelly ridges. Yields well on deep, well-drained loams and sandy loams. Slopes steeper than 1 in 25 should be planted on the contour, and every precaution taken to prevent soil erosion.

Cultivation — Propagation by seed is not difficult, but seedlings are variable in production and nut characteristics, and of little value for commercial plantings. Freshly harvested nuts are best for germination, but require 30 to 90 days before germination. Propagation is usually by cuttings, marcottage, and side-tongue grafts. Root-stocks for grafting are readily grown from seed by ordinary nursery means. Grafting in Macadamia is more difficult than in most nut trees, due to hardness of wood. Best results are obtained when seedling root-stocks are side-wedge grafted with selected scions. After-care of graft is similar to that practiced in other trees. Budding is much less satisfactory than grafting. The most suitable time for transplanting young trees to orchard is from February to April in Australia and in Hawaii, when rainfall is good and sufficient soil moisture available. Taproot should be severed about 30 cm below ground about 6 weeks before time to transplant, to allow fibrous roots to develop. Roots are very susceptible to exposure and should not be allowed to dry out. Grafted trees should be planted with the union well above ground level and watered immediately. Since trees have a tendency to grow tall, young trees, when about 75 cm tall, should be topped little by little to produce a few evenly spaced limbs, thus developing a strong, rounded symmetrical tree. Little pruning is required in bearing trees except to discourage leaders, to reduce lateral growth, to let in light, and to make cultural and harvesting operations more favorable. Pruning should be done toward the end of winter after the crop is harvested. Macadamia grows best in soils with a good supply of humus. Farm-yard manure may be added, and green manure crops can be grown between trees in summer. Under orchard conditions, regular applications of fertilizer are required, as a 8:10:5 formula, at a rate of .45 kg per tree per year of age, maximum of 4.5 kg. Fertilizer should be applied in early spring just before trees make new growth and start flowering. Zinc deficiencies seem to be a problem with this tree — the symptoms being small, yellowish or slightly mottled leaves which are bunched together, crop retardation, and poor shoot growth. The condition corrected by application of foliar spray in early spring after the first flush of growth, at a rate of 4.5 kg zinc sulfate, 1.3 kg soda ash (or 1.7 kg hydrate lime) in 100 gal water. However, spray is effective at any period of year if symptoms are obvious. Since root system is rather close to surface, shallow cultivation for weed control should be practiced. Summer cover crops, e.g., cowpeas, and autumn green manure crops may be grown between trees until harvest time. Grazing cattle on weeds and grass in orchards has the advantage of adding animal manure.

Harvesting — Nuts mature 6 to 7 months after flowering and should be allowed to ripen on the trees. Usually the nuts fall to the ground when mature. In some cvs, nuts remain on trees and must be removed with rakes. After harvesting, nuts are dehusked, usually with an improvised corn-sheller, washed, placed on wire trays for about 6 weeks to dry out, graded, and shipped to market. Machinery for cracking shells has been designed for processing purposes, in addition to several efficient hand-operated crackers, which produce a kernel undamaged. Kernels which are broken during cracking are used by confectioners. Shelled kernels deteriorate rather quickly unless kept in vacuum-sealed jars. Processed nuts when roasted and slightly salted keep extremely well.[118,240]

Yields and economics — Most trees begin bearing fruit at 6 to 7 years, while other trees must be 10 to 15, vegetatively propagated trees bearing earlier. Yield records vary widely, depending on strain characteristics and environmental factors. Macadamia has great commercial potential in the tropics and makes an excellent door-yard tree. In addition to production of nuts in Australia, production in Hawaii in 1970 amounted to 5750 tons. Presently, production is being developed in South Africa, Paraguay, Costa Rica, Jamaica, Samoa, and Zimbabwe.[158,278]

Energy — According to Saleeb et al.,[290] nuts of *M. integrifolia* and *M. tetraphylla* are equal in oil content, with an iodine value of 75.4 and 71.8, respectively. They describe a method for partially extracting the oil (6 to 14% of the weight of intact oven-dry kernels), rendering them more attractive, digestible, and less fattening, while diverting 14% of the weight to oil production. In Australia yields are estimated at about 45 kg per tree annually; in Hawaii, at 135 kg per tree. New cultivars are known to yield as much as 3.75 tons/ha, averaging 1 ton of kernels, which should contain more than 700 kg oil/ha renewably (oil makes up 65 to 75% of the kernel).

Biotic factors — Macadamia trees are attacked by *Gloeosporium sp.* (Blossom blight) and *Macrophoma macadamiae*. Nematodes isolated from trees include: *Helicotylenchus dihystera, Rotylenchus erythrinae,* and *Xiphinema americanum*. In Hawaii, the Southern green stink-bug is a serious problem, damaging about 10% of the seed.[217]

MADHUCA LONGIFOLIA (L.) Macbr. (SAPOTACEAE) — Mahua, Illupei Tree, Mawra Butter Tree
Syn.: *Madhuca indica* J. F. Gmel., *Bassia longifolia* L.

Uses — Mahua is valued for its edible flowers and oil-bearing seeds. Fresh flowers are extremely sweet, less so when dried, having a flavor resembling that of figs. Rich in vitamins, the flowers are eaten fresh or dried and cooked with rice, grains or shredded coconut, fried or baked into cakes, or ground into flour and used in various foodstuffs. A large portion of the crop of flowers is made into syrup containing ca. 60% sugar, suitable for making jams, sweetmeats, or as a honey substitute, for production of alcohol (with average yields of 90 gals of 95% alcohol per ton of dried flowers), for making vinegar, or distilled liqueurs and wine. Molasses sugar of good quality is made from mahua. Syrup is used by natives of Bastar (in Madhya Pradesh) instead of brown sugar. Flowers, and spent flowers after fermentation, are used as feed for livestock. The flesh of animals fed on mahua flowers has a delicate flavor. Pressed cake of corollas is used as fertilizer. Mahua cake has insecticidal and piscicidal properties. Because the saponin present in it has a specific action against earthworms, it is applied to lawns and golf greens. Used, along with *Acacia concinna,* as a hairwash in South India. Seeds, with 50 to 60% fat content, are the source of Mahua Oil or Tallow Mawra Butter, used for manufacturing soaps and candles, and when refined, used as butter. Oil has poor keeping quality. Used for edible and cooking purposes in some rural areas. Refined oil is also used in the manufacturer of lubricating greases and fatty alcohols, and as a raw material for the production of stearic acid. Wood is durable, lasting exceptionally well under water, planes well, and takes a good finish, but is difficult to saw, and has a tendency to split or crack. Wood is used for building purposes, as door and window frames, beams, and posts, furniture, sports goods, musical instruments, oil and sugar presses, boats and ship-building, bridges, well construction, turnery, agricultural implements, drums, carving, and has been tried for railway sleepers. The bark contains 17% tannin and is used for dyeing and tanning. Mahua berries are eaten raw or cooked, and are eaten by cattle, sheep, goats, monkeys, and birds. Sometimes used as green manure.[24,70,278]

Folk Medicine — According to Hartwell,[126] the flowers are used in folk remedies for abdominal tumors. Reported to be anodyne, antidote, astringent, bactericide, carminative, demulcent, emetic, emollient, expectorant, insecticide, lactagogue, laxative, piscicide, refrigerant, stimulant, and tonic, mahua is a folk remedy for bee-sting, bilious conditions, blister, blood disorders, breast ailments, bronchitis, cachexia, cholera, colds, consumption, cough, diabetes, dysuria, ear ailments, eye ailments, fever, fistula, gingivitis, headaches, heart problems, intestinal ailments, itch, leprosy, orchitis, phthisis, piles, pimples, rheumatism, skin ailments, smallpox, snakebite, suppuration, tonsillitis, tuberculosis, tumors of the abdomen, and wounds.[91] The gummy juice is used for rheumatism, the bark decoction as an astringent and emollient, and as a remedy for itch; root, bark, leaves, and flowers for snakebite, the flowers for scorpion sting.[165] Mahua is considered to be astringent, stimulant, emollient, demulcent, and nutritive in Ayurvedic medicine. Bark used to treat rheumatism, ulcers, itches, bleeding and spongy gums, tonsillitis, leprosy, and diabetes. The emollient oil is used in skin diseases, rheumatism, bilious fevers, burning sensations, headaches; being laxative, it is useful in habitual constipation, piles, and hemorrhoids; and is used as an emetic. Used in winter for chapped hands. Roots are applied to ulcers, bleeding tonsillitis, rheumatism, diabetes mellitus, and spongy gums. Medicinally, flowers are reported to be cooling, aphrodisiac, demulcent, galactagogue, expectorant, nutritive, tonic, and carminative, are considered to be beneficial in heart diseases, bronchitis, coughs, wasting diseases, burning sensation, biliousness, and ear complaints; dried flowers used as a fomentation in orchitis. Fried flowers are eaten by people suffering from piles. Mahua flowers show antibacterial activity aginst *Escherichia coli*. The edible honey from the flowers is reported to be used for eye diseases. Liquor made from the flowers used as an astringent and a tonic. Mahua leaves are astringent, used in embrocations. Fruit used for bronchitis, consumption, and blood diseases; seeds are galactagogue.[24,70,165,278]

Chemistry — Per 100 g, the inflorescence (ZMB) is reported to contain 5.0 g protein, 1.8 g fat, 89.0 g total carbohydrate, 1.6 g fiber, 4.2 g ash, 130 mg Ca, and 120 mg P. Per 100 g, the leaf (ZMB) is reported to contain 9.1 g protein, 3.9 g fat, 79.4 g total carbohydrate, 19.0 g fiber, 7.6 g ash, 1460 mg Ca, and 210 mg P.[89] An insoluble gum from incisions on the trunk contains 48.9% gutta, 38.8% resin, and 12.3% ash. Bark contains 17% tannin.[278] The wood contains naphthaquinone, lapachol, and alpha- and beta-lapachones; the essential oil from the fruit pulp contains ethyl cinnamate, alpha-terpineol, and a sesquiterpene fraction. Myricetin and myricetin-3-O-L-rhamnoside has been isolated from the leaves.[24] In addition, *The Wealth of India*[70] reports 51.1% fatty oil, 8.0% protein, 27.9% N-free extract, 10.3% fiber, and 2.7% ash in an analysis of the seed kernel. Senaratne et al.[302] report the fatty acid components of the seed oil to be 23% palmitic, 15% stearic, 46% oleic, 14% linoleic, and traces of linolenic acids. The glyceride structure of the oil is reported to be 1% dipalmitostearins, 1% oleo-dipalmitins, 27% oleo-palmitostearins, 41% palmito-dioleins, and 30% stearodioleins. The *Wealth of India*[70] reports the values are trace trisaturated, 47% monounsaturated-disaturated, 36% mono-saturated-diunsaturated, and 17% tri-unsaturated. Per 100 g, the corollas are reported to contain 18.6% moisture, 4.4% protein, 0.5% fat, 72.9% total sugars, 1.7% fiber, 2.7% ash, 140 mg P, 140 mg Ca, and 15 mg Fe; magnesium and copper are present. The sugars are identified as sucrose, maltose, glucose, fructose, arabinose, and rhamnose. Corollas also contain 39 IU carotene, 7 mg ascorbic acid, 37 μg thiamine, 878 μg riboflavin, and 5.2 mg niacin per 100 g. Folic acid, pantothenic acid, biotin, and inositol are also present. Corollas also contain an essential oil, anthocyanins, betaine, and salts of malic and succinic acids. The ripe fruits, per 100 g, are reported to contain 73.64% moisture, 1.37% protein, 1.61% fat, 22.69% carbohydrates, 0.69% mineral matter, 45 mg Ca, 22 mg P, 1.1 mg Fe, 512 IU carotene, and 40.5 mg ascorbic acid; tannins are present. The oil contains 22.7% ethyl cinnamate, 3.5% alpha-terpineol, and 67.9% sesquiterpene and sesquiterpene alcohol. The green leaves contain 78.95% moisture, 19.60%

organic matter, 0.43% N, 1.45% mineral matter, 0.43% potash (K_2O), 0.087% phosphoric acid (P_2O_5). and 0.10% silica. Analysis of samples of coagulum from incisions made in the bark show 12.2 to 19.9% caoutchouc, 48.9 to 75.8% resin, and 11.9 to 38.9% insolubles.[24,70,278,302]

Toxicity — According to Burkill,[56] there is a saponin or sapo-glucoside in the seeds which has a destructive action on the blood. Awasthi et al.[24] report the presence of a bitter glucosidic principle from mahua seed that was shown to possess digitalis-like action on frog heart. Over-consumption of mahua flowers is reported to cause vomiting and stomach disorders.;2[4]

Description — Large deciduous tree, 13 to 17 m tall, with a short trunk and numerous spreading branches forming a dense rounded crown. Leaves elliptic to linear-lanceolate, 8 to 20 cm long, 3 to 4.5 cm wide, tapering to base, glabrous when mature, clustered at ends of branches. Flowers small, in dense clusters of 30 to 50 at ends of branches; corolla tubular, 1.5 cm long, yellowish to cream-colored, thick, fleshy, globe-shaped, enclosed at the base in a velvety chocolate-brown calyx. Fruit an ovoid berry up to 5 cm long, yellow when ripe. Seeds 1 to 4, yellow to brown, ovoid, shining, 2.5 to 3 cm long, kernel about 70% by weight of seed and containing 35 to 40% of a greenish grease (fat-oil). Trees shed their leaves in February, and flowers appear in March and April, at which time the ground beneath the trees is carefully cleared. Flowers March to April; fruits May to June.[278]

Germplasm — Reported from the Hindustani Center of Diversity, mahua, or cvs thereof, is reported to tolerate drought, frost, insects, poor soil, slope, savanna, and waterlogging.[82] According to *The Wealth of India,*[70] van Royen revised the taxonomy and nomenclature of the genus *Madhuca* of the Malaysian area. He merged *M. indica* and *M. longifolia* under the latter name and distinguished two varieties, var. *longifolia* and var. *latifolia.*[70]

Distribution — Native to southern India. Although it grows spontaneously in some parts, it is extensively cultivated throughout India and Sri Lanka.[278]

Ecology — Ranging from Warm Temperate Moist through Tropical Very Dry to Moist Forest Life Zones, mahua is reported to tolerate annual precipitation of 7.0 to 40.3 dm (mean of 4 cases = 17.7), annual temperatures of 24.2 to 27.5°C (mean of 4 cases = 25.4°C), and pH of 5.0 to 7.5 (mean of 3 cases = 6.6).[82] Mahua, usually drought-resistant, is especially suited for dry or waste lands where little else will grow. Trees thrive on dry, stony ground in all parts of India, and are protected by the natives. Trees are frost-hardy, but do suffer from severe conditions. It is sometimes found in waterlogged or low-lying clayey and shallow soils. Requires full sun and is readily suppressed by shade. When cut in dry season, plants coppice well, but not during the rainy season.[278]

Cultivation — In southern India, trees are frequently cultivated as an avenue tree. Seeds may germinate naturally during the rainy season, soon after falling, where earth is washed into small hollows. Subsequent growth is slow, but is favored by sunlight. For artificial propagation, seeds are sown directly or for transplant. Fresh seeds are sown in July and August, in prepared lines or patches. Transplanting may be risky due to the long, delicate taproots. In India, seeds are sown directly in deep containers or the seedlings transplanted into them from the nursery during the first rainy season a few weeks after germination. Young trees are frequently intercropped with annual crops, at least during the first 10 to 15 years.[278]

Harvesting — Under favorable soil and climatic conditions, mahua trees begin to bear fruit in 8 to 10 years after planting, and continue to do so for over 60 years. Corollas fall in great showers in early morning to the previously cleaned ground, from about the end of March until the end of April. They are collected by women and children and spread out on mats to dry in the sun, shrinking to about one-half their weight and turning reddish-brown. Sometimes flowers are collected before they drop; in some places it is the practice to remove only the corolla, leaving the pistil to ripen to a fruit. Harvest period is 7 to 10 days for a single tree. Flowers, when dried, are sold to distilleries, where they are immersed in water for about 4 days, allowed to ferment and thereafter distilled. The spirit, somewhat similar to Irish whiskey, has a strong, smokey, and rather fetid flavor, improved by aging, producing a strong palatable drink. One ton of dried flowers produces ca. 90 gal of 95% ethyl alcohol. Fruits may occur in alternate years. Fruits fall from tree when ripe or may be dropped by shaking the branches. Season for collecting is short, from May to June, but may be extended until December in southern India. Seeds are separated from the smooth chestnut-brown pericarp by bruising, rubbing, or subjecting them to moderate pressure. Then they are dried and shelled to get the kernel, these constituting the Mahua seed of commerce. Mahua oil is extracted by cold expression; the yield of oil, depending on the efficiency of equipment, varies from 20 to 43%, the highest gotten when extracted by solvents. In Central India, kernels are pounded, boiled, wrapped in several folds of cloth, and then the oil is expressed. Fresh Mahua Oil from properly stored seeds is yellow with a disagreeable odor. In warmer areas, the oil is a liquid; in cold weather or areas, it solidifies to a buttery consistency. Mahua cake from seeds is used as a manure, alone or mixed with mineral fertilizers, or made into a compost with sawdust, cane trash, or bagasse, about 3 months being required for nitrification of the cake. Quantities (1,000 to 1,750 tons) of this compost are exported from India to Sri Lanka and Britian annually. Mahua cake also has insecticidal and piscicidal properties, and is applied to lawns and golf courses against earthworms.[278]

Yields and Economics — Trees require about 20 years to attain full production of flowers and seeds; an average tree producing from 90 to 125 kg of flowers per year. Mahua is essentially a forest crop. Still, the total amount of seeds collected in the forest is less than from trees in semi-cultivated areas. An estimated 7 million trees in India produce about 100,000 tons of seed per year. India is the principal producer of all products of mahua, and the bulk of the crop is consumed locally. Some products are exported to Belgium, Germany, France, and Britain. Indian mills convert 15,000 to 30,000 tons of seeds into oil annually.[278]

Energy — A good fuelwood, it is hard and heavy, specific gravity approximately 0.95 to 0.97. Pruning, perhaps amounting to 2 to 4 MT ha, could be used for firewood. Sapwood has a calorific value of 4,890 to 4,978 calories (8,802 to 8,962 Btu); heartwood, 5,005 to 5,224 calories (9,010 to 9,404 Btu). Seed oil (20 to 43%) could be used for diesel substitution, the press-cake converted to power alcohol. Assuming 100 trees per ha and 100 kg flowers per tree, one might expect 900 gallons (>20 barrels) ethanol per hectare.

Biotic factors — Trees are damaged by loranthaceous parasites. Mahua trees are affected by several fungi: *Scopella (Uromyces) echinulata* (rust), *Polystictus steinheilianus* (white spongy rot), *Fomes caryophylli* (heart rot of stems), and *Polyporus gilvus* (root and butt rot). Leaves are eaten by caterpillars: *Achaea janata, Anuga multiplicans, Bomboletia nugatrix, Metanastria hyrtaca,* and *Rhodoneura* spp.; *Acrocercops* spp. (leaf-miners); the bark is destroyed by *Odonotermes obesus, Coptotermes ceylanicus,* and *Kalotermes* sp. (white ants) and *Xyloctonus scolytoides* (bark borers); sapwood of dead trees is damaged by *Schistoceros anabioides* and *Xylocis tortilicornis* (ghoon borers).[278] Also attacked by the sapsucker *Unaspis acuminata.*[70] In addition, Browne[53] lists Angiospermae: *Dendrophthoe falcata;* Lepidoptera: *Ophiusa janata.*

MORINGA OLEIFERA Lam. (MORINGACEAE) — Horseradish-Tree, Benzolive Tree, Drumstick-Tree, Sohnja, Moringa, Murunga-Kai
Syn.: *Moringa pterygosperma* Gaertn., *Moringa nux-ben* Perr., *Guilandina moringa* L.

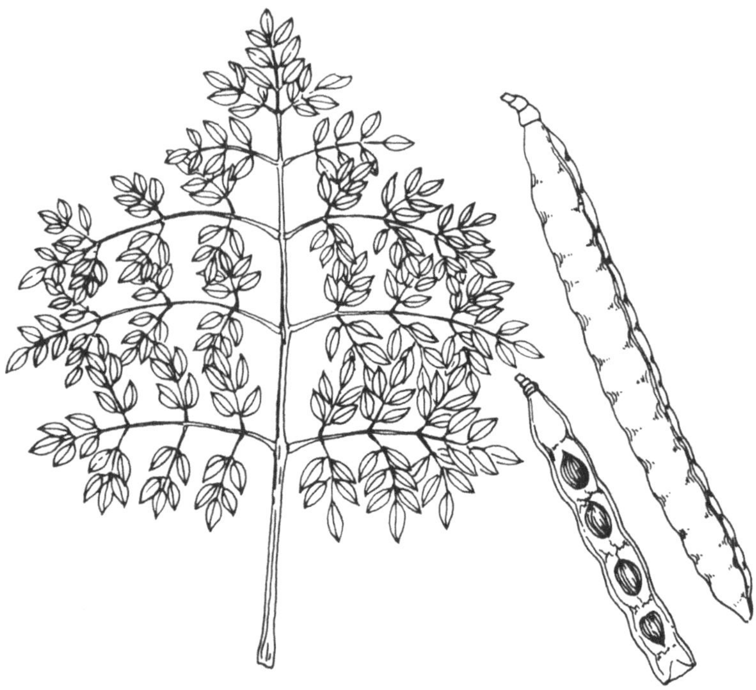

Uses — Described as "one of the most amazing trees God has created".[383] Almost every part of the *Moringa* is said to be of value for food. Seed is said to be eaten like a peanut in Malaya. Thickened root used as substitute for horseradish. Foliage eaten as greens, in salads, in vegetable curries, as pickles and for seasoning. Leaves pounded up and used for scrubbing utensils and for cleaning walls. Flowers are said to make a satisfactory vegetable; interesting particularly in subtropical places like Florida, where it is said to be the only tree species that flowers every day of the year. Flowers good for honey production. Young pods cooked as a vegetable. Seeds yield 38 to 40% of a nondrying oil, known as Ben Oil, used in arts and for lubricating watches and other delicate machinery. Haitians obtain the oil by crushing browned seeds and boiling in water. Oil is clear, sweet and odorless, said never to become rancid (not true, according to Ramachandran et al.).[384] It is edible and used in the manufacture of perfumes and hairdressings. Wood yields blue dye. Leaves and young branches are relished by livestock. Commonly planted in Africa as a living fence (Hausa) tree. Ochse[238] notes an interesting agroforestry application; the thin crown throws a slight shade on kitchen gardens, which is "more useful than detrimental to the plants". Trees planted on graves are believed to keep away hyenas and its branches are used as charms against witchcraft. In Taiwan, treelets are spaced 15 cm apart to make a living fence, the top of which is lopped off for the calcium- and iron-rich foliage.[383] Bark can serve for tanning; it also yields a coarse fiber. Trees are being studied as pulpwood sources in India. Analyses by Mahajan and Sharma[385] indicate that the tree is a suitable raw material for producing high alpha-cellulose pulps for use in cellophane and textiles. In rural Sudan, powdered seeds of the tree *Moringa oleifera* are used to purify drinking water by coagulation. In trials, the powder was toxic to guppies (*Poecilia reticulata*), protozoa (*Tetrahymena pyriformis*), and bacteria (*Escherichia coli*), and it inhibited acetylcholinesterase. It had no effect on coliphages, lactic dehydrogenase, or invertase, and the equivalent of cotyledon powder up to 1000 mg/liter had no mutagenic effect on salmonella. Pericarp had no effect. Powdered cotyledon at 5 mg/liter affected oxygen uptake of *T. pyriformis,* 30 to 40 mg/liter disturbed locomotion of guppies, and the 96-H LC_{50} for guppies was 196 mg/liter. Toxic effects may have been due to 4(alpha-1-rhamnosyloxy) benzyl isothiocyanate, a glycosidic mustard oil. The toxin seemed not to be a danger to the health of man, at least not in the concentrations present during the use of the seeds for nutrition, medicine, or water purification.[386] For the low-turbidity waters of the Blue Nile, only a quarter seed per liter of water is required, for moderately turbid water, half a seed, and for fully turbid, 1 to 1.5 seeds per liter. Such seed are hulled, crushed, and reduced to a powder.[387]

Folk Medicine — According to Hartwell,[126] the flowers, leaves, and roots are used in folk remedies for tumors, the seed for abdominal tumors. Reported to be abortifacient, antidote (centipede, scorpion, spider), bactericide, cholagog, depurative, diuretic, ecbolic, emetic, estrogenic, expectorant, purgative, rubefacient, stimulant, tonic, vermifuge, and vesicant — horseradish-tree is a folk remedy for adenopathy, ascites, asthma, baldness, boils, burns, catarrh, cholera, cold, convulsion, dropsy, dysentery, dysuria, earache, epilepsy, erysipelas, faintness, fever, gout, gravel, hematuria, hysteria, inflammation, madness, maggots, neuralgia, palsy, pneumonia, rheumatism, scabies, scrofula, scurvy, skin ailments, snakebite, sores, spasms, splenitis, sterility (female), syphilis, toothache, tumors, ulcers, vertigo, wounds, and yellow-fever.[91] The root decoction is used in Nicaragua for dropsy. Root juice is applied externally as rubefacient or counter-irritant. Leaves applied as poultice to sores, rubbed on the temples for headaches, and said to have purgative properties. Bark, leaves, and roots are acrid and pungent, and are taken to promote digestion. Oil is somewhat dangerous if taken internally, but is applied externally for skin diseases. Bark, regarded as antiscorbutic, exudes a reddish gum with properties of tragacanth; sometimes used for diarrhea. Bitter roots act as a tonic to the body and lungs, and are emmenagogue, expectorant, mild diuretic, and stimulant in paralytic afflictions, epilepsy, and hysteria. Other medicinal uses are suggested in Kirtikar and Basu,[165] Morton,[224] and Watt and Breyer-Brandwijk.[322]

Chemistry — Per 100 g, the pod is reported to contain 86.9 g H_2O, 2.5 g protein, 0.1 g fat, 8.5 g total carbohydrate, 4.8 g fiber, 2.0 g ash, 30 mg Ca, 110 mg P, 5.3 mg Fe, 184 IU Vitamin A, 0.2 mg niacin, and 120 mg ascorbic acid, 310 μg Cu, 1.8 μg I. Young pods contain indoleacetic acid and indole acetonitrile.[388] Leaves contain 75 g H_2O, 6.7 g protein, 1.7 g fat, 14.3 g total carbohydrate, 0.9 g fiber, 2.3 g ash, 440 mg Ca, 70 mg P, 7 mg Fe, 110 μg Cu, 5.1 μg I, 11,300 IU Vitamin A, 120 μg Vitamin B, 0.8 mg nicotinic acid, 220 mg ascorbic acid, and 7.4 mg tocopherol per 100 g. On a ZMB, leaf curries contain 25.8 ppm thiamin, 7.26 ppm riboflavin, and 35 ppm niacin.[389] If ascorbic acid is the target, leaves should be gathered before flowering and consumed quickly. Estrogenic substances, including the antitumor compound, beta-sitosterol, and a pectinesterase are also reported. Leaf amino acids include 6.0 g arginine per 16 g N, 2.1 histidine, 4.3 lysine, 1.9 tryptophane, 6.4 phenylalanine, 2.0 methionine, 4.9 threonine, 9.3 lucine, 6.3 isoleucine, and 7.1 valine. Pod amino acids include 3.6 g arginine per 16 g N, 1.1 g histidine, 1.5 g lysine, 0.8 g tryptophane, 4.3 g phenylalanine, 1.4 g methionine, 3.9 g threonine, 6.5 g leucine, 4.4 g, isoleucine, and 5.4 valine. Seed kernel (70 to 74% of seed) contains 4.08 g H_2O, 38.4 g crude protein, 34.7 g fatty oil, 16.4 g N free extract, 3.5 g fiber, and 3.2 g ash. Seeds contain 100 ppm Vitamin E and 140 ppm beta-carotene.[390] The seed oil contains 9.3% palmitic, 7.4% stearic, 8.6% behenic, and 65.7% oleic acids among the fatty acids. (Myristic and lignoceric acids have also been reported.) The cake left after oil extraction contains 58.9% crude protein, 0.4% CaO, 1.1% P_2O_5 and 0.8% K_2O. Gums exuding from the trunks contain L-arabinose, D-galactose, D-glucuronic acid, L-rhamnose, and D-xylose.[391] Pterygospermin ($C_{22}H_{18}O_2N_2S_2$), a bactericidal and fungicidal compound, isolated from Moringa has an LD_{50} subcutaneously injected in mice and rats of 350 to 400 mg/kg body weight. It might serve as a fruit- and vegetable preservative. In low concentrations, it protects mice against staphylococcus infections.[70] Root-bark yields two alkaloids: moringine and moringinine. Moringinine acts as a cardiac stimulant, produces rise of blood-pressure, acts on sympathetic nerve-endings as well as smooth muscles all over the body, and depresses the sympathetic motor fibers of vessels in large doses only. The root alkaloid, spirochin, paralyzes the vagus nerve, hinders infection, and has antimycotic and analgesic activity. In doses of 15 g, the root bark is abortifacient.[187]

Description — Short, slender, deciduous, perennial tree, to about 10 m tall; rather slender with drooping branches; branches and stems brittle, with corky bark. Leaves feathery, pale-green, compound, tripinnate, 30 to 60 cm long, with many small leaflets, 1.3 to 2 cm long, 0.6 to 0.3 cm wide, lateral ones somewhat elliptic, terminal one obovate and slightly larger than the lateral ones; flowers fragrant, white or creamy-white, 2.5 cm in diameter, borne in sprays, with 5 sepals, 5 petals; stamens yellow. Pods pendulous, brown, triangular, splitting lengthwise into 3 parts when dry, 30 to 120 cm long, 1.8 cm wide, containing about 20 seeds embedded in the pith, pod tapering at both ends, 9-ribbed. Seeds 1 to 2 cm wide, dark-brown, with 3 papery wings. Main root thick. Fruit production in March and April in Sri Lanka.

Germplasm — Reported from the African and Hindustani Centers of Diversity, Moringa or cvs thereof is reported to tolerate bacteria, drought, fungus, laterite, mycobacteria, and sand.[82] Several cvs are grown: 'Bombay' is considered one of the best, with curly fruits. Others have the fruits 3-angled or about round in cross-section. In India, 'Jaffna' is noted for having fruits 60 to 90 cm, 'Chavakacheri murunga' 90 to 120 cm long. (2n = 28.)

Distribution — Native to India, Arabia, and possibly Africa and the East Indies; widely cultivated and naturalized in tropical Africa, tropical America, Sri Lanka, India, Mexico, Malabar, Malaysia, and the Philippine Islands.

Ecology — Ranging from Subtropical Dry to Moist through Tropical Very Dry to Moist Forest Life Zones, Moringa is reported to tolerate annual precipitation of 4.8 to 40.3 cm (mean of 53 cases = 14.1) annual temperature of 18.7 to 28.5°C (mean of 48 cases = 25.4) and pH of 4.5 to 8. (mean of 12 cases = 6.5). Thrives in subtropical and tropical climates, flowering and fruiting freely and continuously. Grows best on a dry sandy soil, but grows "in all types of soils, except stiff clays".[70] Drought resistant.

Cultivation — In India, the plant is propagated by planting limb cuttings 1 to 2 m long, from June to August, preferably. The plant starts bearing pods 6 to 8 months after planting, but regular bearing commences after the second year. The tree bears for several years.

Harvesting — Fruit or other parts of the plant are usually harvested as desired, according to some authors; but in India, fruiting may peak between March and April and again in September and October. Seed gathered in March and April and oil expressed.

Yields and economics — I feel, from personal observations, that Moringa's biomass and pod production should approach that of Prosopis growing in the same habitat. A single tree, 3 years old, can yield more than 600 pods per year, or up to 1000.[70] A single fruit will have ca. 20 seeds, each averaging 300 mg, suggesting a seed yield of 6 kg per tree, an oil yield conservatively of 2 kg per tree. Such could be very useful in poor developing countries which import vegetable oils. I would suggest a target yield of about 10 MT pods per hectare. Horseradish-tree is grown locally in India, Sri Lanka, and elsewhere, and is consumed as a local product, either ripe or unripe.

Energy — According to Verma et al.,[328] "saijan" is a fast-growing tree being planted in India on a large scale as a potential source of wood for the paper industry. At Fort Meyers, Florida, trees attain ca. 5 m height 10 months after seed is planted.[383] It seems doubtful that the wood and seed oil could both be viewed as fountains of energy. According to Burkill,[56] "The seeds yield a clear inodorous oil to the extent of 22 to 38.5%. It burns with a clear light and without smoke. It is an excellent salad oil, and gives a good soap . . . It can be used for oiling machinery, and indeed has a reputation for this purpose as watch oil, but is now superseded by sperm oil.'' Sharing rather similar habitat requirements with the jojoba under certain circumstances, perhaps it might be investigated as a substitute for sperm whale oil like jojoba. Growing readily from cuttings, the ben oil could be readily produced where jojoba grows. Coming into bearing within two years, it could easily be compared to jojoba in head-on trials. I recommend such.

Biotic factors — Fruitflies (*Gitona* spp.) have infested the fruits which then dried out at the tip and rotted.[392] Leaves of young plants and freshly planted stumps are attacked by several species of weevils (*Myllocerus discolor* var. *variegatus, M. 11-pustulatus, M. tenuiclavis, M. viridanus* and *Ptochus ovulum*). Also parasitized by the flowering plant, *Dendrophthoe flacata*. Fungi which attack the horseradish-tree include: *Cercospora moringicola* (Leaf-spot), *Sphaceloma morindae* (Spot anthracnose), *Puccinia moringae* (rust), *Oidium* sp., *Polyporus gilvus,* and *Leveillula taurica* (Papaya powdery mildew).[393]

NELUMBO NUCIFERA Gaertn. (NELUMBONACEAE) Sacred Lotus, Lotus Root, Indian Lotus, Hasu

Syn.: ***Nymphaea nelumbo* L., *Nelumbo nelumbo* (L.) Karst., *Nelumbium nelumbo* (L.) Druce, *Nelumbium speciosum* Willd.**

Uses — Rhizomes and seeds of the sacred lotus are frequently used for food, especially in the Orient. The small scale-like leaves on the rootstock, up to 30 cm long, are used as food in some countries. Plants are grown by Chinese and Japanese for the edible tubers, which are used much like sweet potatoes, roasted, steamed, or pickled. In China, a type of arrowroot is prepared from the rhizomes. Leaves may be eaten raw as a vegetable in salads. Fruits can be eaten after the seeds are removed. Flowering stalks are eaten as a vegetable. Seeds are usually boiled or roasted after removing the bitter-tasting embryo, or eaten raw.[209,278,283]

Folk medicine — According to Hartwell,[126] the lotus is used in folk remedies for corns, calluses, and tumors, and/or indurations of the abdomen, cervix, ear, limbs, kidney, liver, and spleen. In China, the leaf juice is used for diarrhea or decocted with licorice for sunstroke or vertigo. Flowers decocted, alone or with roses, for premature ejaculation. Floral receptacle decocted for abdominal cramps, bloody discharge, metrorrhagia, and non-expulsion of amniotic sac. Fruit decocted for agitation, fever, heart, and hematemesis. Seed used for diarrhea, enteritis, insomnia, metrorrhagia, neurasthenia, nightmare, spermatorrhea, splenitis, leucorrhea, and seminal emissions. The nourishing seeds are believed useful in preserving health and strength, and promoting circulation. Root starch given for diarrhea, dysentery, dyspepsia, the tonic paste applied to ringworm and other skin ailments. Plant refrigerant in smallpox, said to stop eruptions. Antidote to alcohol and mushroom. Honey from bee visitors is considered tonic; used for eye ailments. The embryo is used for cholera, fever, hemoptysis, spermatorrhea. Knotty pieces of rootstock used for epistaxis, dysentery, hematemesis, hematochezia, hematuria, hemoptysis, and metrorrhagia. Cotyledons believed to promote virility and alleviate leucorrhea and gonorrhea. Stamens said to purify the heart, permeate the kidneys, strengthen the virility, blacken the hair, make joyful the countenance, benefit the blood and check hemorrhages; for hemoptysis, spermatorrhea.[90] According to Kirtikar and Basu,[165] nearly every part of the plant has a distinct name and economic use. Ayurvedics use the whole plant to give tone to the breast, and to correct biliousness, fever, nausea, strangury, thirst, and worms. They use the root for biliousness, body heat, cough, and thirst, the stem for blood disorders, leprosy, nausea, and strangury, the leaves for burning sensations, leprosy, piles, strangury, and thirst, the flower for biliousness, blood defects, cough, eyes, fever, poisoning, skin eruptions, and thirst, the "aphrodisiac" anthers in bleeding piles, diarrhea, inflammations, mouth sores, poisoning, thirst, and as a uterine sedative, the fruit for blood impurities, halitosis, and thirst, the "aphrodisiac" seeds for burning sensations, diarrhea, dysentery, leprosy, nausea, and to strengthen the body, and the honey as an excellent tonic, useful in eye diseases. Yunani employ the diuretic root in chest pain, leucoderma, smallpox, spermatorrhea, and throat ailments, the flower for bronchitis, internal ailments, thirst, and watery eyes, and as a tonic for the brain and heart, the seeds for chest complaints, fevers, leucorrhea, menorrhagia, and as a uterine tonic.[165]

Chemistry — Per 100 g, the seed (ZMB) is reported to contain 318 to 390 calories, 16.6 to 24.2 g protein, 1.0 to 2.7 g fat, 70.2 to 76.2 g total carbohydrate, 2.5 to 13.1 g fiber, 4.5 to 5.2 g ash, 139 to 330 mg Ca, 298 to 713 mg P, 6.1 to 7.1 mg Fe, 17.4 to 49.0 mg Na, 942 to 1665 mg K, 0 to 35 μg beta-carotene equivalent, 0.65 to 0.75 mg thiamine, 0.18 to 0.26 mg riboflavin, 1.9 to 7.8 mg niacin, and 0 to 44 mg ascorbic acid. The rhizome (ZMB) contains 16.7 mg protein, 0.6 g fat, 74.1 g total carbohydrate, 4.9 g fiber, 6.8 g ash, 370 mg Ca, 1.36 mg thiamine, 0.37 mg riboflavin, 12.96 mg niacin, and 93 mg ascorbic acid.[89] Saline extracts bacteriostatic. Extracts show antitumor activity, vindicating its herbal anticancer reputation. Liriodenine is active in the KB tumor system, oxoushinsunine, cytotoxic; nuciferine and nornuciferine, antispasmodic. Anonaine, armepavine, demethylcoclaurine, gluconic acid, isoliensinine, liensenine, liriodenine, lotusine, D-N-methylcoclaurine, neferin, nelumboside, N-nornuciferine, nornuciferine, nuciferine, pronuciferine, quercitin, and roemerine are reported.[86,90] Hagers Handbook[187] mentions quercetin, isoquercitrin, leucocyanidin, and leucodelphinidin from the leaves, quercetin, isoquercitrin, luteolin, glucoluteolin, kaempferol, and robinin in the petals and stamens. Seeds contain the active beta-sitosterol and related esters, as well as glutathione, the embryo containing methylcorypalline (a coronary dilator[141]), luteolin-7-glucoside, rutin, and hyperoside. Raffinose and stachyose have been isolated from the rhizome, (+)catechin, (+)-gallocatechin, neochlorogenic acid, gallocatechin, leucocyanidin, and leucodelphinidin from the roots.[187] Hsu et al.[141] add the cardiotonic alkaloid higenamine. Is it a wonder that a chemistry set like this is considered sacred in some parts of the world?

Description — Perennial rhizomatous herbaceous aquatic, from a stout, creeping rootstock 10 to 20 m long, branching, bearing numerous scale-like leaves as well as foliage leaves, with milky juice; leaves blue-green with a silvery sheen, waterproof, peltate, circular, up to 90 cm in diameter, concave, on petioles up to 1 m long above water, margins raised upwards, the leaf-stalks and flower-stalks 1 to 2 m tall, hollow, with small scattered prickles; flowers borne singly at ends of stalks, opening on three successive days before fading, fragrant, extending above the leaves on long cylindrical stems; flowers 10 to 26 cm in diameter, sepals 4 to 5, green caducous, inserted at base of receptacle, petals numerous, rose-red to white, free, obovate, obtuse, 8 to 12 cm long, 3 to 7 cm broad, anthers linear, yellow, 15 to 20 mm long, the filaments linear, 7 to 25 mm long; receptacle spongy, in fruit in 10 cm high and wide, flat, the nuts (seeds) embedded within; nuts 2.0 cm by 1.3 cm, ovoid to ellipsoidal, brown to blackish, protruding like knobs, without endosperm, with a hard pericarp. Flowers June to August.[165,278]

Germplasm — Reported from the Near Eastern Center of Diversity, sacred lotus, or cvs thereof, is reported to tolerate bacteria, frost, and waterlogging.[82] Many varieties are cultivated in various parts of the world. Some of the best known cvs are *album grandiflorum; album plenum* ('Shiroman', with double white flowers 30 cm across); *kermesinum* (light rose); *kinshiren* (white shaded pink); *osiris* (deep rose); *pulchrum* (dark rosy-red); *pekinese rubrum* (rosy-carmine); *roseum* (rosy-pink); *plenum* (large and double); *pygmaeum* (dwarf). Seeds known to be 200 years old have been germinated from collections in dry Gobi Desert lakes, plants of these are now being grown in the Kenilworth Aquatic Gardens in Washington, D.C.[278] Priestley and Posthumus[268] describe viable Manchurian seed radiocarbon dated as over 450 (ca. 466) years old. (2n = 16.)

Distribution — Native from the southern border of the Caspian Sea to Manchuria, south throughout the warmer parts of India, Pakistan, China, Iran, Japan, and Australia. It is cultivated in some Mediterranean countries and is naturalized in Rumania. It was commercially introduced in the U.S. about 1876; it has now become naturalized.[278]

Ecology — Ranging from Warm Temperate Dry to Moist through Tropical Very Dry to Moist Forest Life Zones, sacred lotus is reported to tolerate annual precipitation of 6.4 to 40.3 dm (mean of 11 cases = 14.2), annual temperature of 14.4 to 27.5°C (mean of 11 cases = 19.6°C), and pH of 5.0 to 7.5 (mean of 10 cases = 6.2).[82] Lotus thrives with plenty of sunshine and rich soil. The rhizomes grow in mud at the bottom of water, 60 to 90 cm deep. They require a minimum winter temperature above freezing. A good soil would contain two parts loam and one part well-decayed manure. Once set, the plants flower freely. Unless the roots are frozen, they are not harmed by the cold.[278]

Cultivation — Sacred lotus may be propagated from seed, sown in shallow pans of sandy soil, immersed in water tanks heated to 15°C. Seedlings are allowed to grow in the seed pans until large enough to plant out in tubs or ponds. When seeds are sown directly in ponds or pools, they are rolled in a ball of clay and dropped in the water. The hard seeds germinate better if scarified by boring or filing. Plants may be propagated by sections of the rhizomes placed in large tubs or pools, indoors or outdoors. Divisions of the tubers may also be used similarly. From 30 to 45 cm of compost is placed in a vessel, or tubs may be filled with soil and submerged so that the soil surface is 18 to 30 cm below water level. Planting should be in spring when weather has definitely warmed. Plants will grow in ponds or larger bodies of water, as well as in tubs or half-barrels. Tubers may also be planted in late spring just before they start new growth, in rich soil in the bottom of a pond, in water 30 to 90 cm deep. If rhizomes are covered with sufficient water to prevent them from freezing (about 90 cm), they will over winter satisfactorily. If water is not deep enough to prevent the rhizomes from freezing, the pool should be drained in the fall, the tubs removed to a cellar or some place where the temperature is maintained about 1 to 8°C, or the plants should be covered with 1 m or so leaves, hay, or straw and left outdoors for the winter.[278,283]

Harvesting — Parts are harvested when available or needed.[278]

Yields and economics — Commercially, only the rhizomes are sold in shops and markets in southeastern Asia.[278] Duke[82] reports rhizome yields of 4.6 MT/ha.

Energy — This aquatic plant seems better viewed as an edible ornamental rather than a vigorous biomass candidate. I don't find it recommended (like the water hyacinth and cattail, for example) by the champions of aquatics for energy.

Biotic factors — Sacred lotus is attacked by several fungi: *Alternaria nelumbii, A. tenuis, Cercospora nelumbii, Fusarium bulbigenum, Gloeosporium nelumbii, Macrosporium nelumbii, Myrothecium roridum, Phoma nelumbii, Phyllosticta nelumbonis, Physoderma nelumbii,* and *Sclerotium rolfsii.* It is also attacked by *Bacillus nelumbii.*[186,278]

NYPA FRUTICANS Wurmb. (ARECACEAE) — Nipa Palm

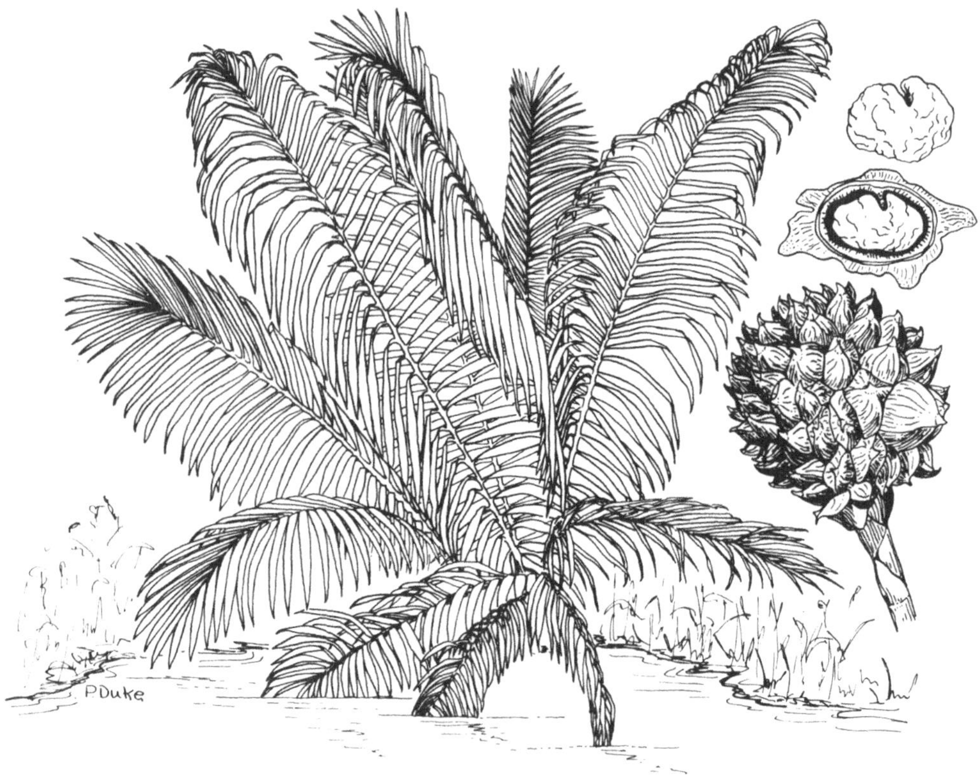

Uses — Menninger[209] summarizes that the palm supplies roofing, thatching, baskets, matting, cigarette wrappers, fuel, alcohol, sugar, toddy, and other products. Also useful for stabilizing soils in tidal terrain. The nut is jelly-like at first, becoming nutty, and finally so hard as to require grating or pounding for eating raw.[209] The tender palm hearts are eaten as a vegetable. Leaves are much valued for thatching, basketry, and mats. Umbrellas, sun-hats, raincoats, mats, and bags are made from the leaves in the Philippine Islands. Midribs are used for making coarse brooms and as fuel. Young unexpanded leaves are used as cigarette wrappers. Leaflets, with 10.2% tannin and 15.2% hard-tans are used for tanning leather. When fishing, fishermen submerge nipa leaves in the sea to attract fish. Salt is obtained by burning the roots or leaves and leaching the ash. The ash is used, with wood-tar, in blackening teeth. Sap is used for making jaggery, sugar,[267] alcohol, and vinegar. Arrows are made from the petioles in the Mentawai Islands.[56,70]

Folk Medicine — Reported to be intoxicant, nipa palm is a folk remedy for centipede bites, herpes, sores, toothache, and ulcers.[91] The sugar is used in a tonic prescription. The stem-bud has been used in making a charmed preparation to counteract poison.[56]

Chemistry — Of 18% of solids in the fresh sap, 17% was found to be sucrose, 1/2% ash. The increase in total carbohydrates in the kernels was from 71 to 78%, between the time they were removed for sugar-tapping (3 months) and at maturity (4 months). Leaves contain 10% tannin.[56] Fresh nipa sap contains ca. 17% sucrose and only traces of reducing sugars. Vinegar (from sap fermented ca. 2 weeks) contains 2 to 3% acetic acid. Immature seeds contain ca. 70% starch. Leaflets contain ca. 10.2% tannin and 15.2% hard-tans.[70]

Description — Gregarious palm, the rootstock stout, branched, covered with the sheaths of old leaves, leafing and flowering at the ends of the branches. Leaves pinnatisect; 4.5 to

9 m long; leaflets linear-lanceolate, 1.2 to 1.5 m long, the sides reduplicate in vernation. Spadix 1.2 to 2.1 m long, terminal, erect in flower, drooping if fruit. Flowers monoecious, male in catkin-like lateral branches of the spadix, female crowded in a terminal head, perianth glumaceous. Male flowers minute, surrounded with setaceous bracteoles; sepals linear with broad truncate inflexed tips, imbricate; petals smaller; stamens 3; filaments connate in a very short column; anthers elongate, basifixed; pistillode 0. Female flowers much longer than the male; sepals 6, rudimentary, displaced; staminodes 0; carpels 3, connate, tips free with an oblique stigmatic line; ovules 3, erect. Fruit large, globose syncarp, 30 cm in diameter, of many obovoid, hexagonal, 1-celled, 1-seeded carpels, 10 to 15 cm long, with pyramidal tips and infra-apical stigmas; pericarp fleshy and fibrous; endocarp spongy and flowery; seed erect, grooved on one side; testa coriaceous, viscid within, adherent to the endocarp; hilum broad; endosperm horny, equable, hollow; embryo basilar, obconic.[165]

Germplasm — Reported from the Indochinese-Indonesian Center of Diversity, the nipa palm, or cvs thereof, is reported to tolerate heavy soils, salt, and tidal waterlogging. (2n = 16.)

Distribution — India south to Australia and New Guinea,[249] in tidal mud from the mouth of the Ganges to Australia.[209] Introduced in the mangroves of South Nigeria, where it has run wild.[350] Reported to have grown successfully in brackish waters of southern Florida.

Ecology — Estimated to range from Subtropical Dry to Wet through Tropical Dry to Moist Forest Life Zones, nipa palm is estimated to tolerate annual precipitation of 5 to 45 dm, annual temperature of 21 to 27°C, and pH of 6.5 to 8.5. Often gregarious in mangrove swamps and tidal forests, growing best in alluvial deposits of clayey loam with sufficient salt.[70]

Cultivation — Cultivated in Sumatra for wine and foliage production. Reproduces naturally by seed and detached portions of rhizome. It may attain 2 m height during its first year.[70] Management consists of thinning natural stands to 2,500 to 3,500 palms per ha, 1.5 to 2 m apart. Periodic pruning to maintain 7 to 8 leaves if favorable to sap production. Other authors suggest much wider spacings, 380 to 750 trees per ha. Bangladesh nursery results are best where submerged at least 230 min/day.

Harvesting — Nuts are harvested as needed. The palm is ready for wine tapping after the second flowering, when about 5 years old. Tapping may continue 50 years or more. If the plant bears more than one spadix, one is topped, the other removed. Sap collection is continued for about 3 months.

Yields and economics — The average yield of sap per plant is 43 ℓ. According to McCurrach,[207] one hectare of nipa will yield 8,000 gals of sweet syrup, inexpensive source of sugar, vinegar, and particularly alcohol. Nipa production is rural-based and labor intensive, though probably less so than other alcohol plants.

Energy — On Bohol Island in the Philippines, a mini-distillery was set up to evaluate potential for the production of ethanol from the nipa palm. Sap of the nipa contains ca. 15% sugar, which can be collected from mature fruits stalks after cutting off the head. With care, this can be repeated over an extended period of time, yielding up to 40 ℓ per tree per season. This translates to a projected 30,000 ℓ juice per hectare. Cultivated palms may produce as much as 0.46 ℓ per tree per day, equivalent to ca. 8,000 ℓ alcohol per ha per year.[267] Halos[121] states that nipa is a better alcohol producer on a hectare basis than sugarcane or coconut, comparing better with sweet potato. In 1919, 2 1/4 million gallons (more than 50,000 barrels) alcohol were produced from nipa palm. Midribs of the leaves are sometimes used for fuel.

Biotic factors — Grapsid crabs are the worst pests of young nipa palms. Pollinated by Drosophila flies.

ORBIGNYA COHUNE (Mart.) Dahlgren ex Standl. (AREACEAE) — Cohune Palm
Syn.: *Attalea cohune* Mart.

Uses — Seeds are source of Cohune Oil, a nondrying oil, considered finer than that of coconut, used in food, as illuminant, and in the manufacture of soap. Very young buds, or cabbage, consumed as a vegetable. Young leaves used to make hats and other apparel, and for thatching.[278] Pole-like rachis of the leaf used for forming the framework of huts. Large quantities of nuts were once used in England for preparing charcoal used in gas masks.[209] Fruits made into sweetmeats and used as fodder for livestock. Trunk used for building. Sap used for winemaking and for making intoxicating beverage.[278,313]

Folk medicine — Reported to be poisonous.[91]

Chemistry — Per 100 g, the seed (ZMB) is reported to contain 6.9 g protein and 52.2 g fat. The tissue removed from the seed contained 1.2 g protein and 0.5 g fat.[92]

Toxicity — "It was said that if too much of the nut was eaten, constipation and sometimes death might result."[58]

Description — Tall monoecious palm 16 to 20 m tall; trunk to 30 cm thick, spineless, usually ringed, covered with old leafbases. Leaves with petioles flat above, rounded below, fibrous at base; blade up to 10 m long, erect, pinnate with 30 to 50 pairs of leaflets; leaflets 45 cm or less long, stiff, dark-green; flower-stalks from lowest leaves, in woody spathe. Flowers small; staminate flowers fall as spathe opens; anthers slender, pale, contorted and spirally twisted. Fruit 7.5 cm long, ovoid, in large grape-like clusters. Flowers February.[278]

Germplasm — Reported from the Middle American Center of Diversity, cohune palm, or cvs thereof, is reported to tolerate limestone, poor soil, sand, slope, savanna, and waterlogging. 2n = 32.[82]

Distribution — Native to wet Atlantic lowlands of Central America from Mexico to Honduras and Belize; grown south to Panama and northern South America.[278,315]

Ecology — Ranging from Subtropical Dry to Moist through Tropical Day Forest Life Zones, cohune palm is reported to tolerate annual precipitation of 6.4 to 40.3 dm (mean of 5 cases = 18.3), annual temperature of 21.3 to 26.5°C (mean of 5 cases = 24.1°C), and pH of 5.0 to 8.0 (mean of 3 cases = 6.9). Thrives in tropical swamps and uplands, or in tropical greenhouses, where night temperatures are not below 15.5°C; occurs from sea-level to 600 m altitude, and appears on all types of soils, including marls, limestones, granites, and slate-derived soils, as well as shales and mudstones. Grows in small congested patches. Occurs also along large streams, on upland sites, on hills and in valleys, preferring rich pockets of soil.[82,278]

Cultivation — Lacking basal shoots, the palm is propagated by seed, in rich soil containing loam, manure, and sand in proportions of 3-1-1. Seeds retain their viability for ca. 6 months. Seeds should be planted about 5 cm deep and watered freely. Spacings between trees should allow about 100/ha.[278]

Harvesting — When freed of competing vegetation, lianas, and epiphytes, each palm bears prolifically. In natural habitat, trees generally do not bear fruit until crown is free in the canopy.[278]

Yields and economics — Yields vary; often nuts are not available enough to supply an oil-mill economically. Large supplies of nuts are not readily available and accessible. Fruits or nuts are exported from Central America for soapmaking.[278]

Energy — Although not so promising as the babassu for oil production, the germplasm of the cohune may contribute to building a bigger genetic base for other oleiferous species. Specific gravity of the wood is 0.868 to 0.971.[324]

Biotic factors — The following fungi cause diseases in this palm: *Achorella attaleae*, *Gloeosporium palmigenum*, and *Poria ravenalae*.[186,278] Bruchid beetles may damage the seeds, destroying both embryo and endosperm.

ORBIGNYA MARTIANA Barb. Rodr., *ORBIGNYA OLEIFERA* Burret, *ORBIGNYA SPECIOSA* (Mart.) Barb. Rodr. (ARECACEAE) — Babassu

Uses — Babassu kernels taste, smell, and look like coconut meat, but contain more oil. The oil can be used for the same purposes as coconut oil, for margarine, shortening, general edibles, toilet soap, fatty acids, and detergents. Unlike many palm oils, the babassu oil does not quickly turn rancid. Babassu oil is rich in "practically all of the elements needed in the manufacture of plastics, detergents, emulsifiers, and many related materials" (H. G. Bennett, as quoted in Balick[29]). The protein- and oil-rich seed cakes are suitable for animal feed. The endocarp is a good fuel. Leaves are used for thatching. Palm hearts are also eaten.

Folk medicine — The oil is used in medicinal salves.

Chemistry — Atchley[23] cites analyses with 9.4 to 16.2% protein, fat content of 0.2 to 62.9% oil — the higher oil figure possibly representing fruit rather than seed. NAS[229] notes that fruit oil may be as high as 72%. Pesce[152]compares the analysis of the coconut with babassu (Table 1). Mesocarp runs 16.3 to 17% moisture, 1.5 to 4.9% fatty material, 63.8 to 71.3% starch, 0.0 to 0.8% sugar, dextrim cellulose 2.05%, 3.12 to 3.19% nitrogenated

Table 1
BABASSU KERNELS AND COCONUT COPRA

	Babassu (*Orbignya martiana*) (%)	Coconut (*Cocos nucifera*) (%)
Moisture	4.21	3.80
Oil	66.12	66.00
Protein	7.17	7.27
Digestible carbohydrates	14.47	15.95
Woody fiber	5.99	4.55
Ash	2.03	2.43

From Johnson, D. V., Ed. and Transl. (Original by Pesce, C.), *Oil Palms and Other Oilseeds of the Amazon* Reference Publications, Algonac, Mich., 1985, 199. With permission.

Table 2
CHEMICAL COMPOSITION AND PROPERTIES OF COCONUT AND BABASSU OIL

Fatty acids	Coconut oil (%)	Babassu oil (%)
Saturated		
Caproic	0.0—0.8	0.0—0.2
Caprylic	5.5—9.5	4.0—6.5
Capric	4.5—9.5	2.7—7.6
Lauric	44.0—52.0	44.0—46.0
Myristic	13.0—19.0	15.0—20.0
Palmitic	7.5—10.5	6.0—9.0
Stearic	1.3	3.0—6.0
Arachidic	0.0—0.4	0.2—0.7
Unsaturated		
Oleic	5.0—8.0	12.0—18.0
Linoleic	1.5—2.5	1.4—2.8

From Eckey, E. W., *Vegetable Fats and Oils*, Reinhold Publishing, New York, 1954. With permission.

material, 1.2% ash, and 0.3 to 11.4% undetermined. The press-cake has 11.6% moisture, 6.5% oil, 19.8% protein, 40.0% digestible carbohydrates, 16.5% woody fiber, and 5.6% ash.[152] Eckey[366] compares the coconut oil with that of babassu (Table 2).

Description — Tall, erect, smooth-stemmed palm. Leaves erect-declined, large, elegant, recurved at the flexuous apex; leaflets long, rigid, proximate, oblique-acuminate, disposed in a vertical plane. Spadix large, ramose, pendent; branches rigid, bracted, dense; female spadices with many sessile flowers on branches and male flowers abortive, small in the apices; in male spadices, flowers with small calyx, petals two, rarely three, biquadridentate; curved inward, overlapped; stamens 24, aggregated in groups of eight; loculus of anthers irregularly coiled and twisted. Female flowers much larger, ovoid-oblong, bibracted, ferruginous tomentose; sepals broadly oblong, obtuse-careened-acuminate; petals slightly smaller, oblong, with irregularly serrated margins, at the protracted apex tri-dentate; androecia abortive, half the number of petals; stigmas 3 to 6. Drupe large, oblong, conical, pointed, enveloped almost half-way, at the base ferruginous-tomentose and at the apex albo-tomentose, haloed, 3 to 6 seeded.[201]

Germplasm — Reported from the Brazilian Center of Diversity, babassu, or cvs thereof, is reported to tolerate alkalinity, sand, savanna, and waterlogging, perhaps even brackish water.[201] Taxonomically confusing, the literature has contradictory references to *O. martiana, O. oleifera,* and *O. speciosa* as the true "Babassu". The taxon *oleifera* "prefers a drier, semi-deciduous forest".[29]

Distribution — Babassu ranges from 3 to 10°S latitude and 40 to 70°W longitude in Brazil.

Ecology — Estimated to range from Tropical Dry to Wet through Subtropical Dry to Wet Forest Life Zones, babassu is estimated to tolerate annual precipitation of 15 to 60 dm, annual temperature of 23 to 29°C, and pH of 4.5 to 8.0. Babassu grows best in alkaline or neutral soils, under average rainfall and good drainage; but it is found in areas of high to low rainfall, dry to swampy conditions, and generally in siliceous soils. It occurs as isolated specimens and in solid stands, but principally in mixed hardwood forests, except in Maranhao and Piaui, on the Pantanal of Mato Grosso and in local areas in some river valleys, where it may form dense forests.[82,201] The day I spent on the bus crossing Maranhao and Piaui was dominated by panoramas of babassu.

Cultivation — Mostly harvested from the wild, like Brazil-nuts and cashews. While plantations have been established, little has been done to examine the variability of wild trees for use in breeding and selection programs.[298]

Harvesting — Slow to mature, babassu may start yielding at 8 years, rising to 12 years, and bearing for 75 years or longer. While the palm flowers year round, it does not always set fruit. In Brazil, fruits ripen from July to November, then fall to the ground. After collection, the fruit is usually dried in the sun to facilitate removing the kernel from the shell.[229] With an axe and mallet, capable natives can shell up to 8 kg kernels a day, but are more likely to average 4 to 5 kg a day.

Yields and economics — At an Office of Technology Assessment in 1980, Duke adduced incredible figures stating that some babassu trees were reported to yield more than a ton of fruit per year. Of the fruit, 10% is kernel, 50% (to 68%) of which is oil, indicating a yield of ca. 40 kg oil per tree, or a barrel of oil for every four trees.[84] Assuming a 63 to 70% oil content per kernel, Balick[29] suggests a possible maximum of ca. 63 kg oil per tree per year, indeed a living "oil-factory". Though individual trees are reported to produce 1000 kg nuts a year, palms on cultivated plantations have yielded 1,500 kg/ha nuts. The fruit weighs 150 to 200 g and may contain 3 to 8 kernels containing 60 to 70% oil and constituting 10% of the fruit's weight. The kernel is surrounded by a pulp that is 10% starch, enclosed by a hard woody shell nearly 12 mm thick. The pulp constitutes 20% of the weight of the fruit.[229] American imports peaked in 1945 at nearly 45,000 tons in a year when Brazil harvested more than 70,000 tons. In 1974, Brazil produced >200,000 tons babassu kernels worth ca. 500 million cruzeiros. Babassu is probably the only species that could replace coconut in the production of olein and stearin.[229] Babassu, covering nearly 15 million swampy hectares in the Amazon and employing nearly 100,000 people in Brazil, has been recommended for further study and use by the NAS. Back in 1957, Markley[200] noted, "It is probably the largest vegetable oil industry in the world wholly dependent on a wild plant, developed from an indigenous cottage industry and still capable of further expansion." Markley[201] gives details of the historical production and value of the Brazilian crop. Pinto[254] tabulates data for 1940 to 1949.

Energy — As early as 1951, Pinto[254] noted, "The shells and husks have proved to be a source of fuel and when distilled may yield useful hydrocarbon products and also carbon suitable for gas absorption. The whole nut is occasionally used for the production of oily smoke in the curing of wild rubber; also, buttons are made from the shells." Michael Balick[31] says, "the babassu palm is one of the best sources of fuel in the form of charcoal or coke. Babassu charcoal burns with a lower content of sulfur, and in some cases has more volatile material than certain mineral coals." In Brazil during World War I, the nuts were found equivalent to coal in heat content, and the husks were easily converted to coke.[29] Analyzing 62 kinds of biomass for heating value, Jenkins and Ebeling[149] reported a spread of 19.92 to 18.83 MJ/kg, compared to 13.76 for weathered rice straw to 23.28 MJ/kg for prune pits. On a percent DM basis, the husks contained 79.71% volatiles, 1.59% ash, 18.70% fixed carbon, 50.31% C, 5.37% H, 42.29% O, 0.26% N, 0.04% S, and undetermined residue. Assuming 250 babassu trees per hectare, Pinto[254] projects a potential production of 34,932,040 MT of kernels and (using his 65% figure) >22 million MT oil (or more than 55 million barrels per year). This is about 15 times 1974 production of ca. 220,000 MT and 1978 production of ca. 240,000 MT. During World War II, liquid fuels were derived from babassu, which burned easily and cleanly in diesel engines. Residues can be converted to coke and charcoal. Clearly, this and other oil palms deserve further study as potential energy sources.[81]

Biotic factors — The tree is sometimes attacked by beetles. *Pachymerus nucleorum* often destroy the fallen fruits.

PACHIRA AQUATICA Aubl. (BOMBACACEAE) — Saba Nut, Malabar Chestnut, Provision Tree, Maranhau Nut

Uses — According to Sturtevant,[317] the roasted nuts taste like chestnut, no nut being better than this nut cooked with salt. Not all nut-eaters would agree. Young leaves and flowers are also used as a vegetable. The seeds contain 50 to 58% oil, with an aroma suggesting licorice or fenugreek. Panamanians and/or Colombians make a breadstuff from powdered roasted seed. Uphof[324] suggests that seeds of large fruited types are used as cacao substitutes. Choco witch doctors are said to use the seeds as a narcotic (but I'm not sure that, in fact, they do). Bark yields a yellow dye used to tint sails, fishing nets, and lines.

Folk medicine — Saba nut is a folk remedy for eye ailments and inflammations. Guatemalans use the bark and immature fruits for liver afflictions. Bark, which has demonstrated antibiotic activity, is used for diabetes in Panama.[80]

Chemistry — Per 100 g (ZMB), the seed of *Pachira macrocarpa* is reported to contain 560 calories, 16.9 g protein, 41.4 g fat, 37.9 g total carbohydrate, 13.1 g fiber, 3.7 g ash, 87.7 mg Ca, 302.3 mg P, 4.0 mg Fe, 76.1 mg Na, 700 mg K, 1300 μg beta-carotene equivalent, 0.03 mg thiamine, 0.06 mg riboflavin, 4.02 mg niacin, and 25.4 mg ascorbic acid.[89] Seeds contain 58% fat. The seed fats of a Congo specimen contained 46% palmitic, 43% oleic, and 11% linoleic acids. Those of a Sudanian specimen contained 50.7% palmitic and stearic, 40.8% oleic, and 8.5% linoleic. Those from South America contained 56% palmitic, 3% stearic, 7.5% oleic, and 5% linoleic acids. There is also a report of 26.5% cyclopropenoid acids in the seed fat.[128] Bark contains 2.7% tannin.

Description — Evergreen tree to 23 m high and 70 cm dbh, often buttressed; outer bark hard, planar, thin, with weak distant vertical fissures; inner bark thick, reddish, marbled with white. Leaves palmately compound, glabrous; stipules ovate, ca. 1 cm long; petioles to 24 cm long, often ribbed, swollen at both ends; leaflets 5 to 7(9), oblong-ovate to elliptic, caudate-acuminate to apiculate at apex, tapered to an acute base and decurrent on petiolule, 5 to 29 cm long, 3 to 15 cm wide, whitish-lepidote especially below. Flowers sweetly aromatic, usually solitary in upper axils; pedicels stout, 1 to 5.5 cm long; calyx more or less tubular, truncate, the lobes obscure; petals 5, valvate, linear, greenish-white to brown, 17 to 34 cm long, ca. 1.5 cm wide, curled outward at anthesis, stellate-puberulent outside, glabrous to villous inside; stamens many, scarlet in apical third, white basally, erect to spreading, slightly shorter than petals, variously united in small clusters basally to middle, the clusters finally uniting with staminal column; anthers horseshoe-shaped, dehiscing by straightening; ovary broadly ovoid, ca. 1 cm long; style colored like stamens but several cm longer; stigma of 5 tiny lobes. Capsules reddish-brown, elliptic, oblong-elliptic, or subglobose, shallowly 5-sulcate, mostly to 20(30) cm long and 10(12) cm wide, the valves 5, densely ferrugineous outside, appressed-silky-pubescent within; seeds usually 2 or 3 per carpel, irregularly angulate, mostly 3 to 4.5 cm long at maturity, brown, buoyant, embedded in solid, white, fleshy mesocarp.[67]

Germplasm — Reported from the Latin America Center of Diversity, saba nut, or cvs thereof, is reported to tolerate drought and waterlogging. The genus apparently contains only one more species, the very similar *Pachira insignis*.[281]

Distribution — Native to the Americas, Mexico to Peru and Brazil, but cultivated in Angola and the Congo, Florida, and the West Indies. According to Robyns,[281] it ranges from southern Mexico through Central America to Ecuador, northern Peru and northern Brazil; often cultivated throughout tropical America, in some isles of the Antilles, in Africa and Asia.

Ecology — Ranging from Tropical Moist to Wet through Premontane Moist to Wet Forest Life Zones, saba nut is estimated to tolerate annual precipitation of 20 to 50 dm, annual temperature of 22 to 28°C, and pH of 6 to 8.5. Rather pure stands occur, rather typical of Tropical Moist and Wet Forests in Panama.[67] Apparently confined to riverine and swamp situations in my experience. The seeds may germinate while floating, striking root when they lodge on soils.

Cultivation — Menninger[209] says it is grown commercially in the Congo, but I know of no cultivation, except as a curio, here in America.

Harvesting — Trees as short as 2 m may begin flowering and fruiting.[224] In Panama, flowering all year though concentrated in February to April; most of the fruits mature from March to August. New leaves appear around May.

Yields and Economics — With precocious fruiting, the tree may produce many large fruits, with many large seeds.

Energy — With more than 50% oil, seeds might be viewed as an oilseed candidate for fresh-water and slightly brackish swamps in the tropical moist to wet forest life zones.

Biotic factors — *Pachira insignis* is listed as an important alternative host to *Steirostoma breve* (Cocoa beetle), major cocoa pest in tropical South America and the Caribbean Islands.[167]

PAULLINIA CUPANA Kunth ex H.B.K. (SAPINDACEAE) — Guarana, Uabano, Brazilian Cocoa
Syn.: *Paullinia sorbilis* Mart.

Uses — Guarana is a dried paste, chiefly of crushed seeds, which may be swallowed, powdered, or made into a beverage. It is a popular stimulant in Brazil among natives who grate a quantity into the palm of hand, swallow it, and wash it down with water. Taste is astringent and bitterish, then sweetish. A refreshing guarana soft drink is made in Brazil similar to making the ordinary drink, but sweetened and carbonated. Odor is similar to chocolate. Cultivated by the Indians and seed made into a paste, sold in two grades. Said to be used also in cordials and liqueurs (fermented with cassava). Brazilian Indians make a breadstuff from pounded seeds. Tyler[323] notes that Coca-Cola — Brazil uses guarana in a carbonated beverage it markets there. I enjoyed it with rum at the airport in Rio. ''Zoom'', a rather tasty beverage, has been promoted as a ''cocaine'' substitute. Menninger calls it ''the most exciting nut in the world''. Erickson et al.[97] mention the product ''guarana flor'', a flour extracted from burned flowers.

Folk medicine — A nervine tonic and stimulant, the drug owes its properties to caffeine. Used for cardiac derangements, headaches, especially those caused by menstrual or rheumatic derangements, intestinal disorders, migraine and neuralgia. Action is sometimes diuretic, and used for rheumatic complaints and lumbago. Said also to alleviate fever, heat stress, and heart ailments. With words like aphrodisiac, diet, narcotic, and stimulant associated with guarana in the herbal literature, it is little wonder that the herb has excited curiosity among avante-garde Americans. Promotional literature states that guarana outsells Coke in Brazil, suggesting that Amazon natives sniff the powdered seeds, and stating (wrongly or rightly) that guarana decreases fatigue and curtails hunger. However, Latin Americans used the plant mainly as a stimulant and for treating chronic diarrhea and headache.[86,180] People accustomed to guarana swear "that it improves health, helps digestion, prevents sleepiness, increases mental activity", and many whisper that it also improves sexual activity, but "it might act as a limiting factor to fertility"[255] (Pio Correa, as quoted in Menninger[209]).

Chemistry — Indians in South America also made an alcoholic beverage from the seeds along with cassava and water. Guarana contains guaranine, an alkaloid similar to theine of tea and caffeine of coffee; about 2.5 to 5% caffeine; and 5 to 25% tannin, as catechutannic acid. An 800 mg tablet of "Zoom" is said to contain ca. 60 mg caffeine.[323] Adenine, catechin, choline, guanine, hypoxanthine, resin, saponins, theobromine, theophylline, timbonine, and xanthine are reported, in addition to the caffeine.[180]

Toxicity — May be quite high in caffeine (possibly the highest of any plant).[86] Dysuria often follows its administration. Has been approved for food use (§172.510). In humans, caffeine, 1,3,7-trimethylxanthine, is demethylated into three primary metabolites: theophylline, theobromine, and paraxanthine. Since the early part of the 20th century, theophylline has been used in therapeutics for bronchodilation, for acute ventricular failure, and for long-term control of bronchial asthma. At 100 mg/kg theophylline is fetotoxic to rats, but no teratogenic abnormalities were noted. In therapeutics, theobromine has been used as a diuretic, as a cardiac stimulant, and for dilation of arteries. But at 100 mg, theobromine is fetotoxic and teratogenic.[86] Leung[180] reports a fatal dose in man at 10,000 mg, with 1,000 mg or more capable of inducing headache, nausea, insomnia, restlessness, excitement, mild delirium, muscle tremor, tachycardia, and extrasystoles. Leung also adds "caffeine has been reported to have many other activities including mutagenic, teratogenic, and carcinogenic activities; . . . to cause temporary increase in intraocular pressure, to have calming effects on hyperkinetic children . . . to cause chronic recurring headache . . . ".

Description — Large, woody, evergreen perennial, twining or climbing vine to 10 m tall, usually cultivated as a shrub; leaves small, pinnate, 5-foliolate, alternate, stipulate, 10 to 20 cm long, the petiole 7 to 15 cm long, flowers in axillary racemes, yellow; the sepals 3 to 5, 3 mm long, petals 3 to 5 mm long, hairy; fruit a 3-valved capsule with thin partitions, in clusters like grapes, pear-shaped, 3-sided; seed(s) globose or ovoid, about the size of a filbert, purplish-brown to brown or blackish, half enclosed in the aril, flesh-colored, white, yellow, or red, easily separated when dry. Germination cryptocotylar, the eophylls unifoliolate.[278]

Germplasm — Reported from the South American Center of Diversity, guarana, or cvs thereof, is reported to tolerate a pronounced dry season.[82]

Distribution — Native to the Brazilian Amazon Basin, especially in the region of Maues, in the valley of the Papajoz River, below Manaos,[284] in the upper regions of the Orinoco Valley in southwestern Venezuela, and in the Moist Evergreen Forests of northern Brazil. It has been reported in parts of Uruguay and was introduced in Sri Lanka and France (1817) from South America.[278] It seems to be thriving at the New Alchemist's outpost in Gandoca, Costa Rica (TMF).

Ecology — Ranging from Tropical Dry to Moist through Subtropical Moist to Wet Forest Life Zones, guarana is reported to tolerate annual precipitation of 10 to 24 dm and annual temperature of 23 to 27°C.[82] Guarana grows naturally in deep acidic oxisols, where there is a pronounced dry season from June to September. Flowering commences at the end of the rainy season. The plant does not tolerate soil compaction. Although guarana was originally a swamp creeper in the moist evergreen forests of the Amazon, it has been more successfully grown on well-drained black sandy soils. Plants do not do well when cultivated on yellow clay soils.[82,278]

Cultivation — Guarana is obtained from both wild and cultivated plants.[278] Pio Correa,[255] however, states that the plants are never found wild.[255] Since seeds require about 3 months to germinate, cultivated guarana is usually propagated by shoots. Young shoots are spaced about 7 m apart, and a triangular bower is built over each plant to provide support for the climbing vine. Or seedlings may be spaced at 4 × 4 m (625 plants per ha) or 3 × 3 m (1,100 plants per ha). Young plants should be shaded. Leguminous ground covers are often established between the plants (*Pueraria, Vicia*). It has been suggested as an intercrop for *Bactris gasipaes*. It can be planted among growing cassava plants. Planting is usually in February and March. Once established, plants require practically no care, except for weeding.[97,278]

Harvesting — Plants begin to flower and produce a small quantity of fruit when about 3 years old. Production increases with maturity, and vines live about 40 years. Fruits, like clusters of grapes, are hand-picked in October, November, and December, after they have ripened. As soon as the berries are harvested, they are thoroughly soaked in water, passed over a sieve to remove the seeds from the white pulp (aril) that surrounds them and the seeds placed in the sun to dry. After drying, seeds are immediately baked or roasted for half a day to prevent fermentation, which sets in rapidly after the fruit is picked from the vine. Seeds are roasted over a slow fire in clay ovens, skillfully, so that all seeds are equally toasted and not burned. Roasted seeds, removed from the ovens, are separated from their dry paper-thin shells by rubbing in the palm of the hands or by placing them in sacks and beating them with clubs. Then the kernels are macerated with mortar and pestle. The coarse powder produced is mixed with a little water and kneaded into a paste which is shaped into cylindrical sticks or loaves ca. 2.5 cm in diameter and 12 to 30 cm long, weighing about 225 g each (about 1/2 lb). These "cakes" are dried and smoked for about 60 days in an open-fire drying house, where they require a dark chocolate-brown color and a metallic hardness. Crude guarana is sold on the market in this form, which will keep for many years.[97,278]

Yields and economics — A mature guarana shrub or vine averages 1.3 to 5.0 kg/year seed, occasionally yielding 9 kg; still, yields run only 77 to 175 kg dried seed per hectare.[97] In past decades, Brazil produced about 80 MT of guarana paste annually, and exported about 50 MT. Herbal interests may have stimulated trade since then. About 6,000 ha are now cultivated, much of it in the Brazilian county of Maues, which produces ca. 80% of the world's supply as of 1980.[97,278]

Energy — Lacking biomass data for this species, I will suggest that the pods, as residue, might equal or exceed in quantity the biomass of the harvested seeds. The pulp and aril probably represent less biomass, also a waste product. Prunings might be used for fuelwood.

Biotic factors — The most severe fungus known to attack guarana is *Colletotrichum guaranicola* Alb., which attacks the foliage and inflorescence. "Black speckle", caused also by *Colletotrichum* sp., can be controlled by such compounds as benomyl, captafol, macozeb, and methyl thiophanate. *Fusarium decemcellarare* Brick. (so-called "trunk gall"), causing a proliferation of buds resulting in large masses of nonproductive tissue, can kill the plant. A red root rot is caused by *Ganoderma philippi* (Bres. & P. Henn.) Bres., causing yellowing of the foliage, gradual decline, often followed by death. Pollination is by insects, primarily bees and wasps; ants are also numerous.[97]

PHYTELEPHAS MACROCARPA Ruiz and Pav. (ARECACEAE) — Ivory Nut Palm, Tagua

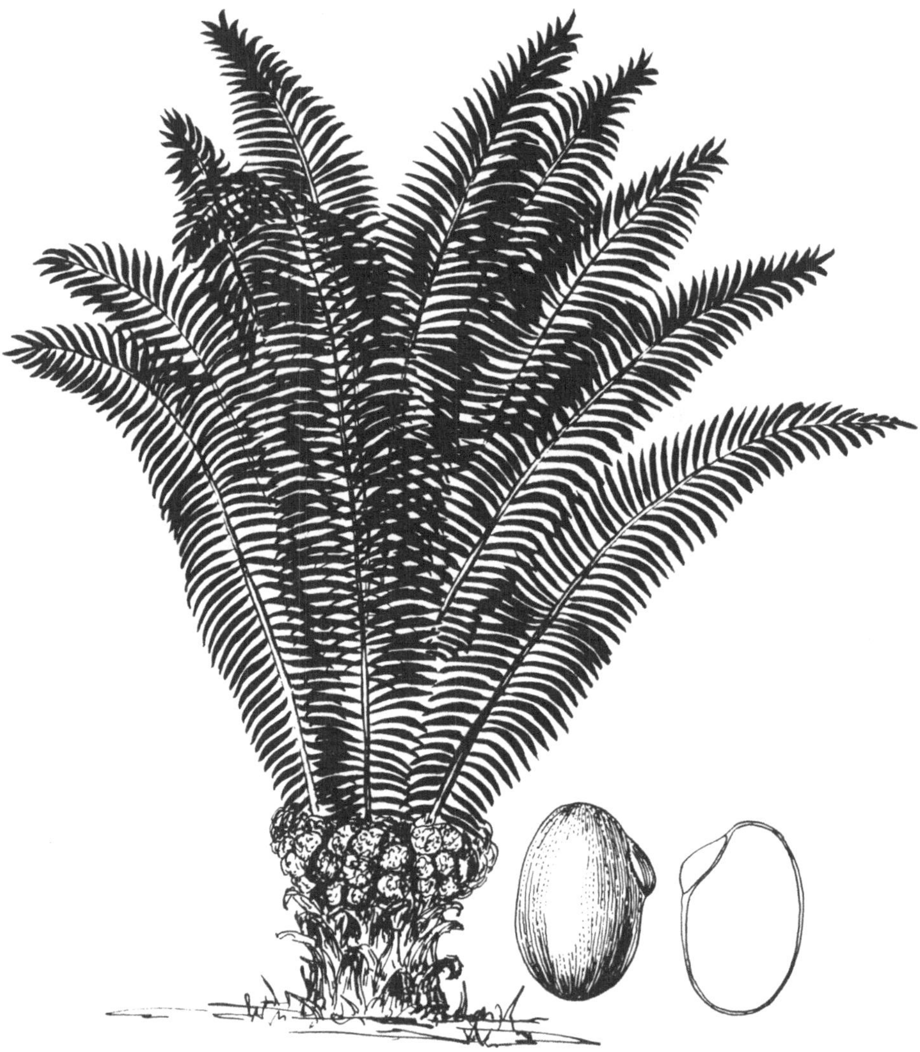

Uses — In Ecuador, they have "commercialized" the hard, compact, heavy, brilliant seeds, so highly valued for their thousand uses, and industrial applications, especially in the button industry. The cabbage is quite edible, usually cooked, and the young fruits make a beverage said to be just as good as coconut water; older fruits become thicker and more mucilaginous or gelatinous, at which time it may be spooned out as a custard. Finally, they harden as the "vegetable ivory." More recently, it has been used in polishing compounds for the metals finishing industry. The roots are boiled to make a beverage. Refuse from the button "turnerys" can be made into cattle food and it is less legitimately used as a coffee substitute, probably after scorching. According to Gohl,[110] ivory nut meal can be used for all classes of livestock without any particular restrictions. Durable leaves used for thatch, the stems are split and used for flooring. Empty spathes have been used as very durable broom heads.

Folk Medicine — A liquid prepared by boiling the roots is considered diuretic in Ecuador.

Chemistry — Per 100 g, the seeds contain (ZMB) 5.3 g protein, 1.6 g fat, 91.6 g total carbohydrate, 9.3 g fiber, and 1.5 g ash.[110] Seeds may contain 40% Mannan A and 25%

Mannan B. Mannan A yields on hydrolysis 97.6% mannose, 1.8% galactose, and 0.8% glucose. Mannan B yields 98.3% mannose, 1.1% galactose, and 0.8% glucose. Nuts are said to contain the alkaloid *phytelephantin*. Personal correspondence reveals that it is *the* raw material for the preparation of the sugar D-mannose. D-mannan has shown some antitumor activity.

Description — Acaulescent or short-stemmed diocecious palms to 20 m tall, 70 cm DBH. Leaves pinnate, to 4 m long, 15 to 30 in the rosette, leaflets in a single plane, to 70 cm long. Male flowers in elongate cluster to 2 m long. Female flowers in heads to 50 cm long with perianth, the tepals to 30 cm long. Ovary 4- to 6-locular; style with 4 to 9 long lobes.[187]

Germplasm — Reported from the Tropical American Center of Diversity, tagua, or cvs thereof, is reported to tolerate rocky soil, shade, and temporary waterlogging.

Distribution — Panama to Brazil, Venezuela, and Ecuador. Although a species of mature forest, it is often left to stand in cleared pastures and banana plantations.

Ecology — Estimated to range from Subtropical Moist to Rain through Tropical Moist to Rain Forest Life Zones, tagua is estimated to tolerate annual precipitation of 20 to 110 dm, annual temperature of 22 to 28°C, and pH of 4.5 to 8.0. Sometimes gregarious; said to "prefer" naturally drained or porous soils, but flourishes on some rocky terrain and in clay alluvial terraces. Ranges from sea level to 1800 m above sea level.

Cultivation — Tagua is rarely cultivated. There have been a few plantations started in Ecuador, mountaineers merely scattering the seeds and weeding them, perhaps thinning them occasionally. Seeds begin to germinate in 3 to 4 months. Young plants may need protection from the sun.

Harvesting — A tagua may mature in 10 years, faster than commonly believed,[2] starting flowering only at 14 to 15 years (Burkill[56] says they start fruiting at 6 years), such that the fruit appears to arise from the ground. Then the females produce fruits "uninterruptably" every subsequent year, a palm lasting for centuries in the mountains. Fruiting occurs throughout the year. When collectors are in too big a hurry, they may destructively fell the tree, which kills it, unable to coppice. The unripe fruits thus obtained, are artificially matured under organic matter, becoming the "blond" tagua, as opposed to the "dark" or "black" tagua. Leaves to be used for thatching are first fermented for 8 to 15 days.

Yields and economics — Well-developed palms produce 15 to 16 mazorcas (clusters), each of which weighs 8 to 15 (to 19) kg. Twelve inflorescences will yield 100 pounds of seeds with their shells, or ca. 60 pounds of shelled seed. Seeds may weigh up to 240 g. Burkill figures that each tree produces 45 to 100 kg fruits per yr for 50 to 100 years. Back in 1948, Acosta-Solis[2] noted that a good price was about $0.70/100 lb. In 1928, Esmeraldas Ecuador exported more than 1000 tons of seeds, 1929 being the highest year ever, with nearly 2200 tons exported, dwindling down to 500 tons by 1941, and almost nothing after that. In New York, in 1941, the Esmeraldas tagua was worth only ca. $2.00/100 lb, a mere $0.10/kg.[2]

Energy — *Phytelephas microcarpa* is said to produce a valuable oil.[331]

Biotic factors — In Ecuador, a coleopteous larva attacks the stem, destroying the pith, and often killing the tree; superficially this resembles the larva of *Rhynchophorus palmarum*. *Dryocoetes* sp. (Coleoptera) may attack the fruit.[2]

PINUS EDULIS Engelm. (PINACEAE) — Piñon, Pine Nut, Nut Pine, Silver Pine
Syn.: *Pinus cembroides* var. *edulis* (Engelm.) Voss and *Caryopitys edulis* Small.

Uses — The State Tree of New Mexico, this species furnishes the piñon nuts or Indian nuts of commerce. Piñon nuts are evident in the firepots of the Gatecliff Shelter, Nevada, carbon-dated at 6000 years. Nuts (seeds) considered main article of subsistence by Indians of California, Nevada, and Utah, eaten raw or, more frequently, roasted. Nuts have a rather disagreeable flavor but are highly nutritious, rich in protein. Seeds are smaller but tastier than those of the single-leaf piñon. In spring, buds at ends of limbs, inner bark, and core of cone (which is something like cabbage stalk when green) are eaten. Wood is mainly used for fuel and fenceposts; infrequently the tree-form is used for lumber of fair quality. The piñon wood was also used in Indian construction. The pitch was used as a glue for waterproofing jugs, as a black dye for blankets, and to repair pottery. Navajo smeared piñon pitch on a corpse prior to burial. Hopi dabbed it on their foreheads to protect them against sorcery. Navajo used it for incense.[278,283]

Folk medicine — According to Hartwell,[126] the pitch is used in folk remedies for tumors of the fingers and external cancers. Reported to be antiseptic and suppurative, the plant is used as a folk remedy for boils, bugbites, laryngitis, myalgia, pneumonia, sores, sore throat, swellings, syphilis, and wounds. Various parts of the plant are used medicinally by Indians: crushed nuts for treatment of burns or scalds; smoke from burning branches for coughs, colds, and rheumatism; and pitch for sores and wounds. Fumes of burning pitch were inhaled by Indians for headcold, cough, and earache.[91,278]

Chemistry — Per 100 g, the "nut" is reported to contain 714 calories, 3.0 g H_2O, 14.3 g protein, 60.9 g fat, 18.1 g total carbohydrate, 1.1 g fiber, and 2.7 g ash.[89]

Description — Straggling tree, forming a broad, pyramid-shaped crown in young trees and later becoming round-topped, to 15 m tall, usually smaller; diameter to nearly 1 m; trunk often crooked and twisted; bark irregularly furrowed and broken into small scales. Leaves mostly 2 to a fascicle, sometimes with varying proportions of 1- or 3-needled fascicles, 2 to 4 (5) cm long, sharp-pointed, margins entire, sheaths of the fascicles deciduous, the odor of the crushed foliage fragrant. Staminate cones about 6 mm long, yellow, soon fading. Ovulate cones subterminal or lateral, 2 to 5 cm long, nearly as wide, ovoid, usually brown at maturity, short-stalked, the scales becoming thickened, 2 to 6 mm long, 4-sided, knobbed at the apex, the dorsal umbo inconspicuous; seeds large, 10 to 16 mm long, brown, wingless, thick-shelled; cotyledons 6 to 10.[68]

Germplasm — Reported from the North American Center of Diversity, piñon, or cvs thereof, is reported to tolerate severe climatic conditions, including low relative humidity, very high evaporation, intense sunlight, low rainfall, hot summers, slope, weeds, and alkaline soil.[82] Piñon, or pine nuts, refer to the seeds of several pine species which grow along the western area of North America from British Columbia southward into Mexico. *Pinus cembroides* Zucc., or Mexican nut-pine, occurs in mountains of central and northern Mexico and extends northward into New Mexico and southeastern Arizona; it is a tree to 20 m tall with needles in fascicles of 3, bright green, 2.5 to 5 cm long, and seeds more or less cylindrical to obscurely triangular and somewhat compressed at apex, about 1.5 to 2 cm long. *Pinus monophylla* Torr., or single-leaf piñon, occurs from Utah and Nevada, south to Baja California and Arizona; it is a tree up to 7 m tall with the needles occurring singly, or rarely in pairs, rather pale glaucous green, about 3.5 cm long, and seeds 1.3 to 1.5 cm long and oblong. Hybrids between *P. edulis* and *P. monopylla* are produced naturally, especially in Utah; such trees have both 1 or 2 needles per fascicle, and other anatomical features of the leaves are intermediate between the two species. Artificial hybrids have also been produced with similar characteristics. *P. edulis* var. *albo-variegata* Hort. has white leaves mixed with the green leaves.[68,69,176]

Distribution — Dry rocky places in the Colorado Plateau region of southwest Wyoming, Utah, western Colorado, extreme western tip of Oklahoma, western Texas, New Mexico, adjacent Chihuahua, Mexico, and eastern Arizona.[68]

Ecology — Ranging from Warm Temperate Thorn to Wet through Subtropical Moist Forest Life Zones, piñon, or cvs thereof, is estimated to tolerate annual precipitation of 3 to 21 dm (mean of 3 cases = 15), annual temperature of 15 to 21°C (mean of 3 cases = 18), and pH of 5.0 to 8.5 (mean of 3 cases = 5.1).[82] To 1500 to 2750 m elevation. Thrives on high tablelands at elevations from 1,600 to 3,000 m altitude, on shallow, rocky soil, where annual rainfall of 30 to 45 cm and climate is arid. Sometimes forms pure groves but more often grows along with oak, juniper, or yellow pine.[278] Hardy to Zone 5.[343]

Cultivation — Trees or shrubby plants cultivated by Indians as far north as British Columbia. Propagated from seeds scattered over ground. Not apparently cultivated in any orderly fashion. Elsewhere grown as an ornamental. Trees are slow-growing, and often form a compact shrub.[278]

Harvesting — Cone matures in August or September of second season and sheds seed shortly thereafter. Seeds are gathered in quantities in favorable seasons. Indians usually collect nuts from the ground after cones have opened, or beat the nuts loose from their cones with poles. Present-day nut-collectors, who often collect the nuts for recreation and then sell them to local groceries, break off cone-bearing limbs, or tear green cones loose with garden rakes, causing serious damage to trees, thus lowering their productivity. Nuts are dried and sorted much like other nuts. They have unusual keeping qualities and may be stored for as much as 3 years without becoming rancid.[278]

Yields and economics — No data available, as most nuts are collected from wild plants which vary widely in their size and productivity. Trees do not bear regularly nor equally fruitfully. Piñon is considered a staple food for some Indian groups, both for themselves and as an article for selling at markets, especially in New Mexico, Arizona, and Mexico. Prices range to as much as $2.85 per pound in retail groceries.[278]

Energy — Historically, the wood, the cones, the needles, and the pitches and resins of pines have been used as energy sources. Scandinavians have even adapted automobile engines to run on turpentine-like compounds. Although the seeds may run more than 50% oil, they seem better adapted to edible than to energy ends.

Biotic factors — *Agriculture Handbook 165*[4] lists the following as affecting this species: *Arceuthobium campylopodum* Engelm. f. *divaricatum* (western dwarf mistletoe), *Armillaria mellea* (root rot), *Coleosporium crowellii* (needle rust), *C. jonesii* (needle rust), *Cronartium occidentale* (piñon blister rust), *Diplodia pinea* (seedling blight), *Elytroderma deformans* (needle cast, witches'-broom), *Fomes pini* (butt and heartwood rot), and *Hypoderma saccatum* (needle cast).[4] Wild animals also collect the nuts.

PINUS QUADRIFOLIA Parl. ex Sudw. (PINACEAE) — Parry's Pine-Nut, Piñon
Syn.: ***Pinus cembroides* var. *parrayana* (Engelm.) Voss and *Pinus parrayana* Engelm.**

Uses — Nuts (seeds), which are rich in proteins, are used as an important food supply by Mexicans and Indians, in Lower California especially. Seeds are eaten raw or in confections under name of pignolia. Taste is that of piney-flavored peanuts, except that the meat is softer. Dense foliage makes the tree desirable as an ornamental tree in cultivation.[278] Trees also used in environmental forestry, as watershed, and as habitat or food for wildlife.[5]

Folk medicine — According to Hartwell,[126] the ointment derived from the pitch is said to be a folk remedy for external cancers. Duke and Wain[91] report Parry's pine-nut to be a folk remedy for cancer.

Chemistry — No data available.

Description — Evergreen trees to 12.3 m tall, with thick, spreading branches forming a pyramid, eventually becoming round-topped and irregular. Needles stout, in fascicles of 4, not over 3.5 cm long, pale glaucous green, incurved, irregularly deciduous, mostly falling the third year. Cones subglobose, chestnut-brown, lustrous, 3.5 to 5 cm broad, broadly ovate, compact until mature; scales thick, pyramidal, conspicuously keeled, umbo with minute prickle. Seeds few, large, dark red-brown, mottled, about 1.3 cm long; shell thin and brittle. Fruit matures in August or September of second season. Hybridizes with *P. monophylla*, single-leaf piñon, from border of U.S. into Baja California, Mexico.[278] Fruit green before ripening; yellowish or reddish-brown when ripe. Flowers June; cone ripens in September; seeds dispersed September to October.[5]

Germplasm — Reported from the Middle America Center of Diversity, Parry's pine-nut, or cvs thereof, is reported to tolerate drought, heat, poor soil, and slope.[82]

Distribution — Native at low elevations of southern California and northern Baja California, Mexico. Not hardy northward.[278] Most abundant of piñon pines. There are very dense and extensive stands in the Sierra Juarez and the Sierra San Pedro Martir, which produce tremendous quantities of piñon nuts.[339]

Ecology — Ranging from Warm Temperate Wet through Subtropical Moist Forest Life Zones, Parry's pine-nut is reported to tolerate annual precipitation of 10.3 to 21.4 dm (mean of 2 cases = 15.9), annual temperature of 21.2°C, and pH of 5.0 to 5.3 (mean of 2 cases = 5.2).[82] Thrives on arid mesas and low mountain slopes on well-drained soils. Tolerates high temperatures and low rainfall; very drought-resistant.[278]

Cultivation — Trees not known to be in cultivation for the nuts. Sometimes trees are cultivated as ornamentals. Propagated from seed, mainly distributed naturally.[278] First cultivated in 1885. Germination hastened and improved by cold stratification of stored seeds for up to 30 days at 0° to 5°C in a moist medium.[5]

Harvesting — Natives usually collect nuts from the ground after cones have opened, or they beat nuts loose from cones with long poles. Present-day nut-collectors, who often collect the nuts for recreation and then sell them to local groceries, break off cone-bearing branches, or tear green cones loose with garden rakes, causing serious damage to trees, thus lowering their productivity. Nuts have good keeping qualities and unshelled piñon nuts can be stored for 3 years without becoming rancid. Piñon nuts mature in the second season during August and September.[278] There is a 1- to 5-year interval between large seed crops. Seeds are dried for 2 to 8 days. Seeds may be collected by shaking the tree and collecting seeds on a cloth spread on the ground.[5]

Yields and economics — In California, between 820 and 1,200 (average 960) seeds per pound were collected from three samples.[5] Exact yield data are difficult to obtain, as fruiting is uneven, and nearly all piñon nuts are harvested from wild plants, which may be scattered. Nuts form a very important item of the diet for some Mexicans and Indians, especially in Baja California, and are sold in markets from San Diego southward, for as much as $2.85 per lb.[278]

Energy — Historically, the wood, the cones, the needles, and the pitches and resins of pines have been used as energy sources. Scandinavians have even adapted automobile engines to run on turpentine-like compounds. Although the seeds may run more than 50% oil, they seem better adapted to edible than to energy ends.

Biotic factors — This piñon nut tree is attacked by a fungus, *Hypoderma* sp. and may be parasitized by the mistletoe, *Arceuthobium campylopodum*.[278]

PISTACIA VERA L. (PISTACIACEAE) — Pistachio

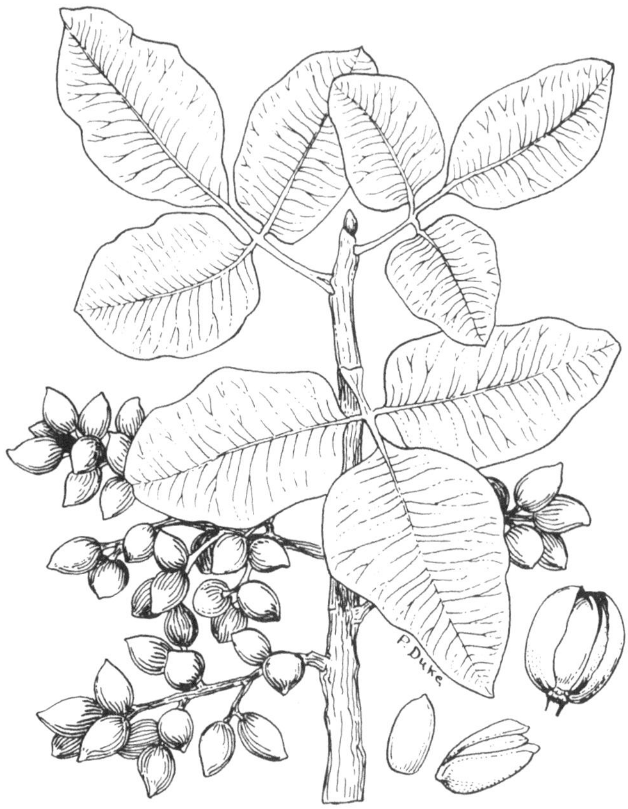

Uses — Pistachio is cultivated for the nut, rich in oil, eaten roasted, salted, or used to flavor confections and ice cream. Arabs call the nut "Fustuk". The outer husk of the fruit, used in India for dyeing and tanning, is imported from Iran. The fruit is the source of a non-drying oil. In Iran, Bokhara Galls of Gul-i-pista, are used for tanning.[278] The nuts are much liked by squirrels and some birds, including bluejays and red-headed woodpeckers. The wood is excellent for carving and cabinet work.[153] In Iran, fruit husks are made into marmalade; they are also used as fertilizer.[70,329,335]

Folk medicine — According to Hartwell,[126] the nuts are said to be a folk remedy for scirrhus of the liver. Reported to be anodyne and decoagulant, pistachio is a folk remedy for abdominal ailments, abscess, amenorrhea, bruises, chest ailments, circulation problems, dysentery, dysmenorrhea, gynecopathy, pruritus, sclerosis of the liver, sores, and trauma.[91] Algerians used the powdered root in oil for children's cough. Iranians infused the fruits' outer husk for dysentery. Lebanese used the leaves as compresses, believing the nuts enhanced fertility and virility.[85] Arabs consider the nuts to be digestive, aphrodisiac, and tonic. They are used medicinally in East India.[278]

Chemistry — Per 100 g, the seed (ZMB) is reported to contain 624 to 627 calories, 19.7 to 20.4 g protein, 56.4 to 56.7 g fat, 20.1 to 20.6 g total carbohydrate, 2.0 g fiber, 2.9 to 3.3 g ash, 138 mg Ca, 528 mg P, 7.7 mg Fe, 1026 mg K, 146 μg beta-carotene equivalent, 0.71 mg thiamine, 1.48 mg niacin, and 0.0 mg ascorbic acid.[89] Galls produced on leaves contain 45% tannin. Tannin contains gallotanic acid, gallic acid, and an oleo-resin, to which the odor is due.[278] Low in sugar (ca. 8%), high in protein (ca. 20%) and oil (>50%). The oil is nearly 90% unsaturated fatty acid (70% oleic and 20% linoleic fatty acid).[153] The edible portion of the nuts contains 9.0 ppm Al, 0.02 As, 0.002 Au, 11 B, 0.1 Ba, 16 Br, 1066 Ca, 0.04 Cd, 408 Cl, 0.2 Co, 0.6 Cr, 0.1 Cs, 33 Cu, 0.1 Eu, 3.8 F, 46 Fe, 0.1 Hg, 51 I, 8639 K, 0.02 La, 0.01 Lu, 949 Mg, 3.4 Mn, 538 Na, 1.1 Ni, 0.8 Pb, 10 Rb, 960 S, 0.05 Sb, 0.004 Sc, 0.1 Se, 0.03 Sm, 0.4 Sn, 10 Sr, 0.4 Th, 3.1 Ti, 0.01 V, 0.1 W, 0.1 Yb, and 30 ppm Zn dry weight. The normal concentration of some of these elements in land plants are 50 ppm B, 14 Ba, 15 Br, 2000 Cl, 0.5 Co, 0.2 Cs, 14 Cu, 3.200 Mg, 630 Mn, 3 Ni, 20 Rb, 3,400 S, 26 Sr, and 0.2 ppm Se dry weight. They were higher in copper, fluorine, iodine, and potassium, and they were equal or higher in europium and thorium than any of the 12 nut species studied by Furr et al.[102] Moyer[226] reports pistachios to contain, per 100 g edible portion, 594 calories, 19.3 g protein, 53.7 g fat, 19.0 g carbohydrates, 5.3% water, 131 mg Ca, 500 mg P, 7.3 mg Fe, 972 mg K, and 158 mg Mn. An analysis of pistachio kernels in the *Wealth of India* gave the following values per 100 g: 5.6% moisture, 19.8% protein, 53.5% fat, 16.2% carbohydrates, 2.1% fiber, 2.8% mineral matter, 0.14% Ca, 0.43% P, 13.7 mg Fe, 240 I.U. carotene (as vitamin A), 0.67 mg thiamine, 0.03 mg riboflavin, 1.4 mg nicotinic acid, no vitamin C, and 626 calories. The fatty acid composition of the oil is 0.6% myristic, 8.2% palmitic, 1.6% stearic, 69.6% oleic, and 19.8% linoleic acids. Galls contain 50% tannins. Both young and mature leaves contain shikimic acid.[70]

Description — Slow-growing, long-lived (700 to 1500 years), small, dioecious bushy tree, to 10 m tall, developing a large trunk with age; branches pendant. Leaves odd-pinnate, the 3 to 11 leaflets ovate, slightly tapering at the base. Flowers dioecious, without petals, brownish-green, small, in axillary racemes or panicles; pedicels bracted at base; staminate flowers with 5-cleft calyx and 5 very short stamens with large anthers; pistillate flowers with 3 to 4 cleft calyx, 1-celled sessile ovary and short 3-cleft style. Fruit a dry, ovoid to oblong, pedicelled drupe, up to 2.5 cm long, reddish and wrinkled, enclosing 2 yellow-green oily cotyledons (kernel). Flowers early summer; fruits August to September.[278]

Germplasm — Reported from the Central Asia Center of Diversity, pistachio, or cvs thereof, is reported to tolerate drought, frost, and heat.[82] Many varieties of pistachio have been developed, because the crop has been grown for several thousand years, most are named after the area in which they were cultivated. Iranian: light-yellow kernel, larger size but lacks oily nut flavor; Sicilian, Syrian, and Turkish: almost green kernel throughout, with good flavor; Afghan and Italian: deep-green kernels prefered for ice cream and pastry. In Syria, district cvs are 'Alemi', 'Achoury', 'Aijimi', 'Aintab', 'Ashoori', 'El Bataury', 'Mirhavy'. In Sicily: 'Trabonella', and 'Bronte'. In California, 13 cvs have been tested: 'Ibrahmim', 'Owhadi', 'Safeed', 'Shasti', 'Wahedi' (largest nuts of any cv). In Turkey: 'Uzun' (nuts 34 to 36 mm long) and 'Kirmizi' (red-hulled, thin-shelled, free-splitting, green-kerneled, containing 20.3% protein and 65.47% oil).[278] Joley[153] reports on cvs being tested at Chico, California. Male cv 'Peters', nearest to a universal pollinator, coincides well with 'Red Aleppo' and 'Trabonella' (early blossoming) and with 'Kerman' (late blossoming). Cultivar 'Chico' provides a supplement to 'Peters'. The first nut-bearing cvs tested at Chico were 'Bronte', 'Buenzle', 'Minassian', 'Red Aleppo', 'Sfax', and 'Trabonella'. The most promising in quality and greeness of kernels are 'Bronte', 'Red Aleppo', and 'Trabonella'. 'Kerman' is liked by importers and processors for its size, crispness, and snap when bitten into and chewed. A sister seedling of 'Kerman', 'Lassen', also produces good quality large-sized nuts.[153] (2n = 30.)

Distribution — Native to Near East and Western Asia from Syria to Caucasus, and Afghanistan, forming pure stands at altitudes up to 1000 m; pistachio has been introduced and is now cultivated in many subtropical areas of the world, such as China, India, the Mediterranean, and U.S. (Arizona, California, and Florida).[278]

Ecology — Ranging from Warm Temperate Dry through Subtropical Thorn to Dry Forest Life Zones, pistachio is reported to tolerate annual precipitation of 4.7 to 11.2 dm (mean of 7 cases = 6.2 dm), annual temperature of 14.3°C to 26.2°C (mean of 7 = 18.8°C), and pH of 7.1 to 7.8 (mean of 4 cases = 7.6).[82] Hardy to Zone 9.[343] Pistachio requires cold winters (to −18°C) and long hot dry summers (to 38°C) to mature. Philippe[251] assumes it requires 600 to 1500 hours below 7°C to meet its chilling requirements. In Iran, it grows at 1200 m elevation on desert plateau. In Turkey and California it grows in the same areas as olives and almonds, but flowers later in the spring than almonds, and is less susceptible to fruit injury. Requires from 30 to 45 cm annual rainfall, any less may need irrigation, but requires less than most other culitvated fruit and nut trees. Soils should be deep, friable, and well-drained but moisture-retaining; the root is deep-penetrating.[278] It can, however, survive in poor, stony, calcareous highly alkaline or slightly acid, or even saline soils.[251]

Cultivation — Trees are difficult to transplant; green seed (nuts) are planted in their permanent place. Other species of *Pistacia* are used as stock upon which to bud pistachio. Care should be taken to select areas for the pistachio orchard which are protected from wind, as in a valley, with less exposure to cold, and with soil relatively free of sand but possessing the ability to retain moisture. After planting, soil should be cultivated periodically for 5 to 7 years, by which time the trees are 2 to 3 m tall. Branches of selected cvs are then bud-grafted (2 buds per tree to insure at least one taking) on new trees. Male varieties shedding pollen during the first half of female blooming period should be selected. In California, male varieties 'Peters' and 'Chico 23' correspond well to female 'Red Aleppo', 'Trabonella', and 'Bronte'. One male tree should be planted to 7 or 8 females. Plantings should be about 9 m apart under irrigation, farther apart without irrigation. Pistachio responds favorably to applications of nitrogen. After grafting, 4 to 6 years are required before trees begin to bear. The trees do not bear fully until they are 20 to 25 years old, and continue to bear for 40 to 60 years or more. Pistachio trees are delicate, and production of nuts is influenced by excess of rain, drought, excessive heat or cold, and high winds.[278,335,337]

Harvesting — Harvest period is August to September in most areas. It is best to harvest the whole tree when most of the crop is ripe. Nuts can be knocked from trees. Clusters of nuts are removed, allowed to dry 3 days on the ground, and beaten or stamped on to separate the nuts from the clusters. They are then put in a tank of water to soak for 12 hr, and then stamped on or beaten to remove the outer green husk. Finally, they are washed and dried in the sun.[153,278]

Yields and economics — Adult trees yield an average of 11.25 kg annually. Three kg unshelled nuts yield 1 kg shelled.[278] Joley[153] reports on the average yield of four pistachio cvs per tree per year since start of production: 'Kerman' 22.45 kg dry weight, 15 years in production; 'Bronte' 11.25 kg dry weight, 14 years in production; 'Trabonella' 6.35 kg dry weight, 12 years in production; and 'Red Aleppo' 4.50 kg dry weight, 12 years in production. For 8- to 15-year old trees in Jordan, Philippe[251] estimates yields at 2 to 8 kg in shell per tree, 200 to 800 kg/ha, for 16- to 30-year-old trees, 8 to 30 kg per tree, 800 to 2,400 kg/ha. In 1976, the yield of American pistachios was 150,000 lbs; in 1979, more than 17 million lbs. Yields of 50 to 150 lbs per tree are reported in California.[329] Duke[82] reports 7 kg fruit per plant. Nuts are marketed mostly unshelled and salted. Soaked in a brine solution, they are quickly dried in the sun or in artificial driers to prevent development of surface mold. Before marketing the shell is cracked for consumer convenience. "Red" pistachios are roasted, salted, and shell is colored with a vegetable dye; "White" pistachios are roasted and shell coated with a mixture of salt and cornstarch; "Naturals" have only salt added after roasting.[278] In 1979, the revolution in Iran caused the world's main pistachio supply to disappear, which in turn caused prices to rise from $1.25/lb in 1978 to $2.05/lb in 1980. In 1982, the American crop of 43 million pounds of pistachios was valued at more than $60 million. Indications are that the U.S. crop will be 70 to 80 million pounds by the 1990s, eventually topping 120 million pounds.[329]

Energy — The wood has a specific gravity of 0.9179 to 0.9200,[324] and is said to make an excellent fireplace wood.[153] Analyzing 62 kinds of biomass for heating value, Jenkins and Ebeling[149] reported a spread of 19.26 to 18.06 MJ/kg, compared to 13.76 for weathered rice straw to 23.28 MJ/kg for prune pits. On a percent DM basis, the shells contained 82.03% volatiles, 1.13% ash, 16.85% fixed carbon, 48.79% C, 5.91% H, 43.41% O, 0.56% N, 0.01% S, 0.04% Cl, and undetermined residue.

Biotic factors — Pollination is by wind or air drift. Many insects are serious crop-destroying pests and should be controlled; an aphid (*Anapleura lentisci*) is one such pest. Numerous fungi, causing serious damage, attack pistachio: *Alternaria tenuissima, Asteromella pistaciarum, Cladosporium herbarum, Cylindrosporium garbowskii, C. pistaciae, Cytospora teretinthi, Fomes rimosus, Fusarium roseum, F. solani, Monilia pistacia, Ozonium auriconium, Papulospora* sp., *Phellinus rimosus, Phleospora pistaciae, Phyllactinia suffulta, Phyllosticta lentisci, P. terebinthi, Phymatotrichum omnivorum, Phytophthora parasitica, Pileolaria terebinthi, Pleurotus ostreatus, Rhizoctonia bataticola, Rosellina necatrix, Septogloeum pistaciae, Septoria pistaciae, S. pistaciarum, S. pistacina, Stemphylium botryosum, Tetracoccosporium* sp., and *Uromyces terebinthi.* The *Phytophthora* causes footrot via damage to cambium; *Septoria* spp. cause defoliation and CuS spray should be used; *Phomopsis* and *Fusarium* attack the female flowers; a virus causing rosettes is serious in Asia; mistletoe attacks the trees; and the following nematodes have been isolated from pistachio: *Heterodera marioni, Meloidogyne* sp., and *Xiphinema* index.[278] The roots are very susceptible to root-knot nematodes.[153] Verticillium wilt is the primary threat, according to Vietmeyer.[329] Rice et al.[279] report epicarp lesion symptoms being reproduced on apparently healthy pistachio fruit clusters exposed to field-collected adult leaf-footed bugs, *Leptoglossus clypealis.* Two species of leaf-footed bugs, *Leptoglossus clypealis* and *L. occidentalis,* and at least four species of stink-bugs in the genera *Thyanta, Chlorochroa,* and *Acrosternum* produced similar external and internal damage to pistachio fruits. Other fruit symptoms, not associated with insects, were panicle and shoot blight, endocarp necrosis, and stylar-end lesion. Several species of smaller plant bugs in the family Miridae, including *Lygus hesperus* and *Calocoris norvegicus,* caused epicarp lesion symptoms.[279]

PITTOSPORUM RESINIFERUM Hemsl. (PITTOSPORACEAE) — Petroleum Nut (English), "Hanga" (Philippine)

Uses — Called petroleum nuts because of the fancied resemblance of the odor of the fruit's oil to that of petroleum, the fruits, even green ones, burn brilliantly when ignited. Hence, they are used like torch nuts or candlenuts for illumination in the bush. Dihydroterpene ($C_{10}H_{18}$) is used in perfumes and medicines. Heptane (C_7H_{16}) is a component of gasoline, and has been suggested as a possible component of paint and varnish.[15]

Folk medicine — The fruit is used as a panacea by Philippine traditionalists — especially, however, for abdominal pain. The oleoresin is used to treat muscular pains and skin diseases.[249] The nut decoction is used for colds. Crushed nuts are mixed with coconut oil as a relief for myalgia. Altshul[10] quotes from a 1947 Sulit herbarium specimen, "Petroleum gas extracted from the fruit is medicinal for stomach-ache and cicitrizant." Hurov[142] says the fruit is used to treat rheumatism, muscle pains, and wounds.

Chemistry — The volatile oil of the fruit is reported to contain "dihydroterpene and heptane, which is a cardiac glycoside".[249] The Horticultural and Special Crops Laboratory at Peoria analyzed an accession of fruit, and identified, from its "squeezings", constituents passing through a gas chromatographic column, heptane (about 45% of the elutents) nonane, alpha-pinene or beta-ocimene, beta-pinene, myrcene, and unidentified materials. The essential oil (8 to 10% of fruit weight) contains myrcene (40%) and alpha-pinene (38%) in

± equal quantities (oil of *P. undulatum* contains mostly limonene). The two components n-heptane (5%) and n-nonane (7%) are minor components.

Description — Aromatic tree to 30 m tall, but probably smaller in its elfin forest habitat (perhaps even epiphytic); fruiting when only 6 to 12 m tall. Leaves aromatic, coriaceous, entire (possibly evergreen), thickest above the middle, pinnately nerved, with a short acumen at the tip. Flowers fragrant, white, clustered on the stems. Fruits average 25 mm in diameter (12 to 43). Each fruit has 5 to 72 seeds (average 31), the seeds ranging from 1 to 4 mm, averaging 3 mm. The seeds are about as close to hexahedral and prismatic as any I have seen, being quite angular, black to blackish-gold, often still surrounded by a gummy or resinous endocarp.

Germplasm — The FORI Director in the Philippines is actively collecting superior germplasm in the high mountains of Bontoc and Benguet where they abound, especially in elfin forests.[15,326]

Distribution — In the Philippines, petroleum nut is locally known in Benguet as apisang, abkol, abkel, and langis; in the Mountain Province, dael and dingo, and in Abra, sagaga. It abounds in Mt. Pulis, Ifugao, and is reported from the head-waters of the Agno and Chico River Basins. Also in the Bicol Provinces, Palawan, Mindoro, Nueva Ecija, and Laguna Provinces. It is being cultivated at FORI's Conifer Research Center, Baguio City.[15,326]

Ecology — Petroleum nut is reported to range from 600 to 2,400 m elevation, usually in elfin or Benguet Pine Forest. Average of 7 climatic data sites where the *Pittosporum* grows was close to 1,000 m, the range from ca. 550 to 2,000 m. Whether or not it can stand frost, dry heat, and drought is questionable. Frequently, species of elfin forests have very narrow ecological amplitudes and do not thrive in other vegetation types. Results of transplants and trials are unavailable to me now. Reportedly, seed were introduced once, at least to Hawaii. Thanks to Professors Ludivina S. de Padua, S. C. Hales, and Juan V. Pancho of the Philippines, we now have a fairly good idea of the ecosystematic amplitudes of the *Pittosporum,* an energy plant that has captured the imagination of many. Professor de Padua checked off all the climatic data points (from our climatic data base) at which *Pittosporum resiniferum* was growing, prior to its widespread introduction for potential energy studies elsewhere in the Philippines. Ranging from Tropical Dry to Moist through Subtropical Forest Life Zones, the petroleum nut grows where the annual precipitation ranges from 15 to more than 50 dm (mean of 36 cases = 27 dm), annual temperature from 18 to 28°C (mean of 17 cases = 26°C). Of 17 cases where both temperature and rainfall data were available to us, 13 would suggest Tropical Moist Forest Life Zone, three would suggest Tropical Dry, and one would suggest Subtropical Rain Forest Life Zones.

Cultivation — Seeds and cutting can be used to propagate the tree. Seeds may lose their vitality rather rapidly. According to Juan V. Pancho (personal communication, 1982), "from my experience, the seed lost its viability after one month storage."

Harvesting — Currently, seeds are harvested from the wild.

Yields and economics — A single fruit yields 0.1 to 3.3 mℓ, averaging about 1.3 mℓ oil. In general, the bigger the fruit, the larger the seed, and the greater the oil content.[326] It is reported[15] that a single tree from Mount Mariveles, Bataan, yielded 15 kg green fruits, which yielded 800 cm^3 of oil. The residue, ground up and distilled with steam, yielded 73 cm^3 more. Another report gave 68 g per kg fresh nuts, suggesting about 1 kg oil per tree yielding 15 kg.[15] Currently, seeds are being sold at $2.00 per gram in 5-gram lots (ca. 40 seeds per g) by the FORI Seed Officer, Forest Research Institute College, Laguna, Philippines.

Energy — The plant was discovered as a hydrocarbon source just after 1900. Based on the previous paragraph, it seems it would take 1,000 trees per ha to get one MT oil per hectare from the fruits. Perhaps the resin in the leaves, twigs, etc. would equal or exceed this; figures are not yet available. The oil derived from the fruits is quite sticky and rapidly turns resinous when laid thin. In an open dish, it burns strongly, although with a sooty flame.[15] C. A. Arroyo[18] notes that for home use as fuel, "the husk of African oil palm nuts could be much better than the petroleum nut that emits sooty smoke and strong smell." President Marcos was said to encourage each Philippine farmer to plant five trees in the hopes that they could obtain 300 ℓ of oil therefrom, per year. I saw nothing about this at the Philippine exhibit at the World's Fair in June 1982. However, if yields of 60 ℓ of oil per tree are possible, the tree should certainly be examined! In the January 1981 issue of *Canopy International*, Generalao[108] lists petroleum nut at the top of a long list of potential oil seeds including *Pongamia pinnata, Sterculia foetida, Terminalia catappa, Sindora supa, Calophyllum inophyllum, Canarium luzonicum, Aleurites moluccana, Aleurites trisperma, Mallotus philippensis, Barringtonia asiatica, Sindora inermis, Pithecellobium dulce, Tamarindus indica, Chisocheton cumingianus, Jatropha curcas*, and *Euphorbia philippensis* to help the Philippines solve their energy problem (importing 85%). Presidential Decree 1068 declares the imperative acceleration of research on energy alternatives. Editorial notes in *Canopy International* suggest that the flammable element in petroleum nut is volatile, evaporating quickly like acetone.[15] Some chemists believe admixing another element will stabilize the compound. One Hurov seed catalog[142] has very optimistic notes about the plant: "The Gasoline Tree produces masses of apricot-sized orange fruits which when cut and touched with a match leap into flame and burn steadily. The fruits contain 46% of gasoline type components (heptane, dihydroterpene, etc.), which are found in extensive networks of large resin canals. If planted, the estimated yield would be about 45 tons of fruit or 2500 gallons of 'gasoline' per acre per year."[15,142,326]

Biotic factors — No data available.

PLATONIA ESCULENTA (Arr. Cam.) Rickett & Stafl. (CLUSIACEAE) — Bacury, Bacuri, Pakuri, Parcouril, Piauhy, Wild Mammee (Guyana)
Syn.: *Aristoclesia esculenta* (Arr. Cam.) Stuntz; *Platonia insignis* Mart.

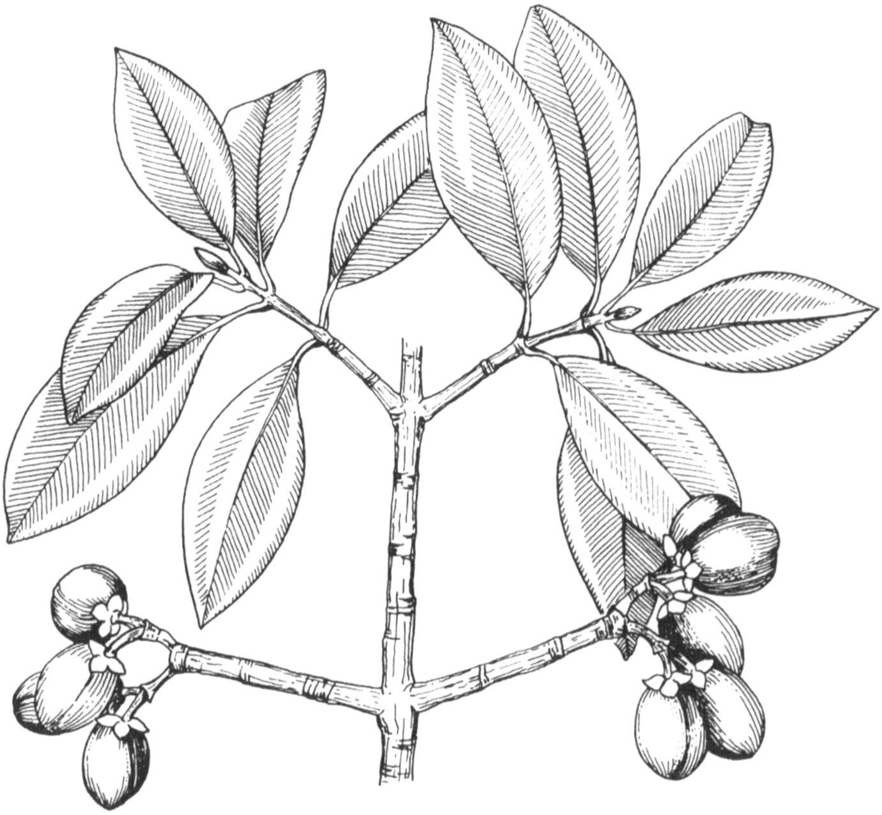

Uses — Seeds are the source of Bacury Kernel Oil, a nondrying oil used in the manufacture of candles and soaps. Fruits are used for pastry and preserves, and are highly esteemed in the Amazon region for the delicious pulp from the large fruit, used in sweets and ice cream. Extracts of the fruit are toxic to black carpet beetles, but not to the larvae of *Aedes* and *Anopheles*. A yellow gum resin secreted by the bark is used in veterinary medicine. Wood, brownish-yellow, turning black upon exposure to air, is durable, resistant to insect attacks; resilient, rather fine-grained, easy to work, taking on a lustrous finish; used for flooring, planks, fancy wood-work, and construction of buildings. Presently, wood is used for making rum barrels, and cases or crates for shipping bananas; it is also excellent for cabinet work and carpentry. Bark is white, exfoliating, fibrous, used for cordage, and yielding a black viscous resin used for caulking boats.[278,324]

Folk medicine — No data available.

Chemistry — This is one of the few outstanding exceptions to the generally evenly distributed glyceride structure of solid seed fats. In 1945, it was reported to have the unusually high melting point of 51 to 52°C and contain 24% fully saturated glycerides, although its component acids were approximately 56% saturated (palmitic and stearic) with 39% oleic and 4% linoleic acid. Component acids reported are myristic 1.0, palmitic 55.1, stearic 6.4, arachidic 0.3, hexadecenoic 3.2, oleic 31.7, and linoleic 2.3%, and the component glycerides: fully saturated 20 (tripalmitin 15), oleodipalmitin 38, oleopalmitostearin 17, palmitodiolein 19, stearodiolein 6%. Apart from the fully saturated glyceride content, the rest of the fat is constituted on the usual lines, and bacury fat thus resembles laurel kernel fat in that it is only the fully saturated glycerides which are abnormal. It is possible, but of course not in any way proven, that such departures from the normal are caused by certain acids (in this instance, palmitic) being produced in the seed at some stage of its development in much greater proportions than the average content of the acid in the total seed fat at maturity; if so, the departure from normality would be more apparent than real. Elsewhere, Hilditch and Williams report among the saturated fatty acids 1.2% C_{14}, 57.2% C_{16}, 6.0% C_{18}, and 0.2% C_{20} or above. Among the glycerides, 19% were trisaturated, 55% were disaturated, 26% monosaturated.[128]

Description — Large trees with yellowish sap; trunk straight, cylindrical, 50 to 55 cm in diameter (up to 1.3 m), free of branches 20 to 25 m up, with indistinct, low, thick butresses; cortex dark-gray, with deep vertical cracks 1 to 3 cm apart, or with large scales 5 to 25 mm thick; crown broad, flattened, with thick straight slightly slanting branches; twigs straight and stout; entire plant glabrous except the inflorscence. Leaves remote at ends of branches and in single pairs at ends of short lateral branchlets; petioles 1 to 2 cm long, margined; blade elliptic, obovate or oblong to oblong-lanceolate, apex and base rounded, acute or slightly acuminate, up to 15 cm long, coriaceous, glossy above, midrib flat or impressed above, prominent to strongly prominent beneath; primary veins prominent on both sides, connected by a submarginal vein. Inflorescence 1- to 3-flowered, terminating the leaf-bearing branchlets; peduncle absent; pedicels 1 to 3 cm long, their bases surrounded by a series of deltoid bracts 3 to 4 mm long, leaving transverse scars; flowers minutely pulverulent; sepals ovate to broad semi-orbicular, 6 to 8 mm long; petals elliptic, 3.5 to 4 cm long, pink outside, white inside; bud reddish. Fruit globose, 5 to 7.5 cm long, green, turning yellow; mesocarp edible, often containing only 1 seed, rarely more, of pleasant flavor. Flowers September to November; fruits March to May.[278]

Germplasm — Native to the South American Center of Diversity, bacury tolerates sand and some waterlogging. Mors and Rizzini[222] state "it would be an ideal object of study for plant breeders, who could increase the pulpy part at the expense of the very large seeds". Oilseed specialists might breed in the other direction.

Distribution — Native to Brazil (Para, Maranhao, Ceara, Goyaz, Amazon) and Guyana.[278]

Ecology — Estimated to range from Subtropical Dry to Wet through Tropical Dry to Wet Forest Life Zones, perhaps tolerating annual temperatures of 18 to 25°C, annual precipitation of 5 to 40 dm, and pH of 4.5 to 8.0. On sandy, dry plains and in marshy regions, growing scattered in tropical environment.[278]

Cultivation — Trees grow naturally from seeds in the forests, and the tree is not known to be cultivated.[278]

Harvesting — Trees are cut from the forest for timber. Fruits and seeds are collected by natives and sold at trading centers. Most of the products are used locally by the natives.[278]

Yields and economics — No yield data available. Seeds and fruits are sold at local markets in Brazil, French Guyana, and Surinam. Also lumber, dyewood, and fiber are sold in some markets.[278]

Energy — Prunings and falling leaves might provide 5 to 10 MT dry matter per ha per yr, which could be diverted to energy production, for direct combustion or conversion into alcohol or methane. With no yield data on the nuts, I cannot speculate as to how much renewable oil, resin, and fuelwood this produces.

Biotic factors — No serious pests or diseases have been reported for this tree.

PRUNUS DULCIS (MILL.) D.A. WEBB (ROSACEAE) — Almond

Uses — Almonds are cultivated for the nuts, used in candies, baked products, and confectioneries, and for the oils obtained from the kernels. Oil is used as a flavoring agent in baked goods, perfumery and medicines. Benzaldehyde may be used for almond flavoring, being cheaper ($1.54/kg) than almond oil ($5.28 to $6.60/kg).[351] Much valued in the orient because it furnishes a very pleasant oil. In Tuscany, almond branches are used as divining rods to locate hidden treasure. Modern English Jews reportedly still carry branches of flowering almonds into the synagogue on spring festival days. There is the legendary story of Charlemagne's troops' spears (almond) sprouting in the ground overnight and shading the tents the next day. As essential oils go, there is only bitter almond oil. Sweet almond oil is used for cosmetic creams and lotions, although in a crisis, it might conceivably be used as an energy source. The gum exuded from the tree has been used as a substitute for tragacanth.[85]

Folk Medicine — According to Hartwell,[126] the seed and/or its oil are used in folk remedies for cancer (especially bladder, breast, mouth, spleen, and uterus), carcinomata, condylomata, corns, indurations and tumors. Reported to be alterative, astringent, carminative, cyanogenetic, demulcent, discutient, diuretic, emollient, laxative, lithotriptic, nervine, sedative, stimulant and tonic, almond is a folk remedy for asthma, cold, corns, cough, dyspnea, eruptions, gingivitis, heartburn, itch, lungs, prurigo, skin, sores, spasms, stomatitis, and ulcers. The kernels are valued in diet, for peptic ulcers. It is no surprise that the seeds and/or oil (containing amygdalin or benzaldehyde) are widely acclaimed as folk cancer remedies, for all sorts of cancers and tumors, calluses, condylomata, and corns. Lebanese extract the oil for skin trouble, including white patches on skin; used throughout the Middle East for an emollient; also for itch. Raw oil from the bitter variety is used for acne. Almond and honey was given for cough. Thin almond paste was added to wheat porridge to pass gravel or stone. It is believed by the Lebanese to restore virility. Iranians make an ointment from bitter almonds for furuncles. Bitter almonds, when eaten in small quantity, sometimes produce nettle-rash. When taken in large quantity, they may cause poisoning. Ayurvedics consider the fruit, the seed and its oil aphrodisiac, using the oil for biliousness and headache, the seed as a laxative. Unani use the seed for ascites, bronchitis, colic, cough, delirium, earache, gleet, hepatitis, headache, hydrophobia, inflammation, renitis, skin ailments, sore throat, and weak eyes.[91,85]

Chemistry — Per 100 g, the seed is reported to contain 547 to 605 calories, 4.7 to 4.8 g H_2O, 16.8 to 21.0 g protein, 54.1 to 54.9 g fat, 17.3 to 21.5 g total carbohydrate, 2.6 to 3.0 g fiber, 2.0 to 3.0 g ash, 230 to 282 mg Ca, 475 to 504 mg P, 4.4 to 5.2 mg Fe, 4 to 14 mg Na, 432 to 773 mg K, 0 to 5 μg beta-carotene equivalent, 0.24 to 0.25 mg thiamine, 0.15 to 0.92 mg riboflavin, 2.5 to 6.0 mg niacin, and traces of ascorbic acid. According to WOI, the seeds contain 5.8 mg/100 g Na, 856 K, 247 Ca, 257 Mg, 4.23 Fe, 0.14 Cu, 442 P, 145 S and 1.7 Cl. About 82% of the P is in phytic acid. Seeds contain 0.45 ppm folic acid, 150 mg/kg alpha-tocopherol and 5 mg/kg gamma-tocopherol. The chief protein is a globulin, amandin, which contains 11.9% arginine, 1.6% histidine,, 0.7% lysine, 2.5% phenylalanine, 4.5% leucine, 0.2% valine, 1.4% tryptophane, 0.7% methionine, and 0.8% cystine. The approximate fatty acid composition of the oil is 1% myristic, 5% palmitic, 77% oleic, and 17% linoleic.Sweet almond oil from Kashmir showed 0.2% myristic, 8.9% palmitic, 4.0% stearic, 62.5% oleic, and 24.4% linoleic. The essential oil is 81 to 93% benzaldehyde, close kin to laetrile. The hulls (fleshy pericarp) contain: 7.5% moisture, 25.6% total sugars, 7.2% reducing sugars, 4.4% tannin, 2.6 to 4.7% protein, 1.6% starch, 2.4% pectin, 1.1 to 1.2% ether extract, 12.6% crude fiber, and 4.6 to 6.3% ash.[70] The gum which exudes from the trunk hydrolyses into 4 parts L-arabinose, 2 parts D-xylose, 3 parts D-galactose, and 1 part D-glucuronic acid. The edible portion of the nuts contain 3.2 ppm Al, 0.02 As, 0.001 Au, 18 B, 2.6 Ba, 20 Br, 2720 Ca, 0.02 Cd, 28 Cl, 0.2 Co, 1.7 Cr, 0.1 Cs, 14 Cu, 0.1 Eu, 1.3 F, 54 Fe, 0.04 Hf, 0.1 Hg, 0.1 I, 6346 K, 0.03 La, 0.01 Lu, 2297 Mg, 14 Mn, 0.3 Mo, 20 Na, 1.6 Ni, 0.4 Pb, 13 Rb, 3420 S, 0.1 Sb, 0.003 Sc, 0.02 Se, 960 Si, 0.1 Sm, 0.7 Sn, 16 Sr, 0.03 Ta, 0.2 Th, 3.5 Ti, 0.02 V, 0.1 W, 0.1 Yb, 32 ppm Zn dry weight. The normal concentration of some of these elements in land plants are 50 ppm B, 14 Ba, 15 Br, 2000 Cl, 0.5 Co, 0.2 Cs, 14 Cu, 3.200 Mg, 630 Mn, 3 Ni, 20 Rb, 3,400 S, 26 Sr, and 0.2 ppm Se dry weight. They were higher in calcium and chromium than any of the 12 nut species studied by Furr et al.[102]

Description — Tree to 10 m tall, the alternate leaves lanceolate to oblong lanceolate, minutely serrate. Flowers solitary, white to pink, actinomorphic, 20 to 50 mm broad, appearing with or before the foliage. Fruit an oblong drupe 30 to 60 mm long, pubescent, the tough flesh splitting at maturity to expose the pitted stone; endocarp thin or thick; seed flattened, longovoid, the seed coat brown.

Germplasm — Reported from the Central Asian and Near Eastern Centers of Diversity, almond or cvs thereof is reported to tolerate drought, frost, high pH, heat, mycobacteria, nematodes, slope, and wilt.[82] 'Cavaliera' is very early, 'Nonpareil' early, 'Ferragnes' medium, 'Marcona' late, and 'Texas' very late. (2n = 16.)

Distribution — Widely distributed in cultivation now, the sweet almond is said to have wild types in Greece, North Africa, and West Asia. Almond was cultivated in China in the 10th Century BC, in Greece in the 5th Century BC.

Ecology — Ranging from Cool Temperate Moist to Wet through Subtropical Thorn to Moist Forest Life Zones, almond is reported to tolerate annual precipitation of 2.0 to 14.7 dm (mean of 11 cases = 7.5) annual temperature of 10.5 to 19.5°C (mean of 11 cases = 14.8) and pH of 5.3 to 8.3 (mean of 7 cases = 7.3). Almond does well in the hot, dry interior valleys of California, where the nuts mature satisfactorily. The leaves and nuts are less subject to attack by disease-causing fungi in the hot, dry climate than under cooler and more humid conditions. It has a low winter chilling requirement. Because of this low chilling requirement (or short rest period), and the relatively low amount of heat required to bring the trees into bloom, the almond is generally the earliest deciduous fruit or nut tree to flower, hence extremely subject to frost injury where moderately late spring frosts prevail. Almonds need ample rainfall or irrigation water for maximum production of well-filled almond nuts. Trees have been planted in certain areas where supplies of water are inadequate for other fruit or nut crops; however, yields of nuts were low. In general, conditions favoring peach production will also favor almonds. The almond tree has been successfully grown on a wide range of soils. It is a deep-rooted tree and draws heavily on the soil, which should be deep, fertile, and well drained. Sandy loams are best. Since sandy soils are often deficient in plant food elements, careful attention must be paid to proper fertilization of the trees. Almond trees have high N and P requirements. Sandy soils are easy to cultivate, and cover crops are comparatively easy to grow on them provided they are properly fertilized.

Cultivation — In India, trees are raised from seedlings, the seeds usually having a chilling requirement. Seeds are sown in nurseries, the seedlings transplanted after about one year. For special types, as in the U.S., scions are budded or grafted on to bitter or sweet almond, apricot, myrobalan, peach, or plum seedlings. Trees are planted 6 to 8 m apart and irrigated, in spite of their drought tolerance. Application of nitrogenous and/or organic fertilizers is said to improve yield. Trees should be pruned to a modified leader system. All types are self-sterile, so cvs or seedlings should be mixed.

Harvesting — Fruits occur mainly on shoot spurs, which remain productive up to five years. Bearing trees may be pruned of surplus branches to about 20% of the old-bearing wood. Tree exhibiting decline may be severely cut back at the top. In India, the trees bear from July to September. Fruits are harvested when the flesh splits open exposing the stone. The flesh is then removed from the stones manually or by machine.

Yields and economics — In 1971, commercial almond production in the U.S. was centered in California, which produced more than 99% of the domestic marketed nuts. California's production of in-shell nuts during the 1960s nearly tripled. It reached about 140,000 in-shell tons in 1970. Only sweet almonds are grown commercially. Imports, largely from Spain and Italy, vary widely from year to year, ranging from about 280 to 1,700 tons on the in-shell basis for the past 7 years. The U.S. imported 67,252 kg of bitter almond oil worth $271,600 in 1981, 354 kg from Canada worth $1,300, 48,470 kg from France worth $221,300, 998 kg from Haiti worth $2,600, 17,400 kg from Spain worth $46,000, and 30 kg from Switzerland worth $400. On August 2, 1982,[351] posted prices were ca. $7.70/kg of natural bitter almond, and $2.64/kg of sweet almond. Dealers in bitter almond oil include:

Berge Chemical Products, Inc.
5 Lawrence Street
Bloomfield, NJ 07003

Florasynth, Inc.
410 E. 62nd Street
New York, NY 10021

Hagelin & Co., Inc.
241 Cedar Knolls Road
Cedar Knolls, NJ 07927

International Sourcing, Inc.
555 Route 17 S.
Ridgewood, NJ 07450

Dealers in sweet almond oil include:

Berje Chemical Products, Inc.
5 Lawrence Street
Bloomfield, NJ 07003

Lipo Chemicals, Inc.
207 Nineteenth Avenue
Paterson, NJ 07504

Mutchler Chemical Co., Inc.
99 Kinderkamack Road
Westwood, NJ 07675

PPF Norda Inc.
140 Rt. 10
East Hanover, NJ 07936

Energy — According to *The Wealth of India*,[70] average California yields are ca. 400 kg/ha, but they attain over 1,200 kg/ha. However, for Baluchistan, WOI reports 2,375 kg/ha, basing this on an optimistic yield of 7.3 kg for each of 325 trees per ha. Yields of 2 to 3 kg per tree seem more realistic; Duke,[82] however, reports seed yields of 3000 kg/ha. With an oil yield of 50 to 55%, it is easy to project oil yields of 1500 kg/ha. With recommended pruning to 20% of the old-bearing wood, several MT firewood should be available from the pruning. Analyzing 62 kinds of biomass for heating value, Jenkins and Ebeling[149] reported a spread of 20.01 to 18.93 MJ/kg, compared to 13.76 for weathered rice straw to 23.28 MJ/kg for prune pits. On a percent DM basis, the orchard prunings of almond contained 76.83% volatiles, 1.63% ash, 21.54% fixed carbon, 51.30% C, 5.29% H, 40.90% O, 0.66% N, 0.01% S, 0.04% Cl, and undetermined residue. The hulls, showing a spread of 17.13 to 18.22 MJ/kg, contained 71.33% volatiles, 5.78% ash, 22.89% fixed carbon, 45.79% C, 5.36% H, 40.60% O, 0.96% N, 0.01% S, 0.08% Cl, and undetermined residue. The shells, with a spread of 18.17 to 19.38 MJ/kg, contained 73.45% volatiles, 4.81% ash, 21.74% fixed carbon, 44.98% C, 5.97% H, 42.27% O, 1.16% N, 0.02% S, and undetermined residue.

Biotic factors — Prominent diseases in India include "shot hole" caused by *Clasterosporium carpophilum* (Lev.) Aderh., "white spongy rot" due to *Fomes lividus* Kl, "brown patchy leaf rot" due to *Phyllosticta prunicola* (Spiz) Sacc., "brown rot" due to *Sphaerotheca pannosa* (Wallr.) Lev. and a mosaic disease due to virus; all plague the almond. The chrysomelid *Mimastra cyanura* Hope and the almond weevil *Myllocerus laetivirens* Marshall feed on the leaves. The San Jose scale *Quadraspidiotus perniciosus* Comstock is a minor problem. The almond moth *Ephestia cautella* Wlk. infests shelled almonds and dried apricot, currant, date, fig, peach, and plum.

QUERCUS SUBER L. (FAGACEAE) — Cork Oak
Syn.: *Quercus occidentalis* Gay

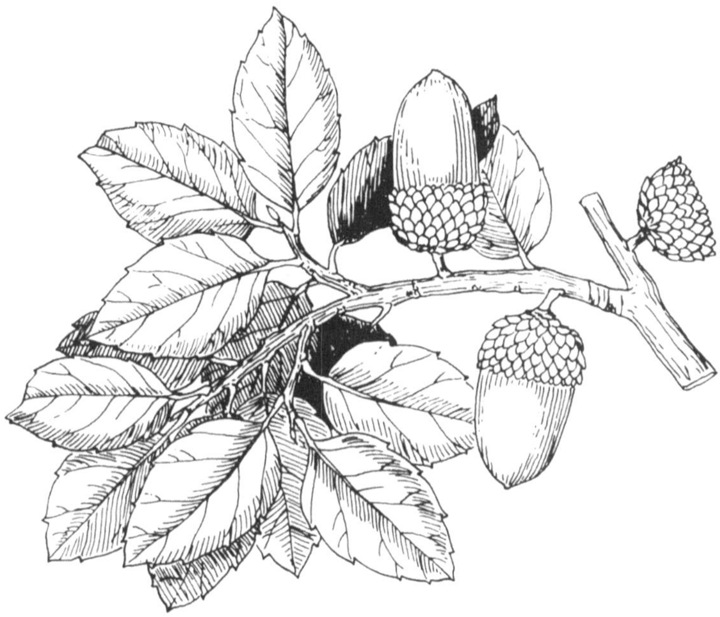

Uses — Bark provides the cork of commerce, used for bungs and stoppers for bottles and other containers, life preservers, mats, ring buoys, floats, shoe inner-sole liners, artificial limbs, sealing liners for bottle caps, novelties, switch-boxes, household appliances and friction rolls, gaskets of various types for automobiles, electric motors, polishing wheels, cork-board, and for insulation, acoustical, and machinery isolation purposes. It is also used in the manufacture of linoleum. The hard wax extracted from the cork waste is used for making shoe pastes.[70,278] Acorns provide forage for hogs, and the orchards are profitably grazed as well by sheep and goats.[310] Acorns may be eaten, especially when roasted, in cases of necessity.[209] Acorns of all oaks can be converted into "edible nuts", but in the bitter species much work is involved, compared to the "sweet oaks" like *Quercus prinos*.

Folk medicine — No data available.

Chemistry — Age, growing conditions, and grades of the bark determine the chemical composition of cork. A good specimen conforms to the following values: 3 to 7% moisture, 20 to 38% fatty acids, 10 to 18% other acids, 2.0 to 6.5% tannins, 1.0 to 6.5% glycerin, 12.6 to 18.0% lignin, 1.8 to 5.0% cellulose, 4.5 to 15.0% ceroids (waxes, stearins, etc.), 0.1 to 4.0% ash, 8 to 21% other substances. Suberin, the characteristic constituent of cork, is composed mainly of high-molecular polymerides of hydroxy fatty acids, the major component being phellonic acid (22-hydroxy docosanoic). Other fatty acids present are phloionic (9,10-dihydroxy octadecanediotic), phloionolic (9,10,18-trihydroxy octadecanoic) and its stereoisomer (m.p. 133°), cis- and trans-9-octadecenoic, 18-hydroxy-9-octadecenoic, and several unidentified acids. Crude cork wax contains cerin (chief constituent), friedelin, steroids, acids, etc.[70] Suberin is a mixture containing several acids, including phloionic acid ($C_{18}H_{34}O_6$), acid XX ($C_{18}H_{32}O_4$), phloionolic acid ($C_{18}H_{36}O_5$), acid XVIII ($C_{18}H_{34}O_3$), acid V ($C_{18}H_{34}O_4$), phellonic acid ($C_{22}H_{44}O_3$), and phellogenic acid ($C_{22}H_{42}O_4$). The cork wax is a mixture of esters and triterpenes (cerin $C_{30}H_{50}O_2$; friedelin $C_{30}H_{50}O$; betulinic acid, betulin, and suberindiol $C_{28}H_{46}O_2$), also tannin, phlobaphen, cellulose, ligin, cyclitol, and vanillin. Thus, synthetic vanilla could be a by-product of the cork industry.[187] The bark contains much silica.[215]

Toxicity — Exposure to the bark is reported to produce a respiratory disorder, suberosis, which starts with rhinitis, cough, and dyspnea, and then proceeds to chronic bronchitic changes or extrinsic allergic alveolitis.[215]

Description — Large, subtropical, evergreen tree, to 20 m tall, the trunk circumference to 10 m, with thick, corky bark; twigs tomentose. Leaves 3 to 7 cm long, ovate-oblong, sinuate-dentate, dark-green above, gray-tomentose beneath; midrib sinuous; petiole 8 to 15 mm long. Male flowers in aments, female flowers in small clusters on short twigs. Fruit ripening in the first year in spring-flowering trees, but some trees flower in autumn and ripen their fruits late in the following summer; involucral scales long and patent, the lower usually shorter and more appressed.[278]

Germplasm — Reported from the Mediterranean Center of Diversity, cork oak, or cvs thereof, is reported to tolerate drought, high pH, poor soil, and sand.[82] Highly variable, with only one type differing sufficiently to be regarded as a subspecies, i.e., *Q. occidentalis*, differing principally in its slower maturing acorns, known from the Iberian Peninsula, southeastern France, and Corsica. Individual clones have been selected and cultivated in many areas, including the U.S.[278,288,289] (2n = 24.) Among the American oak species, acorns of chestnut oak and white oak are most likely to serve as nuts.

Distribution — Native and forming extensive forests (in the past) from northwestern Yugoslavia, west to Spain and Portugal, the islands of the western Mediterranean and north Africa (Morocco and Algeria). Introduced and cultivated for cork in eastern India, Japan (southern islands), and in southern California. Trees also planted from New Jersey to Florida and westward to California for experimental purposes in the late 1940s.[278]

Ecology — Ranging from Cool Temperate Moist through Tropical Dry Forest Life Zones, cork oak is reported to tolerate annual precipitation of 3.1 to 13.5 dm (mean of 10 cases = 8.2), annual temperature of 9.7 to 26.5°C (mean of 10 cases = 16.3), and pH of 4.9 to 8.2 (mean of 9 cases = 6.9).[82] Hardy to Zone 7.[343] Subtropical climate is essential for good bark formation. Trees have withstood temperatures of − 18°C in South Carolina. In general, a mean annual temperature of not less than 5°C with range of not lower than 2°C and maximum mean annual temperature of 21°C is best for growth. About 57% of cork is grown in the 18 to 21°C region. Trees are quite drought-resistant and do not require irrigation after the first few years. Will grow well with 2.5 to 10 dm annual rainfall; optimum is 5 to 10 dm/year. Grows best in neutral or slightly acid, sandy, well-drained, soils. Trees grow from sea-level up to 1,300 m. Though granitic, clay, or slate soils are suggested,[278] Smith[310] says, "the poorer the soil, the better the cork".

Cultivation — Best method of planting is by direct seeding. Ripe acorns are planted in groups of 4 or 5 (about 625 groups per ha), each group in a shallow furrow covered to a depth of 1.3 cm. The stand is later thinned so that one plant remains at each site. Seeds may be germinated in seed-beds and transplanted later, but the seedlings should not be disturbed after the taproot has become established. Viability of seeds is short, but can be lengthened by wet cold storage at 0.5 to 1.5°C.[278] Requiring no stratification, the seeds show 73 to 100% germination after 20 to 30 days at 27°C day and night temperatures.[5] Trees may also be grafted on both evergreen or deciduous native oaks. Techniques for clonal cuttings have been worked out.[310] Older saplings should be thinned to avoid shading. At age 50, trees should be thinned to ca. 500 per ha; at 75 years to about half that number; at 120 years, there should be about 100 per ha. With such reduction, overcrowding is avoided and cork production per ha is relatively stabilized.[278]

Harvesting — Cork of commercial value is not produced by trees less than 30 years old. Since transplanting of saplings and small trees should not be attempted, and direct seeding is practiced, it is impossible to bring trees into production in less than 30 years. First stripping of bark may occur when the tree is about 20 years old. This virgin bark or mascalage is rough and coarse and of little commercial value. Its removal stimulates the growth of cork so that during the succeeding 2 or 3 years, much of the cork is produced. In Algeria, this virgin or male bark is put back in place around the tree and held there by wires for 2 years or so, thus protecting the new bark that is forming. This growth gradually decreases in rate until after about 9 years scarcely any further thickening of the bark is perceptible, and at the end of that period, the second stripping takes place. The second and all subsequent strippings produce bark of commercial value. At around 120 years, decline sets in. Replanting should follow. Harvest is rotated, with only a certain number of trees stripped each year. Each tree is stripped, usually at 9-year intervals, but intervals may vary from 6 to 12 years, depending on the conditions of growth. If pruning is necessary, trees should not be stripped until 3 years after pruning. In North Africa, bark is stripped in winter; in other areas, in spring, when the sap is rising to make bark removal easier. Cork stripping requires considerable skill. Bark must be removed without injuring the inner-most layer, which must remain to continue growth.[278] Acorns may be borne at age 12, with good crops every 2 to 4 years.

Yields and economics — Mature trees yield good quantities of cork for 150 to 200 years in the Mediterranean region. Trees yield about 1.3 kg of cork per stripping, in California. In the Mediterranean, each tree yields from 20 to 240(to 300) kg at each stripping, depending on age and size of tree.[278] Trees are stripped at intervals of 9 to 12 years. About 12 ℓ of acorn will yield a kg of pork.[310] In Portugal, a cork oak forest is said to produce 34 kg/ha pork compared to 68 for a *Quercus ilex* forest. "Lard from acorn-fed hogs is said not to harden; hence they are sometimes finished on corn for hardening the fat."[310] Portugal is the largest producer of cork, supplying 46.2% of the world's tonnage from 33.8% of the total hectarage. There are about 69,000 ha of Portuguese cork oak forests, mainly in the south-central portion of the country.[278] Smith[310] reports 400,000 ha in Portugal producing annually 240 kg/ha. Between 1931 and 1948, cork was varying widely in price, from $30 to $600/ton. Bigger and better trees can yield a ton in one stripping, following another ton 12 years earlier. English owners of cork estates in Portugal estimate that acorns alone produce 1/2 to 2/3 of Portuguese pork. The USDA once said "one gallon of acorns is equal to ten good ears of corn." Pigs may graze the grass and acorns while sheep and goats may graze the bushes and shrubs.[310]

Energy — Felled trees and bigger prunings make excellent charcoal. With low energy input on tough terrain, this seems to be an energy-efficient land-holding scenario yielding cork, firewood, pork, and land stability.

Biotic factors — The following fungi have been reported on the cork oak: *Armillariella mellea, Ascochyta irpina, Aspergillus terreus, A. wentii, Auricularia mesenterica, Chalara quercina, Clitocybe olearia, Coccomyces dentatus, Coriolus pergamenus, C. versicolor, Cyphella candida, Cytospora microspora, Daedalea biennis, Diatrypella quercina, Endothis gyrosa, Ganoderma applanatum, Hirneola auricula judae, Hypoxylon mediterraneum, Irpex deformis, Ithyphallus imperialis, I. impudicis, Lenzites quercina, Leptoporus adustus, L. dichrous, Leucoporus brumalis, Merulium tremellosus, Mucor ramannianus, Mycoleptodon ochraceum, Panus conchatus, Peniphora corticalis, Phellinus igniarius, P. torulosus, Pholiota cylindricea, P. spectabilis, Phoma quercella, Physalospora elegans, Phytophthora cinnamomi, Pleurotus lignatilis, P. ostreatus, Polyporus giganteus, Poria vaporaria, Propolis faginea, Radulum quercinum, Schizophyllum commune, Sebacina crozalsii, Septoria ocellata, S. quercicola, Sphaerotheca lanestris, Stereum fuscum, S. gausapatum, S. spadiceum, Tomentella fusca, T. rubiginosa, Tomentellina bombycina, Trametes campestris, T. cinnabarina, T. serialis* var. *resupinata, Ungulina fomentaria, U. ochroleuca, Uredo quercus, Volvaria bombycina, Vuilleminia comedens, Xanthochrous cuticularis, X. ribis.* The following nematodes have been isolated from this oak: *Caconema radicicola* and *Heterodera marioni.*[186,278]

RICINODENDRON HEUDELOTII Pierre (EUPHORBIACEAE) — Manketti Nut, Sanga Nut, Essang Nut, Ojuk Nut
Syn.: *Ricinodendron africanum* Muell. Arg.

Uses — Source of essang oil, seeds contain 35 to 55% oil, which has been recommended in the drying oil industries. The nuts are consumed as food after boiling. Dried kernels are ground and cooked with food, e.g., in the Cameroons. The kernels only account for ca. 30% of the fruit, the hard shell is difficult to remove. Wild animals, including elephants, are fond of the fallen fruits, leading hunters to lie in wait beneath the trees. Ashes of the wood are used for salt and in the preparation of soap and indigo. Williams[340] describes the use of this species as living telegraph poles. Stakes 6 to 10 m long are cut and placed in holes. During the rainy season, the stake quickly strikes root. Wires are placed on the poles 6 or more meters above the ground as soon as they are firm. Branches tend to sprout only at the summit, rarely interfering with the wires. The wood, quite light, has been suggested as a substitute for balsa. Easily carved, it is used for utensils, masks, musical instruments, boxes, coffins, etc. The hard seeds are used, like marbles, in games, rattles, etc. The very light sawdust is suitable for life jackets and pith helmets.[71,146,399]

Folk medicine — Nigerians use the root-bark, with pepper and salt, for constipation. On the Ivory Coast, the decoction is drunk for dysentery. Pounded and warmed bark is applied locally for elephantiasis. The bark infusion is used in Liberia to relieve labor pains and prevent miscarriage, in the belief that it prevents sterility. The pulped bark prevents abortion. The bark decoction is used for gonorrhea; the leaf decoction as a beverage or bath in calming fever.[71,146]

Chemistry — The seed fatty acids of *R. africanum* include ca. 50% eleostearic acid with ca. 25% linoleic-, 10% oleic-, and 10% saturated acids.[128] The seed, seed shell, and latex, containing a resin, are used for diarrhea and gonorrhea.[332]

Description — Fast-growing, deciduous tree to 33 m or more high and up to 2.5 m girth; buttresses very short, branches whorled. Leaves alternate, hairy when young, with stellate hairs, digitately lobed, the 3 to 5 leaflets up to 25 × 15 cm, sessile, obovate-elliptic, acuminate, narrowed to base, with 10 to 16 pairs of lateral nerves, petioles up to 20 cm long, stipules persistent and leaf-like; flowers paniculate (December to April in Africa), the inflorescence yellow-tomentose, white, falling readily. Fruits 3-celled, ca. 2.5 cm in diameter. Seeds ovoid, rich in oil.

Germplasm — From the Africa Center of Diversity, the essang nut seems to tolerate savanna, second growth, slopes, and weeds.

Distribution — Widespread in tropical Africa. Fast-growing native of the secondary forests of the Belgian Congo and possibly also of Nigeria.[340] Irvine[146] describes it as common in fringing, deciduous, and secondary forests, from Guinea to Angola and the Belgian Congo to Sudan, Uganda, and East Africa.

Ecology — With no ecological data available to me, I speculate that this species ranges from Subtropical Thorn Woodland to Moist through Tropical Thorn Woodland to Moist Forest Life Zones, tolerating annual precipitation of 3 to 25 dm, annual temperature of 23 to 29°C, and pH of 6 to 8.[82] According to Williams,[340] it requires a wet, humid climate.

Cultivation — Coppicing and rooting readily, this tree is often planted as cuttings for vine stakes, living fences, and telephone poles.

Harvesting — Said to bear fruit in its 7th to 10th year.[71]

Yields and economics — Irvine describes the nut yields as prolific.

Energy — The wood does not make good firewood, but "it is much used for fuel". Seeds yield 45 to 47% oil which could be used for energy, but because of the high husk/kernel ratio, the fruits yield only ca. 14% oil.

Biotic factors — According to a forester quoted by Menninger,[209] elephants eat the fruits greedily, and "seed will not germinate until it has spent a week in the elephant", but even the elephant's digestive system barely affects the fruit and the enclosed kernel. "The natives of Rhodesia, therefore, follow the elephant, recover the hard-shelled nuts where they have been dropped, clean and dry them, then crack the extremely hard shell, and find the contents perfectly delicious. This story is a bit grizzly, but it is part of the nut story."[209] The fungus *Fomes lignosus* is reported to attack this species.[186]

RICINODENDRON RAUTANENII Schinz (EUPHORBIACEAE) — Mogongo Nut, Manketti

Uses — A much-prized species with edible fruits that are a staple food of Africans and Bushmen, who eat them raw (fresh or fried), cooked, or fermented into a beer. The thin, fleshy portion, under the tough skin, may be eaten raw or cooked into a sweet porridge. The kernel has a sweet, milky, nutty flavor; eaten raw, pounded and fried, or mixed with lean meat. The seeds can be roasted whole, cracked, and the kernels pounded into a coarse meal, which is eaten dry, with meat, with other roots, or mixed with baobab pulp. It is the main food (constituting half of the vegetable diet) of the Bushmen in the Dobe area. One to three hundred nuts are consumed every day for all but a few months of the year. Also a staple food of elephants. The timber is yellowish, light, and soft, and is used for carving bowls, cups, ashtrays, and ornamental figures of animals and birds. Also used for floats, dart and drawing boards, packing-cases, boxes, toys, insulating material, and coffins. Trees are often hollow and collect life-saving water. It is believed, in South West Africa and Botswana, that this species controls the weather, so that it is never struck by lightning.[246] During one study of the !Kung Bushmen, mongongo nuts contributed 56.7 g protein per day per bushman, compared to 34.5 g from meat, and only 1.9 from other vegetable foods. To the !Kung, the mongongo nut is "basically the staff of life".[124] The light timber is used for furniture, coffins, and an inferior paper.[332]

Folk medicine — The fruits are astringent.

Chemistry — The average daily per capita consumption of 300 nuts weighs ca. 212.6 g but contains the caloric equivalent of 1,134 g cooked rice and the protein equivalent of 396.9 g lean beef. Watt and Breyer-Brandwijk[332] say the fruits contain 7.9% protein and no true starch. Fruits yield 30 to 40% oil, kernels 57 to 63%. Skins of the kernel yield 37% oil. The oil cake has only 0.32 mg vitamin B_1 and 0.7 mg calcium pantothenase per 100 g. The percentages of amino acid in the seed protein are calculated at 2.6% histidine, 4.1% cystine, 7.9% isoleucine, 6.2% leucine, 5.1% lysine, 2.0% methionine, 4.6% phenylalanine, 7.9% threonine, 1.2% tryptophane, and 7.1% valine.[309] The seed fatty acids of the related *R. africanum* ("essang oil") include ca. 50% eleostearic acid with ca. 25% linoleic-, 10% oleic-, and 10% saturated acids.[128] The aromatic fruit contains a gum-resin and 31% saccharose.

Toxicity — The seed coat is nontoxic to rats when constituting 10% of diet, but it is an unsuitable food because of its toughness and indigestibility.[332]

Description — Spreading, deciduous, dioecious tree to 10(to 24) m tall, the trunk to 1 m in diameter; the bark greenish or goldish; twigs and branches, stubby with glabrescent robust young twigs. Leaves alternate, stipulate, digitately compound, with 3 to 7 leaflets; petioles pubescent, to 15 cm long; leaflets broadly lanceolate to ovate, apically blunt or rounded, basally rounded or truncately inequilateral, marginally glandular denticulate, rarely lobulate, 5 to 13 cm long, 2.5 to 9 cm broad; dark-green above, pale below, with stellate hairs on both surfaces, the midribs and veins rufose; petiolules biglandular. Male flowers in slender loose panicles, whitish, the female panicles shorter and few-flowered. Fruits plum-shaped, to 4 cm long, hairy when young, the stone exceedingly hard, containing one or two light-colored kernels.[246]

Germplasm — Reported from the African Center of Diversity, mogongo nut, or cvs thereof, is reported to tolerate sands and savannas.[246,332]

Distribution — Northern southwest Africa, Botswana, Zimbabwe, and Mozambique, and often tropical Africa; grows in groves or forests together on wooded hills and dunes, and always on Kalahari sand. Makes almost pure forest in parts.

Ecology — With no ecological data available to me, I speculate that this species ranges from Subtropical Thorn Woodland to Moist through Tropical Thorn Woodland to Moist Forest Life Zones, tolerating annual precipitation of 3 to 25 dm, annual temperature of 23 to 29°C, and pH of 6 to 8.[82] Tending to flower in spring before rain.

Cultivation — Not normally cultivated.

Harvesting — According to Harlan,[124] women and children are primarily involved in gathering plant materials among the !Kung Bushmen. But adults gather the mongongo nuts. Over a 3-week study period, the Bushmen averaged $2^1/_2$ days a week (average 6 hr work per day) devoted to subsistence activities. Compared to hunting, gathering is a low-risk, high-return enterprise. Fruits ripen ca. February in southern Africa.

Yields and economics — Before the war of 1914-1918, Germans granted a concession to exploit the forests near Tsumeb in Southwest Africa, which were estimated to yield 50,000 tons of nuts per year.[246]

Energy — If there are forests with 50,000 tons[246] of edible nuts therein, the kernels yielding 60% oil, one could theoretically obtain 30,000 tons of oil, and 20,000 tons defatted edible nuts therefrom.

Biotic factors — Fruits greatly relished by elephants.

SANTALUM ACUMINATUM A. DC. (SANTALACEAE) — Quandong Nut, Native Peach

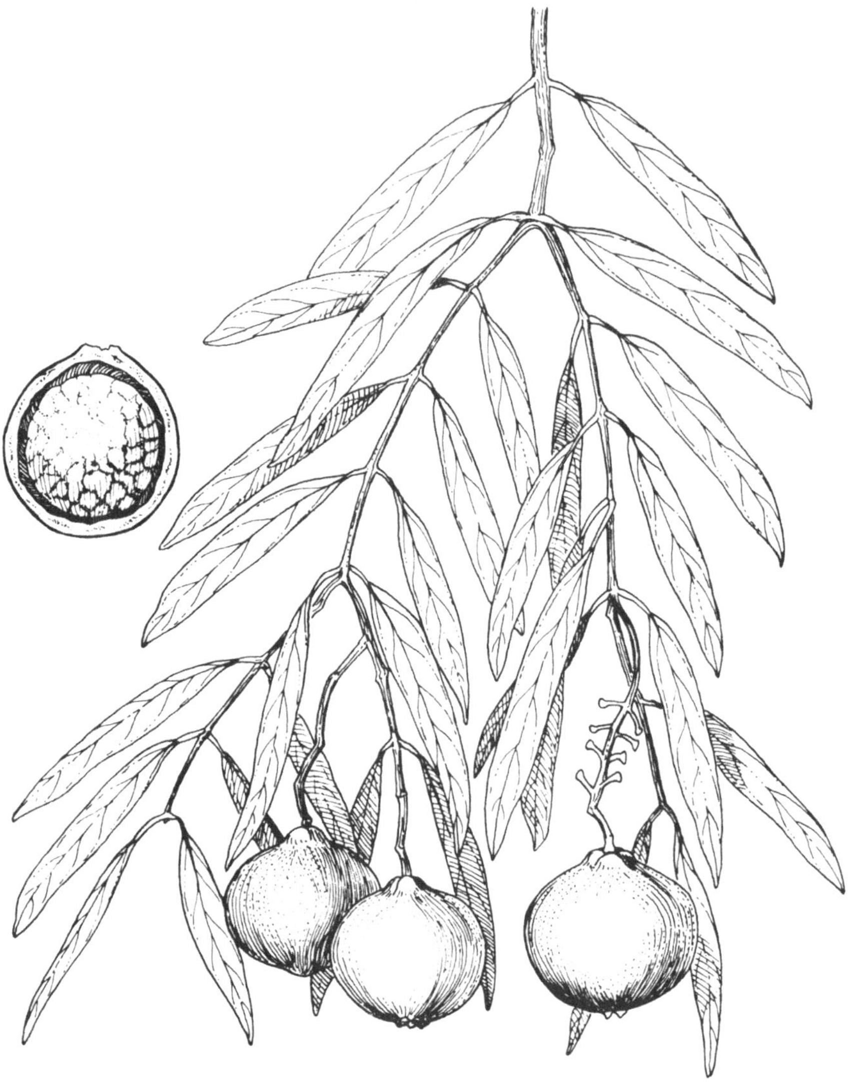

Uses — It's easier for me to remember the popular name Quandong than *Eucarya, Fusanus,* or *Santalum,* the three generic names among which the quandong has been categorized. Both the fruit or ("peach") and nut (or "quandong") are edible. The red flesh is cooked in chutneys, jams, and pies. The nut is said to be quite tasty, slightly roasted, and was a favored food of the aborigines. Pierced with a stick as a candle-nut, the seeds will burn away with a clear light. Nuts are also made into bracelets, necklaces, and other ornaments. The hard, durable close-grained timber is used for cabinet making and wood engraving.[208,283,324]

Folk medicine — The seed oil is used medicinally.

Chemistry — Per 100 g, the fruit is reported to contain 345 kJ, 76.7 g H_2O, 1.7 g protein, 0.2 g fat, 19.3 g carbohydrate, 2.1 g ash, 42 mg Ca, 40 mg Mg, 0.2 mg Zn, 0.2 mg Cu, 51 mg Na, and 659 mg K.[47] Data in Menninger, no doubt reflecting dry nuts, report 60% oil and 25% protein. The fruits are rich in vitamin C. Fatty acids in the seed contain oleic, linoleic, and stearic acids, also santalbinic acid.[187] Some estimates put the "santalbic" content at 40 to 43%. Others say the seed fat is mostly oleic acid except for 3 to 4% palmitic acid.[128] Wood contains 5% essential oil containing nerolidol.

Description — A tall shrub or a tree to 10 m. Leaves opposite, lanceolate, acute, or sometimes when young with a short hooked point, mostly 5 to 7.5 cm long, tapering into a petiole 4 to 6 mm long, coriaceous, with the lateral veins often prominent when old. Flowers rather numerous, in a terminal pyramidal panicle scarcely longer than the leaves. Perianth spreading to ca. 5 mm diameter, the lobes somewhat concave even when open. Free margin of the disk very prominent, broadly rounded between the stamens which curve over the notches. Anthers very short. Style exceedingly short and conical or scarcely any, with deeply 2- or 3-lobed stigma. Fruit globular, 10 to 20 mm in diameter, with a succulent epicarp, and bony pitted endocarp, the perianth-lobes persisting on the top until the fruit is nearly or quite ripe.[39]

Germplasm — Reported from the Australian Center of Diversity, Quandong, or cvs thereof, is reported to tolerate arid conditions and drought.[283,324]

Distribution — Endemic to Australia, especially northern Australia, and the southwest, extending into the desert areas.

Ecology — No data available.

Cultivation — Rosengarten comments on an experimental plantation in Quorn, Australia. Kikuya grass was planted to serve as root host.

Harvesting — Some plantation trees have fruited in the third year.

Yields and economics — Rosengarten sums it up, "Despite its captivating tang, the quandong seems destined to remain a minor Australian nut."[324]

Energy — Serving as candle-nuts, quandongs are so abundant in part of Australia that they might serve as oil-seeds in the future.

Biotic factors — No data available.

SAPIUM SEBIFERUM (L.) Roxb. (EUPHORBIACEAE) — Chinese Tallow Tree, Vegetable Tallow, White Wax Berry

Syn.: ***Carumbium sebiferum*** **Kurz,** ***Croton sebiferus*** **L.,** ***Excoecaria sebifera*** **Muell.,** ***Stillingia sebifera*** **Michx.,** ***Triadica sebifera*** **(L.) Small**

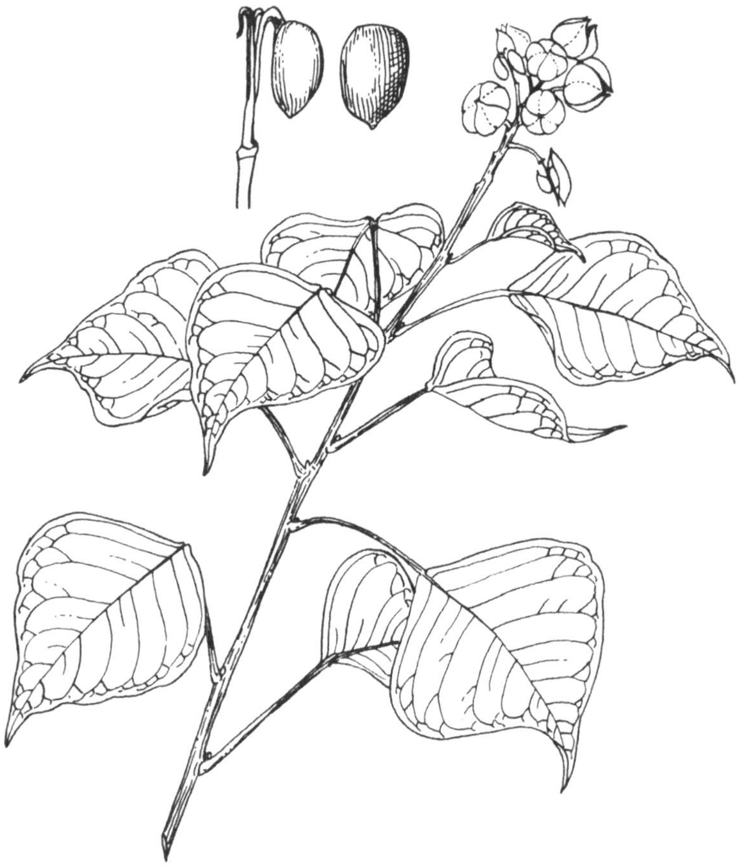

Uses — Chinese Tallow Tree is cultivated for its seeds as a source of vegetable tallow, a drying oil and protein food, and as an ornamental. Fruits yield two types of fats: outer covering of seeds contain a solid fat with low iodine value, known as Chinese Vegetable Tallow; kernels produce a drying oil with high iodine value, called Stillingia Oil. Tallow is used for manufacturing candles, a layer of wax being placed over the tallow body to prevent too-rapid burning; has excellent burning quality, and gives an inodorous, clear, bright flame; also used for making soap, cloth dressing, and fuel. Pure tallow fat is known in commerce as Pi-yu. Oil is used in making varnishes and native paints because of its quick-drying properties, in machine oils and as a crude lamp oil. Pure oil expressed from the inner part of the seeds is known in commerce as Ting-yu. Oil-cakes made from crushed seeds with tallow and oil together is known as Mou-yu. Residual cake, after oil is expressed, is used as manure, particularly for tobacco fields. Wood is white and close-grained, suitable for carving and used for making blocks in Chinese printing; also used for furniture making and incense. Chinese prepare a black dye by boiling leaves in alum water. Tree grows rapidly, develops an attractive crown, and, as leaves turn red in fall, it is cultivated as a shade or lawn tree about houses. It is used as a soil binder along roads and canals. Chinese place an insect on the tree to feed; it lays eggs in the seed, making some of the "jumping beans," because of movements of larvae inside.[261,294,295]

Folk medicine — In Chinese medicine, oil is used as purgative and emetic, not as a usual vegetable oil for humans. Overdose of native medicine probably would cause violent sickness and perhaps death. Additionally, Chinese use the plant as an alexeteric, suppurative, and vulnerary, especially for edema and skin ailments. Decoction of the root bark is used for dyspepsia, considered tonic. Resin from root bark considered purgative. The latex is an acrid and powerful vesicant.[249]

Chemistry — The fatty acid composition of the oil is caprylic, 1.50; capric, 1.00; myristic, 0.97; palmitic, 2.80; stearic, 1.00; oleic, 9.40; linoleic, 53.40; and linolenic, 30.00%. A Hong Kong sample contained 26.8% oil, with: capric, traces; palmitic, 7; stearic, 3; 2,4-decadienoic, 5; oleic, 7; linoleic, 24; and linolenic, 54%. Stillingia oil is considered superior to linseed oil in its drying and polymerizing properties, probably due to the presence of 2,4-decadienoic acid. Seed meal, left after the extraction of oil, possesses a high content of protein, and is a valuable feed and fertilizer. It can be processed into a refined flour, containing 75% protein, fit for human consumption. The amino acid composition of the protein is as follows: arginine, 16.6; aspartic acid, 11.7; cistine, 1.3; glycine, 4.9; glutamic acid, 17.3; histidine, 2.9; leucine, 7.4; lycine, 2.6; methionine, 1.6; tyrosine, 3.7; and valine, 7.8%. The Vitamin B content of the flour compares favorably with that of wheat-flour. The flour, supplemented with lysine and methionine, is reported to be superior to wheat-flour. Ethanol extraction of powdered root bark yielded 0.1% phloracetophenone 2,4-dimethylether, and methanol extraction gave xanthoxylin ($C_{10}H_{12}O_4$). The bark also contains moretenone, moretenol and a new triterpene, 3-epimoretenol (m.p., 223 to 24°). Leaves contain gallic and ellagic acids, isoquercitrin, and tannin (5.5%).[70,133,136]

Description — Small to large deciduous tree, 10 to 13 m tall (in 30 years), often with a gnarled trunk, bark gray to whitish-gray with vertical cracks; stem exudes a milky poisonous juice. Leaves alternate, broad rhombic to ovate, 3.5 to 8.5 cm long, 4 to 9 cm wide, cordate-acuminate at apex, usually round at base, turning orange to scarlet in autumn, falling early in the cold season; petioles 1.5 to 7 cm long, with 2 conspicuous glands at apex and on each side of scale-like bracts. Flowers monoecious, greenish-yellow, in terminal spikes, 5 to 10 cm long. Fruit a capsule, subglobose, 0.95 to 1.7 cm in diameter, 3-valved, with three seeds coated with a white wax. Seeds half-ovate, 0.6 to 1.0 cm long, 0.43 to 0.6 cm wide, 0.5 to 0.77 cm thick, with an acrid penetrating taste. Flowers April to June; fruit ripens September to October.[278]

Germplasm — Of the many cvs grown, more than 100 are found in Taiwan. Two main types are 'Eagle-Claw' and 'Grape', varying according to form of fruit-spikes, fruit-sprigs, fruit-stalks, and maturing period. Native to the China-Japan Center of Diversity, tallow tree is reported to tolerate some frost, grazing, slopes, waterlogging, and weeds. (2n = 36,40.)[82,278]

Distribution — Native to many provinces of central China, especially north of the Yangtze Valley, and Japan. Chinese tallow tree is also cultivated there and on Hainan Island, Hong Kong, Taiwan,[333] and Korea. It has been introduced into Sri Lanka (where naturalized), Indochina, Bengal, India, Sudan, Martinique, southern U.S. (S. California, S. Arizona, and Texas to Florida, north to South Carolina), southern France and Algeria.[278]

Ecology — Ranging from Warm Temperate Dry to Moist through Tropical Dry Forest Life Zones, tallow tree is reported to tolerate annual precipitation of 6.6 to 15.2 dm (to 37) (mean of 6 cases = 11.3), annual temperature of 14.7 to 24.3°C (mean of 6 cases = 18.1), and pH of 5.5 to 7.8 (mean of 5 cases = 6.7).[82] Adapted for growing on canal banks, on steep mountain slopes, granite hills, or sandy beaches, it grows in weakly alkaline soils, saline or strongly acid soils. Said to thrive in alluvial forests, on low alluvial plains, and on rich leaf-molds, growing best in well-drained clayey-peat soils. Favorable climatic conditions are mean air temperatures of 12.5 to 30.1°C, and an annual precipitation from 13 to 37 dm. It is generally a subtropical to warm temperate plant, hardy and able to withstand a few degrees of frost; unripened twigs are susceptible to frost injury. It grows at elevations 100 to 800 m.[70,278]

Cultivation — Propagated by seed, cuttings, layering or top-grafting on seedling stock. Seed usually sown in late autumn or early spring. Seedlings in the first year may grow 0.3 to 0.9 m in height and should be transplanted. When seedlings are about 1 m tall (in the spring of the third year), they should be planted out in permanent areas. Tree grows rapidly, 5 to 8.5 m tall with DBH of 13 to 17 cm in 10 years, and 10 to 13 m tall with DBH 30 to 40 cm in 20 to 30 years. When cultivated, trees are grown in plantations or transplanted to borders of fields or canals, so as not to interfere with the cultivation of the soil. Chinese also make cuttings by breaking small branches and twigs, care being taken not to tear or wound the bark. These are layered and rooted.[278]

Harvesting — Fruits and seeds, about the size of a pea, are harvested by hand in November and December when leaves have fallen. Plants require from 3 to 8 years to bear, but then continue to bear for years, averaging 70 to 100 years. Trees attain full size in 10 to 12 years. Seed can be threshed from the tree and collected by hand (once estimated at less than $.03/kg). Mechanical methods may be readily adapted to the harvest. When fruit is harvested by hand in midwinter, they are cut off with their twigs with a sharp, crescent-shaped knife attached to the end of a long pole, which is held in the hand and pushed upward against the twigs. The capsules are pounded gently in a mortar to loosen the seeds from the shells, from which they are separated by sifting.[278]

Yields and economics — In plantations trees should be planted one rod apart each way, giving 400 trees per hectare, and if trimmed to a convenient size for hand harvesting, would yield 14 MT seed per ha, containing 2.6 MT oil, 2.8 MT tallow, 1.5 MT protein concentrate, 1.1 MT fibrous coat, and 4.5 MT shell. Oil, tallow, and protein meal would bring about $750 per hectare. This yield could increase with age. Scheld et al.[295] report yields of 4,000 to 10,000 kg/ha (in excess of 11,000 kg/ha in VODF Seminar II[293]), and cite estimates of 25 barrels of oil per year as a sustained energy yield. Tallow is separated by placing the seed in hot water, thereby melting the tallow which floats on the surface, or by melting tallow with steam and collecting it when it drips off. Solvent extraction of the tallow from the seed is also used; tallow still adhering to the seed is removed by an alkali treatment. The fairly thick hard shell prevents extraction of the oil inside, so that the seed is crushed and Stillingia Oil is obtained by pressing or solvent extraction. According to one report, seed contains about 20% oil, 24% tallow, 11% extracted meat, 8% fibrous coat, and 37% shell. Yields of Stillingia Oil as high as 53% of the kernel have been reported in some varieties. Seed yields vary with the variety and age-gradations of the trees — a tree averaging at 5 years of age 0.453 kg, at 10 years, 3.379 kg, and at 20 years, 11.989 kg, with yields gradually decreasing after that. White meal, obtained by the extraction of the kernel, has a pleasant nut-like flavor, and contains 76% protein. Flour and protein of Chinese tallow nut contain vitamin B (thiamine). In China and other Oriental countries, as in other regions of the world, large quantities of tallow and oil are extracted annually from this tree. Tallow mills are erected where the tree is extensively grown. In addition to its economic value (from $750/ha for the oil, tallow, and protein), the tree is extensively propagated for ornamental purposes alone in Houston, Texas.

Energy — Coppicing well, the tree grows rapidly, the mean annual girth increment 2.6 to 5.2 cm. The wood, weighing only 513 kg/m^3 is used for fuel. With some tolerance to salt, the tallow trees should be investigated as energy crops for saline situations. Scheld[293] reports standing dry wood mass on 4-year plantations at >40 MT/ha, or more than 10 MT/ha/yr. Princen,[269] assuming an annual oil yield of 25 barrels per hectare, estimates that only 24 million hectares of oilseeds (like Sapium) would be required to produce a replacement for the ca. 8% of our petroleum usage which goes into chemical production. That means 300 million ha could replace all our petroleum usage (ca. 35% of Brazil, 108% of Argentina, 32% of the U.S.). Specific gravity of the wood ranges from 0.37 to 0.48 (mean 0.44) in samples from 18- to 24-year-old trees. Energy values range from 7,226 to 7,835 Btu/lb (mean 7,586). Rapidity of coppicing, taproot production, drought and salt tolerance, and rapid growth rate are attributes leading Scheld and Cowles to regard the tree as a promising biomass candidate (in the warm coastal region of the U.S.) which can be established over large acreages by conventional agricultural planting methods and which can provide woody biomass for direct burning or conversion to charcoal, ethanol, or methanol.[269,293,294,295]

Biotic Factors — Flowers are favored by honey-bees, and fruits are readily eaten by birds, including domestic fowl. It has been considered a desirable plant for bird-food. The tree is remarkably free of insect pests. The root-knot nematode, *Meloidogyne javanica,* has been reported.[382] Fungi known to attack this tree include: *Cercospora stillingiae, Clitocybe tabescens, Dendrophthoe falcata, Phyllactina corylea, Phyllosticta stillingiae,* and *Phymatotrichum omnivorum.*[186,278]

SCHLEICHERA OLEOSA (Lour.) Merr. (SAPINDACEAE) — Lac Tree, Kusum Tree, Malay Lac-Tree, Honey-Tree, Ceylon Oak
Syn.: *Schleichera trijuga* Willd., *Pistacia oleosa* Lour.

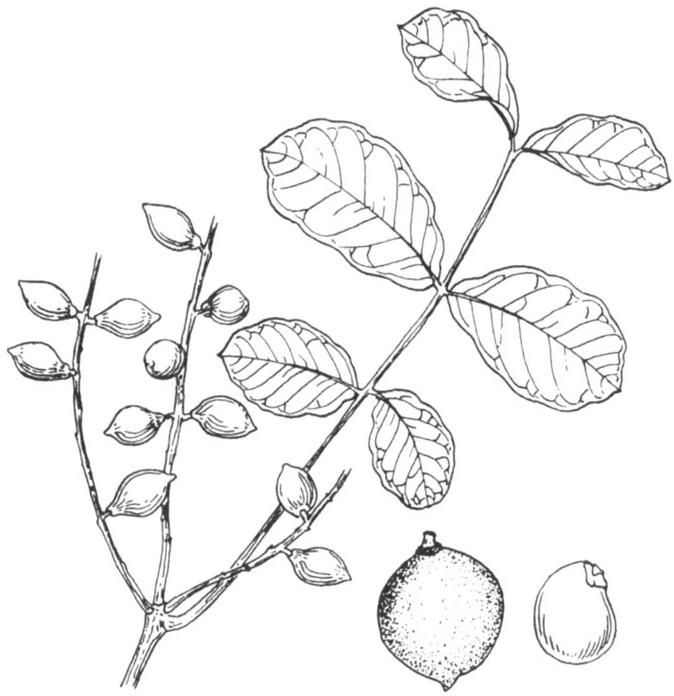

Uses — Seeds of the Lac tree are source of Macassar Oil, used in ointments, for candles, for illumination, as a lubricant for machinery, and in Madura for Batik work. Seeds yield about 40% of an edible oil or fat, sometimes used for culinary purposes and as a hair oil. Seeds also are eaten raw or roasted. Unripe fruits are pickled, and fruit may be eaten when other food supplies are scarce. The ripe fruits, often eaten during the summer, have whitish pulp and pleasant, acidic taste. Young leaves are eaten with rice. Young shoots are eaten; they are also lopped for fodder. Combined with wheat-straw and rape-cake, they make good roughage. Wood is close-grained, very hard, heavy, resistant to moisture, whitish with heartwood light reddish-brown, taking a fine finish, and used for making mortars, pestles, axles and hubs, felloes, and stocks of cart wheels, agricultural implements, such as yokes, plows, and teeth of harrows, shafts, violin bows,, screw rollers in sugar mills, in cotton and oil presses, tool handles for hammers, axes, and picks. Treated lumber is used for construction, cabinet work, beams, rafters, purlilns, trusses, posts, sleepers, and for wagon building. In addition, it is used for road paving, block flooring in mills and warehouses, pit-props, side-props in shafts and galleries in mines. Bark is employed in tanning; flowers yield a dye. Trees serve as host for lac insects.[70,137,278]

Folk medicine — Reported to be anodyne, cyanogenetic, larvicide, and refrigerant, lac tree is a folk remedy for acne, backache, burns, fever, inflammation, itch, malaria, neuralgia, pleurisy, pneumonia, rheumatism, skin problems, and sores.[91] The bark is reported to cure leprosy, skin diseases, inflammation, and ulcers. The unripe fruit is heating to the body, heavy to digest, causes biliousness, astringent to the bowels. The ripe fruit is digestible, astringent to the bowels, heating, appetite stimulant. The seeds are tonic, increase appetite, cure biliousness. The oil is considered a tonic, stomachic, anthelmintic, purgative, cure for skin diseases and ulcers. The astringent bark is used as a cure for the itch when rubbed on with oil. Oil of the seeds is used as a stimulating agent for the scalp, both cleansing it and promoting the growth of hair. The oil is also used as a purgative and as prophylactic against cholera; used externally in massage for rheumatism, for the cure of headaches; for skin disease. Powdered seeds are applied to ulcers of animals and for removing maggots. Bark is applied to swollen glands and ripening boils.[165] Bark is also used for pain in the back and loins, inflammation, and ulcers.[70]

Chemistry — Seeds are reported to contain 0.3% HCN; the oil is reported to contain 1.6% palmitic-, 10.0% stearic-, 19.7% arachidic-, 0.9% palmitoleic-, 52.2% oleic-, 8.5% gadoleic-, and 4.0% C_{22}-acid. The oil-cake is reported to contain 5.57% moisture, 22.31% protein, 48.53% fat, 14.43% soluble carbohydrates, 5.39% fiber, 3.40% soluble mineral matter, 0.37% sand, 3.08% phosphoric acid (P_2O_5), and 1.3% potash (K_2O). Green leaves are reported to contain (ZMB) 10.37% crude protein, 1.93% ether extract, 32.34% crude fiber, 49.21% N-free extract, 2.42% Ca, 0.71% P, 5.09% gallo-tannic acid. The bark is reported to contain 9.4% tannin.[70] Another source reports cotyledons to contain 65 to 70% oil, with the glycerides composed of lauric-, palmitic-, arachic- (25%), oleic- (ca. 70%), butyric-, and lignoceric-acid, and traces of benzaldehyde and hydrocyanic acid, and the bark to contain 7% tannins.[187]

Toxicity — Presumably due to the presence of hydrocyanic acid, the seed and seed oil induce symptoms similar to irritant poisons (giddiness, dilation of pupils, and syncope, sometimes death).[70]

Description — Trees 15 to 40 m tall, mostly gnarly and crooked, slow-growing; stems furrowed; branches thin, finely short-hairy to subglabrous, leafing and flowering in early spring. Leaves alternate, without stipules, 20 to 40 cm long, paripinnate; leaf-rachis sparingly finely hairy, 5 to 14.5 cm long; leaflets 4 to 8, opposite, obovate-lanceolate, 2.5 to 25 cm long, 1.6 to 11 cm broad, the lowest pair the smallest, obtuse or shortly acuminate, entire, coriaceous, glabrescent; young leaves purple; petiolules very sparingly finely hairy to glabrous, 1 to 3 mm thick. Inflorescence 1.5 to 13 cm long, on pedicels 2.5 mm long, finely short-hairy, the racemes glabrous, apiculate, smooth or spinose; calyx glabrous or nearly so, about 1.5 mm in diameter, the segments erect, triangular, acute; disk glabrous, ovary thinly pilose, style persistent, after anthesis indurate. Fruit broadly ellipsoid, glabrous with thin, hard pericarp, indehiscent, 1.6 to 2.5 cm long, 1-seeded; seed with a large chalaza; aril pulpy, subacid, edible. Flowers spring; fruits fall; January to December in Java.[278]

Germplasm — Reported from the Hindustani Center of Diversity.[82] Lac tree is reported to tolerate shade, frost, and drought. Seedlings should be protected in early stages as they are frost-tender.[70]

Distribution — Native and distributed all over Southeast Asia, from the sub-Himalayan region to Nepal, and central and south India, Sri Lanka, Malaya, Burma, Timor, and Java. Cultivated in many areas, e.g., near Calcutta and in Java. Introduced in southern California.[278]

Ecology — Ranging from Tropical Dry to Moist Forest Life Zones, lac tree is reported to tolerate annual precipitation of 9 to 15 (to 30) dm and annual temperature of 24 to 25°C.[82] Lac trees occur in tropical moist to wet evergreen and semi-evergreen forests and in moist deciduous teak forests in India, as well as in dry deciduous forests. Trees are not particular about soil structure or content. It grows best below 1,000 m altitude, in nature growing up to 600 m in teak forests. Optimum temperature should be above 24°C, with precipitation varying from 9 to 10 dm to 30 dm or more per annum.[278] Lac tree is common on well-drained boulder deposits, frequently in large numbers along ravines or on the edges of terraces in the sub-Himalayan tract and the outer hills. Common on sides of ravines on sandstone or on boulder beds in Siwalik range. Scattered near banks of streams in mixed forests in central India. Prefers slightly acidic soils; thrives best on light well-drained, gravelly or loamy soils; occurs on sandy and laterite soils.[70]

Cultivation — Trees propagated by seed and root suckers, either naturally in the forest or under cultivation. Seeds viable only for a short period, but can be stored for 1 year in gunny sacks or 2 years in sealed tins. Seeds are started in seed-beds, and young trees planted out when 0.5 to 1 m tall. Once established, no special care is required. No special fertilizers or soil pH are needed. Wild trees and those grown for boundaries may also be used for lac-trees. When cultivated, trees are planted about 275/ha. Stump-planting seems to give better results in moist climates. Stumps, with ca. 4 cm shoot and 23 cm root, are prepared from seedlings which have attained 7 to 13 mm diameter. Trees should be protected from grazing and weeded regularly for the first few years. Trees will tolerate only light pruning; apical pruning is better than surface pruning.[70,278]

Harvesting — Seeds are harvested in the autumn. Collectors climb trees and cut off fruit-bearing branches. Fruits are depulped by keeping them in a heap for 2 to 4 days and rubbing the decaying pericarp off with the hands. Seeds are then washed, dried, and stored. Kusum bears a good crop of lac every second or third year.[70,278]

Yields and economics — Average annual production of stick-lac varies from 1 to 1.5 kg per tree, to as much as 9 to 18 kg from well-cultivated trees.[278] The quantity of lac produced per tree varies with the size of the crown and the vigor of the shoots. Average seed yields of about 28 to 37 kg in one season are reported, which translates to 7 to 13 kg of easily expressed oil.[70] India, Sri Lanka, and Java are the principle producers of the lac, and the U.S. is the main consumer. Lac from trees from India and Sri Lanka command the highest price.[278]

Energy — The very heavy wood, specific gravity approximately 0.91 to 1.08, makes good fuel and excellent charcoal. Sapwood has a calorific value of 4,950 calories (8,910 Btu); heartwood, 4,928 calories (8,872 Btu). Kernels (60 to 65% of the fresh fruit; 15.3% of dried fruit) contain 59 to 72% oil, although yields are only 32 to 35% oil by boiling decorticated seeds, 25 to 27% with ox-driven presses. With 275 trees per hectare, there could be 1,925 to 3,575 kg oil per ha. The oil is used for candles and for illumination; the oil-cake is also used as fuel.[70]

Biotic factors — Monkeys and birds eat the seeds, thus interfering in their collection for use for oil. The fungus *Meliola capensis* is known to attack trees, and *Dendrophthoe falcata* sometimes parasitizes it.[4,278] Browne[53] lists the following as affecting lac tree: Fungi — *Corticium salmonicolor, Rosellinia bunodes*. Coleoptera — *Holotrichia serrata, Myllocertus cardoni, Xyleborus fornicatus, Xylosandrus morigerus*. Hemiptera — *Laccifer lacca*. Lepidoptera — *Ascotis selenaria, Catachrysops strabo, Cusiala raptaria, Dasychira grotei, Ectropis bhurmitra, Heliothis armigera, Hyposidra successaria, H. talaca, Rapala iarbus, Thalassodes falsaria*. In addition, *The Wealth of India*[70] reports *Rosellina bunodes* (stem blight), *Polyporus weberianus* (yellow-cork-rot), *Daedalea flavida* (white spongy rot), *Hexagonia apiaria* (white spongy rot), *Irpex flavus* (white fibrous rot). *Serinetha augus* attacks the seed. *Laccifer lacca*, the lac insect, is considered the most important insect attacking the tree.[70]

SCLEROCARYA CAFFRA Sond. (ANACARDIACEAE) — Marula Nut, Caffir Marvola Nut

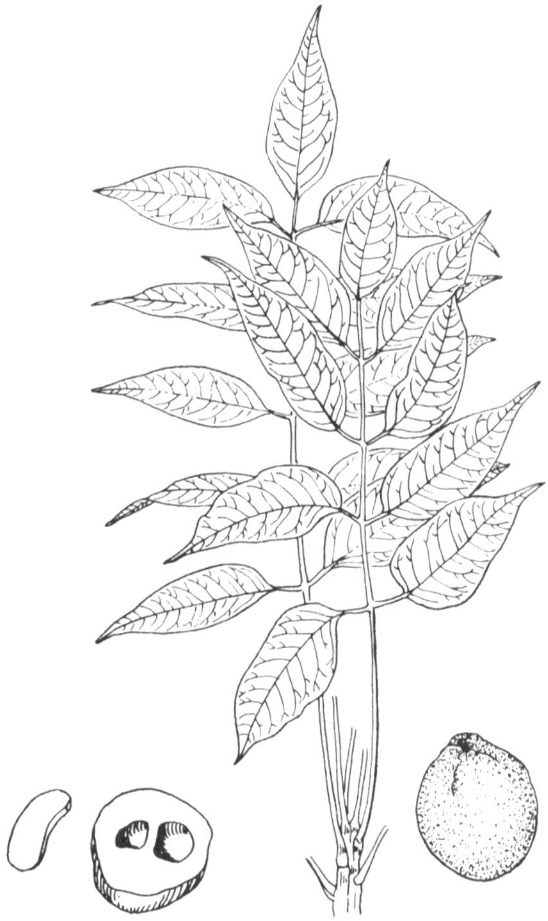

Uses — Tree is important for shade and shelter as well as food to a variety of animals. Fruits (or kernels, or both) edible, yet said to serve as an insecticide. Kernels of stones have a delicious nut-like flavor, and are eaten raw, or dried and ground and added to soups or stews. Fruits, the size of plums, have a pleasant flavor and are a source of food for parrots and mammals. With a turpentine aroma, the fruit is juicy, tart, and thirst-quenching. Fruit juice, boiled down, yields jelly or syrup used as sweetening agent. Fruit is also used by natives to make a fermented beverage which is intoxicating. Elephants and monkeys apparently become drunk from eating fermenting fruits. Seeds, extracted with difficulty, are oily, nutrituous and high in vitamin C. Kernels contain about 60% oil, extracted by boiling and used to preserve and soften skin shirts by Zulu women. Oil is used to treat meat which is to be kept for up to a year. Oil is also used for cooking and as a base for cosmetic red ochre. Pedi use the ground up kernels for making a porridge, the embryo as a condiment, and the leaf as a relish. Bark is used to make a bitter brandy tincture, and is the source of a red dye. Gum from the bark is mixed with soot and used for ink. Wood, pinkish white, often with a greenish tinge, changing to a brown-red on exposure, is fairly soft, fairly durable, saws well, and takes nails, and is used for making fruitboxes, canoes, furniture, panelling, utensils, troughs, stamping blocks, structures, spoons, bowls, dishes, and drums. Leaves are browsed by many animals; elephants eat the bark and roots.[86,246,278,280,332]

Folk medicine — Reported to be astringent, marula is a folk remedy for diarrhea, dysentery, malaria, and proctitis. The bark decoction is used for diarrhea, dysentery, and malaria, and to clean out wounds. The leaf juice is applied to gonorrhea. Europeans in South Africa take the bark decoction both for the cure and prevention of malaria (but experiments have not confirmed antimalarial activity). Zulu use the bark decoction to prevent gangrenous rectitis. Fruits are believed to serve both as an aphrodisiac and contraceptive for females. (African cattle, having partaken of too much fruit, have been observed to become both aggressive and infertile.)[86] Europeans and Africans use the bark as a prophylactic and to treat malaria, the steam for eye disorders. Because of their abundant fruits, the trees are widespread fertility charms in Africa. The bark is thought to control the sex of unborn children; bark of the male tree is administered if a son is desired, and of a female tree if a daughter is desired.[246,332]

Chemistry — Per 100 g, the fruit (ZMB) is reported to contain 361 calories, 6.0 g protein, 1.2 g fat, 90.4 g total carbohydrate, 6.0 g fiber, 2.4 g ash, 72.3 mg Ca, 229 mg P, 1.2 mg Fe, 0.36 mg thiamine, 0.60 mg riboflavin, 2.41 mg niacin, and 819 mg ascorbic acid. The seed (ZMB) is reported to contain, per 100 g, 629 calories, 25.6 g protein, 59.8 g fat, 9.6 g total carbohydrate, 2.8 g fiber, 5.0 g ash, 149 mg Ca, 1299 mg P, 0.4 mg Fe, 0.04 mg thiamine, 0.12 mg riboflavin, and 0.73 mg niacin.[89] Per 100 g, the fruit (APB) is reported to contain 30 calories, 91.7 g H_2O, 0.5 g protein, 0.1 g fat, 7.5 g total carbohydrate, 0.5 g fiber, 0.2 g ash, 6 mg Ca, 19 mg P, 0.1 mg Fe, 0.03 mg thiamine, 0.05 mg riboflavin, 0.2 mg niacin, and 68 mg ascorbic acid. The seed (APB) is reported to contain 604 calories, 3.9 g H_2O, 24.6 g protein, 57.5 g fat, 9.2 g total carbohydrate, 2.7 g fiber, 4.8 g ash, 143 mg Ca, 1248 mg P, 0.4 mg Fe, 0.04 mg thiamine, 0.12 mg riboflavin, and 0.7 mg niacin. Bark contains 3.5 to 10% tannin, leaves 20% tannin, a trace of alkaloids, and 10% gum. Fruits contain citric and malic acid, sugar, and 54 mg vitamin C per 100 g. Seed oil (53 to 60%) contains ca. 55 to 70% oleic acid. The pattern of the amino acids (particularly rich in arginine, aspartic acid, and glutamic acid) in the mean differ only slightly from that in human milk and eggs.[86] The juice contains 2 mg vitamin C per gram (*South Africa Digest*, March 5, 1982).

Toxicity — One source lists it as a narcotic hallucinogen(?). In 1972, a flurry of newspaper articles heralded the propensity of pachyderms to get pickled on the fruit. Elephants, baboons, monkeys, warthogs, and humans may overindulge in Kruger Park (South Africa).[86]

Description — Small to large much-branched dioecious, deciduous tree, up to 20 m tall, with rounded crown with a spread of 10 m; trunk 30 to 90 cm in diameter; bark pale, nearly smooth, peeling in disk-shaped flakes,which leave circular depressions. Leaves alternate, crowded toward apex of stem, up to 30 cm long, compound with 3 to 8 pairs of opposite leaflets; leaflets long-petiolulate, ovate or elliptic, blue-green; serrate on margin in juvenile plants but smooth in older plants, glabrous, 3.7 to 5 cm long 2.5 to 3.3 cm wide, base acute, cuspidate. Flowers unisexual, male and female on different trees; male flowers in terminal reddish spikes or racemes, with 12 to 15 stamens, inserted around a fleshy, depressed, entire disk; sepals 4, dark-crimson; petals 4, pinkish; female flowers long-peduncled, borne singly or 2 or 3 together at ends of young shoots (rarely flowers are fully bisexual); usually only female trees bear fruits, but frequently terminal flowers of male inflorescences may develop fruits; ovary subglobose, 2- to 3-locular. Fruit a fleshy, obovoid, 2- to 3-celled, yellow drupe, each cell containing a seed, and each cell with an "eye" to permit the embryo to grow out of the shell. Seed or stone about 2.5 cm long, 1.5 cm wide, weighing 3 to 4 g. Flowers August; fruits December to March in South Africa.[278]

Germplasm — Reported from the Africa Center of Diversity, marula, or cvs thereof, is reported to tolerate drought, heat, insects, and sand.[82]

Distribution — Native to South Africa, particularly to Natal and Transvaal, but widespread in hotter drier regions, Bechuanaland and tropical Africa, north to Sudan and Ethiopia, established at Miami, Florida.[278]

Ecology — Ranging from Subtropical Dry through Tropical Dry to Wet Forest Life Zones, marula is reported to tolerate annual precipitation of 6.7 to 43.0 dm (mean of 3 cases = 21.6 dm), annual temperature of 20.6 to 27.4°C (mean of 3 cases = 24.1°C), and pH of 6.1.[82] One of the more common trees in the savannas of the Transvaal. It does not tolerate frost. Thrives especially in hot dry regions, and is rarely found in higher rainfall areas. Occurs mainly in woodlands from the coast up to about 700 m altitude, on sandy soils or occasionally on sandy loams. Reported as growing in savanna grasslands where annual elevation and rainfall are as follows: Mozambique — 200 to 900 m, 630 to 1000 mm; South Africa-Mozambique — 300 to 1000 m, 250 to 500 mm; South Zimbabwe — 450 to 1000 m, 380 to 640 mm; Angola — 80 to 1000 m; 600 to 710 mm; South Africa — 600 to 1500 m, 380 to 640 mm. In Malagasy, it occurs in areas with 1,000 to 1,500 mm precipitation.[278]

Cultivation — Common in the wild, marulas have grown very slowly under experimental conditions, but grow quickly in natural conditions. Seeds germinate readily; the hard stones should be sown intact. Trees may be propagated by truncheons, 10 to 12.5 cm thick, which root freely if laid in during early spring. Trees grow fairly rapidly and are drought-resistant when once established.[278] A project to breed marula was scheduled to begin in 1982 by the Department of Horticulture at the University of Pretoria (*South Africa Digest,* March 12, 1982).

Harvesting — Trees are said to bear fruit more copiously than related species. Fruits are collected from the ground or by climbing the trees. Natives regard these as the greatest delicacy and store them carefully. A gift of marula kernels is valued as a mark of highest friendship among natives.

Yields and economics — Trees are very plentiful in the forests where they grow spontaneously, and fruits are collected as needed. One tree yields up to 2 tons of fruit (*South Africa Digest,* March 5, 1982); 30 g of fruit produces 1 ℓ of marula beer (*South Africa Digest,* March 12, 1982). From a single tree, 91,000 fruits have been reported.[246] Kernels consist of nearly 88% hard shell, 12% kernel, the kernel yielding ca. 50% oil. Within the fruit, the shell contains the small oily kernel that burns with a steady flame.[278] Because of its local economic importance, trees are usually preserved by Bantu and others, even on cultivated land. In Transvaal also, the trees are protected.[246,278]

Energy — With two tons of fruit possible per tree, one might possibly obtain more than 6,000 ℓ of beer, distilling down to possibly 300 ℓ ethanol per tree, or 3,000 liters assuming 100 trees per hectare. The hard nut endocarp could be converted to charcoal, the kernel yielding 50% oil. Sap of the tree could also be converted to ethanol. Prunings and by-products could be used in pyrolysis.

Biotic factors — Fungi known to attack marula are *Cercospora caffra* and *Gloeosporium sclerocaryae.* Trees are host of a small beetle (*Polydada*) of which the highly poisonous grubs are used by Bushmen as an ingredient for arrow poisons. Mopane Caterpillars also grow on the tree, and are eaten after roasting by Bantu and Bushmen.[278] Water, which runs off the trunk and crown into holes — usually where a branch has broken off — is used by mosquitoes for breeding. Larvae of *Gonimbrasia belina* sometimes breed on marula. Butterflies and the green lunar moth breed on the foliage. Wood is very liable to blue discoloration through fungi and beetle attacks.[246]

SIMMONDSIA CHINENSIS (Link) C. Schneid. (BUXACEAE) — Jojoba

Uses — *Simmondsia* is unique among plants in that its seeds contain an oil which is a liquid wax. Oil of *Simmondsia* is obtained by expression or solvent extraction. It is light-yellow, unsaturated, of unusual stability, remarkably pure, and need not be refined for use as a transformer oil or as a lubricant for high-speed machinery or machines operating at high temperatures. The oil does not become rancid and is not damaged by repeated heating to temperatures over 295°C or by heating to 370°C for 4 days; the color is dispelled by heating for a short time at 285°C, does not change in viscosity appreciably at high temperatures, and requires little refining to obtain maximum purity. Since Simmondsia oil resembles sperm whale oil both in composition and properties, it should serve as a replacement for the applications of that oil. The CMR[351] reports that a new oil from the fish known as orange roughy is "attempting to make inroads on the jojoba and sperm whale markets". Jojoba oil can be easily hydrogenated into a hard white wax, with a melting point of about 73 to 74°C, and is second in hardness only to carnauba wax. The oil is a potential source of both saturated and unsaturated long-chain fatty acids and alcohols. It is also suitable for sulfurization to produce lubricating oil and a rubber-like material (factice) suitable for use in printing ink and linoleum. The residual meal from expression or extraction contains 30 to 35% protein and is acceptable as a livestock food. Seeds were said to be palatable and were eaten raw or parched by Indians. Recent studies suggest they are toxic. They may also be boiled to make a well-flavored drink similar to coffee, hence the name coffeeberry. It is an important browse plant in California and Arizona, the foliage and young twigs being relished by cattle, goats, and deer, hence such names as goatnut.[86]

Folk medicine — Indians of Baja California highly prized the fruit for food and the oil as a medicine for cancer and kidney disorders. Indians in Mexico use the oil as a hair restorer. According to Hartwell,[126] the oil was used in folk remedies for cancer. Reported to be emetic, jojoba is a folk remedy for cancer, colds, dysuria, eyes, head, obesity, parturition, poison ivy, sores, sore throat, warts, and wounds. Seri Indians applied jojoba to head sores and aching eyes. They drank jojoba-ade for colds and to facilitate parturition.[85,126]

Chemistry — I was amazed to see, in searching through my massive files on jojoba, that I had no conventional proximate analysis. It was not even included in two of my most treasured resources, Hager's Handbook and *The Wealth of India*.[70,187] Perhaps this is due to the relative novelty of interest and the unique situation that the seed contains liquid wax rather than oil, sort of unusual for the conventional analyses.[85] Verbiscar and Banigan[327] approximated a proximate analysis, some of which follows: per 100 g, the seed is reported to contain 4.3 to 4.6 g H_2O, 14.9 to 15.1 g protein, 50.2 to 53.8 g fat, 24.6 to 29.1 g total carbohydrate, 3.5 to 4.2 g fiber, and 1.4 to 1.6 g ash. The amino acid composition of deoiled jojoba seed meal is 1.05 to 1.11% lysine, 0.49% histidine, 1.6 to 1.8% arginine, 2.2 to 3.1% aspartic acid, 1.1 to 1.2% threonine, 1.0 to 1.1% serine, 2.4 to 2.8% glutamic acid, 1.0 to 1.1% proline, 1.4 to 1.5% glycine, 0.8 to 1.0% alanine, 1.1 to 1.2% valine, 0.2% methionine, 0.8 to 0.9% isoleucine, 1.5 to 1.6% leucine, 1.0% tyrosine, 0.9 to 1.1% phenyalanine, 0.5 to 0.8% cystine and cysteine, and 0.5 to 0.6% tryptophane. Detailed analyses of the wax esters, free alcohols, and free acids, are reported in NAS.[229] Per 100 g jojoba meal, there is 1.4 g lysine, 0.6 g histidine, 1.9 g arginine, 2.6 aspartic acid, 1.3 threonine, 1.3 serine, 3.2 glutamic acid, 1.5 proline, 2.4 glycine, 1.1 alanine, 0.6 cystine, 1.5 valine, 0.1 methionine, 0.9 isoleucine, 1.8 leucine, 1.1 tyrosine, and 1.2 g phenylalanine. The two major flavonoid constituents of the leaves are isorhamnetin 3-rutinoside (narcissin) and isorhamnetin 3,7-dirhamnoside.[86]

Toxicity — Simmondsin, a demonstrated appetite-depressant toxicant is contained in seeds, 2.25 to 2.34%; seed hulls, 0.19%; core wood, 0.45; leaves, 0.19 to 0.23%; twigs, 0.63 to 0.75%; and inflorescence, 0.22%. Three related cyanomethylenecyclohexyl glucosides have also been isolated from the seed meal. The acute oral LD_{50} for crude jojoba oil to male albino rats is higher than 21.5 mℓ/kg body weight. Strains of *Lactobacillus acidophilus* can ameliorate this toxicity.

Description — Leafy, xerophytic, long-lived (100 to 200 years), evergreen dioecious shrub, ca.0.5 to 1 m tall in the wild, but occasionally to 6 m tall; leaves thick, leathery, bluish-green, oblong, opposite, 2.5 to 3.5 cm long, entire; flowers apetalous, the female ones usually solitary in the axils, the male ones clustered with 10 to 12 stamens per flower; female flowers with 5 greenish sepals, soft and hairy; the flowers on different plants, male and female plants about equal in nature; fruits ovoid, usually dehiscent, with 1 to 3 peanut-sized, brown seeds each, the endosperm scanty or absent; seeds about 750 to 5,150/kg, about 50% oil.[278]

Germplasm — Reported from the Middle American Center of Diversity, jojoba, or cvs thereof, is reported to tolerate alkali, drought, heat, high pH, and slope.[82] Yermanos[348] describes a monoecious strain which may lead to self-pollinating cvs. (n = 52, 56, ca.100.)

Distribution — Native to areas of northern Mexico, Lower California, on the Islands off the coast of California, New Mexico, and Arizona. It inhabits the mountains bordering the Salton Sea basin in the Colorado Desert in California, and the southern portion of San Diego County. In Arizona, it is found in the mountains around Tucson, near Phoenix, and north of Yuma. In nature, it grows between 600 and 1500 m elevation in the desert, down to sea level near the coast, between latitudes 25° and 31° N. There is a major effort underway in the U.S., Mexico, and Israel to domesticate jojoba; e.g., there are reports that is has been planted in Argentina, Australia, Brazil, Costa Rica, Egypt, Haiti, Israel, Paraguay, Rhodesia, the Sahel, and South Africa. The Israeli examples are bearing fruit. We are anxious to hear more success stories. There seems to be no major difficulty in growing the plant in frost-free, arid, subtropical, and tropical zones, but not many success stories have materialized.

Ecology — Ranging from Warm Temperate Desert (with little or no frost) to Thorn through Tropical Desert Forest Life Zones, jojoba is reported to tolerate annual precipitation of 2 to 11 dm, annual temperature of 16 to 26°C, and pH of 7.3 to 8.2.[82] Jojoba is usually restricted to well-drained, coarse, well-aerated desert soils that are neutral to alkaline, with an abundance of phosphorus. It grows best where the annual rainfall exceeds 30 cm, but does exist where less then 12.5 cm occurs. Where rainfall is ca.75 mm, the jojoba grows to ca.1 m tall; where rainfall is 250 to 400 mm, it may attain 5 m. It tolerates full sun and temperatures ranging from 0 to 47°C. Mature shrubs tolerate temperatures as low as −10°C, but seedlings are sensitive to light frosts just below freezing.[278]

Cultivation — Jojoba seeds retain nearly 99% germinability after 6 months, and 38% after 11 years stored in an open shed. Germination is good in alkaline sands at temperatures of 27 to 38°C. Seedlings are frost sensitive. Field seeding can be done with a modified cotton planter. Seedlings need two or three irrigations during the first summer and must be protected from animals. Weeding is recommended after each irrigation. Adventitious roots may form on 50 to 80% of the cuttings treated with growth-promoting substances. Plants could start producing seeds in 5 years, but full production would not be attained for 8 to 10 years. Using a 2 × 4 m spacing in planting would permit the planting of about 500 female and 50 male pollinating plants per hectare. Apomictic plants are known, lessening the need for male nonfruiting plants in the orchard. Suggested methods for planting include: *Close spacing*, ca.15 cm apart, resulting in hedge-rows, with the seeds planted in flat borders or in a slightly depressed ditch so as to keep them moist until they germinate (ca.10 to 14 days). Male plants should be thinned out to about a 5-1 ratio, finally allowing about 2,500 plants per hectare, with possible annual yields of 2.5 MT/ha seed. *Propagation* by cuttings from selected shrubs could increase seed and/or oil yields. Generally, flowering nodes and leaf nodes alternate, but some plants flower at nearly all nodes; some plants produce more than one flower per node. Transplanted seedlings survive readily, if the roots are pruned. Hence, cuttings could be made in a nursery for later transplanting in the field. The more efficient spacing for this method of planting is in rows 4 m apart, and the bushes in the rows 2 m apart. Male bushes should be interspersed throughout the grove (about 1,500 female and 250 male plants per hectare), possibly yielding ca. 2.75 MT/ha seed. When softwood cuttings were treated with IBA, 4 mg/g of talc, they rooted 100% in 38 days.[278]

Harvesting — In the wild, the only method for harvesting has been hand-collecting from under the plants, since mature seeds fall from the bush. Under cultivation, hedge-row, or orchard-like plantations, without undergrowth, seeds could be raked from under the bushes and then picked up by suction. Pruning the lower branches might be advantageous if this method be used. A device could be designed to pick the seeds from the bush prior to the time of falling. Cost of harvesting would depend on the method.[278]

Yields and economics — Buchanan and Duke[367] accept a figure near 2,250 kg/ha for yields of jojoba. Individual plants may yield 5 kg (dry weight) seeds and more, of which

50% (43 to 56%) by weight is a colorless, odorless liquid wax commonly called "jojoba oil".[230] Yermanos[347] suggested that a 5-year-old orchard should yield about 825 kg of nuts per hectare, increasing to 4,125 kg/ha in the 12th year, suggesting a renewable "oil" yield of ca.2 MT/year. Such yields may be optimistic, even for well-managed plantations. Estimates of the amount of wild nuts available each year range from 100 million to 1 billion pounds, the plants growing over 100 million acres in California, Mexico, and Arizona. Usually plants in cultivation yield oil in 6 to 7 years; the Israelis report their best specimens yield 2 or more kg of seed in the 4th year; wild plants yield about 1 kg of nuts per year, and cultivars should yield twice that amount or more. The seeds contain up to 50% "oil". In 1958, long before the whale oil became endangered, the value of Simmondsia "oil" as a hard wax was estimated at $.55/kg. Because of the present demand for the wax and oil, jojoba is being considered as a noncompetitive crop, that could replace wheat and cotton in Texas and southern California, with as much as the yield from 70,000 hectares being absorbed by industry. The Chemical Marketing Reporter[351] stated that jojoba prices doubled in 6 months to $200/gal. The cost of establishing a plantation can vary from $3,000/ha on land with irrigation available to $5,600/ha on rough desert terrain.[377] Once established, maintenance costs are low — only ca. $200/year. One hectare can yield 1,125 to 2,250 kg oil per year. (Recent prices have approached $50/kg, suggesting to the uninitiated yields of $100,000/ha, right up there with the hyperoptimistic ginseng yields. In either case, a wait of at least 5 years for the first return might seem interminable. Prices have gone down considerably since this was sarcastically written.)

Energy — With 641 plants per hectare, the aerial phytomass (over 6% of total phytomass) was 1,573 kg/ha and annual productivity only 327 kg/ha.[48] Daugherty et al.[72] were optimistic, but not so optimistic as Yermanos about jojoba oil yields. They projected ca.500 kg/ha oil for jojoba, ca.nearly 100 for cottonseed, ca.200 for flaxseed, ca.250 for soybean, and nearly 300 for safflower (based on 10-year averages for the conventional oilseeds, speculation for jojoba).

Biotic factors — One fungus (*Sturnella simmondsiae* Bonar) occurs on the leaves, calyxes, and peduncles, but little damages the plant in this country. *Phytophthora parasitica* and *Pythium aphanidermatum* may cause root rot in jojoba plantations. Cuttings are sensitive to *Alternaria tenuis*, seedlings to *Sclerotium bataticola* and *Fusarium oxysporum*. A scale insect that inhabits the leaves also is not detrimental. There is a harmful pest, probably a microlepidopoterous insect, that destroys a large part of the wild crop by consuming the very young ovules. One spraying at the proper time might eliminate this damage. The scale *Situlaspis yuccae* and the unique mealybug *Puto simmondsia* have been reported.[4,278,347]

TELFAIRIA OCCIDENTALIS Hook.f. (CUCURBITACEAE) — Fluted Pumpkin, Oyster Nut

Uses — Young shoots and leaves are used as a pot-herb. Leaves are much sought after by sheep and goats. Seeds are eaten and are said to have a pleasant almond-like flavor, but the bitter seed-coat must be discarded. Seeds are boiled and eaten or put in soups, or used as the source of a nondrying oil for native cookery and soap-making. Seeds are also used for polishing native earthenware pots. Dry shell of the fruit is sometimes used for utensils. Dried seeds are powdered and used to thicken soups. Dried fiber from macerated stems is used like loofa for paper.[55,239-278]

Folk medicine — No data available.

Chemistry — Per 100 g, the seed (ZMB) is reported to contain 579 calories, 21.9 g protein, 48.0 g fat, 25.1 g total carbohydrate, 2.3 g fiber, 5.6 g ash, 89.6 mg Ca, and 610 mg P. Per 100 g, the leaf (ZMB) is reported to contain 346 calories, 21.2 to 21.3 g protein, 12.9 to 13.2 g fat, 51.5 to 52.0 g total carbohydrates, 12.5 to 12.8 g fiber, and 13.9 to 14.0 g ash.[89] Burkill[55] reports that the oil contains 37% oleic acid, 21% palmitic acid, 21% stearic acid, 15% linoleic acid. Seeds contain a trace of alkaloid, while none has been detected in the roots.[55] On a wet weight basis, the pulped leaves contain 11 mg beta-carotene, juice 9, fiber 1, supernatant 1, and wet LPC 8 mg beta-carotene per 100 g. Under stored conditions, the LPC lost 82% beta-carotene and 58% xanthophyll over 12 months.[368] Ca.70% of the total N and 63% of the protein N was extracted; the potential protein extractability is ca.90%. The oil, by weight, contains 16% palmitic-, 3% stearic-, 23% oleic-, 23% linoleic- and 19% alpha-eleostearic-acids. Seeds of *T. occidentalis* contain fairly large amounts of alpha-eleostearic and no linolenic glycerides, while the seed fat of *T. pedata* derives from the usual mixture of saturated, oleic, linoleic, and linoleic acids.[128]

Description — Perennial, dioecious liana, up to 33 m long; stems herbaceous, ribbed, glabrous or pubescent, becoming thickened when old. Leaves petiolate, 3- to 5-foliolate; median leaflet elliptic, acuminate, acute, tapered into the petiolule, entire or shallowly sinuate-toothed, glabrous or sparsely hairy or punctate, 3-veined from near base with 2 well-developed ascending lateral veins, 6 to 17 cm long, 3 to 10 cm broad; lateral leaflets similar, with petiolules 0.2 to 2 cm long, petiole 1.9 to 8 cm long, pubescent; probracts 5 to 8 mm long. Male flowers in racemes 10 to 30 cm long, the bracts 2.5 to 8 mm long, 1.5 to 3 mm broad, pedicels 8 to 35 mm long, receptacle-tube campanulate, 2.5 to 3.5 mm long, densely glandular-hairy inside above; lobes triangular, glandular-dentate, 2 to 4 mm long; petals about 2.5 cm long, 1.2 cm broad, white with dark-purple marks at base inside, or creamy white with red-purple spot (eye); stamens 3, anthers coherent in center of flower; female flowers stalked. Fruit pale glaucous green or whitish with waxy bloom when ripe, flesh yellowish, ellipsoid, tapering at both ends, rather sharply 10-ribbed, up to 60 (to 90) cm long. Seeds numerous, very broadly and asymmetrically ovate, 3.2 to 3.6 cm long, 3.3 to 3.7 cm broad, and 1.0 to 1.2 cm thick; testa smooth with endocarpic fibrous sheath poorly developed or absent.[278,394] Flowers and fruits year-round.[278]

Germplasm — Reported from the African Center of Diversity, the fluted pumpkin, or cvs thereof, is reported to tolerate drought, low pH, poor soil, and shade.[82] Very similar to the following species, which is the commercial source of true oyster nut oil.[55] The true oyster nut has purplish-pink flowers, whereas the fluted pumpkin has white flowers with a purplish eye. (2n = 24.)

Distribution — Native to tropical Africa from Sierra Leone to Angola, the Congo Area; Fernando Po, Uganda.[278,394] Introduced to tropical America.

Ecology — Ranging from Warm Temperate Moist through Tropical Dry to Wet Forest Life Zones, the fluted pumpkin is reported to tolerate annual precipitation of 13.6 to 22.8 dm (mean of 2 cases = 18.2), annual temperature of 24.4 to 26.2°C (mean of 2 cases =

25.3°C), and pH of 5.0 to 5.0 (mean of 2 cases = 5.0).[82] Thrives best in closed-forest country,[55] ca.1,200 m above sea level.[394] Apparently best adapted to a hot, humid climate (e.g., TMF), common in littoral hedges, and lowland rain-forests up to about 1,200 m.[278] Its occurrence at the edges of forests may be the consequence of previous cultivation. It thrives in plantings in Talamanca, Costa Rica.

Cultivation — Cultivated in some places, especially S. Nigeria and by some tribes in Ghana. Grown on stakes or trained up trees.[55] Propagated by seeds either planted near trees upon which to climb, or more often allowed to sprawl over the ground, as is done in Nigeria. Once established, plants are perennial for several years. Grows well in any good garden soil where there is plenty of heat and moisture.[278]

Harvesting — Leaves and shoots are picked continuously as the plant grows.[55] Fruits are collected whenever ripe and needed. No special season, as plants flower and fruit yearround, and the fruits are gradually ripened throughout the year.[278]

Yields and economics — Often cultivated for the seeds by natives in West Tropical Africa, East Tropical Africa, and Southeast Asia; and probably elsewhere in the hot, humid tropics. Mainly used for the seeds as a vegetable and for oil, and the stem for the fibers for making paper.[278]

Energy — This plant climbed up trees in Talamanca like kudzu does in tropical America, and fruited copiously. Its relatively high seed-oil content suggests that this is as promising an energy species as China's *Hodgsonia*. No doubt the foliage could provide LPC (leaf protein concentrate) and the seeds oil, with the residues being used as by-products for energy production.

Biotic factors — No serious pests or diseases have been reported.[278]

TELFAIRIA PEDATA (Sm. ex Sims) Hook.f. (CUCURBITACEAE) — Oyster nut, Zanzibar Oilvine, Telfairia nuts, Jikungo

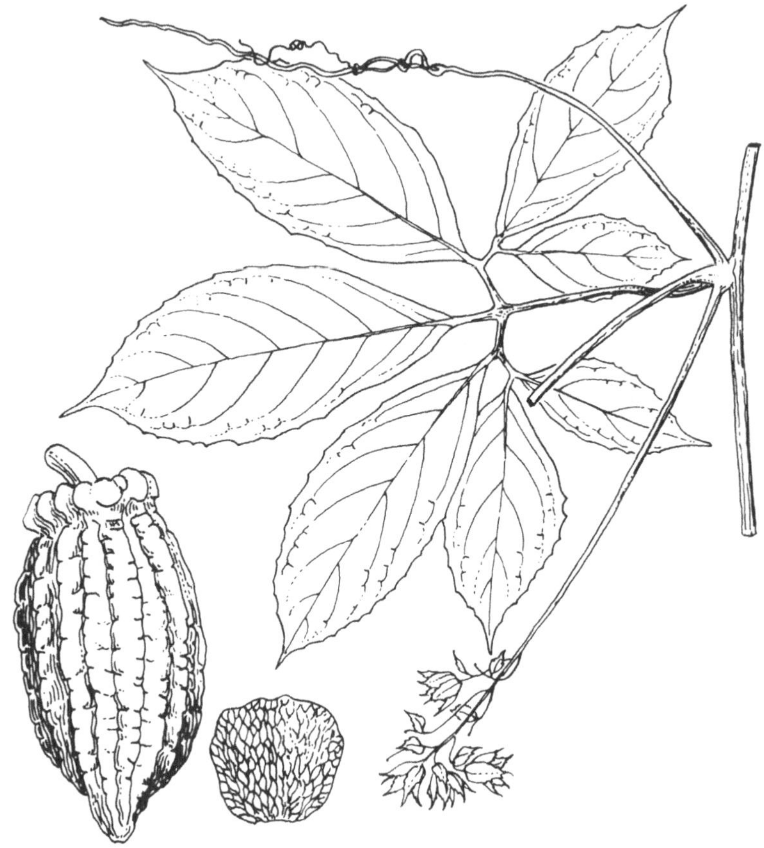

Uses — Oyster nut is cultivated for its edible seeds[394] and oil yield (about 62%). The fruits are used in soups, and the nuts are used in confectionery and chocolates, either alone or as a partial substitute for Brazil nuts or almonds, and are quite palatable fresh or roasted, as well as pickled. The seeds are the source of Castanha Oil, used in manufacture of soaps, cosmetics, salad dressings, paints, and candles. One quote from an unpublished W. E. Bailey typescript, "Possibly the oil can be converted into explosives, just as the Germans have done with Romanian soybeans.[26] The oil is almost indistinguishable from olive oil. The nuts may be pounded, cooked in water, and eaten as a cereal (porridge). The kernel has a high vitamin content, and residue from the kernel after the oil has been extracted can be used for livestock feed.[278] However, Watt and Breyer-Brandwijk[332] describe the seed-cake as "useless for stock feeding on account of its bitterness".

Folk medicine — Medicinally, oyster nuts have laxative properties, and women in Usamabar eat the nut immediately after childbirth to cause early contraction of the pelvis, increase the flow of milk, and insure an early return of their strength so they can return to normal duties in a day or two.[278] East Africans use the seed oil for stomach ailments and rheumatism, the leaf as a bitter tonic. Chagga use the seed as a puerperal tonic and lactagogue. The plant reportedly has taenifuge properties, especially the seed.

Chemistry — Per 100 g, the seed (ZMB) is reported to contain 31.1 g protein, 66.2 g fat, 2.7 g ash, 10.5 mg Ca, 596 mg P, and 4.3 mg Fe.[89] Per 100 g, the kernel (51 to 60% of seed) is reported to contain 4.4 g H_2O, 29.7 g protein, 63.3 g fat, 2.6 g ash, 10 mg Ca,

570 mg P, and 4.1 mg Fe.[70] An unpublished London Fruit Exchange report on file in the Germplasm Introduction and Evaluation Laboratory, gives 6.56% moisture, 19.63% protein, 36.02% fat, 28.45% N-free extract, 7.3% fiber, and 2.04% ash. The oil is yellowish with a brownish fluorescence, practically odorless, with a low acid value, and possesses a pleasant, slightly sweet taste. Somewhat viscous, it is liquid at room temperature, deposits stearine on standing, saponifies readily, and contains stearic, palmitic, and telfairic acids, as well as about 27% protein (as compared to 40% in soybeans).[278] The shell, especially the bast, contains abundant tannin and a bitter crystalline substance. Seed husks contain three antitumor compounds, Cucurbitacin B, D, and E, as well as tannin.

Toxicity — Watt and Breyer-Brandwijk attribute headaches to eating the fruits.[89,278,332]

Description — Perennial dioecious, herbaceous vine, to 30 m long; the stem herbaceous, ribbed, glabrous. Leaves alternate, digitate, 5- to 7-foliolate, the leaflets lanceolate, elliptic or narrowly ovate or obovate, penninerved, obscurely sinuate-toothed, to 13 × 6 cm. Male flowers pinkish purple, in racemes on long stems, opening in sequence, female flower single on shorter stem. Fruit a green gourd-like ellipsoid pepo, 32 to 45 cm long, 16 to 25 cm in diameter, bluntly 10-ribbed, weighing up to 30 kg, filled with a dense fleshy pulp in which seeds are embedded (difficult to separate seed from pulp). Seeds 60 to 200, to 35 mm in diameter, kidney-bean shaped, rich in oil, tasting like almond; kernel protected by two shells, the outer tough, fibrous, the inner hard and brittle; outer shell removed by peeling or burning, the inner one splits with a blow, sometimes a machine known as a belt sander is used to open the nuts.[278,394]

Germplasm — Reported from the African Center of Diversity, oyster nut, or cvs thereof, is reported to tolerate drought, high pH, laterite, poor soil, and shade.

Distribution — Native to East Tropical Africa, especially in Mauritius, Zanzibar, Tanzania, Pemba, and Mozambique. Cultivated throughout the area; especially in Kenya, Masai District, Ngong, and formerly in the Mascarene Islands.[278,394]

Ecology — Ranging from Subtropical Dry to Moist through Tropical Dry Forest and wetter Life Zones, oyster nut is reported to tolerate annual precipitation of 5.2 to 15.3 dm (mean of 4 cases = 11.1), annual temperature of 8.4 to 24.2°C (mean of 4 cases = 17.4°C), and pH of 5.5 to 7.0 (mean of 3 cases = 6.3).[82] Oyster nut grows at the edges of forests, enveloping the trees with its branches, while its trunk frequently attains a diameter of 45 cm. In Africa, it ranges from 0 to 1100 m altitude in lowland rain forest and riverine forest.[394] It grows well in a sheltered position with an eastern exposure, but without strong winds or cold temperatures. It requires medium loams with good drainage, is deep-rooted and drought resistant. It grows well up to 2000 m elevation in Kenya and Tanzania.[278]

Cultivation — Oyster nut is propagated by seeds, which should be planted within 3 months, as the oil dries out of the kernel, causing deterioration of the germ. Seeds, after being soaked in water for 5 days, are planted in a nursery. They germinate in about 21 days. When the seedlings are about 5 cm tall (2 to 3 months old), they are transplanted to the base of trees which they will climb over and often kill — a fast grower, exceeding 20 m in 15 months, if not pruned. The nursery offers protection to the seeds and small plants which are eaten by insects and wild animals; also, the plants are easier to water in dry seasons. Female plants are readily rooted from cuttings. If seed is sown directly in the field, 880 seeds per hectare, at 2 m apart, in double rows, spaced 4 m apart, is recommended. Seed should be planted at half their eventual spacing, since there is no way of distinguishing between the male and female plants until flowering takes place; 10 to 15 male vines needed per hectare. Sometimes trellises are used, these 2 m high, erected 4 to 5 m apart, and connected for the double rows of plants which are trained in opposite directions. This method is expensive, mainly due to the cost of the trellises, and is suitable only to mountainous regions where the posts would not be attacked by white ants. Green manures, compost, or barnyard manure should be used freely from the time of planting. Also bone and fish manures are used, these promoting good growth and fruiting. Lime is used to help control nematodes. Vines should be kept weed free for the first year or so after planting on trellises. After that, the plants will take care of themselves.[278]

Harvesting — The crop begins to bear in 2 years, and continues for 20 to 25 years. However, the plants will die out the third year in a poor soil. About 4 months are required from flower to mature fruit. Plants produce 1 to 2 crops yearly, and may bear almost mature fruits while they are flowering. The fruits are picked by hand as they are needed.[278] Nuts are soaked in water for about 8 hr in 3 changes of water to remove bitterness.

Yields and economics — Assuming 160 vines per ha, a conservative 10 fruits per vine, each fruit with 140 nuts, each weighing ca.12 g, the hectare could yield 2685 kg per seed. Average yields of the nuts are 1000 to 2000 kg/ha. The oil content of the seed is about 62% of its weight, or approximately 35% of the entire weight of the whole nut. This would suggest an oil yield up to 700 kg/ha. Dr. T. W. Whitaker (personal communication, June 1982) suggests that this should be a promising species, but not so exciting as the Asian *Hodgsonia* of the cucurbit family. USDA germplasm teams to China should negotiate for some of this subtropical species.[278]

Energy — From the descriptions, the oyster nut would appear to have aerial biomass attributes similar to or higher than our American weed, kudzu, often over 10 MT DM/ha. One vine reached 12 m tall and 5 cm in diameter in 15 months.

Biotic factors — The major fungi attacking oyster nut are *Armillaria mellea, Colletotrichum* sp., *Didymella lycopersici*, and *Oidiopsis taurica*. Virgin forests should be thoroughly burned before planting to prevent disease. The main nematode, *Heterodera marioni,* is controlled by the natives using a lime dressing, as the seeding stage is most often attacked. In Kenya, the major pests are ground squirrels and porcupines, which dig up recently planted seed, and bucks and grasshoppers, which eat the sprouting seed. Mealy bugs, taken from coffee trees and put on oyster nut vines, died.[278]

TERMINALIA CATAPPA L. (COMBRETACEAE) — Indian Almond, Myrobalan, Badam, Almendro, Bengal almond, Kotamba, Tropical Almond

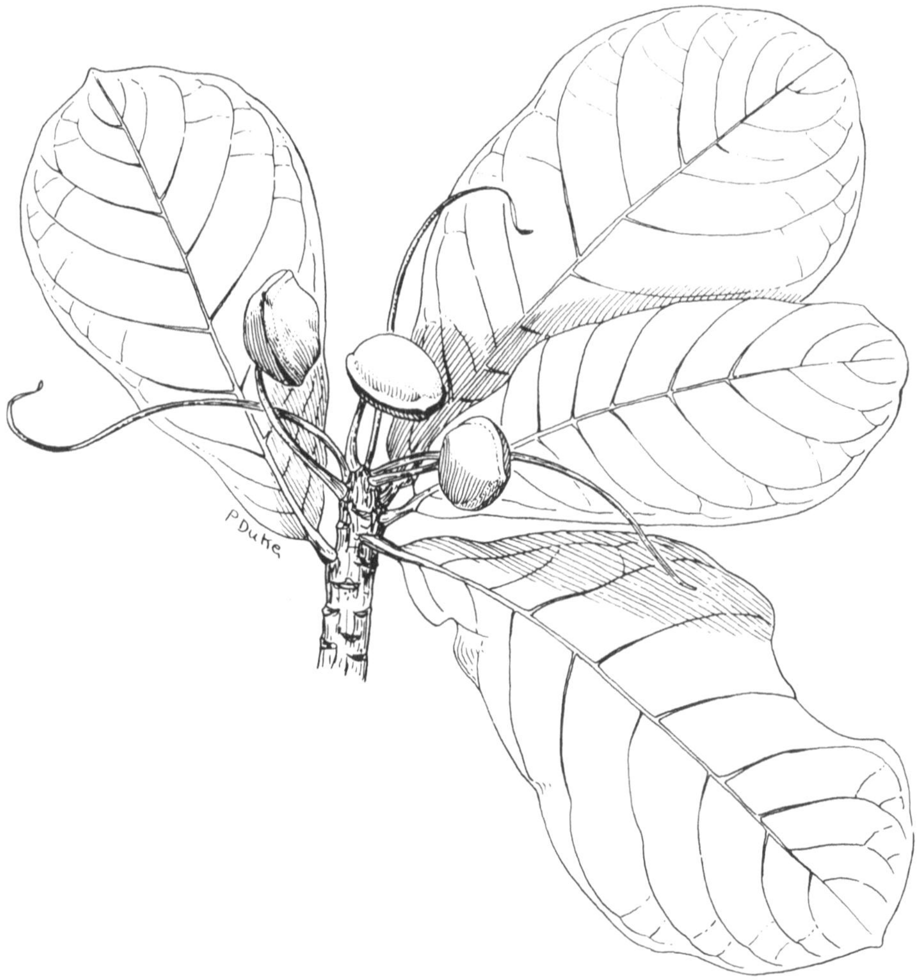

Uses — Indian almond is widely planted in the tropics and subtropics for ornamental, shade, timber purposes, and for the edible nuts. It is cultivated mainly for the edible kernels, used as substitute for almonds (e.g., in Chinese "chicken and almonds" in Trinidad). Kernels contain 50 to 55% colorless oil of excellent flavor, like almond oil in flavor, odor, and specific gravity, highly esteemed in the Orient. Seeds may be eaten raw. Leaves are the food of Tasar Silkworms, and are used as wrapping paper for small shop articles. Roots, bark, and fruits are used in tanning. Fruits are a source of a black dye used in some parts of eastern India to color teeth black. Wood chips in water give a yellow dye. Trees contain a gum, which is the source of a black dye, a source of ink, and a cosmetic. Oil is used as a substitute for groundnut- (Arachis), cottonseed- (Gossypium), and silk-cottonseed- (Bombax) oils. Flowers yield a nectar.

Folk medicine — Reported to be anodyne, astringent, cardiotonic, collyrium, diuretic, emetic, lactagogue, pectoral, purgative, sedative, stimulant, sudorific, tonic, and vermifuge, Indian almond is a folk remedy for arthritis, bugbites, colic, condylomata, cough, diarrhea, dysentery, ear ailments, eruptions, fever, gastritis, glossitis, headache, hemoptysis, insomnia, leprosy, lumbago, neuroses, pyorrhea, rheumatism, scabies, skin ailments, sore throat, stomach-ache, stomatitis, swellings, thrush, ulcers, wounds, and yaws.[91,249] Ayurvedics consider the fruits antibilious, antibronchitic, aphrodisiac, and astringent. In southern India, the juice of the young leaves is put in an ointment for leprosy, scabies, and other skin diseases; also used for colic and headache. Indochinese use the leaves with *Dacrydium* chips and nutgrass rhizomes for dysentery; the fruit, with beeswax, for foul ulcers and hematochezia. Indonesians apply the leaves to swollen rheumatic joints, using the kernel for a laxative and lactagogue. Philippines use the leaf juice, cooked with the kernel oil, for leprosy; and rubbed onto the breast for pain and numbness; or applied to rheumatic joints. Red leaves are believed vermifuge. In the Solomon Islands, leaves, bark, and fruit are used for yaws.[249] Nigerians apply the leaves, macerated in palm oil, for tonsilitis. Cubans take the leaf or fruit decoction for hemoptysis, adding crushed leaves to the bath for skin rash. Haitians take the bark decoction for bilious fevers. Costa Ricans used the bark decoction for crushed nipples and uterorrhagia. Brazilians take the bark decoction for asthma, diarrhea, dysentery, and fever. Colombians take the seed emulsion as pectoral.[224] The root bark is given for diarrhea and dysentery in French Guiana, the stem bark for bilious fevers. Mexicans make a powder from the stems for condylomata.[126]

Chemistry — Per 100 g, the seeds are reported to contain 574 to 607 calories, 2.7 to 6.0 g H_2O, 19.1 to 25.4 g protein, 52 to 56 g fat, 14.9 to 17.2 g total carbohydrate, 1.8 to 14.6 g fiber, 2.4 to 4.0 g ash, 32 to 497 mg Ca, 789 to 957 mg P, 2.4 to 9.2 mg Fe, 70 mg Na, 784 mg K, 0.32 to 0.71 mg thiamine, 0.08 to 0.28 mg riboflavin, 0.6 to 0.7 mg niacin, and 0 mg ascorbic acid. According to Leung, Butrum, and Chang,[181] 94% of the as-purchased nut is refuse, the husk only containing 35 calories, 0.4% moisture, 1.2 g protein, 3.2 g fat, 1.0 g total carbohydrate, 0.1 g fiber, 0.2 g ash, 2 mg Ca, 47 mg P, 0.6 mg Fe, 4 mg Na, 47 mg K, 0.02 mg thiamine, traces of riboflavin, and niacin, and no ascorbic acid. Amino acid values are given as 14.7 arginine, — cystine, 1.7 histidine, 3.4 isoleucine, 7.4 leucine, 2.3 lysine, 7.2 aspartic acid, 24.3 glutamic acid, 4.0 alanine, 6.3 glycine, 4.2 proline, 4.1 serine, 0.9 methionine, 4.2 phenylalanine, 2.9 threonine, — tryptophane, 3.2 tyrosine, and 4.8 valine.[225] Unfortunately, the refuse figures do not add up to 100. Air-dried kernels contain 3.51% moisture, 52.02% fat, 25.4% protein, 14.6% fiber, 5.98% sugars (as glucose). The seed oil contains 1.62% myristic-, 55.49% palmitic, 6.34% stearic-, 23.26% oleic-, and 7.55% linoleic-acids. The oil-cake (7.88% N) contains 8% albumin, 15% globulin, negligible prolamine, and 7.5% gluten. The shell contains ca. 25% pentosans, and hence, is a good source for making furfural. The leaves and fruits contain corilagin, gallic acid, ellagic acid, and brevifolin carboxylic acid, whereas the bark and wood contain ellagic acid, gallic acid, (+)catechin, (−)epicatechin, and (+)leucocyanidin.[70]

Description — Handsome, spreading, pagodiform, deciduous tree, medium- to large-sized, 13 to 27 m tall, 1 to 1.5 m diameter, with horizontal whorls of branches about 1 to 2 m apart; bark smooth, brownish-gray; leaves opposite, simple, leathery, green, turning red before falling, shining, shedding leaves twice a year (February and September), 12 to 30 cm long, 7.5 to 15 cm wide, obovate, tip rounded or somewhat acute, base narrowed, slightly auriculate, petioles about 2.5 cm long; flowers small, greenish-white, arranged crowded in short spikes 15 to 20 cm long, arising in axils of leaves, malodorous; stamens 10 to 12, in staminal flowers towards the apex; fruits yellow-green or reddish, hard, an angular drupe, size of a plum, slightly compressed on 2 sides, broadly oval in outline, elliptical and 2-winged in transverse section, 3.5 to 7 cm long, with thin fleshy pericarp, edible, but with a hard corky interior; seeds slender, pointed, oblong elliptical, 3 to 4 cm long, 3 to 5 mm thick. Germination phanerocotylar, the cotyledons convolute. Flowers June to August, fruits June to November, bearing two crops of fruit annually before dropping leaves.[225,278]

Germplasm — Reported from the Indochina-Indonesia Center of Diversity, Indian almond, or cvs thereof, is reported to tolerate full sunlight, high pH, laterite, lime, low pH, mine-spoil, poor soil, salt spray, sand, shade, slope, waterlogging, and wind.[82,225,232]

Distribution — Indigenous to Andaman Islands and islands of Malay Peninsula, now widely cultivated in the tropics of the Old and New Worlds. Extensively planted in tropical India and Sri Lanka, in West Africa from Senegal to Cameroons, Madagascar, Malaysia, and East Indies. Now pantropical.[278]

Ecology — Ranging from Subtropical Dry to Moist through Tropical Very Dry to Wet Forest Life Zones, Indian almond is reported to tolerate annual precipitation of 4.8 to 42.9 dm (mean of 92 cases = 17.7), annual temperature of 20.4 to 29.9 (mean of 66 cases = 25.2), and pH of 4.5 to 8.78 (mean of 13 cases = 6.1).[82] Though it grows well in sand or shingle, it also thrives in marl and permeable siliceous limestone. It volunteers only in loose sand, muck, or marl.[225] Tolerant of sand and salt, it has been used to stabilize beaches. Indian almond thrives in coastal forests in most tropical areas, from sea level to 1,000 m altitude, preferring coastal soils or light loamy soils. It has been recommended for tropical land soils. According to Morton,[225] it grows equally well in medium shade or full sun, and is highly wind resistant.

Cultivation — Propagated exclusively from seeds, which remain viable for at least one year. In India, whole fruits, exhibiting 25% germination, are planted. Seeds germinate in 2 to 4 weeks. The tree is extensively planted for the red foliage, as few other trees in the tropics develop colored foliage. The tree competes well with weeds.[225,278]

Harvesting — Rotations of 10 to 15 years are average. Fruits are harvested as they ripen. They have a very hard shell, which is easier to crack after the nuts are dry, often cracked between stones. In India, there are two crops a year, spring (April to May) and fall (October to November). There is more-or less constant fruiting in the Caribbean. Perhaps the crop would be desirable to harvest if mechanical means of cracking and cleaning the nuts were devised. Kernels yield nearly 55% oil by extraction and 35% by expression.[278]

Yields and economics — Trees may attain 6 m height in 3 years. A 10-year-old plantation is expected to yield 2.25 to 3.6 MT/ha/year.[233] Grown as a shade tree for cardamon, Indian almond contributed annually 9,300 kg/ha leaf mulch.[369] In Jamaica, nuts run $0.02 to $0.10 each, normally selling for $0.05 each in 1976.[225]

Energy — The wood (sp. gr. 0.59) is often employed as fuel. Erroneously equating *Terminalia catappa* a synonym of *Bucida buceras*, Cannell[61] suggests that the annual litterfall is only 1.7 MT/ha in the Guanica Forest of Puerto Rico, the current annual increment only 2 MT for a forest with 2,160 trees >5 cm DBH, averaging 7.8 m, basal area of 10.7 m^2/ha and standing aerial biomass of 39.1 MT/ha, 36.9 in wood, bark, and branches, 1.7 in fruits and foliage.

Biotic factors — Browne[53] lists the following fungi as affecting this species: *Cercospora catappae, Diplodia catappae, Fomes durissimus, F. fastuosus, Myxormia terminaliae, Phellinus gilvus, Phyllosticta catappae, Polyrhizon terminaliae, Sclerotium rolfsii,* and *Sphaceloma terminaliae.* Also listed are *Dendrophthoe falcata* (Angiospermae); *Amblyrhinus poricollis, Apoderus tranquebaricus, Araccerus fasciculatus, Oncideres cingulata* (Coleoptera); *Coccus hesperidum, Saissetia coffeae, S. nigra* (Hemiptera); *Acrocercops erioplaca, A. ordinatella, A. supplex, A. terminaliae, Antheraea paphia, Dasychira mendosa, Euproctis scintillans, Lymantria ampla, Metanastria hyrtaca, Parasa lepida, Sclepa celtis, Trabala vishnou, Trypanophora semihyalina* (Lepidoptera); and *Rhipiphorothrips cruentatus, R. karna* (Thysanoptera). In India, parakeets steal much of the crop. According to Reed[278] the flowers yield a nectar for honey, which is difficult to collect by bees. In addition, he lists the fungi *Cercospora catappae, Gnomia* sp., *Harknessia terminaliae, Phomopsis terminaliae, Polyporus calcutensis,* and *Sclerotium rolfsii.* It is also attacked by the nematode, *Rotylenchus reniformis.*[186,278] For Puerto Rico, Stevenson[380] lists *Fusiococcum microspermum, Rhytidhysterium rufulum,* and *Trametes corrugata.*

TRAPA NATANS L. and other species (TRAPACEAE) — Water-Chestnut, Jesuit Nut, Water Caltrops

Uses — Water-chestnuts are used as a nut, fresh or roasted, made into a flour, served as a cooked vegetable, or made into a confection, candied much as true chestnuts in Europe. According to Rosengarten,[283] they have been consumed in central Europe since neolithic time. Fresh or boiled nuts are good in salads, having a floury texture and an agreeable nutty flavor. Nuts are often made into rosaries. Roast seed are sometimes used as a coffee substitute. Since water-chestnuts resemble water hyacinths, it has been suggested that they might be used to supplant the water hyacinth, an economic approach to biological control.

Folk medicine — Reported to be alterative, astringent, refrigerant, and tonic, various species of *Trapa* are used in folk remedies for anasarca, bronchitis, cancer, cough, diarrhea, dropsy, fever, flux, rinderpest, and sunstroke.[91] In Japan, the fruits are used in folk remedies for esophageal, gastric, gastrointestinal, lung, stomach, and uterine cancers.[126] Ayurvedics use fruits of *T. bispinosa* (figured) for biliousness, blood disorders, erysipelas, fractures, fatigue, inflammations, leprosy, strangury, and urinary disorders. Yunani, who consider the fruit aperitif, aphrodisiac, and febrifuge, use the fruit for bad teeth, biliousness, bronchitis, fever, lumbago pain, sore throat, and thirst. Cambodians use the infusion of the rind of the fruit for asthenia due to malaria or some other type of fever.[165]

Chemistry — Per 100 g, the fruit of *T. bispinosa* is reported to contain 348 calories, 12.2 g protein, 1.2 g fat, 82.7 g total carbohydrate, 2.4 g fiber, 3.9 g ash, 160 mg Ca, 339 mg P, 3.6 mg Fe, 62.5 mg Na, 1345 mg K, 0.0 μg beta-carotene equivalent, 0.39 mg thiamine, 0.18 mg riboflavin, 5.95 mg niacin, and 20.8 mg ascorbic acid. The seed of *T. bispinosa*, per 100 g, is reported to contain 15.7 g protein, 1.0 g fat, 79.7 g total carbohydrate, 2.0 g fiber, 3.7 g ash, 66.7 mg Ca, 500 mg P, 2.7 mg Fe, 163 mg Na, 2166 mg K, 0.17 mg thiamine, 0.23 mg riboflavin, 2.00 mg niacin, and 30.0 mg ascorbic acid. Per 100 g, the fruit of *T. natans* is reported to contain 11.9 g protein and 1.0 g fat.[89] *The Wealth of India*[70] reports that the kernels contain: moisture, 70.0; protein, 4.7; fat, 0.3; fiber, 0.6; other carbohydrates, 23.3; and mineral matter, 1.1%; calcium 20; phosphorus, 150; and iron, 0.8 mg/100 g. Other minerals reported are copper, 1.27; manganese, 5.7; magnesium, 38; sodium, 49; and potassium, 650 mg/100 g. Iodine (50.6 μ/100 g) is also present. The vitamin contents are thiamine, 0.05; riboflavin, 0.07; nicotinic acid, 0.6; and vitamin C, 9 mg/100 g; vitamin A, 20 IU/100 g. Kernels contain 15.8 mg/100 g oxalates (dry wt). Beta-amylase and much phosphorylase have been reported in the kernels. The nutritive value of flour, prepared from dried kernels, is as follows: moisture, 10.6; protein, 8.0; fat, 0.6; and minerals, 2.6%, calcium, 69; phosphorus, 343; iron, 2.8; and thiamine, 0.44 mg/100 g. The starch, isolated from the flour, consists of 15% amylose, 85% amylopectin.[70] According to Hager's Handbook, the nut (*T. natans*) contains 37% water, 8 to 10% crude protein, 0.7% fat, 1.3% crude fiber, 49% N-free extract (52% starch, 3.2% dextrose). The fruit husk contains 10% tannin.[187]

Description — Hardy aquatic annual or perennial herbs, rooted in the mud, with unbranched stems 0.5 to 2 m long. Plants usually floating with submerged sessile leaves, the lowest opposite, the others alternate, pinnatifid, often functioning as roots; floating leaves in a large rosette, often beautifully variegated, rhombic to nearly orbicular, glabrous above, pubescent at least along the veins beneath, about 7.5 cm in diameter, petioles to 17 cm long, pubescent, often with a fusiform swelling. Flowers solitary, tetramerous, in axils of floating leaves, borne centrally on short stalks above the surface of the water, small inconspicuous, 1 to 2 cm across, white; sepals narrowly triangular, keeled accrescent and indurated in fruit, persistent and forming 2, 3, or 4 horns; petals white, about 8 mm long, caducous. Nut solitary, indehiscent, 2 to 3.5 cm long, 2 to 5.5 cm wide; roots abundant, much-branched. Flowers June to July; fruits autumn.[278]

Germplasm — Reported from the China-Japan Center of Diversity, water-chestnut, or cvs thereof, is reported to tolerate weeds and waterlogging.[82] Although many species and varieties have been described, I am inclined to accept the opinion of *The Wealth of India*,[70] "the more prevalent view seems to be that *Trapa* is a monotypic genus represented by *T. natans* Linn, a polymorphic species". Great variation is found in size of fruit and in number and development of the horns. Some variations seem to be due to edaphic factors, as abnormally high calcium or low potassium and nitrogen concentrations of the water in which they grow.[278] The related *T. bicornis*, the Chinese Ling, is locally important as a food crop. *T. bispinosa* is widely cultivated in India and Kashmir, as the "Singhara Nut". (2n = 36, 40, 48).[82,209,283]

Distribution — Native to central and eastern Europe and Asia, water-chestnuts have been used for food since Neolithic times. They were introduced in 19th century America. The plants spread and became established in the eastern U.S., often choking waterways or crowding out other plants.[278]

Ecology — Ranging from Cool Temperate Moist to Wet through Subtropical Moist Forest Life Zones, water-chestnut is reported to tolerate annual precipitation of 4.3 to 13.2 dm (mean of 5 cases = 8.1), annual temperature of 8.3 to 21.0°C (mean of 5 cases = 11.4°C), and pH of 5.9 to 7.2 (mean of 3 cases = 6.7).[82] Hardy to Zone 5; average annual minimum temperature of −23.3 to −20.6°C (−10 to −5°F).[343] *Trapa natans* is more hardy than the

Ling (*T. bicornis*). The former thrives in ponds and lakes, along slow streams and in stagnant waters, growing best in nutrient-rich but not strongly calcareous waters. It is mainly temperate in climatic requirements.[278]

Cultivation — Water-chestnut is propagated by seed, which must be kept in water before they are sown. They lose their power to germinate quickly if out of water. Seeds are sown in mud. Plants grown in pools or tubs in eastern North America with 5 to 10 cm of loamy soil and filled with water. Plants may also be simply laid on the surface of the water, and they adapt themselves to the situation.[278]

Harvesting — Harvesting the fruits (nuts or seeds) is by hand-picking, sometimes by boat, depending on the size of the field or pond.[278] In India, fruits are ready for harvest about 3 weeks after flowering, i.e., from September to December (to February). At first nuts are harvested once every 2 weeks, then every week, and then nearly every day from November onward.

Yields and economics — Biomass yields of 10 MT/ha seem reasonable. Yields of singhara nut run 4.8 to 6.2 MT/ha. *The Wealth of India* reports yields of 1,760 to 4,440 (to 13,200) kg nut per ha.[70]

Energy — In Japan, the maximum biomass in a floating water-chestnut community was 3 MT/ha at two seasonal peaks, dipping below 1 MT/ha between peaks. But the total dead material may add up to nearly 8 MT/ha, indicating annual biomass potential (life expectancy of the leaves averaged less than 1 month).[370]

Biotic factors — The following fungi are known to attack water-chestnut: *Septoria trapaenatantis* and *Trichoderma flavum*.[278] A leafspot, caused by *Bipolaris tetramera*, seriously affects India's crop. Captan is reported to control the spread. The Singhara beetle, *Galerucella birmanica* is an important widespread pest, controlled in India with 5% BHC.[70] Dusting tobacco or Pyrodust 4000 at 44 kg/ha kills adults and grubs. Chironomid larvae, feeding on petioles and pedicels, may induce malformation of the fruits. *Haltica cyanea*, the blue beetle, feeds and breeds on the leaves. *Bagous trapae* damages soft submerged stems. The aphid *Rhopalosiphum nympheae* occurs on upper leaves, sometimes in large numbers, and often in company with the coccinellid beetles *Pullus nobilus* and *P. piescens*. Larvae of *Nymphula gangeticalis* excavate shelters in the swollen petioles. *Bagous vicinus* and *Nanophyes rufipes* also bore into the petiole.[70,186,278]

TRECULIA AFRICANA Decne. (MORACEAE) — African Breadfruit, African Boxwood, Okwa, Muzinda, Ukwa

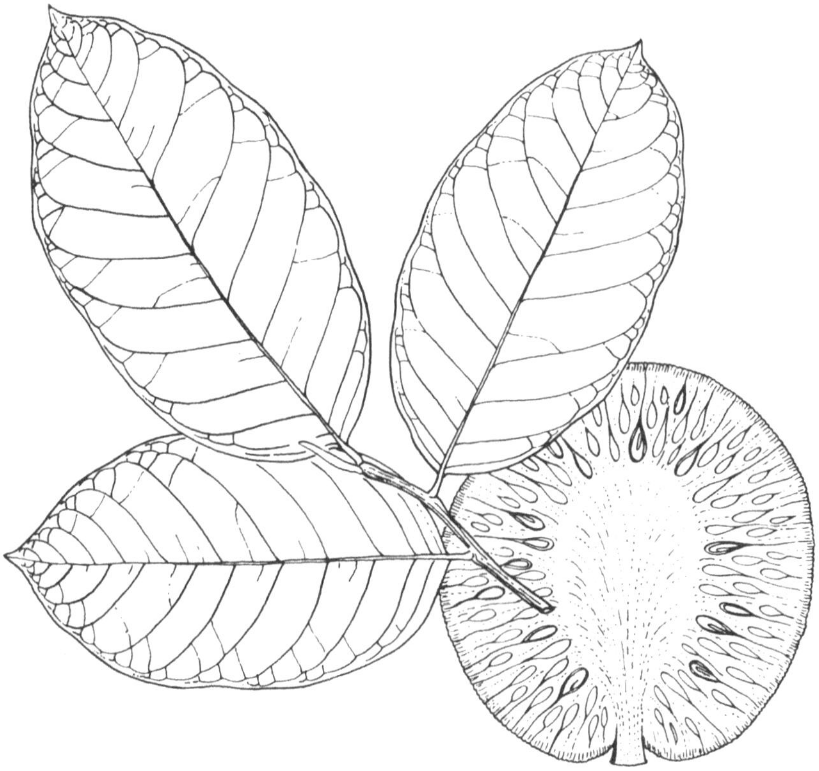

Uses — Seeds are removed from the pulp of African breadfruit by macerating with water, and then eaten cooked, or ground into a meal or flour, or used in soups. Conversely, the seeds can be roasted until the testa becomes brittle for easy removal, the cotyledons then consumed. According to Makinde et al.,[198] the seeds are widely consumed only among the Igbo of Nigeria. "Almond Milk" is a beverage made from this meal. Seeds may be roasted or boiled, peeled and eaten as a dessert nut, or fried in oil. Seeds, with a groundnut flavor, also yield an edible oil. Seeds or oil are put in soaps. Seeds are also used to flavor alcoholic beverages. Heartwood is golden-yellow or yellow-brown (though the very narrow sapwood is yellow-white), very dense and heavy, faintly elastic and flexible, of fine even structure; usable for furniture, wood carving, inlay work and turnery; timber usually marketed as African Boxwood.[278]

Folk medicine — Reported to be laxative, tonic, and vermifuge, African breadfruit is a folk remedy for cough, fever, leprosy, neck ailments, tooth extraction, roundworms, and swelling.[91] Nigerians use the bark decoction for constipation and coughs. Medicinally, a root decoction is used as febrifuge and vermifuge, or drunk as a tonic after illness. It is used for roundworms in children. Bark is used for coughs and as a laxative, and for leprosy.[278]

Chemistry — Per 100 g, the seed (ZMB) is reported to contain 415 calories, 13.9 g protein, 6.2 g fat, 77.5 g total carbohydrate, 1.8 g fiber, 2.4 g ash, 140 mg Ca, and 349 mg P.[89] Seeds contain ca.4 to 7% total lipids, Makinde et al.[198] reporting 5% oil, 13% crude protein. Of the protein extracted, glutelins constituted 53.3%, 23.8% albumins, and 33.8% globulins. Makinde et al.[198] give the amino acid composition shown in Table 1. Table 2[198] compares defatted ukwa protein with other foods. Remember that defatted seeds are not directly comparable to the usual seed analysis (defatted ukwa seeds contain 19%, cf. 13%

Table 1
AMINO ACID COMPOSITION OF *T. AFRICANA* (UKWA) PROTEIN[198]

Amino acids	mg/16 g of N	Amino acids	mg/16 g of N
Aspartic acid	105	Isoleucine	56
Threonine	52	Leucine	74
Serine	67	Tyrosine	56
Glutamic acid	137	Phenylalanine	76
Proline	47	Lysine	62
Glycine	72	Histidine	38
Alanine	40	Ammonia	18
Half-cystine	8	Arginine	79
Valine	61	Tryptophan	2
Methionine	9		

Table 2
ESSENTIAL AMINO ACID CONTENT OF DEFATTED UKWA SEED PROTEIN COMPARED TO SOME OTHER SEED PROTEINS, MAIZE, AND EGG[198]

	mg of amino acid per 16 g of nitrogen							
Amino acids	Defatted Ukwa seed	FAO/WHO (1973)	NAS/NRC (1980)	Harosoy soybean	Cowpeas	Whole maize	Egg	Kidney beans
Histidine	38		17	26	29	23	24	26
Isoleucine	56	40	42	42	40	40	63	42
Leucine	74	70	70	80	76	196	88	81
Lysine	62	55	51	65	68	25	67	67
Total SAA	17	35	26	10	10	19	60	9
Total arom. AA	132	60	73	49	53	44	99	53
Threonine	52	40	35	37	37	47	51	42
Valine	61	50	48	46	48	54	68	51
Tryptophan	2	10	11	18	14	6	34	15

CP for whole seed). The seed fat contains 24.1% palmitic-, 11.7% stearic-, 46% oleic-, and 18% linoleic-acids. Edet et al.[395] report the seeds to contain 7.8% moisture in terms of wet weight; and in terms of dry weight, 13.4% protein, 18.9% fat, 1.4% fiber, 2.1% ash, 58.1% carbohydrate, 3.0% oxalate, and per 100 g, 7 mg Na, 184 mg Mg, 18 mg Ca, 585 mg K, 382 mg P, 3.9 mg Cu, 1.6 mg Fe, 0.20 mg Cr, 7.5 mg Zn, 6.0 mg beta-carotene, 0.5 mg thiamin, 0.3 mg riboflavin, 45 mg ascorbic acid.

Toxicity — Sap of the male tree is caustic and toxic, and if applied on cotton to a carious tooth, will cause it to fall out. No evidence supports the idea that leaves falling into waterholes are poisonous to horses.[278]

Description — Unbuttressed medium-to-large tree, up to 27 m tall and 3 m in girth, bole cylindrical or squarish, fluted at base up to 7 m, bark pale-gray, smooth, latex white; branchlets purple-gray, pithy. Leaves alternate, simple, glabrous, glossy above, elliptic to ovate-elliptic, 20 to 25 cm long, 7.5 to 12 cm wide, sometimes larger, apex shortly pointed, base unequally rounded, petiole very short. Flowers dioecious, male and female flowers in separate inflorescences; male flower-heads globular, 5 cm in diameter, brownish-yellow, very shortly pedunculate, stamens 3. Fruits spherical, up to 45 cm in diameter, and 16 kg in weight, subsessile on the trunk and main limbs, covered with coarse, spine-like tubercles, becoming yellow-brown and soft when ripe. Seeds very numerous, over 1,500 per fruit, smooth, ellipsoid, buried in spongy pulp, ca.1.25 cm long. Flowers January to February fruits February to March (Africa).[146]

Germplasm — Reported from the African Center of Diversity, African breadfruit, or cvs thereof, is reported to tolerate drought, low pH, and waterlogging.[82]

Distribution — Native of West Africa (Guinea, Ivory Coast, Gold Coast, Nigeria, Cameroons, Sierra Leone, Ghana), from Senegal to Angola, Uganda and Nile Land.[278]

Ecology — Ranging from Subtropical Moist through Tropical Dry to Moist Forest Life Zones, African breadfruit is reported to tolerate annual precipitation of 13.6 to 24.1 dm (mean of 3 cases = 18.3), annual temperature of 23.5 to 26.6°C (mean of 3 cases = 25.4°C), and pH of 5.0 to 5.3 (mean of 2 cases = 5.2).[82] Evergreen and deciduous forests. Tree of tropical forests in comparatively dry zones and in villages where planted. Soil under the tree is moist throughout the dry season from condensation. It is usually found near streams or in swampy forests.[146]

Cultivation — Frequently planted in villages and about homesteads.[278]

Harvesting — No data available.

Yields and economics — Uses for the fruit and lumber are mainly local but widespread in Tropical Africa.[278]

Energy — The wood is used for firewood.

Biotic factors — Fruits are eaten by antelopes and large forest snails.

VIROLA SEBIFERA Aubl. (MYRISTICACEAE) — Virola Nut, Red Ucuuba

Uses — The seeds are the source of Virola fat, a nutmeg-scented fat which soon becomes rancid. It is used for making aromatic candles and soaps. Seeds are pierced onto sharp sticks as candle-nuts. The light, soft, pale-brown wood is easy to work but sap stains badly. It is considered suitable for boxes, crates, concrete forms, plywood, and cheap interior construction. Duke[78] notes that some of the economic uses (including narcotic uses) ascribed to this species may be due to confusion with other species in the taxonomically perplexing genus. The jungle names ucachuba, ucahuba, ucauba, uchuhuba, ucuiba, and ucuuba are some of many possible orthographic variants.

Fold medicine — The liniments made from *V. sebifera* are used in folk remedies for tumors. Reported to be a fumitory, the virola is a folk remedy for fever.[91] Brazilians use the fat as a poultice and for rheumatism. The kino-like resin is used for aphtha, angina, caries, and erysipelas. Homeopathically, it is used for abscesses, furuncles, lymphadenitis, and pyodermy. As a tea, the leaves are used for colic and dyspepsia.

Chemistry — Fatty acids of the nuts contain 5 to 13.3% lauric acid, 66.6 to 73% myristic acid, 8.9 to 11% palmitic acid, 6.6 to 11% oleic acid, and up to 3.0% linoleic acid.[128] Hager's Handbook[187] lists N,N-dimethyltryptamine and beta-sitosterol for the husk.[187] Lopes, Yoshida, and Gottlieb[190] report lignans from this species, (2R, 3S)-3-(3,4-dimethyoxybenzyl)-2-(3, 4-methylenedioxybenzyl)-butyrolactone was isolated from the seeds and (2R, 3R)-3-(3,4-dimethyoxybenzyl)-2-(3, 4-methylenedioxybenzyl)-butyrolactone, (2R, 3R)-2,3-di-(3,4-dimethoxybenzyl)-butyrolactone, and (2R, 3R)-2,3-di-(3, 4-methylenedioxybenzyl)-butyrolactone were isolated from the pericarp.

Description — Dioecious, often buttressed trees to 40 m, the younger branchlets persistently tomentose or glabrescent. Leaf blades glabrous above, with persistent, ochraceous stalked-stellate hairs below, coriaceous, oblong to elliptic-ovate or obovate, acute to acuminate, cordate, truncate or acute, 10 to 47 cm long, 4 to 15 cm broad; secondary veins 10 to 28 per side, averaging less than 1/cm along the midrib, the tertiary veins rather prominent below; petioles canaliculate, 8 to 25 mm long, 2 to 5 mm broad. Staminate flowers in much-branched panicles; pedicels 0 to 3 mm long; bracts inconspicuous or absent; perianth tardily 3- (to 5-) lobed, 1.3 to 3.0 mm long; anthers 3 (to 5), 0.7 to 1.5 mm long, usually connate to the apex, the infra-antheral portion of the androecium 0.2 to 1.0 mm long. Pistillate flowers solitary or clustered in racemes 3 to 7 cm long, 2 to 7 cm broad; pedicels 1 to 4 mm long; tepals partially connate, with subpinnate ochraceous pubescence; ovary 1-carpellate, with a sessile, obscurely 2-lobed stigma. Fruits 10 to 30 per inflorescence, the velutinous ligneous pericarp ultimately dehiscing longitudinally into 2 valves, subglobose, the aril laciniate.[78] Germination cryptocotylar but epigeal, the eophylls supracotyledonary.[79]

Germplasm — Reported from the South and Central American Centers of Diversity, virola nut, or cvs thereof, is reported to tolerate waterlogging, but not to the extent that *Virola surinamensis* tolerates flooding.

Distribution — Nicaragua to Bolivia, Brazil, and Peru.[78]

Ecology — Ranging from Subtropical (Premontane) Wet to Rain through Tropical Moist to Rain Forest Life Zones, virola nut is estimated to tolerate annual precipitation of 20 to 45 dm, annual temperature of 23 to 27°C, and pH of 6.0 to 8.0.

Cultivation — Rarely, if ever, cultivated. The cryptocotylar seedlings may be moved from beneath the parent tree.

Harvesting — In Panama, Croat[67] speculates that species flowers twice a year, though mature fruits are seen nearly all year. The length of fruit maturation period is unknown.[67] Gordon[114] describes an unusual collecting method in Brazil. The small subspheroid seeds fall to the forest floor in alluvial forest. When the floods come, the seeds float and go downstream, with the flood, to be scooped up with hand-nets by women and children.

Yields and economics — In 1942, Gordon,[114] referring to both *V. sebifera* and *V. surinamensis*, notes that 4,000 to 5,000 tons are harvested per year in Brazil. According to Markley,[200] in Brazil, "Production of oil has varied between 650 and 1,600 MT/year, and, like other soap oils derived from wild plants, production remains static or is declining, maximum production having occurred in 1941."

Energy — Virola candle-nuts are a poor man's source of energy in many tropical developing countries. The trees offer both fire-wood, leaf litter at the rate of ca. 5 MT/ha, and candle-nuts for energy purposes.

Biotic factors — The wood is subject to pinhole borer injury, if cut logs are allowed to lie after cutting in the forest.[9] *Merulius lacrymans* is reported on *V. merendonis.*[186]

VIROLA SURINAMENSIS (Rol.) Warb. (MYRISTICACEAE) — Ucahuba Nut, White Ucuuba

Uses — Seeds are the source of Ucahuba or Ucuiba Butter, a solid resembling Cacao butter. The seeds are threaded onto wooden spikes and used as candle-nuts by various Amerindian groups. The wood, moderately hard, is easily worked.

Folk medicine — Ucahuba is a folk remedy for rheumatism.[91]

Chemistry — The fatty acids of the nut are 0.7% decanoic, 13.0% lauric, 69.7% myristic, 3.0% palmitic, 7.7% oleic, and 5.1% linoleic. Of the saturated fatty acids, 17.6% are C_{12} or below, 72.9% are C_{14}, and 4.4% are C_{16}, for a total of 94.9%. Of the glycerides, 85% are trisaturated, 15% are disaturated, and none are monosaturated. Another breakdown shows 0.7% capric-, 16.9% lauric-, 72.9% myristic, 4.4% palmitic, and 5.1% linoleic-acids.[128]

Description — Dioecious tree, to 30 m or more tall and ca. 60 cm dbh, often moderately buttressed; outer bark coarse, hard, shallowly fissured, reddish-brown; inner bark tan, reddish on its outer surface; branches often spiraled or clustered, extending nearly horizontally; parts when young bearing ferruginous, sessile, stellate, pubescence, glabrate in age; sap red, lacking distinctive odor. Petioles canaliculate, 5 to 10 mm long; leaf blades oblong, acuminate, rounded to acute at base, 9 to 16 cm long, 1.5 to 4.5 cm wide, coriaceous; major lateral veins in 20 to 30 pairs. All parts of inflorescences densely short-pubescent, the trichomes mostly stellate; pedicels ca. 1.5 mm long; perianth ca. 2 mm long, 3- or 4-lobed usually to middle or beyond, the lobes thick, acute to rounded at apex, spreading at anthesis; staminate flowers in fascicles on panicles to 4 cm long; anthers mostly (2)3(6), connate to apex. Pistillate flowers in clusters of 3 to many, in racemes to 5 cm long; ovary 1-carpellate, ± ovate; stigma sessile, 2-cleft. Capsules ovoid-ellipsoid, thick-walled, light-orange, 3 to 3.5 cm long, bearing dense, short, stellate pubescence; valves 2, woody, ca. 5 mm thick, splitting widely at maturity. Seed 1, ellipsoid, ca. 2 cm long, the aril deeply laciniate, red at maturity (white until just before maturity), fleshy, tasty but becoming bitter after being chewed.[67]

Germplasm — Reported from the South and Central American Centers of Diversity, ucahuba nut, or cvs thereof, is reported to tolerate waterlogging. Natives of the Hyalea distinguish *V. surinamensis* as "ucuuba branca" from *V. sebifera* as "ucuuba vermelha".[222] But in the market, both are sold as ucuhuba fat.

Distribution — Costa Rica and Panama to the Guianas and Brazil and the lesser Antilles.[67] Duke[78] did not include *V. surinamensis* in the *Flora of Panama*. In the Brazilian Hyalea, the trees grow along river banks.

Ecology — Ranging from Subtropical (Premontane) Moist to Wet through Tropical Moist to Wet Forest Life Zones, ucahuba nut is estimated to tolerate annual precipitation of 20 to 40 dm, annual temperature of 23 to 27°C, and pH of 6.0 to 8.0.[82]

Cultivation — Not usually cultivated.

Harvesting — In Panama, flowers from June to March (peaking November to February), maturing fruits from February to August.[67] In the Hyalea of Brazil, the fruits, falling into the water (February to July), float and are gathered with nets made out of bark.[222]

Yields and economics — According to information in Mors and Rizzini,[222] a single tree yields ca. 25 kg ucuuba fat per year. According to Markley,[200] in Brazil, "Production of oil has varied between 650 and 1,600 m tons a year, and, like other soap oils derived from wild plants, production remains static or is declining, maximum production having occurred in 1941."

Energy — Candlenuts are a poor man's source of energy in many tropical developing countries. The trees offer fire-wood, leaf litter at the rate of ca. 5 MT/ha, and candle-nuts for energy purposes.

Biotic factors — No data available.

REFERENCES

1. **Abarquez, A. H.,** Pili Management for Resin and Nut Production, *Canopy Int.*, 8(4), 14, 1982.
2. **Acosta-Solis, M.,** Tagua or Vegetable Ivory — A Forest Product of Ecuador, *Econ. Bot.*, *2,* 46, 1948.
3. **Agaceta, L. M., Dumag, P. U., and Batolos, J. A.,** Studies on the Control of Snail Vectors or Fascioliasis: Molluscicidal Activity of Some Indigenous Plants, in Bureau of Animal Industry, Manila, Philippines, *NSDB Technology Journal,* Abstracts on Tropical Agriculture 7. 38008; (2), 30, 1981.
4. Agriculture Handbook 165, Index of Plant Diseases in the United States, Forest Service, U.S. Department of Agriculture, Washington, D.C., 1960, 531.
5. Agriculture Handbook No. 450, Seeds of Woody Plants of the United States, Forest Service, U.S. Department of Agriculture, Washington, D.C., 1974, 883.
6. **Albuquerque, F. C. De, Duarte, M. De L. R., Manco, G. R., and Silva, H. M. E.,** Leaf Blight of Brazil Nut *(Bertholletia excelsa)* caused by *Phytophthora heveae, Pesquisa Agropecuaria Brasileira,* 9(10), 101, 1977.
7. **Alexander, J. A.,** *Artocarpus integrifolia* (L.), the Jack-tree, *J. Hort. Home Farmer,* March, 1910.
8. **Allen, O. N. and Allen, E. K.,** *The Leguminosae,* The University of Wisconsin Press, Madison, 1981, 812.
9. **Allen, P. H.,** *The Rain Forests of Golfo Dulce,* University of Florida Press, Gainesville, 1956, 417.
10. **Altschul, S. von R.,** *Drugs and Foods from Little-known Plants,* Harvard University Press, Cambridge, Mass., 1973, 366.
11. **Anon.,** Oiticica Oil, Georgia Inst. Tech. State Eng. Exp. Sta., Vol. 11, Bull. 13, 1952.
12. **Anon.,** Coir: Economic Characteristics, Trends and Problems, Commodity Bull. Series, No. 44, Food and Agriculture Organization, Rome, 1969, 1.
13. **Anon.,** Vasicine, the Alkaloid of *Adhatoda vasica,* A Novel Oxytocic and Abortifacient, RRL, *Jammu Newsl.*, 6(2), 9, 1979.
14. **Anon.,** Antifeedant Agents from *Adhatoda vasica,* RRL, *Jammu News.*, 7(5), 21, 1980.
15. **Anon.,** Here Comes (Again), *Canopy Int.*, 7(1), 6, 1981.
16. **Anon.,** Arizona Pecan Orchards Thrive on Climate and Mechanization, *Arizona Land People,* September, 1, 1982.
17. **Archibald,** The use of the fruit of *Balanites aegyptiaca* in the control of Schistosomiasis in the Sudan, *Trans. R. Soc. Trop. Med.*, 27(2), 207, 1933.
18. **Arroyo, C. A.,** Fuel for Home Use, *Canopy Int.*, 7(7), 15, 1981.
19. **Ashiru, G. A. and Quarcoo, T.,** Vegetative Propagation of Kola *(Cola nitida* (Vent.) Schott & Endlicher), *Trop. Agric.* (Trinidad), 48(1), 85, 1971.
20. **Ashworth, F. L.,** Butternuts, Siebold (Japanese) Walnuts, and Their Hybrids, in Jaynes, R. A., Ed., *Handbook of North American Nut Trees,* Northern Nut Growers Association, Knoxville, Tenn., 1969, 224.
21. **Ata, J. K. B. A. and Fejer, D.,** Allantoin in Shea Kernel, *Ghana J. Ag. Sci.*, 8(2), 149, 1975.
22. **Atal, C. K.,** *Chemistry and Pharmacology of Vasicine — A New Oxytocic and Abortifacient,* RRL, Jammu, India, 1980, 155.
23. **Atchley, A. A.,** Nutritional Value of Palms, *Principes,* 28(3), 138, 1984.
24. **Awasthi, Y. C., Bhatnagar, S. C., and Nitra, C. R.,** Chemurgy of Sapotaceous Plants: Madhuca Species of India, *Econ. Bot.*, 29, 380, 1975.
25. **Bailey, L. H.,** *Standard Cyclopedia of Horticulture,* McMillan, New York, 1922.
26. **Bailey, W. E.,** Oyster Nut Growing in British East Africa, Report of American Vice Consul, Nairobi, Kenya Colony, 1, 1940.
27. **Baker, H. G. and Harris, B. J.,** Bat Pollination of the Silk Cotton Tree, *Ceiba pentandra* (L.) Gaertn. (Sensu Lato), in Ghana, *J. Afr. Sci. Assoc.*, 5, 1, 1959.
28. **Balandrin, M. F. and Klocke, J. A.,** Medicinal Plants (Letter), *Science,* 229, 1036, 1985.
29. **Balick, M. J.,** Amazonian Oil Palms of Promise: A Survey, *Econ. Bot.*, 33(1), 11, 1979.
30. **Balick, M. J.,** Economic Botany of the Guajibo, I. Palmae, *Econ. Bot.*, 33(4), 361, 1979.
31. **Balick, M. J.,** New York Botanical Garden, Personal Communication, 1984.
32. **Balick, M. J. and Gershoff, S. N.,** Nutritional Evaluation of the *Jessenia bataua* Palm: Source of High Quality Protein and Oil from Tropical America, *Econ. Bot.*, 35, 261, 1981.
33. **Barrau, J.,** The Sago Palms and Other Food Plants of Marsh Dwellers in the South Pacific Islands, *Econ. Bot.*, 13, 151, 1959.
34. **Bartlett, H. H.,** The Manufacture of sugar from *Arenga saccharifera* in Asahan, on the east coast of Sumatra, 21st Annual Report of the Michigan Academy of Science, 155, 1919.
35. **Batchelor, L. D.,** *Walnut Culture in California,* University of California Publ. Circ., 364, Berkeley, Calif., 1945.
36. **Bavappa, K. V. A., Nair, M. K., and Kumar, T. P.,** Eds., *The Arecanut Palm (Areca catechu* Linn.), Central Plantation Crops Research Institute, Kasaragod, India, 1982, 340.

37. **Beard, B. H. and Ingebretsen, K. H.,** Spring planting is best for oilseed sunflower, *Cal. Agric.*, June, 5, 1980.
38. **Beck, L. and Mittmann, H. W.,** Biology of a Beech Forest, 2, Climate, Litter Production, and Soil Litter, *Carolinea*, 40, 65, 1982.
39. **Bentham, G.,** *Flora Australiensis*, 6 vols., L. Reeve and Co., London, 1873.
40. **Blackmon, W. J., Ed.,** Apios Tribune (newsletter), Louisiana State University, Baton Rouge, 1986.
41. **Boardman, N. K.,** Energy from the biological conversion of solar energy, *Phil. Trans. R. Soc.*, London A 295, 477, 1980.
42. **Bobrov, E. G.,** Betulaceae, in *Flora of USSR*, (English ed.), 5, 211, 1970.
43. **Bogdan, A. V.,** *Tropical Pasture and Fodder Plants*, Longman, London, 1977, 475.
44. **Bolly, D. S. and McCormack, R. H.,** Utilization of the seed of the Chinese Tallow Tree, *J. Oil Chem. Soc.*, March, 83, 1950.
45. **Boulos, L.,** *Medicinal Plants of North Africa*, Reference Publications, Algonac, Mich., 1983, 286.
46. **Braden, B.,** Pecan Timber — A Valuable Resource, *Texas Agric. Progress*, Summer, 20, 1980.
47. **Brand, J. C., Rae, C., McDonnell, J., Lee, A., Cherikoff, V., and Truswell, A. S.,** The Nutritional Composition of Australian Aboriginal Bush Foods, 1, *Food Technol. Australia*, 35(6), 293, 1983.
48. **Braun, W. R. H. and Espericueta, B. M.,** Biomasa y Produccion Ecologica de Jojova (*Simmondsia chinensis* Link) en el Deserto Costero de Sonora, *Deserta*, 5(1978), 57, 1979.
49. **Brinkman, K. A.,** *Corylus* L., Hazel, filbert, in Seeds of woody plants in the United States, *Agriculture Handbook 450*, Forest Service, U.S. Department of Agriculture, Washington, D.C., 1974, 343.
50. **Brooks, F. E. and Cotton, R. T.,** Chestnut Curculios, U.S. Department of Agriculture, Tech. Bull., 130, 1929.
51. **Brown, W. H.,** *Minor Products of Philippine Forests*, 2, 120, 122, Manila, 1921.
52. **Brown, W. H. and Merrill, E. D.,** Minor Products of Philippine Forests, Vol. I, Bureau of Forestry, Department of Agriculture and Natural Resources, Bull. 22, 149, Manila, 1920.
53. **Browne, F. G.,** *Pests and Diseases of Forest Plantation Trees*, Clarendon Press, Oxford, 1968, 1330.
54. **Burk, R. F.,** Selenium, in *Nutrition Reviews, Present Knowledge in Nutrition*, 5th ed., The Nutrition Foundation, Washington, D.C., 1984, 900.
55. **Burkill, H. M.,** *The Useful Plants of West Tropical Africa*, Vol. 1, 2nd ed., Royal Botanic Gardens, Kew, 1985, 960.
56. **Burkill, I. H.,** *A Dictionary of Economic Products of the Malay Peninsula*, 2 vols., Art Printing Works, Kuala Lumpur, 1966.
57. **Bush, C. D.,** *Nut Grower's Handbook*, Orange Judd, New York, 1946, 189.
58. **Bye, R. A., Jr.,** An 1878 Ethnobotanical Collection from San Luis Potosi: Dr. Edward Palmer's First Major Mexican Collection, *Econ. Bot.*, 33(2), 135, 1979.
59. **Cameron, H. R.,** *Filbert Blight*, Nut Grower's Society of Oregon and Washington Proceedings, 53, 15, 1967.
60. **Campbell, C. W.,** Tropical Fruits and Nuts, in *CRC Handbook of Tropical Food Crops*, Martin, F. W., Ed., Boca Raton, Fla., 1984, 235.
61. **Cannell, M. G. R.,** *World Forest Biomass and Primary Production Data*, Academic Press, New York, 1982, 391.
62. **Chandrasekar, V. P. and Morachan, V. B.,** Effect of Advanced Sowing of Intercrops and Nitrogen Levels on Yield Components of Rainfed Sunflower, *Madras Agric. J.*, 66(9), 578, 1979.
63. **Cheney, R. H.,** The Biology and Economics of the Beverage Industry, *Econ. Bot.*, 1, 243, 1947.
64. **Child, R.,** *Coconuts*, Longmans, Green & Co., London, 1, 1964.
65. **Collins, T. F. X.,** Review of Reproduction and Teratology Studies of Methylxanthines, FDA By-Lines, No. 2, Food and Drug Administration, Washington, D.C., 1981, 86.
66. **Corley, R. H. V.,** Oil Palm, in *CRC Handbook of Biosolar Resources*, Vol. II, Resource Materials, McClure, T. A. and Lipinsky, E. S., Eds., CRC Press, Boca Raton, Fla., 1981, 397.
67. **Croat, T. B.,** *Flora of Barro Colorado Island*, Stanford University Press, Stanford, Calif., 1978, 943.
68. **Cronquist, A., Holmgren, A. H., Holmgren, N. H., and Reveal, J. L.,** *Intermountain Flora*, New York Botanical Garden, New York, Vol. 1, 1972, 270.
69. **Crutchfield, W. B. and Little, Jr., E. L.,** Geographic Distribution of the Pines of the World, Forest Service, U.S. Department of Agriculture, Washington, D.C., Misc. Publ. 991, map 16, 1966, 9.
70. C.S.I.R. (Council of Scientific and Industrial Research), *The Wealth of India*, 11 vols., New Delhi, 1948-1976.
71. **Dalziel, J. M.,** *The Useful Plants of West Tropical Africa*, The Whitefriars Press, London, 1937, 612.
72. **Daugherty, P. M., Sineath, H. H., and Wastler, T. A.,** Industrial Raw Materials of Plant Origin, IV, A survey of *Simmondsia chinensis* (Jojoba), Engin. Exp. Sta., Georgia Institute of Technology, Atlanta, Bull. No. 17, 1, 1953.
73. **DeRigo, H. T. and Winters, H. F.,** Nutritional Studies with Chinese Water-chestnuts, *Proc. Am. Soc. Hort. Sci.*, 92, 394, 1968.

74. **Dickey, R. D. and Reuther, W.,** Flowering, fruiting, yield, and growth habits of tung trees, *Fla. Agric. Exp. Sta., Bull.* 343, 1940.
75. **Dickson, J. D.,** Notes on Hair and Nail Loss After Ingesting Sapucaia Nuts (*Lecythis elliptica*), *Econ. Bot.*, 23(2), 133, 1969.
76. *Dorland's Illustrated Medical Dictionary,* 25th ed., W. B. Saunders, Philadelphia, Pa., 1974, 1748.
77. **Dorrell, D. G.,** Sunflower *Helianthus annuus*, in *CRC Handbook of Biosolar Resources*, Vol. II, Resource Materials, McClure, T. A. and Lipinsky, E. S., Eds., CRC Press, Boca Raton, Fla., 1981, 105.
78. **Duke, J. A.,** Flora of Panama, Myristicaceae, *Ann. Mo. Bot. Gard.*, 49(3, 4), 526, 1962.
79. **Duke, J. A.,** On Tropical Tree Seedlings, I, Seeds, seedlings, systems and systematics, *Ann. Mo. Bot. Gard.*, 56(2), 125, 1969.
80. **Duke, J. A.,** *Isthmian Ethnobotanical Dictionary*, 3rd ed., Scientific Publishers, Jodhpur, India, 1986, 205.
81. **Duke, J. A.,** Palms as Energy Sources: A Solicitation, *Principes*, 21(2), 60, 1977.
82. **Duke, J. A.,** The quest for tolerant germplasm, in *ASA Special Symposium 32, Crop Tolerance to Suboptimal Land Conditions*, American Society of Agronomy, Madison, Wisc., 1, 1978; (updated as computer file by Dr. A. A. Atchley, Agriculture Research Service, U.S. Department of Agriculture, Beltsville, Md.)
83. **Duke, J. A.,** *Handbook of Legumes of World Economics Importance*, Plenum Press, N.Y., 1981, 345.
84. **Duke, J. A.,** Magic Mountain, 2000 AD, Paper No. 2, 151, in 97th Congress (1st Session), Background Papers for Innovative Biological Technologies for Lesser Developed Countries, (OTA Workshop Nov. 24-25, 1980), U.S. Government Printing Office, Washington, D.C., 1981, 511.
85. **Duke, J. A.,** *Medicinal Plants of the Bible*, Trado-Medic Books, Buffalo, N.Y., 1983, 233.
86. **Duke, J. A.,** *Handbook of Medicinal Herbs*, CRC Press, Boca Raton, Fla., 1985, 677.
87. **Duke, J. A.,** A Green World Instead of the Greenhouse, in *The International Permaculture Seed Yearbook 1985*, P.O. Box 202, Orange, Mass. 01364, 1985, 15.
88. **Duke, J. A.,** Medicinal Plants (Letter), *Science*, 229, 1036, 1985.
89. **Duke, J. A. and Atchley, A. A.,** *Handbook of Proximate Analysis Tables of Higher Plants*, CRC Press, Boca Raton, Fla., 1986, 389.
90. **Duke, J. A. and Ayensu, E. S.,** *Medicinal Plants of China*, 2 vols., Reference Publications, Algonac, Mich., 1985.
91. **Duke, J. A. and Wain, K. K.,** *Medicinal Plants of the World*, Computer index with more than 85,000 entries, 3 vols., 1981, 1654.
92. **Earle, F. R. and Jones, Q.,** Analyses of Seed Samples from 113 Plant Families, *Econ. Bot.*, 16(4), 221, 1962.
93. **Egolf, R. G.,** The Cycads (Cycadaceae) — The Nuts That Dinosaurs Ate, in *Edible Nuts of the World*, Menninger, E. A., Ed., Horticultural Books, Stuart, Fla., 1977, 161.
94. **Ehsanullah, L.,** The Cultivars of the African Oil Palm, *Principes*, 16(1), 26, 1972.
95. **Ereinoff, V. A.,** Sur l'introduction de *Corylus heterophylla*, our noisetier de Siberie dans l'arboriculture francaise, *Rev. Hort.*, 123, 395, 1951.
96. **Erichsen-Brown, C.,** *Use of Plants for the Past 500 Years*, Breezy Creeks Press, Aurora, Ontario, 1979, 512.
97. **Erickson, H. T., Correa, M. P. F., and Escobar, J. R.,** Guarana *(Paullinia cupana)* as a Commercial Crop in Brazilian Amazonia, *Econ. Bot.*, 38(3), 273, 1984.
98. *FAO Production Yearbook*, Vol. 33, Food and Agriculture Organization, Rome, 1980, 309.
99. **Farris, C. W.,** Hybridization of Filberts, in *Handbook of North American Nut Trees*, Jaynes, R. A., Ed., Northern Nut Growers Association, Knoxville, Tenn., 1969, 299.
100. **Farwell, O. A.,** Botanical source of cola nut of commerce, *Am. J. Pharm.*, 94, 428, 1922.
101. **Freire, F., Das, C. O., and Ponte, J. J.,** Da, Meloidogyne incognita on *Bertholletia excelsa* in the State of Para (Brazil), *Boletim Cearense Agronomia*, 17, 57, 1976.
102. **Furr, A. K., MacDaniels, L. H., St. John, Jr., L. E., Gutenmann, W. H., Pakkala, I. S., and Lisk, D. J.,** Elemental Composition of Tree Nuts, *Bull. Environ. Contam. Toxicol.*, 21, 392, 1979.
103. **Galloway, B. T.,** The Search in Foreign Countries for Blight-resistant Chestnuts and Related Tree Crops, U.S. Department of Agriculture, *Dept. Circ.*, 383, 1, 1926.
104. **Garcia-Barriga, H.,** *Flora Medicinal de Colombia*, Universidad Nacional, Bogota, Vol. 1, 1974, 561.
105. **Garcia-Barriga, H.,** *Flora Medicinal de Colombia*, Universidad Nacional, Bogota, Vol. 2, 1975, 538.
106. **Garcia, P. R.,** Pili: A Potential Reforestation Crop of Multiple Uses, *Canopy Int.*, 9(2), 12, 1983.
107. **Gaydou, A. M., Menet, L., Ravelojaona, G., and Geneste, P.,** Energy Resources of Plant Origin in Madasgascar: Ethyl Alcohol and Seed Oils, *Oleagineux*, 37(3), 135, 1982.
108. **Generalao, M. L.,** Those Seemingly Insignificant Plants, *Canopy Int.*, 7(1), 6, 1981.
109. **Gibbs, R. D.,** *Chemotaxonomy of Flowering Plants*, 4 vols., McGill-Queens University Press, London, 1974.
110. **Gohl, B.,** Tropical Feeds, Feed Information Summaries and Nutritive Values, FAO Animal Production and Health Series No. 12, Food Agriculture Organization, Rome, 1981, 529.

111. **Goldblatt, L. A.,** The Tung Industry, II, Processing and Utilization, *Econ. Bot.*, 13, 343, 1960.
112. **Gomez-Pampa, A.,** Renewable Resources from the Tropics, in *Renewable Resources: A Systematic Approach*, Campos-López, E., Ed., Academic Press, New York, 1980, 391.
113. **Gordon, J.,** Testing for blight resistance in American chestnut, *Northern Nut Growers Assoc. Ann. Rep.*, 61, 50, 1970.
114. **Gordon, J. B.,** Vegetable Oils in Brazil, *Soap*, 18(8), 19, 1942.
115. **Graves, A. H.,** Some outstanding new chestnut hybrids, I, *Bull. Torr. Bot. Club, 87*, 192, 1960.
116. **Green, P. L.,** The Production of Oiticica Oil in Brazil, *Foreign Agric.*, 4, 617, 1940.
117. **Grieve, M.,** *A Modern Herbal*, Reprint 1974, Hafner Press, N.Y., 1931, 916.
118. **Grimwood, B. E.,** *The Processing of Macadamia Nuts*, Tropical Products Inst. G66, London, 1971, 1.
119. **Gyawa,** Personal communication, September 14, 1984.
120. **Hadcock, M.,** It Had to be Simple, Cheap, Reliable, and Easy to Clean, *Internat. Ag. Dev.*, 3(1), 16, 1983.
121. **Halos, S. C.,** Nipa for Alcogas Production, *Canopy Int.*, 1981.
122. **Hardman, R. and Sofowora, E. A.,** A Reinvestigation of *Balanites aegyptiaca* as a Source of Steroidal Sapogenins, *Econ. Bot.*, 26(2), 169, 1972.
123. **Hardon, J. J.,** The Oil Palm: Progress Through Plant Breeding in Malaysia, *SPAN*, 27(2), 59, 1984.
124. **Harlan, J. R.,** *Crops and Man*, American Society of Agronomy, Madison, Wisc., 1975, 295.
125. **Hartley, C. W. S.,** *The Oil Palm*, Longmans, Green & Co., London, 1967, 706.
126. **Hartwell, J. L.,** *Plants Used Against Cancer, A Survey*, Reissued in one volume by Quarterman Publications, Lawrence, Mass., 1982, 710.
127. **Harwood, H. J.,** Vegetable Oils as an On the Farm Diesel Fuel Substitute: The North Carolina Situation, RTI Final Report FR-41U-1671-4, Research Triangle Park, North Carolina, 1981, 65.
128. **Hilditch, T. P. and Williams, P. N.,** *The Chemical Constitution of Natural Fats*, Barnes & Noble, 1960, 745.
129. **Hill, L.,** The unforgettable butternut, *Organic Gardening and Farming*, November, 46, 1974.
130. **Hodge, W. H.,** Oil-producing palms of the world: a review, *Principes*, 19(4), 119, 1975.
131. **Holdt, C.,** Identity of the Oiticica Tree and Uniformity of the Oil, *Drugs, Oils, and Paints*, 52, 316, 1937.
132. **Holland, J. H.,** Oiticica *(Licania rigida)*, *Kew Bull. Misc. Inform. 1932*, 406, 1932.
133. **Holland, B. R. and Meinke, W. W.,** Chinese Tallow Nut Protein, I, Isolation, Amino Acid and Vitamin Analysis, *J. Am. Oil Chem. Soc.*, 25(11), 418, 1948.
134. **Holm, L. G., Pancho, J. V., Herberger, J. P., and Plucknett, D. L.,** A *Geographical Atlas of World Weeds*, John Wiley & Sons, New York, 1979, 391.
135. **Holm, L. G., Plucknett, D. L., Pancho, J. V., and Herberger, J. P.,** *The World's Worst Weeds — Distributions and Biology*, East-West Center, University Press of Hawaii, Honolulu, 1977, 609.
136. **Hooper, D.,** *Sapium sebiferum* (The Chinese Tallow Tree), Chinese or Vegetable Tallow, its preparation, uses, and composition, *The Agricultural Ledger*, 11(2), 1904; Calcultta, 11, 1905.
137. **Hooper, D.,** *Schleichera trijuga* (Kusum tree of India), *Dictionary of Economic Products*, Government Printing Office, Calcutta, 6, 950, 1905.
138. **Hooper, D.,** Useful Plants and Drugs of Iran and Iraq, Botanical Series Field Museum, Chicago 9(3), Publication 387, 71, 1937.
139. *Hortus Third, A Concise Dictionary of Plants Cultivated in the United States and Canada*, MacMillan, New York, 1976, 1290.
140. **Hou, D.,** Rhizophoraceae, in *Flora Malesiana*, Series 1, Vol. 5, van Steenis, C. G. G. J., Ed., P. Noordhoff, Republic of Indonesia, 1958, 429.
141. **Hsu, H. Y., Chen, Y. P., and Hong, M.,** *The Chemical Constituents of Oriental Herbs*, Oriental Healing Arts Institute, Los Angeles, 1972, 1546.
142. **Hurov, R.,** *Hurov's Tropical Seeds, Catalog 34*, Honolulu, ca.1983.
143. **Hussey, J. S.,** *Some Useful Plants of Early New England*, The Channings, Marion, Mass., 1976, 99.
144. **Illick, J. S.,** The American walnuts, *Am. For.*, August, 699, 1921.
145. **Ingham, J. L.,** Phytoalexin Induction and Its Taxonomic Significance in the Leguminosae (Subfamily Papilionoideae), in *Advances in Legume Systematics*, Polhill, R. M. and Raven, P. H., Eds., 1981, 599.
146. **Irvine, F. R.,** *Woody Plants of Ghana*, Oxford University Press, London, 1961, 868.
147. **Jackson, B. D.,** *A Glossary of Botanic Terms*, with their derivation and accent, Hafner, New York, 1953, 481.
148. **Jaynes, R. A.,** *Handbook of North American Nut Trees*, Northern Nut Growers Association, Knoxville, Tenn., 1969, 421.
149. **Jenkins, B. M. and Ebeling, J. M.,** Thermochemical Properties of Biomass Fuels, *Calif. Agric.*, 39(5/6), 14, 1985.
150. **Johnson, D.,** Cashew Cultivation in Brazil, *Agron. Mozamb. Lourenco Marques*, 7(3), 119, 1973.
151. **Johnson, D. V.,** Multi-Purpose Palms in Agroforestry: A Classification and Assessment, *Int. Tree Crops J.*, 2, 217, Academic Publishers, Great Britain, 1983.

152. **Johnson, D. V.**, Ed. and Transl., *Oil Palms and Other Oilseeds of the Amazon*, (Original by C. Pesce), Reference Publications, Algonac, Mich., 1985, 199.
153. **Joley, L. E.**, Pistachio, in *Handbook of North American Nut Trees*, Jaynes, R. A., Ed., The Northern Nut Growers Association, Knoxville, Tenn., 1969, 421.
154. **Jones, C., Griffin, G. J., and Elkins, J. R.**, Association of climatic stress with blight on Chinese chestnut in the eastern United States, *Plant Dis.*, 64(11), 1001, 1980.
155. **Joshi, B. C. and Gopinathan, N.**, *Proceedings Ninth Congress ISSCT, (International Society of Sugar Cane (Technologists)* Vol. II, 1956.
156. **Kasapligil, B.**, *Corylus colurna* L. and its varieties, *Calif. Hort. Soc. J.*, 24(4), 95, 1963.
157. **Kasapligil, B.**, A Bibliography on *Corylus* (Betulaceae) with annotations, *63rd Annual Rep. Northern Nut Growers Assoc.*, 107, 1972.
158. **Keeler, J. T., and Fukunaga, E. T.**, The Economic and Horticultureal Aspects of Growing Macadamia Nuts Commercially in Hawaii, Hawaii Agric. Exp. Sta., *Agr. Econ. Bull.*, No. 27, 1, 1968.
159. **Kelkar, N. V.**, The Betel-Nut Palm *(Areca catechu)* and its cultivation in North Kanara, *Poona Agr. Coll. Mag.*, 7(1), 1, 1915.
160. **Kennard, W. C. and Winters, H. F.**, Some fruits and nuts for the tropics, Agriculture Research Service, U.S. Department of Agriculture, Washington, D.C., misc. publ. No. 801, 80, 1960.
161. **Kent, G. C.**, Cadang-cadang of Coconut, *Philippine Agriculturalist*, 37(5-6), 228, 153.
162. **Keong, W. K.**, Soft Energy from Palm Oil and Its Wastes, *Agric. Wastes*, Dept. Envir. Sci., Universiti Pertanian Malaysia, Serdang, Malaysia, 3(3), 191, 1981.
163. **Kerdel-Vegas, F.**, Generalized hair loss due to the ingestion of 'Coco de Mono' *(Lecythis ollaria), J. Invest. Dermat.*, 42, 91, 1964.
164. **Kerdel-Vegas, F.**, The depilatory and cytotoxic action of 'Coco de Mono' *(Lecythis ollaria)* and its relationship to chronic selenosis, *Econ. Bot.*, 20, 187, 1966.
165. **Kirtikar, K. R. and Basu, B. D.**, *Indian Medicinal Plants*, 4 vols. text, 4 vols. plates, 2nd ed., reprint, Jayyed Press, New Delhi 6, 1975.
166. **Kostylev, A. D.**, Pests and diseases of the filbert *(Corylus maxima)* in the central regions along the Kuban River in the Krasnodar Krau, (Russian), Trudy Kubansk, *Sel'skokhog. Inst.*, 7(35), 91, 1962; Zhurn. Biol. 1963, No. 12E282, transl.
167. **Kranz, J.**, Schmutterer, H., and Koch, W., *Diseases, Pests and Weeds in Tropical Crops*, John Wiley & Sons, New York, 1977, 666.
168. **Krochmal, A. and Krochmal, C.**, *A Guide to the Medicinal Plants of the United States*, Quadrangle/The New York Times Book Co., 1973, 259.
169. **Krochmal, A. and Krochmal, C.**, *A Naturalist's Guide to Cooking With Wild Plants*, Quandrangle/The New York Times Book Co., 1974, 336.
170. **Krochmal, A. and Krochmal, C.**, Uncultivated Nuts of the United States, Agric. Info. Bull. 450, Forest Service, U.S. Department of Agriculture, Washington, D.C., 1982, 89.
171. **Lagerstedt, H. B.**, Filbert Propagation Techniques, *Northern Nut Growers Assoc. Rep.*, 61, 61, 1970.
172. **Lagerstedt, H. B.**, High density filbert orchards, *Nut Growers Soc. Oregon and Wash. Prod.*, 56, 69, 1971.
173. **Lagerstedt, H. B.**, Filberts *(Corylus)* , in *Advances in Fruit Breeding*, Moore, J. N. and Janick, J., Eds., 1973.
174. **Lane, E. V.**, Piqui-a — Potential Source of Vegetable Oil for an Oil-starving World, *Econ. Bot.*, 11(3), 187, 1957.
175. **Langhans, V. E., Hedin, P. A., and Graves, C. H., Jr.**, Fungitoxic Substances Found in Pecan *(Carya illinoensis* K. Koch), *Abstr., Proc. Am. Phytopathological Soc.*, 1976, 3, 339, 1978.
176. **Lanner, R. M.**, Natural hybridization among the Pinyon Pines, *Utah Sci.*, 109, 112, 1972.
177. **Lee, J. P.**, Brazil Nuts, *Brazil*, 20(6), 2, 1946; abstr. in *Econ. Bot.*, 1, 239, 1947.
178. **Leenhouts, P. W.**, The genus *Canarium* in the Pacific, *Bish. Mus. Bull.*, No. 216, 1, 1955.
179. **Leenhouts, P. W.**, A Monograph of the Genus *Canarium* (Burseraceae), *Blumea*, 9(2), 275, 1958.
180. **Leung, A. Y.**, *Encyclopedia of Common Natural Ingredients Used in Food, Drugs, and Cosmetics*, John Wiley & Sons, New York, 1980, 409.
181. **Leung, W., Butrum, R. R., and Chang, F. H.**, Part 1, Proximate Composition Mineral and Vitamin Contents of East Asian Foods, in Food Composition Table for Use in East Asia, Food and Agriculture Organization and U.S. Department of Health, Education and Welfare, 1972, 334.
182. **Lever, R. J. A. W.**, The bread fruit tree, in *World Crops*, September, 63, 1965.
183. **Levy, S.**, Agriculture and Economic Development in Indonesia, *Econ. Bot.*, 11(1), 3, 1957.
184. **Lewis, W. H. and Elvin-Lewis, M. P. F.**, *Medical Botany*, John Wiley & Sons, New York, 1977, 515.
185. **Lima, J. A. A., Menezes, M., Karan, M.deQ., and Martins, O. F. G.**, Genera of Pathogenic Nematodes Isolated from the Rhizosphere of Cashew Tree, *Anacardium occidentale*, *Fitossanidad, Brazil*, 1(2), 32, 1975.

186. **Lipscomb, J.**, Card Catalog, Plant Diseases and Nematodes of the World, Unpublished Card File, Systematic Botany, Mycology, and Nematology Laboratory, Agriculture Research Service, U.S. Department of Agriculture, Beltsville, Md. (no date).
187. **List, P. H. and Horhammer, L.**, *Hager's Handbook of Pharmaceutical Practice*, Vols. 2-6, Springer-Verlag, Berlin, 1969-1979.
188. **Little, Jr., E. L.**, *Common Fuelwood Crops: A Handbook for Their Identification*, McClain Printing, Parsons, W. Va., 1983, 354.
189. **Little, Jr., E. L., Woodbury, R. O., and Wadsworth, F. H.**, Trees of Puerto Rico and the Virgin Islands, *Agriculture Handbook No. 449*, Vol. 2, U.S. Department of Agriculture, Washington, D.C., 1974, 1024. (see also Vol. 1, 426).
190. **Lopes, L. M. X., Yoshida, M., and Gottlieb, O. R.**, Dibenzylbutyrolactone Lignans from *Virola sebifera*, *Phytochemistry*, 22(6), 1516, 1983.
191. **Lozovoi, A. D., and Chernyshov, M. P.**, Yield of Sweet Chestnuts *(Castanea sativa*) from the Forests of Krasnodar Territory, USSR, *Rastitel'nye Resursy*, 15(4), 536, 1979.
192. **Luetzelburg, Ph. von**, Contribucao para o contecimento das "Oiticicas" (Brazil), *Bol. da Insp. de Sec.*, 5(2), 5-15, 1936; Abstr. by Dahlgren, B. D., *Trop. Woods*, 50, 1937.
193. **Lutgen, J. R.**, Blight resistant American chestnuts, *Northern Nut Growers Assoc. Ann. Rep.*, 57, 29, 1966.
194. **MacDaniel, L. H.**, Nut Growing, *Cornell Ext. Bull.*, 701, 1, Ithaca, N.Y. (no date).
195. **MacFarlane, N. and Harris, R. V.**, Macadamia Nuts as an Edible Oil Source, in Pryde, E. H., Princen, L. H., and Mukherjee, K. D., Eds., American Oil Chemists Society, Champaign, Ill., 1981, 103.
196. **MacMillan, H. F.**, *Tropical Gardening and Planting*, 3rd ed., Times of Ceylon, Colombo, Sri Lanka, 1925, 594.
197. **Madden, G., Brison, F. R., and McDaniel, J. C.**, *Handbook of North American Nut Trees*, Pecans, in Jaynes, R. A., Ed., Northern Nut-Growers Association, Knoxville, Tenn., 1969, 163.
198. **Makinde, M. A., Elemo, B. O., Arukwe, U., and Pellett, P.**, Ukwa Seed *(Treculia africana)* Protein, 1, Chemical Evaluation of the Protein Quality, *J. Agric. Food Chem.*, 33(1), 70, 1985.
199. **Markley, K. S.**, Mbocaya or Paraguay Cocopalm — An Important Source of Oil, *Econ. Bot.*, 10(1), 3, 1956.
200. **Markley, K. S.**, Fat and Oil Resources and Industry of Brazil, *Econ. Bot.*, 11(1), 91, 1957.
201. **Markley, K. S.**, The Babassu Oil Palm of Brazil, *Econ. Bot.*, 25(3), 267, 1971.
202. **Martin, F. W. and Ruberte, R. M.**, *Edible Leaves of the Tropics*, Antillian College Press, Mayaguez, Puerto Rico, 1975, 235.
203. **Martin, F. W., Ed.**, *CRC Handbook of Tropical Food Crops*, CRC Press, Boca Raton, Fla., 1984, 296.
204. **Masefield, G. B., Wallis, M., Harrison, S. G., and Nicholson, B. E.**, *The Oxford Book of Food Plants*, Oxford University Press, London, 1969, 206.
205. **Mathieu, E.**, Notes on Cola trees in the economic garden, Singapore, *Straits Settlements Bot. Gard. Bull.*, 2, 74, 1918.
206. **Mattoon, W. R. and Reed, C. A.**, Black Walnuts for Timber and Nuts, U.S. Department of Agriculture, Washington, D.C., *Farmer's Bull.*, 1392, 1, 1933.
207. **McCurrach, J. C.**, *Palms of the World*, Harper & Bros., New York, 1960, 21.
208. **McKay, J. W. and Jaynes, R. A.**, Chestnuts, in *Handbook of North American Nut Trees*, Jaynes, R. A., Ed., Northern Nut Growers Association, Knoxville, Tenn., 1969, 264.
209. **Menninger, E. A.**, *Edible Nuts of the World*, Horticultural Books, Stuart, Fla., 1977, 175.
210. *The Merck Index*, 8th ed., Merck & Co., Rahway, N. J., 1968, 1713.
211. **Micou, P.**, Hark! Something Stirs in the Palm Groves, *ICC Business World*, 1985.
212. **Miege, J. and Miege, M. N.**, *Cordeauxia edulis* — A Caesalpiniaceae of Arid Zones of East Africa, *Econ. Bot.*, 32, 336, 1978.
213. **Millikan, D. F.**, Oaks, Beech, Pines, and Ginkgo, in Handbook of North American Nut Trees, Jaynes, R. A., Ed., Northern Nut Growers Association, Knoxville, Tenn., 1969, 336.
214. **Miric, M., Stanimirovic, D., Hadrovic, H., and Miletic, I.**, Lipid Composition of Chestnuts *(Castanea sativa)* from Metohija, *Hrana I Ishrana*, 14(5/6), 219, 1973.
215. **Mitchell, J. S. and Rook, A.**, *Botanical Dermatology*, Greenglass, Ltd., Vancouver, 1979, 787.
216. **Moerman, D. E.**, *American Medical Ethnobotany: A Reference Dictionary*, Garland Publishing, New York, 1977, 527.
217. **Monroe, G. E., Tung Liang, and Cavaletto, C. G.**, Quality and Yield of Tree-Harvested Macadamia Nuts, Agriculture Research Service, U.S. Department of Agriculture, 42, 1, 1972.
218. **Morgan, P.**, Chinese chestnuts and flame hulling, *Northern Nut Growers' Assoc. Ann. Rep.*, 60, 112, 1969.
219. **Mori, S. A.**, Personal communication, 1986.
220. **Mori, S. A., Orchard, J. E., and Prance, G. T.**, Intrafloral Pollen Differentiation in the New World Lecythidaceae, Subfamily Lecythidoideae, *Science*, 209, 400, 1980.

221. **Mori, S. A. and Prance, G. T.,** The "Sapucaia" Group of Lecythis (Lecythidaceae), *Brittonia,* 33(1), 70, 1981.
222. **Mors, W. B. and Rizzini, C. T.,** *Useful Plants of Brazil,* Holden-Day, Inc., San Francisco, 1966, 166.
223. **Morton, J. F.,** The Jackfruit *(Artocarpus heterophyllus* Lam.): its culture, varieties, and utilization, *Fla. State Hort. Soc.,* 78, 336, 1965.
224. **Morton, J. F.,** *Atlas of Mecidinal Plants of Middle America, Bahamas to Yucatan,* Charles C Thomas, Publisher, Springfield, Ill., 1981, 1420.
225. **Morton, J. F.,** Indian Almond *(Terminalia catappa),* Salt Tolerant, Useful, Tropical Tree with "Nut" Worthy of Improvement, *Econ. Bot.,* 39(2), 101, 1985.
226. **Moyer, J.,** Ed., Nuts and Seeds — The Natural Snacks, Rodale Press, Emmaus, Pa., 1973, 173.
227. **Murthy, K. N. and Yadava, R. B. R.,** Note on the Oil and Carbohydrate Contents of Varieties of Cashewnut *(Anacardium occidentale* L.), *Indian J. Agric. Sci.,* 42(10), 960, 1972.
228. **Nambiar, M. C. and Haridasan, M.,** Fertilizing Cashew for Higher Yields, *Indian Farming,* 28(12), 16, 1979.
229. National Academy of Sciences, Underexploited Tropical Plants with Promising Economic Value, National Academy of Sciences, Washington, D.C., 1975.
230. National Academy of Sciences, Products from Jojoba: A Promising New Crop for Arid Lands, National Academy of Sciences, Washington, D.C., 1975, 30.
231. National Academy of Sciences, Topical Legumes: Resources for the Future, National Academy of Sciences, Washington, D.C., 1979, 331.
232. National Academy of Sciences, Firewood Crops, Shrub and Tree Species for Energy Production, Vol. 2, National Academy of Sciences, Washington, D.C., 1980, 92.
233. National Academy of Sciences, Alcohol Fuels — Options for Developing Countries, National Academy Press, Washington, D.C., 1983, 109.
234. National Research Council, Jojoba: New Crop for Arid Lands, New Material for Industry, National Academy Press, Washington, D.C., 1985, 102.
235. **Nihlgard, B. and Lindgren, L.,** Plant Biomass, Primary Production and Bioelements of Three Mature Beech Forests in South Sweden, *Oikos,* 28(1), 95, 1977.
236. **Obasola, C. O.,** Breeding for short-stemmed oil palm in Nigeria, I, Pollination, compatibility, varietal segregation, bunch quality, and yield of F_1 hybrids *Corozo oleifera* × *Elaeis guineensis, J. Niger. Inst. Oil Palm Res.,* 5, 43, 1973.
237. **Obasola, C. O.,** Breeding for short-stemmed oil palm in Nigeria, II, Vegetative characteristics of F_1 hybrids *Corozo oleifera* × *Elaeis* guineensis, *J. Niger. Inst. Oil Palm Res.,* 5(18), 55, 1973.
238. **Ochse, J. J.,** *Vegetables of the Dutch East Indies,* (Reprinted 1980), A. Asher and Co., Hacquebard, Amsterdam, 1931, 1005.
239. **Okoli, B. E. and Mgbeogu, C. M.,** Fluted Pumpkin, *Telfairia occidentalis:* West African Vegetable Crop, *Econ. Bot.,* 37(2), 145, 1983.
240. **Oppenheimer, C. H. and Reuveni, O.,** Rooting Macadamia cuttings, *Calif. Macadamia Soc. Yearbook,* 7, 52, 1961.
241. **Orallo, C. A.,** Canarium: Alternative Source of Energy?, Canopy Int., 7(10), 10, 1981.
242. **Osborn, D. J.,** Notes on Medicinal and Other Uses of Plants in Egypt, *Econ. Bot.,* 22, 165, 1968.
243. **Page, J.,** Sunflower Power, *Science,* P1, July/August, 92, 1981.
244. **Painter, J. H.,** Filberts in the Northwest, in *Handbook of North American Nut Trees,* Jaynes, R. A., Ed., Northern Nut Growers Association, Knoxville, Tenn., 1969, 294.
245. **Pallotti, C.,** The 'Time for a Coca Cola' may not be right, *Industrie delle Bevande,* 6(6), 123, 1977.
246. **Palmer, E. and Pitman, N.,** *Trees of Southern Africa,* 3 vols., A. A. Balkemia, Cape Town, 1972.
247. **Palmer, I. S., Herr, A., and Nelson, T.,** Toxicity of Selenium in Brazil Nuts, *Bertholletia excelsa,* to Rats, *J. Food Sci.,* 47(5), 1595, 1982.
248. **Patro, C. and Behera, R. N.,** Cashew Helps to Fix Sand Dunes in Orissa, *Indian Farming,* 28(12), 31, 1979.
249. **Perry, L. M.,** *Medicinal Plants of East and Southeast Asia,* MIT Press, Cambridge, Mass, 1980, 620.
250. **Peters, C. M. and Pardo-Tejeda, E.,** *Brosimum alicastrum* (Moraceae): Uses and Potential in Mexico, *Econ. Bot.,* 36(2), 166, 1982.
251. **Philippe, J. M.,** *The Propagation and Planting of Pistachio, Cardo, and Other Fruit and Nut Trees,* United Nations Development Programme, 1969.
252. **Phillips, A. M., Large, J. R., and Cole, J. R.,** Insects and Diseases of the Pecans in Florida, *Univ. Fla. Ag. Exp. Sta. Bull.,* 619, 1960.
253. **Pieris, W. V. D.,** *Wealth from the Coconut,* South Pacific Commission, Ure Smith Pty. Ltd., Sydney, Australia, 1955, 1.
254. **Pinto, R. W.,** Babassu — Brazil's Wonder Nut, *Foreign Commerce Weekly,* Vol. XLV, No. 2, 1951.
255. **Pio Correa, M.,** *Dicionario das Plantas Uteis do Brasil,* Vol. 4, Ministerio da Agricultura, Instituto Brasileiro de Desenvolvimento Florestal, 1984, 765.

256. **PIRB,** *Program Interciencia de Rucursos Biologicos,* Asociacion Interciencia, Bogota, Colombia, 1984, 239.
257. **Plakidas, A. G.,** Diseases of Tung trees in Louisiana, Louisiana State University, *Louisiana Bull.,* No. 282, 1, 1937.
258. **Porterfield, Jr., W. M.,** The Principal Chinese Vegetable Foods and Food Plants of Chinatown Markets, *Econ. Bot.,* 5(1), 3, 1951.
259. **Potter, G. F.,** The Domestic Tung Industry, I, Production and Improvement of the Tung Tree, *Econ. Bot.,* 13, 328, 1960.
260. **Potter, G. F. and Crane, H. L.,** Tung Production: Production in the United States, imports and use, U.S. Department of Agriculture, *Farmer's Bull.,* No. 2031, 1, 1957.
261. **Potts, W. M.,** The Chinese Tallow Tree as a Chemurgic Crop, *Chemurgic Dig.,* 5(22), 373, Columbus, Ohio, 1946.
262. **Powell, G. H.,** The European and Japanese Chestnuts in the Eastern United States, Delaware Coll. Age., *Exp. Sta. Bull.,* 42, Newark, 1898.
263. **Prance, G. T.,** A Monograph of Neotropical Chrysobalanaceae, *Flora Neotropica,* Monograph No. 9, Hafner Publishing, New York, 1972, 409.
264. **Prance, G. T. and da Silva, M. F.,** *Flora Neotropica,* Monograph No. 12, Caryocaraceae, Published for Organization for Flora Neotropica by Hafner Publishing, New York, 1973, 75.
265. **Prance, G. T. and Mori, S. A.,** What is Lecythis?, *Taxon,* 26, 209, 1977.
266. **Pratt et al.,** Vegetable Oil As Diesel Fuel, Seminar II, Northern Agricultural Energy Center, Northern Regional Research Center, Peoria, Ill., October 21 and 22, 1981.
267. **Pratt, D. S., Thurlow, L. W., Williams, R. R., and Gibbs, H. D.,** The Nipa Palm as a Commercial Source of Sugar, *Philippine J. Sci.,* 8(6), 377, 1913.
268. **Priestley, D. A. and Posthumus, M. A.,** Extreme Longevity of Lotus Seeds from Pulantien, *Nature,* 299(5879), 148, 1982.
269. **Princen, L. H.,** New Crop Development for Industrial Oils, *J. Am. Oil Seed Chem. Soc.,* 56(9), 845, 1979.
270. **Pryde, E. H. and Doty, Jr., H. O.,** World fats and oils situation, in *New sources of fats and oils,* Pryde, E. H., Princen, L. H., and Mukherjee, K. D., Eds. AOCS Monograph No. 9, American Oil Chemists' Society, Champaign, Ill., 1981, 3.
271. **Puri, V.,** The life-history of *Moringa oleifera* Lamk., *J. Indian Bot. Soc.,* 20, 263, 1941.
272. **Purseglove, J. W.,** *Tropical Crops,* 4 vols., Longman Group, London, 1968-1972, 334.
273. **Pyke, E. E.,** A note on the vegetative propagation of kola (*C. acuminata*) by softwood cuttings, *Trop. Agric. (Trinidad),* 11, 1, 1934.
274. **Quick, G. R.,** Abstract, A summary of some current research in Australia on vegetable oils as candidate fuels for diesel engines, Seminar II, Peoria, abstr., U.S. Department of Agriculture, 1981.
275. **Quijano, J. and Arango, G. J.,** The Breadfruit from Colombia — A Detailed Chemical Analysis, *Econ. Bot.,* 33(2), 199, 1979.
276. **Radford, A. E., Ahles, H. E., and Bell, C. R.,** *Manual of the Vascular Flora of the Carolinas,* University of North Carolina Press, Chapel Hill, 1968, 1183.
277. **Raghaven, V. and Baruah, H. K.,** Arecanut: India's Popular Masticatory — History, Chemistry, and Utilization, *Econ. Bot.,* 12, 315, 1958.
278. **Reed, C. F.,** Information summaries on 1000 economic plants, Typescripts submitted to the U.S. Department of Agriculture, 1976.
279. **Rice, R. E., Uyemoto, J. K., Ogawa, J. M., and Pemberton, W. M.,** New findings on pistachio problems, *Calif. Agric.,* January-February, 15, 1985.
280. **Riley, H. P.,** *Families of Flowering Plants of Southern Africa,* University of Kentucky Press, 1963, 269.
281. **Robyns, A.,** Family 116, Bombacaceae, *Ann. Mo. Bot. Gard.,* 51(1-4), 37, 1964.
282. **Roche, J. and Michel, R.,** *Oleagineux, 4* 205, 1946.
283. **Rosengarten, Jr., F.,** *The Book of Edible Nuts,* Walker and Company, New York, 1984, 384.
284. **Roth, J. H.,** The Guarana Industry of Amazonas, U.S. Cons. Rep., August, 1924, Manaus, Brazil, 1924.
285. **Rudolf, P. O. and Leak, W. B.,** *Fagus* L., Beech, in Seeds of woody plants in the United States, Agriculture Handbook 450, Forest Service, U.S. Department of Agriculture, Washington, D.C., 1974, 401.
286. **Rumsey, H. J.,** *Australian Nuts and Nut Growing in Australia,* Part I, The Australian Nut, H. J. Rumsey and Sons, Dundas, New South Wales, 1927, 120.
287. **Russell, T. A.,** The kola of Nigeria and the Cameroons, *Trop. Agric. (Trinidad),* 32, 210, 1955.
288. **Ryan, V. A.,** *Some Geographic and Economic Aspects of the Cork Oak,* Crown Cork and Seal Co., Baltimore, Md., 1948, 116.
289. **Ryan, V. A. and Cooke, G. B.,** The Cork Oak in the United States, Smithsonian Report for 1948, 355, Washington, D.C., 1948.

290. **Saleeb, W. F., Yermanos, D. M., Huszar, C. K., Storey, W. B., and Habanauskas, C. K.,** The Oil and Protein in Nuts of *Macadamia tetraphylla* L. Johnson, *Macadamia integrifolia* Maiden and Betche, and their F_1 Hybrid, *J. Am. Soc. Hort. Sci.*, 98(5), 453, 1973.
291. **Savel'eva, T. G., Rolle, A. I. U., Vedernikov, N. A., Sadovskaia, G. M., Novoselova, I. O., and Lisovskii, G. M.,** Acid Hydrolysis of Non-Edible Biomass of Cultivated Plants Grown in an Experimental Biological-Technical System of Human Survival, I, Chemical Composition of Polysaccharides of Wheat and *Cyperus esculentus* Possible Raw Material for the Hydrolysis Industry, *Khim Drev*, May/June 1980 (3), 37, 1980.
292. **Sawadogo, K. and Bezard, J.,** The Glyceride Structure of Shea Butter, *Oleagineux*, 37(2), 69, 1982.
293. **Scheld, H. W.,** Vegetable Oil As Diesel Fuel, Seminar II, Northern Agricultural Energy Center, Northern Regional Research Center, Peoria, Ill., October 21 and 22, 1981.
294. **Scheld, H. W. and Cowles, J. R.,** Woody biomass potential of the Chinese Tallow Tree, *Econ. Bot.*, 35(4), 391, 1981.
295. **Scheld, H. W., Bell, N. R., Cameron, G. N., Cowles, J. R., Engler, C. R., Krikorian, A. D., and Shultz, E. B.,** The Chinese Tallow Tree as a Cash and Petroleum-Substitute Crop, in "Tree Crops for Energy Co-Production on Farms", SERI/CP-622-1086, Solar Energy Research Institute, Golden, Colorado, 1980.
296. **Schroeder, H. W. and Storey, J. B.,** Development of Aflatoxin in 'Stuart' Pecans as Affected by Shell Integrity, *Hortscience*, 11(1), 53, 1976.
297. **Schroeder, J. G.,** Butternut *(Juglans cinerea* L.), Forest Service, U.S. Department of Agriculture, FS-223, June, 1972, 3.
298. **Schultes, R. E.,** The Amazonia as a Source of New Economic Plants, *Econ. Bot.*, 33(3), 259, 1979.
299. **Seabrook, J. A. E.,** A Biosystematic Study of the Genus Apios Fabricus (Leguminosae) with special reference to *Apios americana* Medikus, Thesis, University of New Brunswick, Canada, 1973.
300. **Seabrook, J. A. E. and Dionne, L. A.,** Studies on the Genus Apios, *Can. J. Bot.*, 54, 2567, 1976.
301. **Seida, A. A.,** Isolation, Identification, and Structure Elucidation of Cytotoxic and Antitumor Principles from *Ailanthus integrifolia, Amyris* pinnata, and *Balanites aegyptiaca, Diss. Abst. Int.*, B, 39(10), 4843, 1979.
302. **Senaratne, R., Herath, H. M. W., Balasubramaniam, S., and Wijesundera, C. R.,** Investigations on Quantitative and Qualitative Analysis of oils of *Madhuca longifolia* (L.) Macbr., *J. Natl. Agric. Soc. Ceylon*, 19, 89, 1982.
303. **Sengupta, A. and Roychoudbury, S. K.,** Triglyceride composition of *Buchanania lanzen* seed oil, *J. Sci. Food Agric.*, 28(5), 463, 1977.
304. **Serr, Jr., E. F.,** Persian Walnuts in the Western States, in *Handbook of North American Nut Trees*, Jaynes, R. A., Ed., Northern Nut Growers' Association, Knoxville, Tenn., 1969, 240.
305. **Serr, E. F. and Forde, H. J.,** Blackline, a delayed failure at the union of *Juglans regia* trees propagated on other *Juglans* species, *Proc. Am. Soc. Hort. Sci.*, 74, 220, 1959.
306. **Sharma, R. D. and Sher, S. A.,** Plant Nematodes Associated with Jackfruit *Artocarpus heterophyllus* in Bahia, Brazil, *Nematropica*, 3(1), 23, 1973.
307. **Sharma, V. K. and Misra, K. C.,** Productivity of *Shorea robusta* Gaertn. and *Buchanania lanzan* Spreng. at Tropical Dry Deciduous Forests of Varanasi, Biology Land Plants Symposium, Meerut University, Uttar Pradesh, India, 406, 1974.
308. **Slate, G. L.,** Filberts — including varieties grown in the East, in *Handbook of North American Nut Trees*, Jaynes, R. A., Ed., Northern Nut Growers' Association, Knoxville, Tenn., 1969, 287.
309. **Smith, Jr., C. R., Shekleton, M. C., Wolff, I. A., and Jones, Q.,** Seed Protein Sources — Amino Acid Composition and Total Protein Content of Various Plant Seeds, *Econ. Bot.*, 13, 132, 1959.
310. **Smith, J. R.,** *Tree Crops, A Permanent Agriculture*, Devin-Adair Co., Old Greenwich, Conn., 1977, 408.
311. **Sprent, J. I.,** Functional Evolution in Some Papilionoid Root Nodules, in Advances in Legume Systematics, Polhill, R. M. and Raven, P. H., Eds., 1981, 671.
312. **Spurling, A. T. and Spurling, D.,** Effect of various and morganic fertilizers on the yield of Montana tung *(Aleurites montana)* in Malawi, *Trop. Agric.*, 51(1), 1, 1974.
313. **Standley, P. C.,** The Cohune Palm an *Orbignya*, not an *Attalea*, *Trop. Woods*, 30, 1, 1932.
314. **Steger, A. and van Loon, J.,** Das fette Oel der Samen von *Canarium commune* L., *Rec. Trav. Chim. Pays Bas.*, 59, 168, 1950.
315. **Stevenson, N. S.,** The Cohune Palm in British Honduras, *Trop. Woods*, 30, 3, 1932.
316. **Storey, W. B.,** Macadamia, in *Handbook of North American Nut Trees*, Jaynes, R. A., Ed., Northern Nut Growers' Association, Knoxville, Tenn. 1970, 321.
317. **Sturtevant, E. L.,** *Sturtevant's Edible Plants of the World*, Hedrick, U.P., Ed., Dover Publications, New York, 1972, 686.
318. **Tewari, J. P. and Shukla, I. K.,** *Inhibition of Infectivity of 2 Strains of Watermelon Mosaic Virus by Latex of Some Angiosperms*, Geobios, Jodhpur, India, 9(3), 124, 1982.
319. **Thieret, J. W.,** Economic Botany of the Cycads, *Econ. Bot.*, 12, 3, 1958.

320. **Tokay, B. A.,** Research Sparks Oleochemical Hopes, *Chem. Bus.*, September, 1985.
321. **Tuley, P.,** Studies of the production of wine from the Oil Palm, *J. Nigerian Inst. Oil Palm Res.*, 4, 282, 1965.
322. **Tsuchiya, T. and Iwaki, H.,** Biomass and Net Primary Production of a Floating Leaved Plant *Trapa natans* Community in Lake Kasumigaura, Japan, *Jpn. J. Ecol.*, 33(1), 47, 1983.
323. **Tyler, V. E.,** *The Honest Herbal,* Georgia F. Stickley Co., Philadelphia, Pa., 1982, 263.
324. **Uphof, Th., J. C.,** *Dictionary of Economic Plants,* Verlag von J. Cramer, Lehrte, West Germany, 1968, 591.
325. **Vaughn, J. G.,** *The Structure and Utilization of Oil Seeds,* Chapman and Hall, London, 1970, 279.
326. **Veracion, V. P. and Costales, E. F.,** The Bigger, the More, *Canopy Int.*, 7(6), 1981.
327. **Verbiscar, A. J. and Banigan, T. F.,** Composition of Jojoba Seeds and Foliage, *J. Agric. Food Chem.*, 26(6), 1456, 1978.
328. **Verma, S. C., Banerji, R., Misra, G., and Nigam, S. K.,** Nutritional Value of Moringa, *Curr. Sci.*, 45(21), 769, 1976.
329. **Vietmeyer, N.,** American Pistachios, *Horticulture,* September, 32, 1984.
330. **Vogel, V. J.,** *American Indian Medicine,* University of Oklahoma Press, Norman, 1970, 583.
331. **von Reis, S. and Lipp, Jr., F. J.,** *New Plant Sources for Drugs and Foods from the New York Botanical Garden Herbarium,* Harvard University Press, Cambridge, Mass., 1982, 363.
332. **Watt, J. M. and Breyer-Brandwijk, M. G.,** *The Medicinal and Poisonous Plants of Southern and Eastern Africa,* 2nd ed., E. & S. Livingstone, Edinburgh, 1962, 1457.
333. **Wei-Chi Lin, An-Chi Chen, and Sang-Gen Hwang,** An investigation and study of Chinese Tallow Tree in Taiwan *(Sapium sebiferum* Roxb.), *Bull. Taiwan Forestry Res. Inst.*, No. 57, 32, 1958.
334. **Westlake, D. F.,** Comparisons of Plant Productivity, *Biol. Rev.*, 38, 385, 1963.
335. **Whitehouse, W. E.,** The Pistachio Nut — a new crop for the Western United States, *Econ. Bot.*, 11, 281, 1957.
336. **Whitehouse, W. E. and Joley, L. E.,** Notes on the growth of Persian Walnut propagated on rootstocks of Chinese wingnut, *Petrocarya stenoptera, Proc. Am. Soc. Hort. Sci.*, 52, 103, 1948.
337. **Whitehouse, W. E. and Joley, L. E.,** Notes on culture, growth, and training of pistachio nut trees, *Western Fruit Grower,* October, 3, 1951.
338. **Whiting, M. G.,** Toxicity of the Cycads, *Econ. Bot.*, 17(4), 271, 1963.
339. **Wiggins, I. L.,** *Flora of Baja California,* Stanford University Press, Stanford, Calif., 1980, 1025.
340. **Williams, L. O.,** Living Telegraph Poles, *Econ. Bot.*, 13, 150, 1959.
341. **Woodroof, J. G.,** *Tree Nuts: Production, Processing, Products,* AVI, Westport, Conn., 1967, 356.
342. **WuLeung, Woot-Tseun, Butrum, R. R., and Chang, F. H., Part 1,** Proximate composition mineral and vitamin contents of East Asian food, in Food Composition Table for Use in East Asia, Food and Agriculture Organization and U.S. Department of Health, Education and Welfare, 1972, 334.
343. **Wyman, D.,** *Wyman's Gardening Encyclopedia,* MacMillan, New York, 1974, 1222.
344. **Yamazaki, Z. and Tagaya, I.,** Antiviral Effects of Atropine and Caffeine, *J. Gen. Virol.*, 50(2), 429, 1980.
345. **Yanovsky, E. and Kingsbury, R. M.,** *J. Assoc. Off. Agric. Chem.*, 21, 648, 1938.
346. **Yen, D. E.,** Arboriculture in the Subsistence of Santa Cruz, Solomon Islands, *Econ. Bot.*, 28, 247, 1974.
347. **Yermanos, D. M.,** Jojoba — a Brief Survey of the Agronomic Potential, *Calif. Agric.*, September, 1973.
348. **Yermanos, D. M.,** Monoecious Jojoba, in *New Sources of Fats and Oils,* Pryde, E. H., Princen, L. H., and Mukherjee, K. D., Eds., AOCS Monograph No. 9, American Oil Chemists' Society, Champaign, Ill., 1981, 247.
349. **Zeven, A. C.,** The Partial and Complete Domestication of the Oil Palm *(Elaeis guineensis), Econ. Bot.*, 26(3), 274, 1972.
350. **Zeven, A. C. and Zukovsky, P. M.,** *Dictionary of Cultivated Plants and Their Centres of Diversity,* Centre for Agricultural Publishing and Documentation, Wageningen, Netherlands, 1975, 219.
351. *Chemical Marketing Reporter,* Schnell Publishing Company, New York (often cited herein with date only, single articles are almost always anonymous in this tabloid).
352. **Vilmorin-Andrieux, Mm.,** *The Vegetable Garden;* reprinted by The Jeavons-Leler Press, Palo Alto, Calif., 1976, 620.
353. **Elliott, D. B.,** *Roots — An Underground Botany and Forager's Guide,* Chatham Press, Old Greenwich, Conn., 1976, 128.
354. **Fernald, M. L., Kinsey, A. C., and Rollins, R. C.,** *Edible Wild Plants of Eastern North America,* rev. ed., Harper & Bros., New York, 1958, 452.
355. **Roth, W. B., Cull, I. M., Buchanan, R. A., and Bagby, M. O.,** Whole Plants as Renewable Energy Resources: Checklist of 508 Species Analyzed for Hydrocarbon, Oil, Polyphenol, and Protein, Trans. *Illinois Acad. Sci.*, 75(3,4), 217, 1982.
356. **Kalin Arroyo, M. T.,** Breeding Systems and Pollination Biology in Leguminosae, in *Advances in Legume Systematics,* Polhill, R. M. and Raven, P. H., Eds., 1981, 723.

357. **Sadashivaiah, K. N., Hasheeb, A., and Parameswar, N. S.,** Role of different shade tree foliage as organism manure in cardamom plantation, *Res. Bull. Marathwada Agric. Univ.*, 4(1), 11, 1980.
358. **Agrawal, P. K.,** Rate of Dry Matter Production in Forest Tree Seedlings with Contrasting Patterns of Growth, *Acacia catechu, Butea monosperma*, and *Buchanania lanzan*, Biol. Land Plants Symposium, Meerut University, 391, 1974.
359. **Telek, L. and Martin, F. W.,** Okra Seed: A Potential Source for Oil and Protein in the Humid Lowland Tropics, in *New Sources of Fats and Oils*, Pryde, E. H., Princen, L. H., and Mukherjee, K. D., Eds., AOCS Monograph No. 9, American Oil Chemists' Society, Champaign, Ill., 1981, 37.
360. **Duke, J. A., Atchley, A. A., Ackerson, K. T., and Duke, P. K.,** *CRC Handbook of Agricultural Energy Potential of Developing Countries*, 2 vols., CRC Press, Boca Raton, Fla., 1987.
361. **Anon.,** *Agric. Res.*, December, 1978.
362. **da Vinha, S. G. and Pereira, R. C.,** Producao de Folhedo e sua Sazonalidade em 10 Especies Arborea Nativas no Sul da Bahia, *Revista Theobroma*, 13(4), 327, 1983.
363. **Pereira, H.,** *Small Contributions for a Dictionary of Useful Plants of the State of Sao Paulo*, Rothschild & Co., Sao Paulo, 1929, 799.
364. **Castaneda, R. R.,** *Flora del Centro de Bolivar*, Talleres Univ. Nac. de Colombia, Bogata, 1965, 437.
365. **Mori, S. A.,** The Ecology and Uses of the Species of Lecythis in Central America, *Turrialba*, 20(3), 344, 1970.
366. **Eckey, E. W.** *Vegetable Fats and Oils*, Reinhold Publishing, New York, 1954.
367. **Buchanan, R. A. and Duke, J. A.,** Botanochemical Crops, in *Handbook of Biosolar Resources*, McClure, T. A. and Lipinsky, E. S., Eds., CRC Press, Boca Raton, Fla., 1981, 157.
368. **Oke, O. L.,** Leaf Protein Research in Nigeria, in *Leaf Protein Concentrates*, Telek, L. and Graham, H. D., Eds., AVI, Westport, Conn., 1983, 739.
369. **Sadashivaiah, K. N., Haseeb, A., Parameswar, N. S.,** Role of different shade tree foliage as organic manure in Cardamom plantation, *Res. Bull. Marathwada Agric. Univ.*, 4(1), 11, 1980.
370. **Tsuchiya, T. and Iwaki, H.,** Biomass and Net Primary Production of a Floating Leaved Plant *Trapa natans* Community in Lake Kasumigaura Japan, *Jpn. J. Ecol.*, 33(1), 47, 1983.
371. **Duke, J. A.,** Herbalbum, An LP album of Herbal Folk Music, Produced by Vip Vipperman and Buddy Blackmon, Grand Central Studios, Nashville, Tenn., 1986.
372. **Devlin, R. M. and Demoranville, I. E.,** Wild Bean Control on Cranberry Bogs with Maleic Hydrazide, *Proc. Northeastern Weed Science Soc.*, 35, 349, 1981.
373. **Sanchez, F. and Duke, J. A.,** La Papa de Nadi, *El Campesino*, 115(4), 15, 1984.
374. **Walter, W. M., Croom, Jr., E. M., Catignani, G. L., and Thresher, W. C.,** Compositional Study of *Apios priceana* Tubers, *J. Ag. Food Chem.*, January/February, 39, 1986.
375. **Reynolds, B.,** Research Highlights, *Apios Tribune*, 1(1), 7, 1986.
376. **Keyser, H.,** Personal communication, 1984.
377. **Anon.,** Desert Plant May Replace Sperm Oil, *BioScience*, 25(7), 467, 1974.
378. **Joseph, C. J.,** Systematic Revision of the Genus Pilocarpus, 1° Lugar — Premio Esso de Ciencia, ca. 1970.
379. **Roecklein, J. C. and Leung, P. S.,** A Profile of Economic Plants, typescript, College of Tropical Agriculture and Human Resources, University of Hawaii, 1986.
380. **Stevenson, J. A.,** The Fungi of Puerto Rico and the American Virgin Islands, Contribution of Reed Herbarium, Baltimore, No. 23, 743, 1975.
381. **Duke, J. A.,** Herbalbum: An Anthology of Varicose Verse, J. Medrow, Laurel, Md., 1985.
382. **Golden, A. M.,** USDA, Personal communication, 1984.
383. **Price, M.,** Vegetables from a Tree, *ECHO*, 8(4), 1, 1985.
384. **Ramachandran, C., Peter, K. V., and Gopalakrishnan, P. K.,** Drumstick *(Moringa oleifera)* : A Multipurpose Indian Vegetable, *Econ. Bot.*, 34(3), 276, 1980.
385. **Mahajan, S. and Sharma, Y. K.,** Production of Rayon Grade Pulp from *Moringa oleifera, Indian Forester*,
386. **Grabow, W. O. K., Slabbert, J.L., Morgan, W. S. G., and Jahn, S. A. A.,** Toxicity and Mutagenicity Evaluation of Water Coagulated with *Moringa oleifera* Seed Preparations Using Fish, Protozoan, Bacterial, Coliphage, Enzyme, and Ames Salmonella Assays, *Water S.A.*, 11(2), 9, 1985.
387. **Price, M.,** The Benzolive Tree, *ECHO*, 11, 7, 1986.
388. **Iyer, R. I., Nagar, P. K., and Sircar, P. K.,** Auxins in *Moringa pterygosperma* Gaertn. Fruits, *Indian J. Exp. Biol.*, 19(5), 487, 1981.
389. **Girija, V., Sharada, D., and Pushpamma, P.,** Bioavailability of Thiamine, Riboflavin, and Niacin from Commonly Consumed Green Leafy Vegetables in the Rural Areas of Andhra Pradesh in India, *Int. J. Vitam. Nutr. Res.*, 52(1), 9, 1982.
390. **Dahot, M. U. and Memon, A. R.,** Nutritive Significance of Oil Extracted from *Moringa oleifera* Seeds, *J. Pharm. Univ. Karachi* 3(2), 75, 1985.
391. **Bhattacharya, S. B., Das, A. K., and Banerji, N.,** Chemical Investigations on the Gum Exudate from Sanja *Moringa oleifera, Carbohydr. Res.*, 102(0), 253, 1982.

392. **Kareem, A. A., Sadakathulla, S., and Subramaniam, T. R.,** Note on the Severe Damage of *Moringa* Fruits by the Fly *Gitona* sp., *South Indian Hort.*, 22(1/2), 71, 1974.
393. **Ullasa, B. A. and Rawal, R. D.,** *Papaver rhoeas* and *Moringa oleifera*, Two New Hosts of Papaya Powdery Mildew, *Curr. Sci.*, India, 53(14), 754, 1984.
394. **Milne-Redhead, E. and Polhill, R. M., Eds.,** Flora of Tropical East Africa, Crown Agents for Overseas Governments and Administrations, London (Cucurbitaceae, by C. Jeffrey, 1967), 1968, 156.
395. **Edet, E. E., Eka, O. U., and Ifon, E. T.,** Chemical Evaluation of the Nutritive Value of Seeds of African Breadfruit Treculia africana, *Food Chem.*, 17(1), 41, 1985.
396. **Blackmon, W. J. and Reynolds, B. D.,** The Crop Potential of Apios americana — Preliminary Evaluations, *HortSci.*, 21(6), in press.
397. **Kovoor, A.,** The Palmyrah Palm: Potential and Perspectives, FAO Plant Production and Protection Paper 52, 77, 1983.
398. **Hemsley, J. H.,** Sapotaceae, in *Flora of Tropical East Africa*, Milne-Redhead, E. and Polhill, R. M., Eds., 1968, 78.
399. **Eggeling, W. J.,** *The Indigenous Trees of the Uganda Protectorate*, rev. by I. R. Dale, The Government Printer, Entebbe, 1951, 491.
400. **Duke, J. A.,** *Handbook of Northeastern Indian Medicinal Plants*, Quarterman Publications, Lincoln, Mass., 1986, 212.
401. **Frey, D.,** The Hog Peanut, *TIPSY*, 86, 74, 1986.
402. **Marshall, H.H.,** (Research Station, Morden Manitoba, Canada), correspondence with Noel Vietmeyer, 1977.
403. **Duke, J. A.,** The Case of the Annual "Perennial", *Org. Gard.*, submitted.
404. **Polhill, R. M. and Raven, P. H., Eds.,** Advances in Legume Systematics, in 2 parts, Vol. 2 of the Proceedings of the International Legume Conference, Kew, July 24-29, 1978, 1981.
405. **Gallaher, R. N. and Buhr, K. L.,** Plant Nutrient and Forage Quality Analysis of a Wild Legume Collected from the Highland Rim Area of Middle Tennessee, *Crop Sci.*, 24(6), 1200, 1984.
406. **Turner, B. L. and Fearing, O. S.,** A Taxonomic Study of the Genus *Amphicarpaea* (Leguminosae), *Southwest. Nat.*, 9(4), 207, 1964.
407. **Dore, W. G.,** (Ottawa, Ontario, Canada), typescript and correspondence with Noel Vietmeyer, 1978.
408. **Foote, B. A.,** Biology of *Rivella pallida* Diptera Platystomatidae — a Consumer of the Nitrogen-fixing Root Nodules of *Amphicarpa bracteata* Leguminosae, *J. Kans. Entomol. Soc.*, 58(1), 27, 1985.
409. **Schnee, B. K. and Waller, D. M.,** Reproductive Behavior of Amphicarpaea bracteata (Leguminosae), an Amphicarpic Annual, *Am. J. Bot.*, 73(3), 376, 1986.
410. **Serrano, R. G.,** Current Developments on the Propagation and Utilization of Philippine Rattan, *NSTA Technol. J.*, 9, 76, 1984.
411. **Lapis, A. B.,** Some Identifying Characters of 12 Rattan Species in the Philippines, *Canopy Int.*, April, 3, 1983.
412. **Anon.,** Big Break for Rattan, *Canopy Int.*, September, 2, 1979.
413. **Borja, B.,** MNR-FORI Rattan Research, *Canopy Int.*, 5(9), 1, 1979.
414. **Conelly, W. T.,** Copal and Rattan Collecting in the Philippines, *Econ. Bot.*, 3(1), 39, 1985.
415. **Wong, K. M. and Manokaran, N., Eds.,** Proceedings of the Rattan Seminar, Oct. 2-4, 1984, Kuala Lumpur, Malaysia, The Rattan Information Centre, Forest Research Institute, Kepong, Malaysia, 1985.
416. **Garcia, P. R. and Pasig, S. D.,** Domesticating Rattan Right at Your Backyard, *Canopy Int.*, April, 10, 1983.
417. **Monachino, J.,** Chinese Herbal Medicine — Recent Studies, *Econ. Bot.*, 10, 42, 1956.
418. **Dallimore, W. and Jackson, A. B.,** A Handbook of Coniferae and Ginkgoaceae, 4th ed., rev. by S. G. Harrison, Edward Arnold, Ltd., London, 1966, 728.
419. **Balz, J. P.,** Conditions for Cultivation of *Ginkgo biloba*, personal communication, 1981.
420. **Wilbur, R. L.,** The Leguminous Plants of North Carolina, The North Carolina Experiment Station, *Agric. Exp. Station*, 1963, 294.
421. **Degener, O.,** *Flora Hawaiiensis* or *The New Illustrated Flora of the Hawaiian Islands*, 1957-1963 (published by the author).
422. **Weber, F. R.,** *Reforestation in Arid Lands*, VITA Manual Series Number 37E, 1977, 224.
423. **Descourtilz, M. E.,** *Flore Pittoresque et Medicale des Antilles*, 8th ed., Paris, 1829.
424. **Sargent, C. S.,** *The Silva of North America*, Riverside Press, Cambridge, Mass., 1895.
425. **Bedell, H. G.,** *Laboratory Manual — Botany 212 — Vascular Plant Taxonomy*, 1st ed., Illus. by Peggy Duke, Department of Botany, University of Maryland, College Park, 1984, 159.
426. **Little, E. L., Jr. and Wadsworth, F. H.,** Common Trees of Puerto Rico and the Virgin Islands, Agriculture Handbook No. 249, Forest Service, U.S. Department of Agriculture, Washington, D.C., 1964, 548.
427. **Louis, J. and Leonard, J.,** Olacaceae, in *Flore du Congo Belge et du Ruanda-Urundi*, Robyns, W., Ed., Inst. Nat. l'Etude Agron. Congo (INEAC), Brussels, 1948, 249.

428. **Reed, C. R.,** Selected Weeds of the United States, Agriculture Handbook No. 366, Forest Service, U.S. Department of Agriculture, Washington, D.C., 1970, 463.
429. **Agan, J. E.,** Guarana, *Bull. Pan Am. Union*, September 268, 1920.
430. **Little, E. L., Jr.,** Important Forest Trees of the United States, Agriculture Handbook No. 519, Forest Service, U.S. Department of Agriculture, Washington, D.C., 1978.
431. **Bakker, K. and van Steenis, C. G. G. J.,** Pittosporaceae, in *Flora Malesiana*, Vol. 5, Rijksherbarium, Leiden, 1955-1958, 345.
432. **Duke, J. A.,** *Survival Manual II: South Viet Nam*, Missouri Botanical Garden, St. Louis, 1963, 44.
433. **Maiden, J. H.,** *The Forest Flora of New South Wales*, 2 vols., William Applegate Gullick, Government Printer, Sydney, 1904.
434. **Li, H. L. and Huang, T. C., Eds.,** *Flora of Taiwan*, 6 vols., Epoch Publishing, Taipei, 1979.
435. **Fernandes, R. and Fernandes, A.,** Anacardiaceae, in *Flora Zambesiaca*, Vol. 2(2), Exell, A. W., Fernandes, A., and Wild, H., Eds., University Press, Glasgow, 1966, 550.
436. **Cribb, A. B. and Cribb, J. W.,** *Useful Wild Plants in Australia*, William Collins, Ltd., Sydney, 1981, 269.
437. **Petrie, R. W.,** personal communication, August 6, 1987.
438. **Saul, R., Ghidoni, J. J., Molyneux, R. J., and Elbein, A. D.,** Castanospermine inhibits alpha-glucosidase activities and alters glycogen distribution in animals, *Proc. Natl. Acad. Sci. U.S.A.*, 82, 93, 1985.
439. **Snader, K. M.,** National Cancer Institute, personal communication, July 22, 1987.
440. *Threatened Plants Newsletter*, No. 17, November 1986.
441. **Walker, B. D., Kowalski, M., Goh, W. C., Kozarsky, K., Krieger, M., Rosen, C., Rohrschneider, L. R., Haseltine, W. A., and Sodrowski, J.,** *Proc. Natl. Acad. Sci. U.S.A.*, 84, 8121, 1987.
442. **Walker, B. D., Kozarsky, K., Goh, W. C., Rohrschneider, L. R., and Haseltine, W. A.,** Intl. Conf. on AIDS, Washington, D.C., June 15, 1987.
443. **Hutchinson, J. and Dalziel, J. M.,** *Flora of West Tropical Africa*, Vol. 1, part 2, 2nd ed., revised by Keay, R. M. J., 1958, 392.

FIGURE CREDITS

With a master's degree in botany from the University of North Carolina (1956), complemented by 30 years of experience as an illustrator, Peggy K. Duke is excellently qualified to prepare the figures for this handbook. Peggy and I were pleased and amazed at how generous authors and administrators have been with us, at granting permission to use their published illustrations. Thanks to these fine people, as well as several U.S. Department of Agriculture (USDA) public domain publications, and the curators of the collections at the U.S. National Seed Collection, the National Agricultural Library, and the Smithsonian Institution Botany Department, we have been able to piece together illustrations for the majority of genera treated in this book. Our special thanks go to:

M. J. Balick	P. Kumar	P. H. Raven
H. G. Bedell	H. L. Li	C. F. Reed
E. A. Bell	E. L. Little	J. L. Reveal
M. L. Brown	P. M. Mazzeo	A. Robyns
R. G. Brown	S. A. Mori	C. G. G. J. Van Steenis
O. Degener	W. Mors	R. L. Wilbur
E. Forrero	G. W. Patterson	J. J. Wurdack
H. Garcia-Barriga	G. T. Prance	
C. R. Gunn	T. Plowman	

Photographs from the USDA and New York Botanical Gardens collections were consulted in concert with published photographs and illustrations, especially Menninger's and Rosengarten's, in the publications cited at the end of this book. Mrs. Duke confirmed and/or altered details based on seed specimens of the U.S. National Seed Collection, courtesy C. R. Gunn; and herbarium specimens at the University of Maryland, courtesy J. L. Reveal; the Botany Department of the Smithsonian Institution, courtesy J. J. Wurdack; and the U.S. National Arboretum, courtesy P. M. Mazzeo.

FIGURE CREDIT LIST

Scientific name	**Credit (with permission)**
Acrocomia sclerocarpa	Peggy Duke
Adhatoda vasica	After Little[188]
Aleurites moluccana	After Ochse[238] (courtesy A. Asher & Co., Amsterdam)
Amphicarpaea bracteata	After Wilbur[420]
Anacardium occidentale	After Ochse[238] (courtesy A. Asher & Co., Amsterdam)
Apios americana	Peggy Duke
Areca catechu	Peggy Duke
Arenga pinnata	Peggy Duke
Artocarpus altilis	After Degener[421]
Balanites aegyptiaca	Peggy Duke
Barringtonia procera	Peggy Duke
Bertholletia excelsa	Peggy Duke
Borassus flabellifer	After Weber[422] (courtesy F. R. Weber and Volunteers in Technical Assistance (VITA), Rosslyn, Virginia)
Brosimum alicastrum	After Descourtilz[423]
Bruguiera gymnorrhiza	After Little[188]
Buchanania lanzan	After Kirtikar and Basu[165]
Butyrospermum paradoxum	After Hemsley[398] (reproduced with permission of the Director, Royal Botanic Garden, Kew)
Calamus ornatus	Peggy Duke, after Lapis[411]

FIGURE CREDIT LIST (continued)

Scientific name	Credit (with permission)
Canarium indicum	After Kirtikar and Basu[165]
Carya illinoensis	Peggy Duke, after Sargent[424]
Caryocar villosum	After Prance and da Silva[264] (courtesy New York Botanical Garden)
Caryodendron orinocense	After PIRB[256] and Garcia-Barriga[105] (courtesy Universidad Nacional, Bogota)
Castanea mollissima	Peggy Duke in Bedell[425]
Castanospermum australe	Peggy Duke, after Masefield et al.,[204] courtesy Oxford University Press
Ceiba pentandra	Ochse[238] (courtesy A. Asher & Co., Amsterdam)
Cocos nucifera	After Little and Wadsworth[426] and Masefield et al.,[204] courtesy Oxford University Press
Cola acuminata	Peggy Duke
Cordeauxia edulis	Peggy Duke
Corylus americana	Peggy Duke in Bedell[425]
Coula edulis	After Louis and Leonard[427] (redrawn from "Flore du Congo Belgique et du Ruanda-Urundi", Bruxelles, I.N.E.A.C.
Cycas rumphii	After Ochse[238] (courtesy A. Asher & Co., Amsterdam)
Cyperus esculentus	After Reed[428]
Detarium senegalensis	After Weber[422] (courtesy F. R. Weber and Volunteers in Technical Assistance (VITA), Rosslyn, Virginia)
Elaeis guineensis	Peggy Duke, after Masefield et al.,[204] courtesy Oxford University Press
Eleocharis dulcis	Peggy Duke
Fagus grandifolia	Peggy Duke in Bedell[425]
Ginkgo biloba	Peggy Duke in Bedell[425]
Gnetum gnemon	After Ochse[238] (courtesy A. Asher & Co., Amsterdam)
Helianthus annuus	Peggy Duke
Hyphaene thebaica	After Weber[422] (courtesy F. R. Weber and Volunteers in Technical Assistance (VITA), Rosslyn, Virginia)
Inocarpus edulis	Peggy Duke
Jatropha curcas	After Ochse[238] (courtesy A. Asher & Co., Amsterdam)
Jessenia bataua	After PIRB[256] (courtesy Universidad Nacional, Bogota)
Juglans nigra	Peggy Duke in Bedell[425]
Lecythis ollaria	After Prance and Mori[265] (courtesy New York Botanical Gardens
Licania rigida	Peggy Duke
Macadamia spp.	Peggy Duke, after Degener[421]
Madhuca longifolia	Peggy Duke
Moringa oleifera	After Little and Wadsworth[426]
Nelumbo nucifera	Peggy Duke, after Reed[428]
Nypa fruticans	Peggy Duke
Orbignya martiana	Peggy Duke
Pachira aquatica	After Garcia-Barriga[104] (courtesy Universidad Nacional, Bogota)
Paullinia cupana	Peggy Duke
Phytelephas macrocarpa	Peggy Duke
Pinus edulis	After Little[430]
Pistacia vera	Peggy Duke
Pittosporum resiniferum	After Bakker and van Steenis[431] (courtesy Flora Malesiana)
Platonia esculenta	Peggy Duke
Prunus dulcis	After Kirtikar and Basu[165]
Quercus suber	Peggy Duke
Ricinodendron heudelotii	After Eggerling,[399] and *Flora of West Tropical Africa*[443] (reproduced with permission of the Director, Royal Botanic Garden, Kew) (seed of *R. rautaneninii)*
Santalum acuminatum	Peggy Duke
Sapium sebiferum	After Li and Huang[434] (Flora of Taiwan, with permission)
Schleichera oleosa	After Ochse[238] (courtesy A. Asher & Co., Amsterdam)
Sclerocarya caffra	After Fernandes and Fernandes[435] (reproduced with permission of the Director, Royal Botanic Garden, Kew)

FIGURE CREDIT LIST (continued)

Scientific name	Credit (with permission)
Simmondsia chinensis	Peggy Duke
Telfairia pedata	After Jeffrey[394] (reproduced with permission of the Director, Royal Botanic Garden, Kew)
Terminalia catappa	Peggy Duke
Trapa bispinosa	After Kirtikar and Basu[165]
Treculia africana	Peggy Duke
Virola sebifera	Peggy Duke (after Duke[78])

INDEX

A

B

D

F

G

H

I

J

K

L

M

P

Q

R

U

V

Z